Microplastics in Urban Water Management

Microplastics in Urban Water Management

Edited By

Bing-Jie Ni
School of Civil and Environmental Engineering
Centre for Technology in Water and Wastewater
University of Technology Sydney
Sydney, NSW, Australia

Qiuxiang Xu
State Key Laboratory of Pollution Control and Resources Reuse
College of Environmental Science and Engineering
Tongji University
Shanghai, PR China

Wei Wei
School of Civil and Environmental Engineering
Centre for Technology in Water and Wastewater
University of Technology Sydney
Sydney, NSW, Australia

This edition first published 2023
© 2023 John Wiley & Sons, Inc.

All rights reserved. No part of this publication may be reproduced, stored in a retrieval system, or transmitted, in any form or by any means, electronic, mechanical, photocopying, recording or otherwise, except as permitted by law. Advice on how to obtain permission to reuse material from this title is available at http://www.wiley.com/go/permissions.

The right of Bing-Jie Ni, Qiuxiang Xu, and Wei Wei to be identified as the authors of the editorial material in this work has been asserted in accordance with law.

Registered Office
John Wiley & Sons, Inc., 111 River Street, Hoboken, NJ 07030, USA

For details of our global editorial offices, customer services, and more information about Wiley products visit us at www.wiley.com.

Wiley also publishes its books in a variety of electronic formats and by print-on-demand. Some content that appears in standard print versions of this book may not be available in other formats.

Trademarks: Wiley and the Wiley logo are trademarks or registered trademarks of John Wiley & Sons, Inc. and/or its affiliates in the United States and other countries and may not be used without written permission. All other trademarks are the property of their respective owners. John Wiley & Sons, Inc. is not associated with any product or vendor mentioned in this book.

Limit of Liability/Disclaimer of Warranty
In view of ongoing research, equipment modifications, changes in governmental regulations, and the constant flow of information relating to the use of experimental reagents, equipment, and devices, the reader is urged to review and evaluate the information provided in the package insert or instructions for each chemical, piece of equipment, reagent, or device for, among other things, any changes in the instructions or indication of usage and for added warnings and precautions. While the publisher and authors have used their best efforts in preparing this work, they make no representations or warranties with respect to the accuracy or completeness of the contents of this work and specifically disclaim all warranties, including without limitation any implied warranties of merchantability or fitness for a particular purpose. No warranty may be created or extended by sales representatives, written sales materials or promotional statements for this work. The fact that an organization, website, or product is referred to in this work as a citation and/or potential source of further information does not mean that the publisher and authors endorse the information or services the organization, website, or product may provide or recommendations it may make. This work is sold with the understanding that the publisher is not engaged in rendering professional services. The advice and strategies contained herein may not be suitable for your situation. You should consult with a specialist where appropriate. Further, readers should be aware that websites listed in this work may have changed or disappeared between when this work was written and when it is read. Neither the publisher nor authors shall be liable for any loss of profit or any other commercial damages, including but not limited to special, incidental, consequential, or other damages.

Library of Congress Cataloging-in-Publication Data
Names: Ni, Bing-Jie, editor. | Xu, Qiuxiang, editor. | Wei, Wei, editor. | John Wiley & Sons, publisher.
Title: Microplastics in urban water management / edited by Bing-Jie Ni, Qiuxiang Xu, and Wei Wei.
Description: Hoboken, NJ : John Wiley & Sons, 2023. | Includes bibliographical references
   and index.
Identifiers: LCCN 2022034277 (print) | LCCN 2022034278 (ebook) | ISBN
   9781119759348 (hardback) | ISBN 9781119759362 (pdf) | ISBN 9781119759393 (epub) |
   ISBN 9781119759379 (ebook)
Subjects: LCSH: Microplastics. | Water--Pollution. | Municipal water supply--Management. |
   Urban hydrology--Management.
Classification: LCC TD427.P62 M5395 2023 (print) | LCC TD427.P62 (ebook) | DDC 363.738--dc23/
   eng/20221011
LC record available at https://lccn.loc.gov/2022034277
LC ebook record available at https://lccn.loc.gov/2022034278

Cover image: © OsakaWayne Studios/Getty Images
Cover design by Wiley

Set in 9.5/12.5pt STIXTwoText by Integra Software Services Pvt. Ltd, Pondicherry, India

Contents

Notes of Contributors *xiii*
Preface *xvii*

1 Techniques for Microplastics Detection in Urban Water Systems *1*
Xiaowei Li, Man Li, Lulu Liu, and Xiang Huang
1.1 Introduction *1*
1.2 Sample Collection and Separation *1*
1.2.1 Freshwater Samples *1*
1.2.2 Freshwater Sediments *5*
1.2.3 Wastewater Samples *9*
1.2.4 Sludge Samples *12*
1.2.5 Drinking Water Samples *14*
1.3 Sample Purification *16*
1.3.1 Wet Peroxidation *16*
1.3.2 Enzymatic Degradation *17*
1.3.3 Alkaline and Acid Treatment *18*
1.3.4 Influence of Chemical Purification on Microplastic Property *20*
1.4 Sample Identification *25*
1.4.1 Visual Identification *25*
1.4.2 Microscopic Identification *26*
1.4.3 Spectroscopic Identification *27*
1.4.4 Thermal Analysis *29*
1.5 Quantitative Analysis *32*
1.5.1 LD Method *33*
1.5.2 DLS Method *34*
1.5.3 NTA Method *34*
1.5.4 Challenges in Particle Size Analysis *35*
1.6 Quality Control *36*
1.6.1 Internal Deviation *36*
1.6.2 Judgment Error *37*

1.7	Summary and Future Outlooks *37*
	References *37*

2 Occurrence and Removal of Microplastics in Drinking Water Systems *53*
Junyeol Kim, Yongli Z. Wager, Carol Miller, and John Norton

2.1	Introduction *53*
2.2	What Are Microplastics? *55*
2.2.1	Primary and Secondary Microplastics *56*
2.3	The Emergence of Microplastic *57*
2.3.1	Sources *57*
2.3.2	Transformation *59*
2.4	Occurrence of Microplastics in Drinking Water Systems *61*
2.4.1	Abundance *61*
2.4.2	Distribution *64*
2.4.2.1	Size Distribution *65*
2.4.2.2	Morphological Distribution *66*
2.4.3	Composition *67*
2.5	Removal of Microplastics in Drinking Water Systems *68*
2.5.1	Water Treatment Plant *69*
2.5.1.1	Removal of Microplastics by the Overall Process of Water Treatment *69*
2.5.1.2	Removal Rate of Microplastics Depending on the Size *70*
2.5.1.3	Removal Rate Depending on the Type of Microplastics *73*
2.5.1.4	Removal Efficiency Depending on the Composition of Microplastics *74*
2.5.1.5	Removal of Microplastics by Coagulation, Flocculation, and Sedimentation *75*
2.5.1.6	Removal of Microplastics by Filtration *76*
2.5.1.7	Removal of Microplastics by Ozonation *77*
2.5.2	Microplastic Removal in Lab-scale Studies *78*
2.6	Summary and Prospects *80*
	References *82*

3 Occurrence of Microplastics in Wastewater Treatment Plants *91*
Kang Song and Lu Li

3.1	Introduction *91*
3.2	The Abundance and Removal Performance of Microplastics in WWTPs *92*
3.3	The Microplastics Composition in WWTPs *102*
3.3.1	Microplastics Size Distribution *102*
3.3.2	Microplastic Shapes *103*
3.3.3	Microplastic Materials *104*

3.3.4	Microplastic Color	*105*
3.4	Removal of Microplastics in WWTPs and Contribution of Each Process	*106*
3.4.1	Primary Treatment	*106*
3.4.2	Secondary Treatment	*108*
3.4.3	Tertiary Treatment	*108*
3.5	Summary and Future Outlooks	*109*
	References	*110*

4 Effects of Microplastics on Wastewater Treatment Processes *119*
Yan Laam Cheng, Tsz Ching Tse, Ziying Li, Yuguang Wang, and Yiu Fai Tsang

4.1	Biological Treatment Processes	*119*
4.1.1	Conventional Unit Operations and Processes	*119*
4.1.1.1	Suspended-Growth Processes	*120*
4.1.1.2	Attached-Growth Processes	*122*
4.1.1.3	Advanced Wastewater Treatment Processes	*122*
4.2	Interactions Between Sludge and Microplastics	*123*
4.2.1	Activated Sludge	*126*
4.2.2	Aerobic Granular Sludge	*126*
4.2.3	Anaerobic Granular Sludge	*127*
4.3	Effects of Microplastics on Microorganisms and Key Enzymes	*128*
4.3.1	Heterotrophic Bacteria	*128*
4.3.2	Ammonia-Oxidizing Bacteria	*129*
4.3.3	Nitrite-Oxidizing Bacteria	*132*
4.3.4	Key Enzymes	*132*
4.4	Effects on Sludge Stabilization and Dewatering	*134*
4.4.1	Aerobic Digestion	*134*
4.4.2	Dewatering	*136*
4.5	Perspectives	*137*
4.6	Conclusion	*139*
	Acknowledgments	*140*
	References	*140*

5 Microplastics in Sewage Sludge of Wastewater Treatment *147*
Wei Wei, Xingdong Shi, Yu-Ting Zhang, Chen Wang, Yun Wang, and Bing-Jie Ni

5.1	Introduction	*147*
5.2	Occurrence	*149*
5.2.1	Primary Sludge	*150*
5.2.2	Waste-Activated Sludge	*152*

5.2.3 Dewatered Sludge 153
5.3 Effects of Microplastics on Sludge Anaerobic Treatment 155
5.3.1 Methane 155
5.3.2 Short-Chain Fatty Acid 156
5.3.3 Hydrogen 158
5.3.4 Enzyme Activity 158
5.3.5 Microbial Community 159
5.4 Transport of Microplastics from Sludge to Soil and Landfills 160
5.4.1 Transport of Microplastics from Sludge to Soil 160
5.4.2 Transport of Microplastics from Sludge to Landfills 162
5.5 Enhanced Removal of Microplastics from Sludge 162
5.5.1 Thickening and Dehydration 162
5.5.2 Anaerobic Digestion 163
5.5.3 High Temperature Composting 163
5.5.4 Incineration 164
5.6 Summary and Outlook 165
References 166

6 Discharge of Microplastics from Wastewater Treatment Plants 175
Hongbo Chen and Yi Wu
6.1 Introduction 175
6.2 Microplastics Concentrations in Effluent of WWTPs 176
6.2.1 Concentration of Microplastics in Effluent 176
6.2.2 Types of Microplastics in Effluent 176
6.3 Important Source of the Receiving Waters 179
6.3.1 River 180
6.3.2 Lake 182
6.3.3 Sea 184
6.3.3.1 Microplastics on Beaches and Coastal Areas 186
6.3.3.2 Microplastics on the Surface of Ocean Water 186
6.3.3.3 Microplastic Pollution in Polar Regions 186
6.3.4 Sediments 188
6.4 Uptake of Microplastics in Aquatic Organisms 193
6.4.1 Freshwater Organisms 193
6.4.2 Marine Life 195
6.4.3 Soil and Crops 197
6.5 Conclusions and Considerations for Future Work 198
6.5.1 Conclusions 198
6.5.2 Considerations for Future Work 199
Acknowledgments 199
References 199

7	**Microplastics Removal and Degradation in Urban Water Systems** *211*
	Qiuxiang Xu and Bing-Jie Ni
7.1	Introduction *211*
7.2	Use of Separation-based Technology for the Removal of MPs *215*
7.2.1	CFS *215*
7.2.2	Electrocoagulation *219*
7.2.3	Filtration *221*
7.2.4	Membrane Separation *222*
7.2.5	Adsorption *223*
7.3	Photocatalysis Degradation of Microplastics *225*
7.3.1	Zinc Oxide-based Photocatalysis *226*
7.3.2	Titanium Dioxide-based Photocatalysis *227*
7.3.3	Bismuth-based Photocatalysis *229*
7.4	Chemical Oxidation Degradation of Microplastics *229*
7.5	Future Prospects *231*
	References *232*
8	**Microplastics Contamination in Receiving Water Systems** *243*
	Muhammad Junaid and Jun Wang
8.1	Introduction *243*
8.2	Occurrence of Microplastics in Freshwater Resources *247*
8.2.1	River Surface Waters *247*
8.2.2	Lake Surface Waters *248*
8.3	Composition of Microplastics in Freshwater *249*
8.4	Factors Influencing the Aging of Microplastics *251*
8.5	Uptake and Associated Ecological Impacts of Microplastics in Aquatic Organisms *251*
8.5.1	Invertebrates *251*
8.5.2	Waterbirds *253*
8.5.3	Mammals and Megafauna *254*
8.6	Interactions among Microplastics and Microbes (Bacteria) *256*
8.6.1	Microplastic Biofilms: Formation Mechanisms and Characteristics *256*
8.6.2	Factors Affecting Biofilm Formation *257*
8.6.3	Role of Microplastic Biofilms in Genetic Material Transfer *258*
8.6.4	Microplastics as Pathogen Carriers *260*
8.7	Potential Interactions between Microplastics and Humans *260*
8.7.1	Dietary Exposure *260*
8.7.2	Exposure through Inhalation and Dermal Contact *263*
8.7.3	Microplastics' Toxicity in Humans *264*
8.8	Implications and Suggestions *266*
	Acknowledgments *267*
	References *268*

9 Effects of Microplastics on Algae in Receiving Waters 287
Dongbo Wang, Qizi Fu, Xuemei Li, and Xuran Liu
9.1   Introduction 287
9.2   MPs Induced Effect on the Algae: Growth and Populations 289
9.2.1 Effects of MPs on Algae Growth 289
9.2.2 Effects of MPs on Algae Populations 292
9.3   Factors Affecting Toxicity 293
9.3.1 Dosage 293
9.3.2 Size 295
9.3.3 Materials 295
9.4   Combined Effects of MPs with Contaminants towards Algae 296
9.4.1 Antibiotics 297
9.4.2 Heavy Metals 300
9.4.3 Other Emerging Contaminations 301
9.5   Research Gap and Perspective 302
       References 303

10 Effects of Microplastics on Aquatic Organisms in Receiving Waters 315
Gabriela Kalčíková and Ula Rozman
10.1   Introduction 315
10.1.1 Occurrences in Water and Sediment 316
10.1.2 The Concerns about Potential Ecological Risks 318
10.2   Into the Food Chain of Aquatic Animals 319
10.2.1 Accumulation 320
10.2.2 Transfer within the Organizations 321
10.3   Toxicity to Aquatic Organisms 322
10.3.1 Decomposers 322
10.3.2 Producers 323
10.3.3 Consumers 324
10.4   The Sources of Toxicity 326
10.4.1 The Release of Plasticizers and Other Additives 326
10.4.2 The Adsorbed Pollutants 328
10.4.3 Physical Damage 329
10.5   Summary and Outlook 330
        References 331

11 Chemicals Associated with Microplastics in Urban Waters 345
Yali Wang
11.1   Introduction 345
11.1.1 Chemicals in Microplastics and Its Fragments 347

11.1.2	Chemical Additives in Plastic Consumer Products	*356*
11.2	The Release of Chemicals from Microplastics and Environmental Levels	*358*
11.2.1	Phthalic Acid Esters (PAEs)	*359*
11.2.2	Bisphenol A	*362*
	References	*363*

12 Interactions between Microplastics and Contaminants in Urban Waters *373*

Tianyi Luo, Xiaohu Dai, and Bing-Jie Ni

12.1	Introduction	*374*
12.2	Sorption of Contaminants on Microplastics	*375*
12.2.1	Antibiotics	*376*
12.2.2	Heavy Metals	*376*
12.2.3	Organic Pollutants	*380*
12.3	Enrichment of Antibiotic-Resistant Bacteria and Antibiotic Resistance Genes	*383*
12.3.1	Single Selection	*383*
12.3.2	Co-Selection	*384*
12.4	The Effects of Environmental Conditions	*386*
12.4.1	pH	*386*
12.4.2	Temperature	*386*
12.4.3	Salinity	*388*
12.4.4	Weathering/Aging Effect	*388*
12.5	Joint Potential Risks	*390*
12.5.1	For Contaminants Distribution in Aquatic Environment	*390*
12.5.2	For ARGs and ARB Distribution in Aquatic Environment	*391*
12.5.3	For Aquatic Organisms	*392*
12.5.4	For Human Health	*393*
12.6	Conclusion and Recommendations	*395*
	References	*396*

13 Nanoplastics in Urban Waters: Recent Advances in the Knowledge Base *407*

Ilaria Corsi, Elisa Bergami, Ian J. Allan, and Julien Gigault

13.1	Introduction	*407*
13.2	Nanoplastics in the Aquatic Environment	*408*
13.2.1	Nanoplastics or Polymeric Nanoparticles	*408*
13.2.2	Formation Pathways of Nanoplastics	*410*
13.2.3	Source of Nanoplastics	*411*
13.2.4	The Behavior and Environmental Fate of Nanoplastics	*412*

13.2.5	Interaction of Nanoplastics with Contaminants	*414*
13.3	Interactions between Nanoplastics and Aquatic Organisms	*418*
13.3.1	Effects on Aquatic Organisms: From Microalgae to Fish	*419*
13.4	Ingestion of Nanoplastics in Aquatic Organisms	*427*
13.5	Concluding Remarks and Future Recommendation	*429*
	Funding *430*	
	Acknowledgements *430*	
	Competing Interests *431*	
	References *431*	

Index *445*

Notes of Contributors

Ian J. Allan
Norwegian Institute for Water Research (NIVA)
Oslo, Norway

Elisa Bergami
British Antarctic Survey
Natural Environment Research Council
Cambridge, UK

Hongbo Chen
College of Environment and Resources
Xiangtan University
Xiangtan, China

Yan Laam Cheng
Department of Science and Environmental Studies and State Key Laboratory in Marine Pollution
The Education University of Hong Kong, Tai Po
New Territories,
Hong Kong SAR, China

Ilaria Corsi
Department of Physical, Earth and Environmental Sciences
University of Siena
Siena, Italy

Xiaohu Dai
State Key Laboratory of Pollution Control and Resources Reuse
College of Environmental Science and Engineering
Tongji University
Shanghai, PR China

Qizi Fu
College of Environmental Science and Engineering
Hunan University
Changsha, PR China

Julien Gigault
TAKUVIK Laboratory
CNRS/Université Laval
Quebec City, QC, Canada

Xiang Huang
School of Environmental and Chemical Engineering
Organic Compound Pollution Control Engineering
Ministry of Education
Shanghai University
Shanghai, PR China

Muhammad Junaid
Joint Laboratory of Guangdong Province and Hong Kong Region on Marine Bioresource Conservation and Exploitation
College of Marine Sciences
South China Agricultural University
Guangzhou, China

Gabriela Kalčíková
University of Ljubljana
Faculty of Chemistry and Chemical Technology
Ljubljana, Slovenia

Junyeol Kim
Department of Civil and Environmental Engineering,
Wayne State University
Detroit, MI, USA

Man Li
School of Environmental and Chemical Engineering
Organic Compound Pollution Control Engineering
Ministry of Education
Shanghai University
Shanghai, PR China

Xiaowei Li
School of Environmental and Chemical Engineering
Organic Compound Pollution Control Engineering
Ministry of Education
Shanghai University
Shanghai, PR China

Lu Li
State Key Laboratory of Freshwater Ecology and Biotechnology
Institute of Hydrobiology
Chinese Academy of Sciences
Wuhan, China

Xuemei Li
College of Environmental Science and Engineering
Hunan University
Changsha, PR China

Ziying Li
Department of Science and Environmental Studies and
State Key Laboratory in Marine Pollution
The Education University of Hong Kong, Tai Po
New Territories, Hong Kong SAR, China

Lulu Liu
School of Environmental and Chemical Engineering
Organic Compound Pollution Control Engineering
Ministry of Education
Shanghai University
Shanghai, PR China

Xuran Liu
College of Environmental Science
and Engineering
Hunan University
Changsha, PR China

Tianyi Luo
State Key Laboratory of Pollution
Control and Resources Reuse
College of Environmental Science and
Engineering
Tongji University
Shanghai, PR China

Carol Miller
Department of Civil and
Environmental Engineering
Wayne State University
5050 Anthony Wayne Dr.
Detroit, MI, USA

Bing-Jie Ni
School of Civil and Environmental
Engineering
Centre for Technology in Water and
Wastewater
University of Technology Sydney
Sydney, NSW, Australia

John Norton
Energy, Research, & Innovation
Great Lakes Water Authority
Detroit, MI, USA

Ula Rozman
University of Ljubljana
Faculty of Chemistry and Chemical
Technology,
Ljubljana, Slovenia

Xingdong Shi
School of Civil and Environmental
Engineering
Centre for Technology in Water and
Wastewater
University of Technology Sydney
Sydney, NSW, Australia

Kang Song
State Key Laboratory of Freshwater
Ecology and Biotechnology
Institute of Hydrobiology
Chinese Academy of Sciences
Wuhan, China

Yiu Fai Tsang
Department of Science and
Environmental Studies and
State Key Laboratory in Marine
Pollution
The Education University of Hong
Kong, Tai Po
New Territories, Hong Kong SAR,
China

Tsz Ching Tse
Department of Science and
Environmental Studies and
State Key Laboratory in Marine
Pollution
The Education University of Hong
Kong, Tai Po
New Territories, Hong Kong SAR, China

Yongli Z. Wager
Department of Civil and
Environmental Engineering
Wayne State University,
Detroit, MI, USA

Chen Wang
State Key Laboratory of Pollution Control and Resources Reuse
College of Environmental Science and Engineering
Tongji University
Shanghai, PR China

Dongbo Wang
College of Environmental Science and Engineering
Hunan University
Changsha, PR China

Jun Wang
Joint Laboratory of Guangdong Province and Hong Kong Region on Marine Bioresource Conservation and Exploitation
College of Marine Sciences
South China Agricultural University
Guangzhou, China

Yali Wang
Hebei Key Laboratory of Close-to-Nature Restoration Technology of Wetlands
School of Eco-Environment
Xiong'an Institute of Eco-Environment
Institute of Life Science and Green Development
Hebei University
Baoding, Hebei, China

Yuguang Wang
Department of Civil Engineering
University of Nottingham Ningbo China
Ningbo, China

Yun Wang
State Key Laboratory of Pollution Control and Resources Reuse
College of Environmental Science and Engineering
Tongji University
Shanghai, PR China

Wei Wei
School of Civil and Environmental Engineering
Centre for Technology in Water and Wastewater
University of Technology Sydney
Sydney, NSW, Australia

Yi Wu
College of Environment and Resources
Xiangtan University
Xiangtan, China

Qiuxiang Xu
State Key Laboratory of Pollution Control and Resources Reuse
College of Environmental Science and Engineering
Tongji University
Shanghai, PR China

Yu-Ting Zhang
State Key Laboratory of Pollution Control and Resources Reuse
College of Environmental Science and Engineering
Tongji University
Shanghai, PR China

Preface

Plastics are considered to be one of the greatest industrial inventions of the last century. They are widely used and bring great convenience. As global demand grows, global plastic production continues to increase, exceeding 350 billion tons in 2018. However, the convenience of plastics has also sparked a throw-away mentality, leading to a large amount of plastic pollution entering various environments. It is estimated that the total release of plastics will reach 250 million tons by 2025. The discharged plastics in the environment will gradually be decomposed to plastic particles under various environmental stressors such as sunlight, weathering, and erosion. Among them, those with a particle size in the range of 0.1 mm to 5 mm are considered as microplastics. In addition, small artificial plastic particles added to personal products are also a major source of microplastics. Massive usage of plastic products and poor management of plastic waste disposal have resulted in the release of large quantities of microplastics into the environment. Moreover, microplastics are only slowly degradable through weathering and aging; they therefore accumulate and persist in the various environments for years to decades.

Urban waters such as lakes, river, wastewater, drinking water, and others are closely related to human production and life, and are considered to be important paths for microplastics migration and accumulation. Microplastics have been frequently and significantly detected in different urban waters such as wastewater, wastewater treatment plant effluent, surface water, and drinking water. Microplastics have been demonstrated to induce chronic toxicity to living organisms after ingestion, and this potentially toxic effect could be transmitted to humans through the food chain, posing a potential threat to both aquatic species and human health. Therefore, microplastics have aroused increasing concerns, and the problems and its potential risks are considered more serious than plastic pollution.

The objective of this book is to elucidate the current occurrence, fate, and effect of microplastics during urban water management. Specifically, this book emphatically discusses the effect of microplastics on living organisms including microorganisms, algae, and aquatic animals, and the interactions between microplastics and environmental contaminants, providing comprehensive understanding of the environmental risks of microplastics. This book also summarizes relevant methods for detecting, removing, and degrading microplastics, and is expected to provide theoretical guidance for controlling or mitigating microplastics pollution and its environmental risks. In addition, this book also introduces recent advances in nanoplastics – the key decomposition products of microplastics.

As an excellent reference, this book presents in detail the environmental behavior and potential impacts, risks, detection, and removal of microplastics, and provides an outlook for future research directions and developments. We hope that this 13-chapter book will not only provide water scientists and engineers with professional and valuable information about microplastics in urban water management, but more importantly to make the public, especially policy makers, fully aware of the severity of microplastic pollution and its risks in urban water systems.

Finally, the editors sincerely thank all the authors who contributed to this book for their hard work and patience. We solemnly declare that all opinions expressed in this 13-chapter book are those of the authors themselves and not the organizations in which they work.

Bing-Jie Ni
Qiuxiang Xu
Wei Wei

1

Techniques for Microplastics Detection in Urban Water Systems

Xiaowei Li*, Man Li, Lulu Liu, and Xiang Huang

School of Environmental and Chemical Engineering, Organic Compound Pollution Control Engineering, Ministry of Education, Shanghai University, Shanghai, PR China
* Corresponding author

1.1 Introduction

Microplastics analysis in an urban water system mainly contains three steps outlined in Figure 1.1, i.e., sample collection (separation), sample purification (pretreatment or digestion), and sample identification (quantitative analysis), though the methods are not yet standardized. Samples from urban water may use different techniques since microplastics are widely found in various urban water including in freshwater, sediment, wastewater, sewage sludge, and drinking water. Different dimensions in the final analysis also result from the methods for microplastic identification. In addition, potential deviations related to sample contamination and loss are avoided through quality control in most studies.

1.2 Sample Collection and Separation

1.2.1 Freshwater Samples

Sampling methods are classified as three main types, i.e., selective collection, volume-reduced collection, and bulk collection. Different sampling methods are used for microplastics analysis from various kinds of sample matrix, such as water, sediment, or biota. Selective sampling is often used for large microplastics (1–5 mm) on the surface of sediments that are directly collected by the naked eye. Bulk sampling takes all samples to the laboratory without volume reduction and is usually applied for sediment and biological samples. In contrast to bulk sampling, volume-reduced methods transfer concentrated samples to the laboratory [1]. The microplastic content in water samples is often low, and thus their analysis

Microplastics in Urban Water Management, First Edition. Edited by Bing-Jie Ni, Qiuxiang Xu, and Wei Wei.
© 2023 John Wiley & Sons, Inc. Published 2023 by John Wiley & Sons, Inc.

1 Techniques for Microplastics Detection in Urban Water Systems

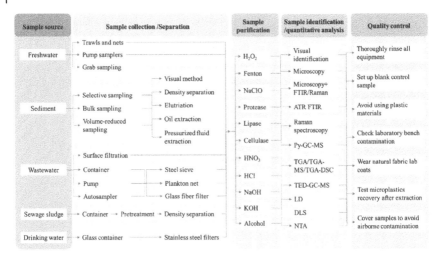

Figure 1.1 The steps and techniques for microplastics analysis in urban water system.

needs large sample volumes [2]. The sample volume is reduced by concentration in the field, which is beneficial for transport. In this section, several main volume-reduced methods for water sample collection are introduced and discussed, such as trawls and net, pump sampling and grab sampling.

(1) Trawls and nets

Trawling nets are often deployed underwater, and towed at low speed over a set distance [3, 4]. Each sampling area is determined by multiplying the width of the trawling net by the towing distance, while the water volume passing through the net is either measured by flowmeter or calculated from the towing distance and net diameter [5]. However, water volume is difficult to determine exactly as the water waves, wind, and boat movement have a continuous effect on the net's immersion depth.

Multiple factors, such as the trawling time, location, length, and tidal strength, have an influence on water sample collection, and thus should receive careful consideration in obtaining a representative sample. In one study, Campanale et al. [6] collected microplastics from the water of the Ofanto River through three plankton nets installed in the center of the river, thereby reducing the space and temporal variation of the samples. If nets are not statically placed, the required trawl distance depends on the microplastic concentration. In addition, further investigation of the effect of the prevailing wind direction with reference to the trawl direction on microplastic content and size collected within trawling nets is needed [7]. In these methods, sample collection nets with a single mesh size are sometimes blocked by suspended materials such as organic matter or phytoplankton [5, 8].

Both the content and size distribution of collected microplastics closely depend on the mesh size of the nets. For example, the number of fiber particles collected through smaller mesh nets might be greater than through nets with a larger mesh. Trawl mesh size often is equal to or greater than 300 μm, frequently causing underestimation of microplastic content in collected samples as a large amount of small microplastics can pass through the nets [9]. Therefore, tandem nets with different mesh sizes are strongly recommended foe investigating small microplastics. Tandem nets can collect sufficient water samples while avoiding net clogging [10]. Dris et al. [11] found that the content of fibers collected by using an 80-μm mesh is 250-times more than that using a 330-μm mesh. Microplastics in surface water from Lake Superior were investigated using a double neuston trawling net (500-μm mesh) by Cox et al. [12]. The results showed no improvement in the double net samples, implying that a single net is enough to gain representative data of microplastic particles in a water sample. Michida et al. investigated microplastics in ocean surface water by comparing a manta net with a neuston net and found a higher content of microplastics using the manta net [13]. The neuston net appears superior in rough water, while a manta net may be fit to collect samples in calm water [10]. A modified basket sampler with nets of different mesh sizes was used to collect the surface, middle, and bottom water of the Austrian Danube River, indicating that a stable equipment carrier plays a vital role [8]. In practice, full information about microplastics in the samples is not available using nets alone, as numerous small microplastics such as fibers pass through the nets. Several net styles used for microplastic sampling are shown in Figure 1.2.

(2) Pump and grab sampling

Microplastic samples are collected through pump sampling, pumping water samples manually or by motor, then filtering with an inline filter. A bucket is applied for water collection and a net is sieved through the sample in the field for grab sampling [15–17]. In grab sampling, a sample bottle is submerged, filled with surface water, and taken for laboratory analysis [18, 19]. Different sample volumes are collected at different depths using pump or grab samplers. The grab sampler has a limited sampling area, possibly causing poor representation, as spatial distribution of microplastics is highly variable. Therefore, samples should be collected in replicates [7]. Pump sampling can take large sample volumes and is more suitable for areas with low microplastic content [20]. The sample volume can vary from 5 mL to 500 L [21]. In addition, more microplastic particles of smaller size are collected in pump and grab sampling, compensating for the limitations of trawling and nets. Comprehensive information on microplastics in

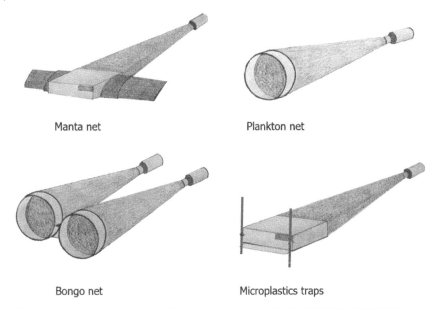

Figure 1.2 Several nets applied for microplastic sampling [14] / MDPI / CC BY 4.0.

water samples can be gained by combining volume-reduced collection, such as net and trawl, with bulk sampling, such as grab collection [22].

(3) Comparison of different sampling methods

Tamminga et al. compared the pump sampling and manta trawling methods in Lake Tollense, Germany [23] and found considerable differences in microplastic content, shape, and size. Fibers and small microplastics are readily collected by pump sampling, while they cannot be sufficiently retained through manta trawl sampling. Additionally, it is necessary to filter a large volume of water during pump sampling to obtain representative results. Large microplastics with lower abundance are more readily collected by manta trawl sampling compared with pump sampling. A positive-skewed distribution with a peak at 500–600 μm in size was found in fibers detected by pump sampling, while the fiber distribution was heterogeneous in the manta samples. Polyethylene (PE) was the dominant microplastic identified by the manta trawl method, while most of the polymer detected in pump samples was polyethylene terephthalate (PET) [23].

Both bulk and volume-reduced sampling were used to assess the microplastic contamination of the Saigon River, Vietnam by Lahens and colleagues [24]. A bulk water sample was collected using a bucket for fiber analysis, while a volume-reduced sample was taken by 300-μm plankton net. Su et al. reported that the average content of microplastics in water samples is lower using bulk sampling than that using plankton-net sampling [25]. Generally, microplastic abundance and size distribution depend on water sample volume. Barrows et al. [26] compared grab and neuston net sampling, and found three orders of magnitude lower amount of microplastics were collected by the net method compared to grab sampling [26]. Karlsson et al. reported that the in situ pump filtration method is superior to manta trawl sampling for point samples because of more accurate volume determination and more filter size options [27]. The trawling method is more suitable for collecting water samples from a larger area and avoids issues associated with patchiness, while areas with a higher abundance of microplastics may be more appropriately collected by pump sampling. Therefore, an effective method for microplastic analysis in a freshwater environment is to combine the volume-reduced net-based sampler such as a trawl net, and bulk sampling such as a pump and grab sampler.

1.2.2 Freshwater Sediments

Microplastics analysis in freshwater sediment samples has not been standardized in terms of sample collection, pretreatment, quantification, and identification [28]. Comparison of the results from different studies is exceedingly difficult [29]. Individual investigations have large dissimilarity in sample size, extraction methods, and reporting units because of the difference in sampling methods [30]. The sampling, pretreatment, isolation, and identification methods for microplastics from sediment are outlined in Figure 1.3.

(1) Sediment sampling

Freshwater sediment can be collected through volume-reduced sampling, bulk sampling, or selective sampling [3]. As the name suggest, in volume-reduced sampling, only a portion of the sample is taken to the laboratory for further analysis after the sample volume is reduced [3]. Bulk sampling refers to methods where all of the collected matrixes from the field are brought back to the laboratory for microplastics detection. Sampling then takes less time and is more standardized [31, 32]. In selective sampling, tweezers are used to directly remove microplastics from sediment samples using naked-eye visualization. This type of sampling is often applied when there is a high content of large microplastics (1–5 mm) that are easily identified [30]. Only 1 of 38 freshwater sediment studies utilized selective sampling for the collection of microplastic pellets [33].

A variety of sampling tools were applied for different freshwater sediment sampling. A stainless shovel, spatula, and spoon were used to take riverbank or lakeshore sediment samples [34], while total microplastic abundance and vertical distribution

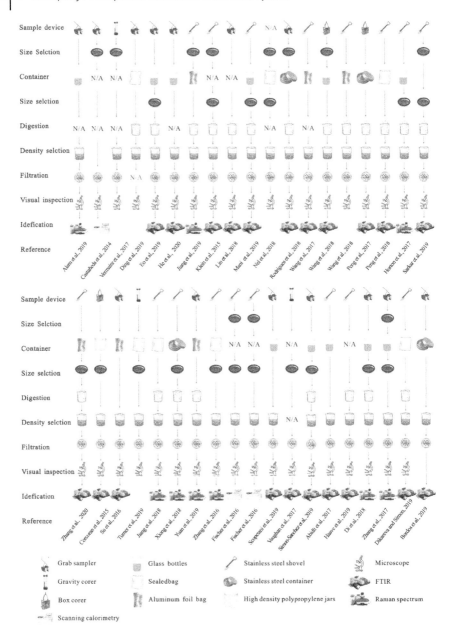

Figure 1.3 Collection, digestion, separation, and inspection methods for microplastics from sediment [28] / with permission of Elsevier.

from sediment samples at the riparian zone or lakeshore were investigated through taking bulk samples from the surface and different depths. The submerged sediment was collected using a vessel or grab sampler such as a Peterson sampler or a Van Veen grab sampler that was dropped to the river or lake bottom [35, 36].

The reporting units depend on the sampling tool [3]. Area is used as a sampling unit in around half of the studies, ranging from 250–930 cm^2. The weight and volume of sediment samples are also applied as sampling units, varying from 0.2–5 kg and 1–3.5 L, respectively. Volume is recommended as the sample unit by MSDF Guidance, as the wet weight of sediments is related to the water content and type of the samples [28]. However, the weight of the dried sediment is used to characterize the abundance of microplastics by the National Ocean and Atmospheric Administration (NOAA).

The collection position, depths, and distance from a human activity center have large influence on microplastic content in sediments [37]. Besley et al. also reported that microplastic abundance is closely related to the collected depth [38]. The average content of microplastics in the samples taken from the top 2–10 cm is lower than that from top 1–5 cm [38]. Specific sampling depth may be required in order to accurately determine microplastic abundance in sediment samples because of uneven vertical distribution [38, 39]. In fact, different studies have used various sampling depths. The samples in some studies were collected from the top 10 cm or deeper, while the majority of collected samples were from the top 2–3 cm or 5 cm of the sediment. Depth was not defined in some studies. Sediment collection from the top 5 cm at least 5 m apart is recommended by the MSFD with a minimum of five replicates [28].

Sampling sites are recommended to be placed at 100 m parallel to the oceanic waterline, according to the standardization protocol for sediment sampling [40]. However, as the range of the ocean is much larger than that of lakeshore or riverbanks in freshwater environments, sampling sites in freshwater sediment may be placed closer to lakes and rivers. The recommended sampling parameters are (1) a collected depth of 5 cm; (2) a collected unit of 30×30 cm area; (3) a sampling tool of metal shovel; and (4) a reserved vessel of glass jar [40]. A box corer or van Veen grabber is recommended for collection of subtidal sediment samples [40]. Stock et al. recommend collecting the sediment sample using a drill corer or a Pürckhauer to prevent interference among samples and avoid sample loss [40].

(2) Sediment sample separation (or extraction)

Microplastics in sediment samples are typically extracted or separated using visual and density-separation methods. Microplastic extraction is extensively carried out through density separation, but the method has some shortcomings. Two new methods, i.e., the oil extraction and the improved pressurized fluid extraction, have been proposed to overcome the limitations of traditional density separation. This section focuses on the introduction to the principles, advantages, and disadvantages of these separation methods [41].

In the visual method, suspended microplastic particles are selected from the collected samples using tweezers by direct or microscope observation. The method is fast, well-suited, and economic, but its disadvantage is inaccuracy because of easy affection by the observer parallax. Hidalgo et al. reported that when microplastics isolated by visual detection are further analyzed using FT-IR spectroscopy, real microplastics account for only 30% [3].

Density separation is another commonly used method to isolate microplastics from sediment samples. In brief, the samples are first dried, then put into an extraction solution such as saturated NaCl [42]. Next, the mixture is stirred using a glass rod for 2 min, then left to settle. Microplastics with lower density than the solution float to the surface and particles with higher density submerge to the bottom. Microplastics are then collected through filtering the upper water. The density extraction method does not require complex equipment and is conveniently and simply operated. However, it is vital to choose a suitable separation solution because of the profound effect of the extraction solution on detected microplastic composition and distribution. Saturated NaCl [43], $ZnCl_2$ [44, 45], and NaI solutions [46] are the most common extraction solutions. For instance, NaCl solution is appropriate for the extraction of low-density microplastics [47], and $ZnCl_2$ or NaI solutions are suitable for the extraction of high-density microplastics. In addition, different extraction solutions have different cost and influence on the environment. Overall, there is no general consensus about which extraction solution gives the best results.

The density extraction method is further optimized in elutriation. In elutriation, an extraction device is used to isolate plastic particles from sediment based on their density difference [48]. Meanwhile, the extraction rate of microplastics can be optimized by improving the flow rate of the elution device and the diameter of the elution column [49]. Low-density microplastics in the water flow upward in the eluting device, and then are captured by a filter screen in flotation equipment.

An oil extraction method has been developed to avoid the problems of the above extraction solutions. Vegetable oil is used as the extraction solution to separate microplastics based on the lipophilicity of the plastics [50]. This method results in a high recovery rate of more than 90% for all types of microplastics. The method takes only 60–120 min to extract microplastics from samples [51]. However, spectral analysis of microplastics is prevented by the oil, so further treatment is required after extraction.

Organic pollutants are often extracted using pressurized fluid extraction. Recently, it was reported that pressurized fluid extraction was also used to isolate microplastics from sediments. In pressurized fluid extraction, the solvents are heated to subcritical temperature and pressure in a pressurized fluid extractor [52]. The method is not limited by microplastic size or density and is suitable for various types of microplastics. It can separate all types of plastics, including PE, polypropylene (PP), polystyrene (PS), and polyvinyl chloride (PVC). In addition, this method is easily automated and simply operated. However, some readily observable characteristics of microplastics are destroyed during the pressurized fluid extraction, making identification and quantitation more difficult

In sum, many methods have been proposed for microplastic extraction in freshwater sediments, but extraction processes still lack standardization. Further study is required to develop an accurate, objective, and standard operating process in which impurities are removed from the sample as much as possible, microplastics are recovered, and the interference of external factors is removed to the maximum extent.

1.2.3 Wastewater Samples

At present, some sampling methods have been put forward to collect wastewater samples for microplastics analysis, including container collection, pumping sampling, and surface filtration [53–58].

(1) Container collection and pumping sampling

Container collection has a straightforward operation for wastewater sampling, but the volumes of collected samples are very limited. Therefore, this method is appropriate to collect the inlet water in WWTPs because of its high content of microplastic particles [59]. was employed In the study of Magnusson and Norén [60], a Ruttner sampler composed of a cylinder was submersed into the wastewater and closed with a plummet. A stainless-steel filter with a mesh size of 300 μm was then used to filter the wastewater sample.

Different from the container collection, pumping sampling coupled with filtration is often employed for collecting effluent samples in WWTPs for the ability to take hundreds of liters of wastewater. As shown in Figure 1.4, a mobile pumping device is used to develop a newly-designed sampling method [61]. This sampling equipment consists of a filter shell of a stainless-steel cartridge filter with a mesh size of 10 μm, and a PVC hose connecting a thickening end, flow meter, and membrane pump. Wastewater is taken to wash the pumping system for 5 min before sampling. Then, the thickening end is set below ca. 10 cm of the surface of wastewater for sampling.

A sampling device (Figure 1.5) consisting of four stainless steel sieves was developed by Ziajahromi et al. [62], and used to collect microplastics in effluents. The reported mesh sizes are 25, 100, 190, and 500 μm, respectively. All mesh sieves are stacked together with the largest one on top. PVC covers are used to protect the stacked sieves and a baffle is put in the inlet for maintaining uniform distribution

Figure 1.4 A sampling method based on mobile pumping device [61] / with permission of Elsevier.

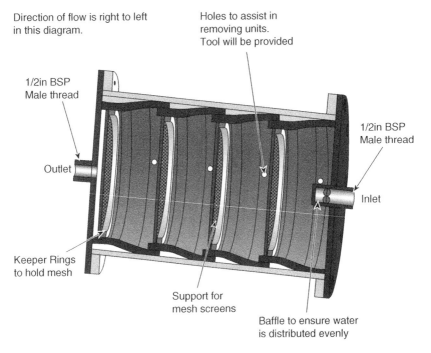

Figure 1.5 A stacked sampling device containing four different stainless steel mesh sieves. Adapted from [62].

of the wastewater samples. Two main advantages of this sampling device are as follows: (1) to classify microplastics in situ according to their size; and (2) to process numerous samples continuously. The recovery efficiency of PS through this sampling method reaches 92% for a 25-μm mesh sieve, and 99% for 500-μm mesh sieve. The method is very effective at trapping microplastic particles of all sizes. In addition, no contamination was found during the sampling process [54].

(2) Surface filtration

Thousands of cubic meters of wastewater can be efficiently collected in surface filtration. As shown in Figure 1.6, a surface filtration [63] was used to skim water at an effluent outfall. Different times were used during the sampling process because of varying flow rate and water quality. However, several limitations exist in surface filtration: (1) it is only suitable for waterfalls; and (2) surface filtration in an open channel inevitably leads to airborne contamination. In addition, surface filtration may underestimate the amount of microplastics, since only low-density microplastics are collected when skimming the water surface [53].

Sample representativeness should be considered during collection, as wastewater has relatively low content of microplastics with heterogeneous spatial and temporal distributions. An increase in sampling volume and collection of 24-h composite samples were attempted to obtain a representative sample [59, 64]. A standardized guide for microplastics sampling needs to be developed, offering efficient and judicious selection of the proper sampling strategy and frequency to reduce sampling errors and enhance data reproducibility. The uncertainty of micropollutants sampling in a wastewater system can be reduced by selecting the sampling mode and frequency based on study aim and flow characteristics [65, 66]. These basic guidelines for micropollutants sampling can be used to design microplastics sampling methods. However, other factors should also be

Figure 1.6 Microplastics sampling using a surface filtration [53, 63].

considered, such as the particle dynamics associated with microplastic density and geometry.

Microplastics collected in wastewater samples are often concentrated by filtration. Therefore, the collected microplastic content is profoundly affected by the mesh and pore sizes of screens and filters [60]. At present, the mesh and pore sizes of the sieves applied are not consistent in reported studies. Mesh sizes can range 1–500 μm [53]. A stack of sieve pans has been also applied for filtration in many studies (Figure 1.5). An increase in the total volume of filtered samples is allowed, and the size categories of microplastics can be distinguished in this method. However, size categorization based on mesh size may be inaccurate. Michielssen et al. [67] reported that the sieves would not be passed through by some particles because of their irregular shapes. However, smaller filters may allow fibers to pass longitudinally because of their specific morphology [61, 62]. A neuston plankton net and manta trawl can also be used to extract microplastics in situ from surface water [39, 68]. Microplastics in freshwater or seawater have been collected using these methods, but these methods have not yet been used to sample microplastics in WWTPs.

Container collection and pumping coupled with filtration are widely used for microplastics collection. Pumping coupled with filtration and surface filtration is applicable for collecting microplastics in effluent samples as these methods can take a large volume of samples [54].

1.2.4 Sludge Samples

(1) Sludge sample collection

Sludge is a complex matrix composed of organic materials, microorganisms, and inorganic particles, and thus its sampling is radically different from wastewater collection [69]. Sludge samples cannot be collected by automatic samplers for 2–24 h like wastewater sampling. A common method of sludge collection is to take a 5–20 g of sludge sample from sludge treatment units [53]. It appears easy to collect sludge samples at one sampling site, but the characteristics of daily influent wastewaters influence sludge sampling. Therefore, microplastic concentrations in sludge samples should be continuously determined at different time frames. The collected sludge sample is often stored at 4°C or −20°C before further microplastic extraction in the laboratory [53, 70–76]. The collected sludge amount and transfer conditions lack standardization and are most likely different in related studies.

(2) Sludge sample pretreatment

Flocs in sewage sludge are stacked with cells because of the presence of extracellular polymeric substances (EPS) [77] that can capture microplastics, thus easily limiting microplastic extraction in the sludge [78]. Therefore, pretreatment methods are used to remove organic matter using various oxidants [69]. A crucial factor is that the oxidant pretreatment has no influence on the integrity of microplastics while oxidizing the organic matter [79]. Some chemicals, such as Fenton's reagent [80–82] and 30–35% H_2O_2 [73, 83], have been used to oxidize organic matter in sludge in some recent studies. Four different pretreatments, i.e., H_2O_2, Fenton's reagents, NaOH, and KOH, were developed by Hurley et al. [80] to remove organic matter in complex solid substrates during microplastic extraction. The results showed that several polymer types undergo surface degradation during NaOH pretreatment. The optimum protocol for microplastic extraction from sludge was Fenton pretreatment followed by density separation. An ice bath is used to control the reaction temperature below 40°C because of the exothermic reaction of Fenton's reagent, and to protect the microplastics [69, 80]. Similarly, Li et al. used 270–500 μm of six different microplastics (PET, polyamide (PA), polymethyl methacrylate (PMMA), PP, PS, and PE) to investigate the microplastic extraction efficiency from four solid matrices, i.e., cattle manure, sewage sludge, soil, and sediment [78]. The lowest extraction efficiency (82.7%) was found in sewage sludge samples, possibly resulting from the presence of EPS preventing microplastic extraction. Therefore, five different pretreatments including H_2O_2 (30%), and Fenton's reagent, HNO_3 or HCl, and NaOH were used to improve microplastic extraction in sewage sludge. First, 30 mL of each reagent was used to treat 30 g of sludge sample (ww) with the addition of each microplastic for 24 h at 60°C. Then, microplastics in the matrices were extracted with saturated NaCl and $ZnCl_2$ solutions, respectively, and the upper water was filtered using a 37-μm stainless steel mesh. The results showed that a raw sludge sample has lower microplastic extraction efficiency than sludge treated by H_2O_2, Fenton's reagent, and 1 M HNO_3. In addition, microplastics were least affected by oxidation procedures with H_2O_2 and Fenton's reagent, while both acidic and alkali pretreatments had a strong effect on PA and PET [78]. Raju et al. [82] used Fenton's reagent to separate microplastics from influent and effluent of wastewater and waste-activated sludge. The sludge samples required doubled exposure time to Fenton's reagent because of their higher organic content compared with the wastewater sample. Then, NaI extraction (1.8 g/cm^3) for waste sludge samples was carried out after the samples settled overnight at ambient temperature. Decreasing sized sieves (2 mm, 1 mm, 500 μm, 250 μm, 125 μm, 53 μm, 38 μm) were used to filter the supernatant liquor of the samples, and microplastics of different sizes were successively extracted. Then, Rose Bengal was used to stain the samples, and nonplastics were marked. Finally, the remaining filtrate of the 38-μm sieve was filtered using 1.5-μm cellulose filters [82]. Edo et al. [84] used 30 mL of H_2O_2 (33%) to pretreat 1 g of sludge samples at 50°C. Saturated NaCl (1.2 g/cm^3) solution was used for density-based extraction after the H_2O_2

treatment, and the solution was then agitated for 24 h and incubated for another 24 h. Finally, the supernatant liquor was filtered. Overall, oxidation with H_2O_2 (30–35%) or Fenton's reagent are the optimal pretreatment procedure for removing sludge organic matter according to this study. During the oxidation process, organic groups in sewage sludge are effectively removed, but microplastics are not degraded. However, before the most efficient oxidation conditions are chosen, an investigation of the removal rate of organic matter is recommended [85].

(3) Sludge sample extraction

The most common method for removing inorganic materials in the sludge samples is based on density separation using salt solution (usually NaCl) [73]. After high-density solution is mixed with sludge samples, denser inorganic materials in the samples sink to the bottom of the solution, while microplastics float in the upper water [69]. The densities of microplastics range 0.90–1.6 g/cm^3 [79]. Not all polymers float to the surface in NaCl solution because of its low density (1.2 g/cm^3), although NaCl is a cheap and environmentally friendly salt. Most polymers can be extracted using $ZnCl_2$ (1.5–1.7 g/cm^3), sodium polytungstate (1.4–1.5 g/cm^3), or NaI (1.6–1.8 g/cm^3) solutions because of their higher density [86]. However, the costs of the density-separation reagents are high, and they are also detrimental to the environment. Kedzierski et al. [87] investigated recovery and reuse of NaI and found no significant difference in density after 10 application cycles (total loss of 35.9%). Rodrigues et al. [88] reported the detection and recovery of all polymer types, and the recovery rate of $ZnCl_2$ solution is higher than 95%. Consequently, there are promising ways to recycle and reuse these chemicals to offset their costs. As a result, denser solutions are recommended for extracting all types of polymers from sludge samples because they are more useful. The last step of pretreatment should be the filtration process. Filters with a proper size are used to collect floating microplastics after density-based separation [89–94].

1.2.5 Drinking Water Samples

(1) Drinking-water sample collection

Microplastic contamination in globally-purchased bottled water was investigated [95–98]. The water is packed in bottles made of various materials like plastic, glass, and carton. The microplastic results are influenced by the diversity of the materials [99]. Samples of tap water were collected around the world by volunteers in another study [100]. In the sampling procedure, after the tap water is run for 1 min, a 500 mL bottle of high-density polyethylene (HDPE) is filled to the overflowing point then emptied. After the fill-and-empty cycle was conducted twice, each bottle was filled a third time and capped [100]. Every bottle was rinsed through the fill- and-empty process before final sample collection.

Influent and effluent samples were collected from drinking water treatment plants (DWTP) located in northwestern Germany [101] and in urban areas in the Czech Republic [102], respectively. Water from individual households was sampled at the meter and at the tap of the distribution water system in the DWTP [101]. Glass containers were used to collect a large number of drinking water samples (Table 1.1), and the samples were then filtered using 3-μm cartridge stainless steel mesh [101] or 5-μm PTFE membrane mesh [102].

Table 1.1 Methods of microplastics analysis in drinking water (N = Number of Studies) [99] / with permission of American chemical Society.

Water sample	Sample amount	Sample collection	Identification method	Comments	
Bottled water (N = 4)	0.5–2 L	Vacuum filtration of samples ($N = 2$)	Micro-Raman spectroscopy	Detection of polymer composition; determination of microplastic particles; detection of small microplastic particles (1 μm)	[95, 96]
		Add the Nile red solution ($N = 1$)	Optical microscope; FTIR spectroscopy	FTIR was only used for confirmation for some particles larger than 100 μm	[98]
		Add the EDTA solution ($N = 1$)	Micro-Raman spectroscopy	Analysis of a small sample area (4.4%)	[97]
tap water (N = 1)	0.46–0.6 L	Add the rose-Bengal solution	Optical microscope for stained particles	False positives might be gotten	[100]
DWTP water (N = 2)	27 L	Digestion of wet peroxide solution	FTIR microscope	The whole filter was analyzed	[101]
	2500 L		FTIR for analysis of particles larger than 10 μm; and Raman microscope for analysis of particles (1–10 μm) by	Each filter analyzed only 25% of the particles	[102]

(2) Drinking-water sample treatment

Compared with wastewater samples, bottled water samples are more easily processed since they do not require chemical digestion, density separation, or centrifugation. Two studies used staining methods [98, 100], while ethylene diamine tetraacetic acid tetrasodium (EDTA) salt was added in another study to avoid precipitation of minerals and facilitate the analysis of microplastic particles [97]. However, the method with EDTA addition for microplastic analysis has not been used or replicated, and thus requires further corroboration. No sample processing was used in two additional studies [95, 96].

Vacuum filtration was used to treat samples in all studies, and then the filters were stored in a sealed Petri dishes for further analysis. The Petri dishes were made of glass [97, 101, 102], polystyrol [95], or other unreported materials [96, 98, 100]. The studies used various types of membrane filters including gold- or aluminum-coated polycarbonate [95, 97], glass fibers [98], or cellulose nitrate [96, 100]. The aperture of the filters ranged 0.4–2.5 μm.

H_2O_2 solution was used to pretreat the collected samples from the DWTPs, removing any possible organic film [101, 102]. Then, 0.2-μm aluminum oxide filters were used to filter the samples. In all studies, any possible airborne pollution was avoided by using a laminar flow box, and any further external contamination was prevented by using a cleaning procedure along with 100% cotton lab coats and glass equipment. All studies used blanks and controls and processed the samples in the laboratory [99].

1.3 Sample Purification

After microplastics are collected and separated from the urban water system, some residual organic matter is often found in the samples because there is a plethora of organic matter in the samples, especially wastewater and sewage sludge. At present, organic matter in the samples is removed through three types of purifications, i.e., wet peroxidation, enzymatic degradation, alkaline and acid treatment.

1.3.1 Wet Peroxidation

The organic matter from samples of the urban water system is often removed using wet peroxidation (WPO). There are many useful oxidants for organic matter, such as H_2O_2, NaClO, and Fenton's reagent.

The most common method for pretreating samples is the NOAA method (48%). This method is designed for collecting samples in marine environments. Wastewater samples can also feasibly be treated by this method because of their large amount of complex matrix and organic particulate matter. Organic matter digestion is carried out using H_2O_2, generating hydroxyl radicals with strong oxidation capacity under Fe(II) solution in the method. Samples are usually then extracted by density separation after digestion. Commonly used solvents include aqueous sodium chloride and zinc chloride. Low-density microplastics float to the surface during density separation, and high-density particles settle to the bottom. The supernatant water is filtered using a filter with sizes ranging 0.7–125 μm. The digestion steps need to be repeated several times to degrade all the organic matter, increasing time and cost of the process compared with other methods [99].

Recent studies have used oxidants such as Fenton's reagent and H_2O_2 to digest samples under different protocols. Studies have used PP, PE, PS, PVC, and PET microplastics to investigate the efficiency of H_2O_2 for organic matter removal using three protocols. A 100:1 ratio of reagents to microplastics is used to assess the effect of H_2O_2 on microplastics, showing that the effects of oxidants are negligible. In the first protocol, 35% H_2O_2 is put into the sample and allowed to stand for seven days at room temperature. Then, water or 80% ethanol solution is used to rinse the residue sample, followed by filtration. In the second protocol, a 30% H_2O_2 solution is added to the sample and left for seven days at 55°C. The residue is then also rinsed with water, followed by filtration. For the third protocol, the addition of 35% H_2O_2 to the biota sample is reacted at 60°C for four days, resulting in nearly 100% removal of organic matter. However, good removal rates are also reported in other studies at lower temperatures (from 25°C to 50°C) [103].

1.3.2 Enzymatic Degradation

Enzymatic degradation is an emerging method for extracting microplastics from organic matter. A mixture of engineered enzymes (e.g., amylase, lipase, chitinase, cellulase, and protease) are used to treat microplastic samples for organic matter degradation [104, 105]. Cole et al. [104] reported the efficiency of digestion methods using HCl, NaOH, and Proteinase-K alone, or in combination with ultrasound. The results showed that the highest digestion efficiency (by weight) of 88.9% is found using enzyme digestion alone. A multistep and enzymatic impregnation method was applied for the treatment of wastewater samples by Mintenig et al. [61]. Protease, lipase, and cellulase were combined with H_2O_2 (35%, in some samples) and sodium dodecyl sulfate (SDS, 5% w/vol), then the mixture was added to the sample for digestion. However, 13 days were required for the entire degradation process. Another study further assessed the effectiveness of removing organic matter using enzymes [106]. Some procedures were improved in the study, such as optimized conditions for incubation, an increased concentration of SDS, a change in buffer composition, and the addition of two optimized enzymes to a sample with a large amount of organic matter. The improved methods will enhance the removal efficiency of the enzyme for organic matter such as polysaccharide and lipids. Various environmental samples (e.g., seawater, freshwater, wastewater, plankton, sediment, and aquatic organ samples) can be digested using the modified protocol.

Enzymatic digestion (such as amylase, chitinase, cellulose, lipase, and proteinase) may be the most promising way to digest organic matter in microplastic samples because it is less hazardous and friendly to microplastics. Cole et al. [104] used Protease-K to treat plankton-rich seawater samples, and 88% digestion efficiency was achieved. An increase in the treatment time, enzymatic concentration, and reacting temperature (50°C) enhanced the efficacy up to more than 97%. The treatment caused no degradation of the microplastic sample. Alternative enzymes have also been developed for bivalve tissue or plankton, sediments, and biota, because of the excessive cost of Protease-K and the complexity of the operation. Some enzymes such as trypsin, collagenase, and papain, all cheaper than Protease-K, were used in the study of Courtene et al. [107]. One of the most economically viable enzymes was also proposed by Catarino et al. [108]. von Friesen et al. [109] investigated the effects of pancreatic enzymatic digestion on identification of 10 polymers through surface functional groups before and after enzyme digestion using FTIR, and no significant changes were found in the microplastics. Compared with other digestion methods, the steps of enzyme digestion vary in different studies, and the application of enzyme digestion for organic matter removal in WWTP samples has been limited. Enzyme digestion of effluent samples from wastewater treatment plants has been carried out using proteases, lipases, and cellulases in another study [61]. Organic material removal and microplastic extraction were satisfactorily carried out in the study, but an additional 10 days was required, and sample loss and contamination may happen in several procedures. Enzymatic digestion needs to be compared with other digestion techniques when used to treat wastewater samples [110]. In addition, enzyme efficiency varies depending on the types of organic matter. Thus, the combination of enzymatic digestion and other extraction approaches such as H_2O_2 and Fenton's reagent are proposed to enhance the purification [111]. Cellulase enzyme was used to degrade fibers in wastewater samples and Fenton's reagent was applied to oxidize other organic matter in the study of Simon et al. [112].

1.3.3 Alkaline and Acid Treatment

Alkali and acids can also be used to digest organic matter of samples. HNO_3 or HCl are often used in acid digestion. Although microplastics have strong oxidation resistance in samples with high organic matter content, melting, yellowing, and destruction of microplastics occur during acid digestion at a temperature above 60°C, decreasing the abundance of microplastics. Alkali digestion is often conducted with KOH or NaOH, and organic matter in samples of sediments and biota can be removed at 60°C for 12–24 h. However, discoloration and degradation of microplastics also occur during alkaline digestion [103].

The polymer and its structure may be damaged during acid treatment with H_2SO_4 or HNO_3, or alkaline digestion. Therefore, it is suggested to use low acid concentration during acid digestion, resulting in a decrease in the removal efficiency of organic matter. Cole et al. [104] found that the application of 1 M HCl led to more than 80% organic matter removal. Enders et al. [113] selected 21 polymers to assess the tolerance of plastics to different reagents during chemical digestions. In the study, an acid mixture of HNO_3 and $HClO_4$ (4:1) was added to the samples for acid digestion, the mixture was left at room temperature for 5 h, then heated to 80°C for 20 min. PA, polyurethane (PU), and black tire rubber elastomers are the most affected by chemical digestion and are dissolved completely. Other incompletely dissolved polymers exhibited some discoloration, such as polycarbonate (PC), PS, and PET. No effect was observed for PP, HDPE, low-density polyethylene (LDPE), polytetraflurorethylene (PTFE), and EVA. Heating during chemical digestion also can have negative effects on the polymers in addition to chemical reagents. Analysis by Raman spectroscopy shows no considerable changes in the polymers other than acrylonitrile butadiene styrene (ABS) after chemical digestion. The spectra of other digested polymers are similar to those of the original polymers. However, some signs of degradation, peak deviation, and weakening of the main peak were found for PS and PVC.

Nylon, HDPE, PS, poly-1,4-butanediol esters, and PVC were soaked in HNO_3 (55%) solution for 1 month in the study of Naidoo et al. [114]. The results showed that all nylons are completely decomposed after 24 hours, with no notable change found for other plastics, implying resistance to acid digestion. Catarino et al. [108] also investigated acid digestion with HNO_3(35%) to the intact structure of plastic and found that the treatment caused a complete dissolution of nylon. A melting effect was also shown in PET and HDPE. The tolerance of polymers to digestion should be fully investigated before digestion is used to treat an actual sample. Acidic reagents can erode certain polymers and lead to the decomposition of certain plastics, resulting in erroneous results during sample counting [110].

Cole et al. [104] reported that microplastics are damaged by 10M of NaOH at 60°C. In addition, low-pH intolerant polymers, such as PA and PS, are destroyed by strongly oxidizing acids such as H_2SO_4 and HNO_3 [48]. Heating or microwave digestion at 110–120°C is often carried out with acid treatment, but some microplastics melt by 90°C [63]. Isopropanol was used for organic matter removal attached to microplastics in only one study, but removal efficiency was not tested [71].

1.3.4 Influence of Chemical Purification on Microplastic Property

Sample digestion may cause an impact on microplastics with varying degrees. The surface area of the polymer decreased using H_2O_2 for chemical purification according to the study of Nuelle et al. [115]. H_2O_2 treatment destroyed PA particles severely at 70°C [80]. However, Tagg et al. [56] demonstrated that the FTIR spectra of digested microplastics were similar to that of pristine microplastics after seven days of exposure to H_2O_2 (30%), indicating that H_2O_2 had no effect on their surface properties. A slight change was found in the size of most microplastics during the H_2O_2 oxidation process [115]. However, the effects of chemical purification on the surface properties of microplastics may be not fully revealed by qualitative analyses such as SEM and FTIR because of low sensitivity and high subjectivity [78].

Changes in mass and size of six microplastics after different chemical purifications are outlined in Table 1.2. The results showed that microplastic mass is less susceptible to chemical digestion than microplastic size. Treatment with 5 M HNO_3 or HCl solution caused an 100% decrease in the PA mass and size, in accordance with the results from previous studies [116, 117]. One likely reason is that peptide bonds in PA molecules hydrolyzed in highly acidic solution. Alkaline purification caused a decrease in the PET mass by 30.2–53.5% and the PET size by 9.4–16.7%. Saponification of the ester bonds of PET occurred in the process of alkali digestion, resulting in a considerable reduction of the PET mass [80].

Carbonyl index (CI) is often applied for quantitative evaluation of microplastic aging [118–120]. As shown in Table 1.3, the CI values of pristine microplastics have a decreasing sequence as follows: PET> PMMA>PA>PS, PP, and PE, indicating that CI values of the polymers change with the chemical composition of microplastics. Compared with the corresponding pristine microplastics, the enhancement of CI values is found in treated microplastics, indicating that aging of the microplastic surface occurred during the chemical purification, in accordance with the previous findings on UV aging [118–120]. Compared with oxidative purifications with H_2O_2 and Fenton's reagent, aging of PP, PS, PA, and PET is more pronounced during alkaline purification. Liu et al. [118] showed that PE and PS aging from Fenton's reagent is lower than from heat-activated $K_2S_2O_8$. The change in CI of PET is higher than that of PMMA and PA, followed by PS, PP, and PE. These results indicate that microplastic aging is affected by polymer type and internal structure. More pronounced aging may happen in polymers with higher molecular weight monomers or oxygen-containing groups [121]. Other results showed that the chemical structures have an effect on the change in CI values of PE, PP, and PS during UV exposure [118, 122].

Table 1.2 Changes in six microplastic mass and size after chemical purifications (mean value ± SD, n=3) [78] / with permission of Elsevier.

Microplastics	30% H_2O_2	Fenton	1 M HNO_3	5 M HNO_3	1 M HCl	5 M HCl	1 M NaOH	5 M NaOH	10 M NaOH
Mass loss (%)									
PE	-1.1 ± 4.7	2.0 ± 1.0	0.3 ± 1.9	2.4 ± 3.7	-1.3 ± 2.2	2.1 ± 3.2	-2.2 ± 1.6	0.3 ± 1.1	-3.1 ± 2.6
PP	-0.4 ± 0.7	0.8 ± 0.2	0.4 ± 0.8	1.8 ± 0.7	0.6 ± 1.4	1.3 ± 2.6	-2.4 ± 3.3	0.4 ± 1.8	-2.4 ± 1.8
PS	-0.5 ± 0.9	0.3 ± 0.1	2.1 ± 0.6	0.2 ± 1.1	-1.8 ± 1.4	-1.0 ± 1.5	-0.9 ± 1.0	0.2 ± 1.1	-2.1 ± 1.8
PA	-0.4 ± 2.0	-0.4 ± 1.1	1.7 ± 2.3	-100	1.1 ± 2.5	-100	-7.1 ± 4.0	-2.4 ± 1.1	-6.5 ± 4.0
PMMA	-1.6 ± 1.5	3.8 ± 2.9	-0.7 ± 1.3	-0.04 ± 2.0	-0.1 ± 1.2	2.1 ± 2.4	-6.1 ± 4.2	-0.01 ± 1.8	-2.5 ± 5.2
PET	-2.7 ± 2.8	0.6 ± 0.7	0.1 ± 2.1	1.9 ± 1.7	0.6 ± 1.4	3.0 ± 2.6	-30.2 ± 6.2	-38.5 ± 5.8	-53.5 ± 7.2
Size loss (%)									
PE	-2.8 ± 1.1	1.8 ± 4.1	4.1 ± 0.3	5.6 ± 0.6	-2.9 ± 6.3	-1.6 ± 1.0	-3.7 ± 0.3	-5.5 ± 4.1	-2.5 ± 1.6
PP	-2.6 ± 3.6	2.6 ± 1.5	0.7 ± 0.2	-3.5 ± 2.0	-6.0 ± 2.4	-6.2 ± 3.6	-3.3 ± 1.9	-2.3 ± 1.6	-4.6 ± 2.1
PS	1.4 ± 0.6	0.9 ± 3.1	-3.2 ± 2.5	-6.0 ± 0.8	-3.3 ± 2.2	-6.9 ± 5.1	-2.3 ± 4.1	-1.3 ± 2.5	-5.4 ± 4.5
PA	-5.2 ± 7.1	2.1 ± 6.3	-7.6 ± 1.7	-100	-4.8 ± 8.1	-100	-11.8 ± 2.1	-8.9 ± 3.5	-9.2 ± 3.1
PMMA	-2.7 ± 1.6	-2.9 ± 1.6	2.6 ± 4.2	-1.2 ± 1.6	-2.3 ± 1.2	-6.8 ± 2.2	-7.7 ± 5.1	-3.2 ± 1.7	-4.4 ± 0.5
PET	-4.6 ± 5.0	-3.4 ± 2.8	-5.6 ± 3.0	-2.4 ± 1.2	1.1 ± 0.9	-4.1 ± 0.2	-9.4 ± 5.8	-13.4 ± 3.9	-16.7 ± 3.5

Table 1.3 Changes in carbonyl index (CI) of six microplastics before and after the chemical purification Adapted from [78].

Microplastics	Virgin	30% H_2O_2	Fenton	1 M HCl	5 M HCl	1 M HNO_3	5 M HNO_3	1 M NaOH	5 M NaOH	10 M NaOH
PE	0.006	0.004	0.000	0.002	0.003	0.002	0.011	0.006	0.013	0.014
PP	0.001	0.004	0.002	0.009	0.024	0.015	0.013	0.018	0.026	0.037
PS	0.003	0.010	0.013	0.007	0.010	0.019	0.023	0.027	0.038	0.056
PA	0.636	0.723	0.470	0.863	ND	1.034	ND	0.905	0.926	1.442
PMMA	3.772	3.284	3.858	5.975	3.136	4.511	4.695	4.481	4.644	4.096
PET	7.338	8.292	7.360	12.788	13.500	10.293	8.737	12.472	17.592	0.423

ND, no data

Chemical purifications also have an impact on the adsorption capacity of microplastics for heavy metals [123]. As shown in Figure 1.7, alkali digestion causes an increase in the adsorption of Cd by six microplastics. Especially, the adsorbed Cd content on PET increases 148-fold after 10 M NaOH treatment. These results indicate that the vector effect of microplastics on contaminants is significantly affected by alkali purification, because more hydroxyl groups on the microplastic surface after alkaline purification cause an increase in the adsorption of cationic metals to balance the surface charges [124]. As shown in Figure 1.8, alkali purification reduces the zeta potential of microplastics, indicating that negative charge increases on the digested microplastic surface. However, oxidation and acid purification have a lower effect on microplastics surface charges than alkaline purification, except for PMMA and PA. A decrease in the adsorption potential of PA is found after acid purification, possibly because of the adsorption of a large number of positive hydrogen ions on the surface [125], resulting in competition with Cd at adsorption sites. The zeta potential of PA after acid purification is higher than that of virgin PA, in accordance with the results of adsorption potential (Figure 1.8). In addition, compared with the corresponding virgin microplastics, a significant increase in CI values also results in higher adsorption potential for Cd after purifications of PMMA and PET (Table 1.3). A positive correlation was found between Cu adsorption capacity and the carbonyl cardinal basis of microplastics based on the study of Yang et al. [126].

Five quantitative indicators of microplastics were analyzed using principal component analysis to understand the effect of chemical purifications. As Figure 1.9A shows, chemical purifications are divided into four categories: P1 (1 M, 5 M, and 10 M NaOH), P2 (5 M HNO_3 and HCl), P3 (1 M HNO_3 and HCl), and P4 (H_2O_2 and Fenton's reagent). Combined with the above results, chemical purifications in

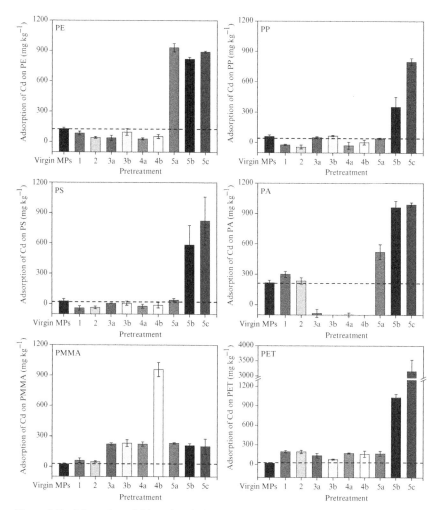

Figure 1.7 Adsorption of Cd on the virgin and treated microplastics. 1, 30% H_2O_2; 2, Fenton's reagent; 3a, 1 M HNO_3; 3b, 5 M HNO_3; 4a, 1 M HCl; 4b, 5 M HCl; 5a, 1 M NaOH; 5b, 5 M NaOH; 5c, 10 M NaOH [78] / with permission of Elsevier.

P1 have a larger effect on microplastics than P2, followed by P3 and P4. In general, physicochemical properties of most microplastics are strongly affected by alkaline purification, especially PET and PA. In contrast, the physicochemical properties of microplastics are less influenced by low-content acid purification than that by high-content ones. As shown in Figure 1.9B, six microplastics are classified into three categories, i.e., M1 (PET, PA, and PMMA), M2 (PS) and M3 (PE and PP). Combined with the above results, the microplastics in M1 have higher resistance

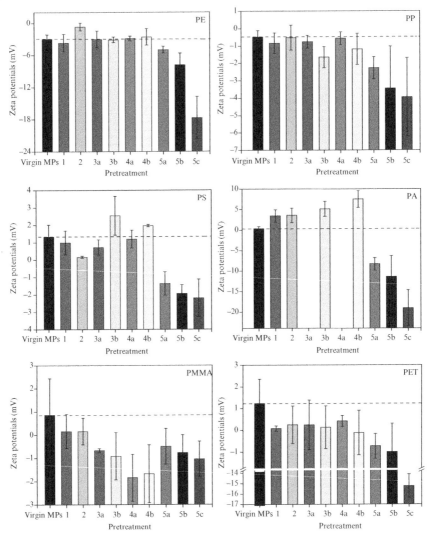

Figure 1.8 Zeta potentials of the virgin and treated microplastics. 1, 30% H_2O_2; 2, Fenton's reagent; 3a, 1 M HNO_3; 3b, 5 M HNO_3; 4a, 1 M HCl; 4b, 5 M HCl; 5a, 1 M NaOH; 5b, 5 M NaOH; 5c, 10 M NaOH. Adapted from [78].

to chemical digestion than M2, followed by M3. Unlike M2 and M3 (PS, PE, and PP), the microplastics in M1 have medium-polar carbonyl groups and heteroatoms, and thus are more liable to hydrolyze because of ester or amide bonds [127]. PS belongs to glassy plastics, while PE and PP belong to rubbery plastics [128, 129] based on the glass transition temperatures. Teuten et al. reported that glassy polymers have dense structures and closed internal nanoscale pores.

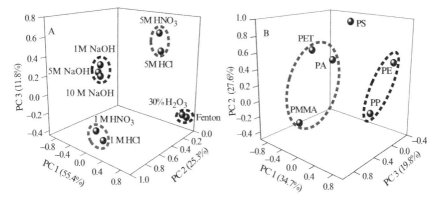

Figure 1.9 Principal component analysis of nine purifications (A) and six microplastic types (B) based on the changes in microplastic physicochemical properties after the purifications [78] / with permission of Elsevier.

Chemical materials are liable to attack the PS carbon backbone during chemical purification because of the presence of nanovoids [129]. Kelkar et al. (2019) found that PS has a lower resistance to chlorination than PP and PE, since the direct erosion by chlorine is promoted in the presence of plastic additives. In sum, the effects of chemical purification depend on the microplastic type, and the physicochemical properties of PP and PE are the least influenced by chemical purification by H_2O_2 and Fenton's reagent.

1.4 Sample Identification

Various techniques are used to determine and identify microplastic particles after sample purification. Microplastics can be detected using a variety of techniques because of various characteristics including color, shape, size, and composition. Visual, spectral, and thermal analysis are the commonly used methods for microplastic identification, but each method and various combination has their respective advantages and disadvantages (Table 1.4).

1.4.1 Visual Identification

Large microplastics of 1–5 mm are reported to be mainly distributed on beaches and to a lesser extent in surface water [131, 132]. Visual sorting and identification of microplastics is usually carried out by the naked eye on a tray using tweezers. This visual method can identify colored plastic fragments and pre-manufactured resin pellets with sizes ranging 2–5 mm [133]. High strandline samples contain a

Table 1.4 Comparison of the commonly used methods for microplastic identification [130] / with permission of The Royal Society of Chemistry.

detection method	Advantages	disadvantages
Microscopy	• Simple, fast, and easy	• No chemical confirmation and polymer composition data • High possibility of false positive and missing small and transparent plastic particles
Microscopy followed by Raman or FTIR spectroscopy	• Polymer composition of major or typical plastic types • Plastic confirmation of subset samples	• Possibility of false positive and missing small and transparent plastic particles
Raman or FTIR spectroscopy	• Detection of down to 1 μm in size • Reduction of false negative data • No possibility of false positive data by chemical confirmation of all the plastic-like particles • Non-destructive analysis • Non-contact analysis (Raman spectroscopy) • Automatic mapping (FPA-reflectance FTIR)	• Expensive instrument • Interference by pigments (Raman spectroscopy) • Laborious work and time consuming for whole particle identification
Thermal analysis	• Synchronous analysis of polymer type and additive chemicals (pyro-GC/MS)	• A few polymer identification (DSC) • Complex data (pyro-GC/MS) • Destructive analysis

large amount of inorganic and organic material interfering with the visual identification of microplastics on the beach. Fuzzy plastic samples ca. 1 mm in size have similar colors to the interferers and may be missed in the classification [130]. However, experts and non-specialists receiving brief training can use this simple and rapid visual method to sort and identify large microplastics [134].

1.4.2 Microscopic Identification

A commonly used identification method is stereo- (or dissecting) microscopy that can investigate polymer particles ranging in the hundreds of micron size, such as neuston net samples [135–139]. Detailed surface texture and structure information to identify fuzzy, plastic-like particles can be viewed through magnified images from the microscope. However, particles less than 100 microns (<100 μm) without color or typical shape are difficult to definitively characterize as microplastics [140]. Since light sediment particles do not separate well from sediment samples, they will affect the microscopic identification of microplastics on filter paper. Additionally, chemical digestion does not completely remove biogenic materials in sediments or neuston net samples and this can also affect microscopic detection. In previous studies, spectral analysis confirmed that the valid identification rate of plastic-like particles is usually less than 80% using microscopy, and that of transparent particles is less than 30% [3, 140, 141]. Synthetic polymers make up just 1.4% of the particles that visually look like microplastics [130]. Many studies have found that fibers are the main or dominant microplastics in water, sediment, and biota [142, 143], but manmade fibers (such as polyester) are very difficult to identify from natural fibers (such as colored cotton) using a microscope alone [140].

High magnification images of plastic-like particles can be supplied by scanning electron microscopy (SEM), thus helping to distinguish microplastics from organic particles [144]. The elemental composition of the same particles can be gained using further analysis with energy-dispersive X-ray spectroscopy (EDS) [145]. EDS helps identify carbon-dominated plastics in inorganic particles. However, SEM-EDS limits the number of samples analyzed, as it the instrumentation is costly and consumes more time and energy. Meanwhile, SEM cannot recognize the color of plastic. Surface characterization and elemental analysis of the plastic particles are recommended to complement this method.

Plastic particles can also be identified under specific circumstances by other advanced microscopy techniques. Polyethylene (PE) particles have been successfully identified using polarized optical microscopy in laboratory accumulation and toxicity experiments [146]. The crystal structure inside the plastic can affect the transmission of polarized light, and diverse types of polymers have differing degrees of crystallinity. For this method, thin microplastics are required to allow enough polarized light to pass through. Microplastic samples on opaque filter paper are not suitable for this method [130].

1.4.3 Spectroscopic Identification

Microplastics recovered from environmental matrices are often identified using infrared spectroscopy. The application of this technique is dual purpose. On the one hand, the misclassification of polymers containing organic matter or other substances in the sample can be avoided; on the other hand, the type of polymer can be identified.

FTIR is one of the simplest and cost-effective techniques [130]. It was used to characterize polymers already identified using fluorescence microscope in some studies [56, 140], while it was directly employed to analyze residues from flotation or digestion in others [147]. In addition, FTIR can be combined with an automated tracking system to detect the presence of microplastics by scanning slices of a sample surface. Atomic Force Microscopy-Infrared (AFM-IR) combines an atomic force microscope with an infrared laser source and uses a pulsed infrared laser to excite molecular vibrations in a sample. The sample is heated rapidly through infrared absorption, leading to rapid thermal expansion, thus causing oscillation of the AFM cantilever beam. The atomic force microscope is used to detect the oscillation, the attenuation law of the oscillation is analyzed by FTIR, and the frequency and amplitude of the oscillation are extracted. Collection of amplitudes and wavelengths gives rise to infrared spectra since the cantilever oscillation is proportional to the local absorption. This method can identify PU nanoplastics in solid matrices based on their characteristic spectra.

FTIR-equipped microscopes and mercury telluride (MCT) single-mode FTIR are two types of FTIR for microplastics analysis. Non-uniform components of microplastics can be analyzed using FTIR-equipped microscopes, while single and batch samples can only be identified using MCT single mode with an ATR tip. Two modes, ATR mode and focal plane array (FPA) detector mode are used to analyze samples using FTIR-equipped microscopy.

Microscopy combined with an ATR tip is used to identify individual particles in ATR mode. Quantization is possible using a micro-FTIR equipped with a FPA detector. Spatially resolved spectra with xyz stage are automatically obtained by the FPA-micro-FTIR. Compared to traditional FTIR that analyzes the whole sampling area using diamond ATR sampling, the FPA detector can detect the particles in the selected grid region through space-resolved spectroscopy.

Moreover, FPA-micro-FTIR combines chemical imaging with spectroscopic analysis and is used to identify heterogeneous microplastic particles. Spatial and spectral information about heterogeneous microplastic particles can be collected by FPA detectors [148]. A large-scale analysis of particles with a diameter up to 47 μm for 9 and 16 h was conducted using FPA-FTIR with 1 and 2 scans per pixel for the environmental analysis, respectively [56]. The spectral correlation and literature were used for large data set analysis after the chemical images obtained by FPA detector were mapped [149]. Heterogeneous microplastic particles can be detected from the samples using FTIR image analysis. Chemical identification was carried out by calculating the correlation between the original spectrum and first derivative of the vector-normalized spectrum using the reference library. The size and number of each microplastic particle were determined from the identified pixels. Then the correct particle recognition was carried out in depth [150]. Reference samples of microplastic identification by FTIR are used to optimize the classification through hierarchical clustering of reference spectra. Therefore, information about size, chemical composition, and shape of the particles were gained [148]. The FPA detector has high lateral resolution, reducing the size detection limit of microplastics to 20 μm.

FPA-micro-FTIR techniques are often used to detect microplastics in wastewater treatment plants. A removal rate of 10–500 μm microplastics is estimated up to 98% using the method [111]. These researchers investigated the number and mass distribution of microplastic particles, such as PE, PVC, PS, PA, in influent and effluent wastewater. However, limitations of accurate estimation of microplastic mass are also highlighted using the FTIR method. FTIR spectroscopy is also used to identify microplastics in drinking water and surface water [151]. FTIR spectroscopy is also used for the identification of microplastic contaminants in bottled water [98]. Microplastics larger than 100 μm were detected at a mean of 10.4 particles/L in bottled water, and PP occupied more than half of the detected polymers (54%) likely from the bottle cap. Plastic particles of 6.5–100 μm were detected by Nile red dye. In combination, FITR and Nile-red-staining optical microscopy detected 325 microplastic particles of more than 6.5 μm. PP and PE, at 1250–4650 n/m^3 and 900–2800 n/m^3 respectively, were found in surface water at each lake sampling site using FTIR [152]. Some researchers reported that particle size detection by FTIR is limited to between 50 μm and 5 mm [4]. The number, type, and shape of microplastics were counted in this study. Microplastics detected in surface water are mainly PE and PP [4]. However, the contents varied with sampling sites.

Raman spectrometers, with wavenumbers in the range 200–2000 cm^{-1}, are also commonly used for detection and characterization of microplastics. Image analysis was also achieved through combining Raman with automatic particle tracking. Matrix volumetric measurements and high barometric pressure of the nebula were automatically scanned as a single-layer slide and microplastics were individually tracked. While tracking, Raman spectroscopic and microscopic image analyses were also automatically conducted. Vis–NIR (visible–near infrared) spectroscopy was used to detect microplastics without extraction. The amount of surface-reflected light with wavelengths in the range 350–2500 nm was measured using portable spectrophotometers, and the percentage of reflections were recorded at each wavelength. Microplastics in the surface level of soil were thus identified. Therefore, to measure an average microplastic content in soil, five or more readings should be collected after the soil is mixed [103].

1.4.4 Thermal Analysis

Thermal analysis differs from vibration spectroscopy (FTIR and Raman) for microplastic identification. Mass spectrometry (MS) is often used to characterize gases produced from thermal decomposition of polymers above 500°C. Compared with Raman and FTIR spectroscopy, the method has the advantage of not requiring multiple sample preparation steps [153]. It also has relatively short analysis time and is independent of microplastic size and shape [154]. Therefore, complex plastic fragments can be identified using a combination of thermal analysis with mass spectrometry.

According to the type of thermal analyzer, thermal analysis can be divided into three types, i.e., Py-GC-MS (pyrolysis-gas chromatography-mass spectrometry), TGA-based (thermogravimetric analysis) methods and TED-GC-MS (thermal extraction-desorption gas chromatography-mass spectroscopy).

(1) Py-GC-MS

Microplastics are detected using Py-GC-MS, based on thermal decomposition of the polymer. After thermal degradation in an inert atmosphere, microplastic fragments are isolated through gas chromatography and analyzed qualitatively and quantitatively by mass spectrometry. Complex structures are readily analyzed using this method. Therefore, this method can analyze microplastics containing organic pollutants. However, the sample weight limitation of this method was 0.5 mg, and the maximum operating temperature was 250 – 300°C. The thin transfer capillary often tends to be contaminated and/or blocked by pyrolysis products of high molecular weight and high boiling point (>300°C), and thus transfer loss occurs [155].

Py-GC-MS can be used to determine microplastics such as PE, PP, and PS [156]. The detection of this method was limited to 1–86 μg/g and the recovery rate of 250 μg/g microplastics was 70–128%. Py-GC-MS has been applied for identification of microplastics, as well as organic and inorganic plastic additives in an ocean environment [157]. This method identified various microplastics such as PE, PS, PP, chlorosulfonated and chlorinated PE, and various organic additives such as dimethyl phthalate, diisobutyl phthalate, and benzaldehyde. Zinc, barium, sulfur, and titanium dioxide nanoparticles were also detected.

After cascade filtration, PY-GC-MS was also used for microplastic characterization from WWTP effluent [158]. In this study, the characteristic pyrolysis products of 2,4,6-triphenyl-1-hexane and 1,14-pentadecadiene were used to quantify the PS and PE, respectively. It is reported that the detection of 2,4,6-triphenyl-1-hexane from PS, and 1,14-pentadecadiene from PE was limited to 0.009 μg and 0.3 μg, respectively.

(2) TGA-based methods (TGA, TGA-MS, TGA-DSC)

TGA can characterize microplastics by decomposing microplastic samples and monitoring mass loss. For microplastic quantification, one heats the sample under an inert environment and then determines the mass loss of the sample. Changes in the sample mass are related to operation time and temperature. The thermobalance is used to measure the mass loss in the TGA system. Microplastic identification in soil environmental samples is often carried out using this method [159, 160]. This method is used to analyze the thermo-decomposed sample, supplying detailed information. Qualitative and quantitative analysis of gas products from microplastic samples are provided through differential scanning calorimetry (DSC) and mass spectrometry. The ion signals are determined by a mass spectrometer detector of a TGA-MS for microplastic identification, and the mass loss of the sample is measured by TGA. TGA-MS has the advantage of accepting relatively large samples but has limitations when interpreting samples with similar mass and degradation temperatures, compared with Py-GC-MS [153]. The endothermic phase change heat flow and peak temperature of different microplastics such as PE and PP in wastewater can be determined by combining TGA with DSC [161]. Other characteristics of the microplastic samples such as particle size, impurities and additives, and degree of polymer branching also affect their characteristic phase transition temperature, causing complex samples that overlap [153, 162, 163]. Therefore, the recommended detectable particle size of microplastics is 200–500 μm. However, direct injection of the gasses evolved from the TGA into the MS detector has an adverse effect on analytical performance by causing ion source deterioration [164].

(3) TED-GC-MS

TED-GC-MS is a method combining thermogravimetry with thermal desorption through a solid phase extraction [165, 166]. TED-GC-MS makes up for the limitations of PY-GC-MS (blocking of thin transfer capillarity and/or contamination of high molecular weight pyrolysis products) by separating the thermal extraction process from the GC-MS system [167]. Compared with PY-GC-MS, TED-GC-MS can analyze a large sample amount (up to 100 mg) in less time. Notably, TED-MS has worse analytical performance than TED-GC-MS, because in TED-MS the gas feeds directly into the MS detector. First, the microplastic sample is pyrolyzed in TGA at temperatures up to 1000°C, then a solid-phase agent such as PDMS is used to adsorb the degradation products. The products are desorbed by increasing the temperature after being transferred to the thermal desorption unit. Then, a chromatographic column is used to isolate the materials, and finally mass spectrometry identifies the separated groups. PP, PE, and PS in biogas plants and rivers can be quantified by TED-GC-MS [168]. The main parameters, including 30 mL/min of purge gas flow and 10°C/min heating rate, were optimized to develop a fast method with good signal intensity [167]. The method enables quantitative and qualitative analysis of polymer blends (PE, PP, and PS) without peak superposition, and has been used successfully for environmental samples. Table 1.5 outlines three thermoanalytical methods of microplastic identification.

1.5 Quantitative Analysis

After various pretreatments extract the microplastics from environmental samples, the microplastic size distribution needs further investigation. The physical parameters of particles ranging from micrometer to nanometer have been directly

Table 1.5 Three thermoanalytical methods used to identify microplastics [169] / with permission of Elsevier.

	Py-GC-MS	TGA-based methods	TED-GC-MS
Principle	• Thermal decomposition of polymer	• Thermal degradation of sample and monitoring weight loss	• Combination of TGA with thermal desorption using solid gas extraction
			• Quantification using MS
Identification	• Characteristic degradation products	• Mass loss during heating	• Mass loss during heating
LOD	• pg–μg < 0.5 mg	• 200–500 μm	• Relatively larger (about 100 mg)
	• >50–100 μm, <1.5 mm		• No size limitation
Advantage	• Complex structure: microplastics with organic and inorganic contaminants are detectable	• Suitable for many samples	• Samples with higher amount
			• Better performance compared with TGA and pyrolysis
			• Reduced analysis time
	• Better Sensitivity		
Limitations	• Lower sample amounts (<0.5 mg)	• Limited by similar degradation temperature and mass	• Complex with similar mass and degradation temperature: difficult to interpret data
		• Direct injection of evolved gas to MS	
Targeted sample	Drinking water	Wastewater	Wastewater

determined using several analysis methods. For example, particle size distribution can be measured using laser diffraction (LD), before which the particles settle for a few hours, or dynamic light scattering (DLS), in which the particles are dispersed in Brownian motion. Another advanced method is nanoparticle tracking analysis (NTA) that can accurately estimate the concentration, size, and size distribution of monodisperse and polydisperse nanoparticles. A variety of fields including environmental engineering, drug delivery, pharmaceuticals, and nanotoxicology widely apply for these techniques for particle analysis. Compared with vibrational spectroscopy (IR and Raman), one of their advantages is that analysis is relatively fast [170, 171]. Each range of particle sizes detected by the three methods such is shown in Figure 1.10. The methods for analysis of microns and nanoparticles are relatively well developed, but few studies focus on their application for identifying characteristics of microplastics. As a complement to vibrational spectroscopy, the principles, advantages, applications, and challenges of these techniques are worth discussing in detail [169].

1.5.1 LD Method

Laser diffraction can be used to analyze particle size distribution by allowing a laser beam to pass through the diffraction pattern of the particles. At the same time, angular variation in light intensity is detected during the above procedure. The particle size distribution is estimated by measuring angular scattering intensity, because particle size is inversely proportional to the diffraction angle. After that, the Fraunhofer approximation (large particles) or Mie theory (all particle

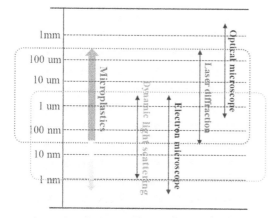

Figure 1.10 Particle size distribution method sorted according to measurement range [169] / with permission of Elsevier.

sizes) is used to estimate the distribution of the volume-weighted sphere-equivalent diameter [172].

1.5.2 DLS Method

The size distribution of small particles in suspension is measured by dynamic light scattering. Variations in scattered light intensity are used to estimate particle size based on Brownian motion of these particles [173]. The measurement range is usually small, ranging one nanometer to a few microns. However, this method may be less accurate all particles are assumed spherical in the analysis.

DLS was used to detect nanoplastics with the range 1–1000 nm in seawater and in ordinary cosmetics such as facial cleansers [174, 175]. DLS detected the presence of nanoplastics (<100 nm) in seawater, resulting from decomposition of marine microplastics by sunlight [176]. A customized in situ DLS instrument found that nanoparticles consisting of PE (90%) and PP (10%) were generated during the photolysis period. The autocorrelation function of DLS characterized the initial small particle formation and losses in millimeter plastic particles decomposed by ultraviolet. Before using a hydraulic model to investigate the migration and retention of nanoplastics and microplastics in rivers, the particle size in the supernatant was measured by DLS. It was also used to investigate the aggregation-sedimentation adhesion efficiency of suspended solids (Kaolin clay) and PS (70, 1050 nm) [177]. Additionally, nanoplastics from fish samples were estimated using DLS with detection limited to 52 µg/g fish [178].

1.5.3 NTA Method

Dispersed particles are tracked through a sequential imaging analysis, and both light scattering and Brownian motion are used in the nanoparticle tracking analysis method. Continuous image analysis of the particles is carried out by a laser scattering microscope combined with a charge-coupled device (CCD) camera. The scattered light of illuminated particles is measured by video microscopy and the behavioral trajectory of particles subjected to Brownian motion is captured in the NTA. The Stokes-Einstein relation is used to estimate particle size according to the three-dimensional motion of spherical particles [179]. A charge-coupled refractive index is combined with laser scattering microscopy to characterize nanoplastics.

Individual particles can be analyzed using NTA based on Brownian motion and tracked through multiple images on video. The concentration of nanoparticles in a solution matrix can be visualized using this method. Both NTA and DLS were used to determine monodisperse PS beads with differing particle sizes in a drug-delivery system. Accurate size distinction was obtained using NTA with unbiased high peak resolution in both mono-and polydisperse samples. Compared with LD and DLS, NTA has higher resolution [180]. However, LD and DLS have more sample sensitivity in the presence of large particles and aggregates compared with NTA [179]. Particles up to 30 nm can be detected, but particles larger than 2000 nm cannot be determined by NTA [181]. However, compared with DLS (several minutes), NTA requires more time (up to 1 h). Particle size distribution of nanoplastics and microplastic in water was analyzed using the method [182, 183]. In a study by Lambert et al., seven microplastics including PE, PP, and five consuming plastics were generated during the decomposition process. NTA was then used to detect their particle size distribution ranging from 30 nm and 60 µm [182, 183]. NTA also characterized the particle size distribution range 30–2000 nm during the decomposition of latex film (PS) [184].

1.5.4 Challenges in Particle Size Analysis

Particle size analyses each have their own advantages, but their application for quantitative analysis of microplastics is hindered by several common challenges. These particle size analyzers calculate the total particle size distribution that can then be used to distinguish microplastics from other contaminant particles [185]. DLS and/or LD can only estimate particle size and distribution of extracted microplastics from the solution matrix. Strict sample pretreatment for the complete removal of all other organic and inorganic pollutants is required before analysis. In addition, Brownian displacement easily leads to differing results between DLS measurement and visual inspection [179]. Average particle size of polydisperse suspension measured by DLS was larger than that by SEM [174], because the scattering of larger particles in a population led to measurement deviation [186]. When a low content of nanoplastics is determined in complex substrates, the accuracy of DLS reduces. Monodisperse particles can be estimated using DLS analysis, but the results may be inaccurate for polydisperse particles [181]. Small-sized particles that are masked by strong scattered light of large particles are detected using DLS [187]. Microplastic particles can be quantified using the NTA. However, information about the size distribution given by NTA will only be helpful in preliminary determination of the characteristics of microplastics in an environmental matrix to help formulate an analysis scheme. Improvements such as a powerful light source or particle tracking software should be investigated to overcome these challenges. A multispectral detection system was evaluated in a recent study, using three lasers to produce polarized, narrow, bright light. Compared to NTA, a wider range of particle sizes of 50–2000 nm is covered in this system. Polydisperse particles are also identified with high sensitivity using this system [181].

When the microplastic content is too low to detect, the sample should be concentrated with extra filtration. This is especially true in samples from surface water and seawater [175]. Quantification of microplastic concentration using particle size analysis may be more accurate than manual counting and vibrational spectroscopy based on the total number of pixels estimated by FPA-m-FTIR. The application of NTA depends on the sample properties. Samples from surface water, ocean, and wastewater treatment plants can be often analyzed using NTA.

1.6 Quality Control

1.6.1 Internal Deviation

Cross-contamination readily occurs in microplastics analysis, affecting the accuracy of microplastic detection, because microplastics come from a wide range of sources such as air, water, synthetic textiles, polymer filters, and the like. Therefore, unnecessary errors can be reduced with preventive measures.

Some measures should be taken during the entire analysis: (1) plastic equipment should be avoided, and non-plastic equipment such as stainless metal or glass is preferred wherever possible; (2) operators should wear 100% cotton clothing and plastic-free gloves rather than synthetic textiles; (3) the solution is pre-filtered to remove interference and the equipment rinsed with filtered ethanol or ultrapure water; (4) negative controls are analyzed, and if microplastic contamination is found, correction is made to samples [188].

To prevent background contamination, either sample water or ultrapure water should be used to wash the sample instrument during sampling. In addition, sample contact with air should be kept to a minimum. It is important that any pump run for several minutes to clear air before sampling urban water samples collected by pump. All tools and containers should be cleaned by ultrapure water. Potential microplastic contamination from the air can be effectively avoided through covering samples. Covering samples leads to more than 90% reduction in contaminants [189].

Loss of microplastics in samples interferes with results. For example, microplastics with smaller-than-filter size or microfibers are leaked during screening; lower extraction volume causes microplastic reservation in sediment; and microplastics attach to container walls. Careful attention can reduce these error rates. An effective isolation device was developed by Xu et al. [190]. Frequent solution transfers were avoided, and good recovery rates were obtained using this equipment. In addition, losses of microplastics can be quantified using a positive control during the analysis processes.

Some factors should be paid attention to in the design of any recovery experiment: (1) all equipment should be covered during the analysis; (2) microplastics of various sizes and types should be included; and (3) sufficient replicates should be carried out [151]. Compared with microplastics collected from WWTPs, microplastics in recovery testing are easier to collect and identify because of their better surface characteristics. Therefore, recovery testing should be optimized through developing a standard surrogate [191].

1.6.2 Judgment Error

Other sources of error include false negatives, false positives, or false count of microplastics, especially in the absence of chemical analysis. Cotton, cellulose, and animal hair all have similar shape to plastic fibers and are easily mistaken for them. In addition, white or transparent microplastics are frequently ignored, as their color is similar to small particles and background paper, leading to an underestimation of microplastic content. Visual identification and additional inspection, such as needle inspection and dyeing, can be combined to identify suspected particles. In addition, some microplastics may be identified as non-plastic by vibrational spectroscopy, especially small microplastics, because of the difference between the standard spectra and sample. Therefore, more time is required for suspect spectra, and sample spectra should be further identified with the help of peer experts. A helpful automatic protocol was developed by Primpke et al. [192], in which the FTIR imaging data of the samples can be analyzed using a designed reference database [150].

1.7 Summary and Future Outlooks

Microplastics in urban water systems have received increasing attention, but detection methods remain elusive. Different collection and purification methods have a profound impact on detection results, so standardized detection methods should are urgently needed. The definition and classification of the forms of microplastics should also be standardized. The ratio of plastic and non-plastic particles to be analyzed is critical. Research results on microplastics are unreliable without chemical composition analysis. Sample quantitation and quality control are key components of microplastics research, and the entire process of sample quantitation and quality control should be standardized. At present, research on microplastics is mainly laboratory simulation, and the effects of various natural factors from the environment on microplastic properties should also be considered.

References

1 Razeghi, N., Hamidian, A.H., Wu, C. et al. (2021). Microplastic sampling techniques in freshwaters and sediments: A review. *Environmental Chemistry Letters* 19 (6): 4655–4655.

2 Huppertsberg, S. and Knepper, T.P. (2018). Instrumental analysis of microplastics-benefits and challenges. *Analytical and Bioanalytical Chemistry* 410 (25): 6343–6352.

3 Hidalgo-Ruz, V., Gutow, L., Thompson, R.C. et al. (2012). Microplastics in the marine environment: A review of the methods used for identification and quantification. *Environmental Science & Technology* 46 (6): 3060–3075.

4 Sighicelli, M., Pietrelli, L., Lecce, F. et al. (2018). Microplastic pollution in the surface waters of Italian Subalpine lakes. *Environmental Pollution* 236: 645–651.

5 Sadri, S.S. and Thompson, R.C. (2014). On the quantity and composition of floating plastic debris entering and leaving the Tamar Estuary, Southwest England. *Marine Pollution Bulletin* 81 (1): 55–60.

6 Campanale, C., Stock, F., Massarelli, C. et al. (2020). Microplastics and their possible sources: The example of ofanto river in southeast Italy. *Environmental Pollution* 258: 113284.

7 Zhang, K., Shi, H., Peng, J. et al. (2018). Microplastic pollution in China's inland water systems: A review of findings, methods, characteristics, effects, and management. *Science of the Total Environment* 630: 1641–1653.

8 Liedermann, M., Gmeiner, P., Pessenlehner, S. et al. (2018). A methodology for measuring microplastic transport in large or medium rivers. *Water* 10 (4): 414.

9 Wang, Z., Su, B., Xu, X. et al. (2018). Preferential accumulation of small (<300 µm) microplastics in the sediments of a coastal plain river network in eastern China. *Water Research* 144: 393–401.

10 Anderson, P.J., Warrack, S., Langen, V. et al. (2017). Microplastic contamination in lake Winnipeg, Canada. *Environmental Pollution* 225: 223–231.

11 Dris, R., Gasperi, J., Rocher, V. et al. (2018). Synthetic and non-synthetic anthropogenic fibers in a river under the impact of Paris megacity: Sampling methodological aspects and flux estimations. *Science of the Total Environment* 618: 157–164.

12 Cox, K., Brocious, E., Courtenay, S.C. et al. (2018). Distribution, abundance, and spatial variability of microplastic pollution in surface waters of lake superior. 47 (5): 1358–1364.

13 Michida, Y., Chavanich, S., Chiba, S. et al. (2019). Guidelines for harmonizing ocean surface microplastic monitoring methods. Chiyoda-ku, Japan: Ministry of the Environment. doi: 10.25607/OBP-513.

14 Campanale, C., Savino, I., Pojar, I. et al. (2020). A practical overview of methodologies for sampling and analysis of microplastics in riverine environments. *Sustainability* 12 (17).

15 Han, M., Niu, X., Tang, M. et al. (2020). Distribution of microplastics in surface water of the lower Yellow River near estuary. *Science of the Total Environment* 707: 135601.

16 Miller, R.Z., Watts, A.J.R., Winslow, B.O. et al. (2017). Mountains to the sea: River study of plastic and non-plastic microfiber pollution in the northeast USA. *Marine Pollution Bulletin* 124 (1): 245–251.

17 Mao, Y., Li, H., Gu, W. et al. (2020). Distribution and characteristics of microplastics in the Yulin River, China: Role of environmental and spatial factors. *Environmental Pollution* 265: 115033.

18 Barrows, A.P.W., Christiansen, K.S., Bode, E.T. et al. (2018). A watershed-scale, citizen science approach to quantifying microplastic concentration in a mixed land-use river. *Water Research* 147: 382–392.

19 Dubaish, F. and Liebezeit, G. (2013). Suspended microplastics and black carbon particles in the jade system, southern North Sea. *Water, Air, and Soil Pollution* 224 (2): 1352.

20 Crawford, C.B. and Quinn, B. (2016). *Microplastic Pollutants*, 1e. Elsevier Limited.

21 Braun, U., Jekel, M., Gerdts, G. et al. (2018). Microplastics analytics: Sampling, preparation, and detection methods. *Discussion Paper, BMBF Research Focus "Plastics in the Environment"*.

22 Fischer, E.K., Paglialonga, L., Czech, E. et al. (2016). Microplastic pollution in lakes and lake shoreline sediments – A case study on lake Bolsena and lake Chiusi (central Italy). *Environmental Pollution* 213: 648–657.

23 Tamminga, M., Stoewer, S.-C., and Fischer, E.K. (2019). On the representativeness of pump water samples versus manta sampling in microplastic analysis. *Environmental Pollution* 254: 112970.

24 Lahens, L., Strady, E., Kieu-Le, T.-C. et al. (2018). Macroplastic and microplastic contamination assessment of a tropical river (Saigon River, Vietnam) transversed by a developing megacity. *Environmental Pollution* 236: 661–671.

25 Su, L., Xue, Y., Li, L. et al. (2016). Microplastics in Taihu Lake, China. *Environmental Pollution* 216: 711–719.

26 Barrows, A.P.W., Neumann, C.A., Berger, M.L. et al. (2017). Grab vs. neuston tow net: A microplastic sampling performance comparison and possible advances in the field. *Analytical Methods* 9 (9): 1446–1453.

27 Karlsson, T.M., Kärrman, A., Rotander, A. et al. (2020). Comparison between manta trawl and in situ pump filtration methods, and guidance for visual identification of microplastics in surface waters. *Environmental Science and Pollution Research* 27 (5): 5559–5571.

28 Yang, L., Zhang, Y., Kang, S. et al. (2021). Microplastics in freshwater sediment: A review on methods, occurrence, and sources. *Science of the Total Environment* 754: 141948.

29 Van Cauwenberghe, L., Devriese, L., Galgani, F. et al. (2015). Microplastics in sediments: A review of techniques, occurrence and effects. *Marine Environmental Research* 111: 5–17.

30 Ivleva, N.P., Wiesheu, A.C., and Niessner, R. (2017). Microplastic in aquatic ecosystems. *Angewandte Chemie International Edition* 56 (7): 1720–1739.

31 Alimba, C.G. and Faggio, C. (2019). Microplastics in the marine environment: Current trends in environmental pollution and mechanisms of toxicological profile. *Environmental Toxicology and Pharmacology* 68: 61–74.

32 Zhang, L., Liu, J., Xie, Y. et al. (2020). Distribution of microplastics in surface water and sediments of Qin river in Beibu Gulf, China. *Science of the Total Environment* 708: 135176.

33 Corcoran, P.L., Norris, T., Ceccanese, T. et al. (2015). Hidden plastics of lake Ontario, Canada and their potential preservation in the sediment record. *Environmental Pollution* 204: 17–25.

34 Abidli, S., Toumi, H., Lahbib, Y. et al. (2017). The first evaluation of microplastics in sediments from the complex Lagoon-Channel of Bizerte (Northern Tunisia). *Water, Air, and Soil Pollution* 228 (7): 262.

35 Alam, F.C., Sembiring, E., Muntalif, B.S. et al. (2019). Microplastic distribution in surface water and sediment river around slum and industrial area (case study: Ciwalengke River, Majalaya district, Indonesia). *Chemosphere* 224: 637–645.

36 Rodrigues, M.O., Abrantes, N., Gonçalves, F.J.M. et al. (2018). Spatial and temporal distribution of microplastics in water and sediments of a freshwater system (Antuã River, Portugal). *Science of the Total Environment* 633: 1549–1559.

37 Qiu, Q., Tan, Z., Wang, J. et al. (2016). Extraction, enumeration and identification methods for monitoring microplastics in the environment. *Estuarine, Coastal and Shelf Science* 176: 102–109.

38 Besley, A., Vijver, M.G., Behrens, P. et al. (2017). A standardized method for sampling and extraction methods for quantifying microplastics in beach sand. *Marine Pollution Bulletin* 114 (1): 77–83.

39 Prata, J.C., da Costa, J.P., Duarte, A.C. et al. (2019). Methods for sampling and detection of microplastics in water and sediment: A critical review. *Trac-trends in Analytical Chemistry* 110: 150–159.

40 Stock, F., Kochleus, C., Bänsch-Baltruschat, B. et al. (2019). Sampling techniques and preparation methods for microplastic analyses in the aquatic environment – A review. *Trac-trends in Analytical Chemistry* 113: 84–92.

41 Zhang, B., Chen, L., Chao, J. et al. (2020). Research progress of microplastics in freshwater sediments in China. *Environmental Science and Pollution Research* 27 (25): 31046–31060.

42 Blettler, M.C., Abrial, E., Khan, F.R. et al. (2018). Freshwater plastic pollution: Recognizing research biases and identifying knowledge gaps. *Water Research* 143: 416–424.

43 Lv, W., Zhou, W., Lu, S. et al. (2019). Microplastic pollution in rice-fish co-culture system: A report of three farmland stations in Shanghai, China. *Science of the Total Environment* 652: 1209–1218.

44 Wen, X., Du, C., Xu, P. et al. (2018). Microplastic pollution in surface sediments of urban water areas in Changsha, China: Abundance, composition, surface textures. *Marine Pollution Bulletin* 136: 414–423.

45 Wen, X., Du, C., Zeng, G. et al. (2018). A novel biosorbent prepared by immobilized Bacillus licheniformis for lead removal from wastewater. *Chemosphere* 200: 173–179.

46 Wang, J., Wang, M., Ru, S. et al. (2019). High levels of microplastic pollution in the sediments and benthic organisms of the South Yellow Sea, China. *Science of the Total Environment* 651: 1661–1669.

47 Fu, Z. and Wang, J. (2019). Current practices and future perspectives of microplastic pollution in freshwater ecosystems in China. *Science of the Total Environment* 691: 697–712.

48 Claessens, M., Van Cauwenberghe, L., Vandegehuchte, M.B. et al. (2013). New techniques for the detection of microplastics in sediments and field collected organisms. *Marine Pollution Bulletin* 70 (1–2): 227–233.

49 Zhu, X. (2015). Optimization of elutriation device for filtration of microplastic particles from sediment. *Marine Pollution Bulletin* 92 (1–2): 69–72.

50 Crichton, E.M., Noël, M., Gies, E.A. et al. (2017). A novel, density-independent and FTIR-compatible approach for the rapid extraction of microplastics from aquatic sediments. *Analytical Methods* 9 (9): 1419–1428.

51 Dong, M., Luo, Z., Xing, X. et al. (2019). Separation of microplastics in soils and sediments used oil extraction protocol. *Research of Environmental Sciences* 1–11.

52 Fuller, S. and Gautam, A. (2016). A procedure for measuring microplastics using pressurized fluid extraction. *Environmental Science & Technology* 50 (11): 5774–5780.

53 Sun, J., Dai, X., Wang, Q. et al. (2019). Microplastics in wastewater treatment plants: Detection, occurrence and removal. *Water Research* 152: 21–37.

54 Hu, Y., Gong, M., Wang, J. et al. (2019). Current research trends on microplastic pollution from wastewater systems: A critical review. *Reviews in Environmental Science and Bio-technology* 18 (2): 207–230.

55 Murphy, F., Ewins, C., Carbonnier, F. et al. (2016). Wastewater treatment works (WWTW) as a source of microplastics in the aquatic environment. *Environmental Science & Technology* 50 (11): 5800–5808.

56 Tagg, A.S., Sapp, M., Harrison, J.P. et al. (2015). Identification and quantification of microplastics in wastewater using focal plane array-based reflectance Micro-FT-IR imaging. *Analytical Chemistry* 87 (12): 6032–6040.

57 Dyachenko, A., Mitchell, J., and Arsem, N. (2017). Extraction and identification of microplastic particles from secondary wastewater treatment plant (WWTP) effluent. *Analytical Methods* 9 (9): 1412–1418.

58 Gündoğdu, S., Çevik, C., Güzel, E. et al. (2018). Microplastics in municipal wastewater treatment plants in Turkey: A comparison of the influent and secondary effluent concentrations. *Environmental Monitoring and Assessment* 190 (11): 626.

59 Talvitie, J., Mikola, A., Setälä, O. et al. (2017). How well is microlitter purified from wastewater? – A detailed study on the stepwise removal of microlitter in a tertiary level wastewater treatment plant. *Water Research* 109: 164–172.

60 Magnusson, K. and Norén, F. (2014). *Screening of Microplastic Particles in and Down-stream a Wastewater Treatment Plant*. Stockholm, Sweden: IVL Swedish Environmental Research Institute.

61 Mintenig, S., Int-Veen, I., Löder, M.G. et al. (2017). Identification of microplastic in effluents of waste water treatment plants using focal plane array-based micro-Fourier-transform infrared imaging. *Water Research* 108: 365–372.

62 Ziajahromi, S., Neale, P.A., Rintoul, L. et al. (2017). Wastewater treatment plants as a pathway for microplastics: Development of a new approach to sample wastewater-based microplastics. *Water Research* 112: 93–99.

63 Carr, S.A., Liu, J., and Tesoro, A.G. (2016). Transport and fate of microplastic particles in wastewater treatment plants. *Water Research* 91: 174–182.

64 Mason, S.A., Garneau, D., Sutton, R. et al. (2016). Microplastic pollution is widely detected in US municipal wastewater treatment plant effluent. *Environmental Pollution* 218: 1045–1054.

65 Ort, C., Lawrence, M.G., Reungoat, J. et al. (2010). Sampling for PPCPs in wastewater systems: Comparison of different sampling modes and optimization strategies. *Environmental Science & Technology* 44 (16): 6289–6296.

66 Ort, C., Lawrence, M.G., Rieckermann, J. et al. (2010). Sampling for pharmaceuticals and personal care products (PPCPs) and illicit drugs in wastewater systems: Are your conclusions valid? A critical review. *Environmental Science & Technology* 44 (16): 6024–6035.

67 Michielssen, M.R., Michielssen, E.R., Ni, J. et al. (2016). Fate of microplastics and other small anthropogenic litter (SAL) in wastewater treatment plants depends on unit processes employed. *Environmental Science: Water Research & Technology* 2 (6): 1064–1073.

68 Li, J., Liu, H., and Paul Chen, J. (2018). Microplastics in freshwater systems: A review on occurrence, environmental effects, and methods for microplastics detection. *Water Research* 137: 362–374.

69 Zhang, Z. and Chen, Y. (2020). Effects of microplastics on wastewater and sewage sludge treatment and their removal: A review. *Chemical Engineering Journal* 382: 122955.

70 Koyuncuoğlu, P. and Erden, G. (2021). Sampling, pre-treatment, and identification methods of microplastics in sewage sludge and their effects in agricultural soils: A review. *Environmental Monitoring and Assessment* 193 (4): 175.

71 Bayo, J., Olmos, S., López-Castellanos, J. et al. (2016). Microplastics and microfibers in the sludge of a municipal wastewater treatment plant. *International Journal of Sustainable Development and Planning* 11 (5): 812–821.

72 Mahon, A.M., O'Connell, B., Healy, M.G. et al. (2017). Microplastics in sewage sludge: Effects of treatment. *Environmental Science & Technology* 51 (2): 810–818.

73 Sujathan, S., Kniggendorf, A.-K., Kumar, A. et al. (2017). Heat and bleach: A cost-efficient method for extracting microplastics from return activated sludge. *Archives of Environmental Contamination and Toxicology* 73 (4): 641–648.

74 Li, X., Chen, L., Mei, Q. et al. (2018). Microplastics in sewage sludge from the wastewater treatment plants in China. *Water Research* 142: 75–85.

75 Liu, X., Yuan, W., Di, M. et al. (2019). Transfer and fate of microplastics during the conventional activated sludge process in one wastewater treatment plant of China. *Chemical Engineering Journal* 362: 176–182.

76 van den Berg, P., Huerta-Lwanga, E., Corradini, F. et al. (2020). Sewage sludge application as a vehicle for microplastics in eastern Spanish agricultural soils. *Environmental Pollution* 261: 114198.

77 Wei, L., Xia, X., Zhu, F. et al. (2020). Dewatering efficiency of sewage sludge during Fe2+-activated persulfate oxidation: Effect of hydrophobic/hydrophilic properties of sludge EPS. *Water Research* 181: 115903.

78 Li, X., Chen, L., Ji, Y. et al. (2020). Effects of chemical pretreatments on microplastic extraction in sewage sludge and their physicochemical characteristics. *Water Research* 171: 115379.

79 Alvim, C.B., Mendoza-Roca, J., and Bes-Piá, A. (2020). Wastewater treatment plant as microplastics release source-quantification and identification techniques. *Journal of Environmental Management* 255: 109739.

80 Hurley, R.R., Lusher, A.L., Olsen, M. et al. (2018). Validation of a method for extracting microplastics from complex, organic-rich, environmental matrices. *Environmental Science & Technology* 52 (13): 7409–7417.

81 Lusher, A.L., Hurley, R., Vogelsang, C. et al. (2017). *Mapping Microplastics in Sludge*. Norway: NIVA.

82 Raju, S., Carbery, M., Kuttykattil, A. et al. (2020). Improved methodology to determine the fate and transport of microplastics in a secondary wastewater treatment plant. *Water Research* 173: 115549.

83 Li, Q., Wu, J., Zhao, X. et al. (2019). Separation and identification of microplastics from soil and sewage sludge. *Environmental Pollution* 254: 113076.

84 Edo, C., González-Pleiter, M., Leganés, F. et al. (2020). Fate of microplastics in wastewater treatment plants and their environmental dispersion with effluent and sludge. *Environmental Pollution* 259: 113837.

85 Okoffo, E.D., O'Brien, S., O'Brien, J.W. et al. (2019). Wastewater treatment plants as a source of plastics in the environment: A review of occurrence, methods for identification, quantification and fate. *Environmental Science: Water Research & Technology* 5 (11): 1908–1931.

86 Silva, A.B., Bastos, A.S., Justino, C.I. et al. (2018). Microplastics in the environment: Challenges in analytical chemistry A review. *Analytica Chimica Acta* 1017: 1–19.

87 Kedzierski, M., Le Tilly, V., César, G. et al. (2017). Efficient microplastics extraction from sand. A cost effective methodology based on sodium iodide recycling. *Marine Pollution Bulletin* 115 (1–2): 120–129.

88 Rodrigues, M., Gonçalves, A., Gonçalves, F. et al. (2020). Improving cost-efficiency for MPs density separation by zinc chloride reuse. *MethodsX* 7: 100785.

89 Turan, N.B., Erkan, H.S., and Engin, G.O. (2020). Microplastics in wastewater treatment plants: Occurrence, fate and identification. *Process Safety and Environmental Protection* 146: 77-84.

90 Bayo, J., Olmos, S., and López-Castellanos, J. (2020). Microplastics in an urban wastewater treatment plant: The influence of physicochemical parameters and environmental factors. *Chemosphere* 238: 124593.

91 Jiang, J., Wang, X., Ren, H. et al. (2020). Investigation and fate of microplastics in wastewater and sludge filter cake from a wastewater treatment plant in China. *Science of the Total Environment* 746: 141378.

92 Campo, P., Holmes, A., and Coulon, F. (2019). A method for the characterisation of microplastics in sludge. *MethodsX* 6: 2776–2781.

93 Priyanka, M. and Saravanakumar, M. (2019). Preliminary examination on isolation of microplastics (MPs) in sewage sludge from the local wastewater treatment plant. *International Journal of Engineering and Advanced Technology* 9 (1): 7514–7516.

94 Leslie, H.A., Brandsma, S.H., van Velzen, M.J. et al. (2017). Microplastics en route: Field measurements in the Dutch river delta and Amsterdam canals, wastewater treatment plants, North Sea sediments and biota. *Environment International* 101: 133–142.

95 Schymanski, D., Goldbeck, C., Humpf, H.-U. et al. (2018). Analysis of microplastics in water by micro-Raman spectroscopy: Release of plastic particles from different packaging into mineral water. *Water Research* 129: 154–162.

96 Wiesheu, A.C., Anger, P.M., Baumann, T. et al. (2016). Raman microspectroscopic analysis of fibers in beverages. *Analytical Methods* 8 (28): 5722–5725.

97 Oßmann, B.E., Sarau, G., Holtmannspötter, H. et al. (2018). Small-sized microplastics and pigmented particles in bottled mineral water. *Water Research* 141: 307–316.

98 Mason, S.A., Welch, V.G., and Neratko, J. (2018). Synthetic polymer contamination in bottled water. *Frontiers in Chemistry* 6: 407.

99 Elkhatib, D. and Oyanedel-Craver, V. (2020). A critical review of extraction and identification methods of microplastics in wastewater and drinking water. *Environmental Science & Technology* 54 (12): 7037–7049.

100 Kosuth, M., Mason, S.A., and Wattenberg, E.V. (2018). Anthropogenic contamination of tap water, beer, and sea salt. *PloS one* 13 (4): e0194970.

101 Mintenig, S., Löder, M., Primpke, S. et al. (2019). Low numbers of microplastics detected in drinking water from ground water sources. *Science of the Total Environment* 648: 631–635.

102 Pivokonsky, M., Cermakova, L., Novotna, K. et al. (2018). Occurrence of microplastics in raw and treated drinking water. *Science of the Total Environment* 643: 1644–1651.

103 Ruggero, F., Gori, R., and Lubello, C. (2020). Methodologies for microplastics recovery and identification in heterogeneous solid matrices: A review. *Journal of Polymers and the Environment* 28 (3): 739–748.

104 Cole, M., Webb, H., Lindeque, P.K. et al. (2014). Isolation of microplastics in biota-rich seawater samples and marine organisms. *Scientific Reports* 4.

105 Loeder, M.G.J., Kuczera, M., Mintenig, S. et al. (2015). Focal plane array detector-based micro-Fourier-transform infrared imaging for the analysis of microplastics in environmental samples. *Environmental Chemistry* 12 (5): 563–581.

106 Loeder, M.G.J., Imhof, H.K., Ladehoff, M. et al. (2017). Enzymatic purification of microplastics in environmental samples. *Environmental Science & Technology* 51 (24): 14283–14292.

107 Courtene-Jones, W., Quinn, B., Ewins, C. et al. (2019). Consistent microplastic ingestion by deep-sea invertebrates over the last four decades (1976–2015), a study from the North East atlantic. *Environmental Pollution* 244: 503–512.

108 Catarino, A.I., Thompson, R., Sanderson, W. et al. (2017). Development and optimization of a standard method for extraction of microplastics in mussels by enzyme digestion of soft tissues. *Environmental Toxicology and Chemistry* 36 (4): 947–951.

109 von Friesen, L.W., Granberg, M.E., Hassellov, M. et al. (2019). An efficient and gentle enzymatic digestion protocol for the extraction of microplastics from bivalve tissue. *Marine Pollution Bulletin* 142: 129–134.

110 Bretas Alvim, C., Mendoza-Roca, J.A., and Bes-Pia, A. (2020). Wastewater treatment plant as microplastics release source- quantification and identification techniques. *Journal of Environmental Management* 255: 109739.

111 Hu, K., Tian, W., Yang, Y. et al. (2021). Microplastics remediation in aqueous systems: Strategies and technologies. *Water Research* 198: 117144.

112 Simon, M., van Alst, N., and Vollertsen, J. (2018). Quantification of microplastic mass and removal rates at wastewater treatment plants applying Focal Plane Array (FPA)-based Fourier Transform Infrared (FT-IR) imaging. *Water Research* 142: 1–9.

113 Enders, K., Lenz, R., Beer, S. et al. (2017). Extraction of microplastic from biota: Recommended acidic digestion destroys common plastic polymers. *ICES Journal of Marine Science* 74 (1): 326–331.

114 Naidoo, T., Goordiyal, K., and Glassom, D. (2017). Are nitric acid (HNO3) digestions efficient in isolating microplastics from juvenile fish? *Water Air and Soil Pollution* 228 (12).

115 Nuelle, M.T., Dekiff, J.H., Remy, D. et al. (2014). A new analytical approach for monitoring microplastics in marine sediments. *Environmental Pollution* 184: 161–169.

116 Avio, C.G., Gorbi, S., and Regoli, F. (2015). Experimental development of a new protocol for extraction and characterization of microplastics in fish tissues: First observations in commercial species from Adriatic Sea. *Marine Environmental Research* 111 (SI): 18–26.

117 Dehaut, A., Cassone, A.L., Frere, L. et al. (2016). Microplastics in seafood: Benchmark protocol for their extraction and characterization. *Environmental Pollution* 215: 223–233.

118 Liu, P., Qian, L., Wang, H. et al. (2019). New insights into the aging behavior of microplastics accelerated by advanced oxidation processes. *Environmental Science & Technology* 53 (7): 3579–3588.

119 Müller, A., Becker, R., Dorgerloh, U. et al. (2018). The effect of polymer aging on the uptake of fuel aromatics and ethers by microplastics. *Environmental Pollution* 240: 639–646.

120 Hüffer, T., Weniger, A.-K., and Hofmann, T. (2018). Sorption of organic compounds by aged polystyrene microplastic particles. *Environmental Pollution* 236: 218–225.

121 Lv, Y., Huang, Y., Kong, M. et al. (2017). Multivariate correlation analysis of outdoor weathering behavior of polypropylene under diverse climate scenarios. *Polymer Testing* 64: 65–76.

122 Song, Y.K., Hong, S.H., Jang, M. et al. (2017). Combined effects of UV exposure duration and mechanical abrasion on microplastic fragmentation by polymer type. *Environmental Science & Technology* 51 (8): 4368–4376.

123 Li, X., Mei, Q., Chen, L. et al. (2019). Enhancement in adsorption potential of microplastics in sewage sludge for metal pollutants after the wastewater treatment process. *Water Research* 157: 228–237.

124 Turner, A. and Holmes, L.A. (2015). Adsorption of trace metals by microplastic pellets in fresh water. *Environmental Chemistry* 12 (5): 600–610.

125 Melo, D.Q., Neto, V.O.S., Oliveira, J.T. et al. (2013). Adsorption equilibria of Cu2+, Zn2+, and Cd2+ on EDTA-functionalized silica spheres. *Journal of Chemical and Engineering Data* 58 (3): 798–806.

126 Yang, J., Cang, L., Sun, Q. et al. (2019). Effects of soil environmental factors and UV aging on Cu 2+ adsorption on microplastics. *Environmental Science and Pollution Research* 26: 23027–23036.

127 Gewert, B., Plassmann, M.M., and MacLeod, M. (2015). Pathways for degradation of plastic polymers floating in the marine environment. *Environmental Science: Processes & Impacts* 17 (9): 1513–1521.

128 Alimi, O.S., Farner Budarz, J., Hernandez, L.M. et al. (2018). Microplastics and nanoplastics in aquatic environments: Aggregation, deposition, and enhanced contaminant transport. *Environmental Science & Technology* 52 (4): 1704–1724.

129 Teuten, E.L., Saquing, J.M., Knappe, D.R.U. et al. (2009). Transport and release of chemicals from plastics to the environment and to wildlife. *Philosophical Transactions of the Royal Society B-biological Sciences* 364 (1526): 2027–2045.

130 Shim, W.J., Hong, S.H., and Eo, S.E. (2017). Identification methods in microplastic analysis: A review. *Analytical Methods* 9 (9): 1384–1391.

131 McDermid, K.J. and McMullen, T.L. (2004). Quantitative analysis of small-plastic debris on beaches in the Hawaiian archipelago. *Marine Pollution Bulletin* 48 (7): 790–794.

132 Gregory, M.R. (1977). Plastic pellets on New Zealand beaches. *Marine Pollution Bulletin* 8 (4): 82–84.

133 Heo, N.W., Hong, S.H., Han, G.M. et al. (2013). Distribution of small plastic debris in cross-section and high strandline on Heungnam beach, South Korea. *Ocean Science Journal* 48 (2): 225–233.

134 Hidalgo-Ruz, V. and Thiel, M. (2013). Distribution and abundance of small plastic debris on beaches in the SE Pacific (Chile): A study supported by a citizen science project. *Marine Environmental Research* 87–88: 12–18.

135 Eriksen, M., Lebreton, L.C., Carson, H.S. et al. (2014). Plastic pollution in the World's Oceans: More than 5 trillion plastic pieces weighing over 250,000 tons afloat at sea. *PloS one* 9 (12): e111913.

136 Desforges, J.-P.W., Galbraith, M., Dangerfield, N. et al. (2014). Widespread distribution of microplastics in subsurface seawater in the NE Pacific Ocean. *Marine Pollution Bulletin* 79 (1): 94–99.

137 Nel, H.A. and Froneman, P.W. (2015). A quantitative analysis of microplastic pollution along the south-eastern coastline of South Africa. *Marine Pollution Bulletin* 101 (1): 274–279.

138 Kang, J.-H., Kwon, O.-Y., and Shim, W.J. (2015). Potential threat of microplastics to zooplanktivores in the surface waters of the Southern Sea of Korea. *Archives of Environmental Contamination and Toxicology* 69 (3): 340–351.

139 Laglbauer, B.J.L., Franco-Santos, R.M., Andreu-Cazenave, M. et al. (2014). Macrodebris and microplastics from beaches in Slovenia. *Marine Pollution Bulletin* 89 (1): 356–366.

140 Song, Y.K., Hong, S.H., Jang, M. et al. (2015). A comparison of microscopic and spectroscopic identification methods for analysis of microplastics in environmental samples. *Marine Pollution Bulletin* 93 (1): 202–209.

141 Eriksen, M., Maximenko, N., Thiel, M. et al. (2013). Plastic pollution in the South Pacific subtropical gyre. *Marine Pollution Bulletin* 68 (1): 71–76.

142 Browne, M.A., Galloway, T.S., and Thompson, R.C. (2010). Spatial patterns of plastic debris along estuarine shorelines. *Environmental Science & Technology* 44 (9): 3404–3409.

143 Lusher, A.L., McHugh, M., and Thompson, R.C. (2013). Occurrence of microplastics in the gastrointestinal tract of pelagic and demersal fish from the English channel. *Marine Pollution Bulletin* 67 (1): 94–99.

144 Cooper, D.A. and Corcoran, P.L. (2010). Effects of mechanical and chemical processes on the degradation of plastic beach debris on the island of Kauai, Hawaii. *Marine Pollution Bulletin* 60 (5): 650–654.

145 Vianello, A., Boldrin, A., Guerriero, P. et al. (2013). Microplastic particles in sediments of Lagoon of Venice, Italy: First observations on occurrence, spatial patterns and identification. *Estuarine, Coastal and Shelf Science* 130: 54–61.

146 von Moos, N., Burkhardt-Holm, P., and Köhler, A. (2012). Uptake and effects of microplastics on cells and tissue of the Blue Mussel Mytilus edulis L. after an experimental exposure. *Environmental Science & Technology* 46 (20): 11327–11335.

147 Maes, T., Jessop, R., Wellner, N. et al. (2017). A rapid-screening approach to detect and quantify microplastics based on fluorescent tagging with Nile Red. *Scientific Reports* 7.

148 Xu, S., Ma, J., Ji, R. et al. (2019). Microplastics in aquatic environments: Occurrence, accumulation, and biological effects. *Science of the Total Environment* 703: 134699.

149 Primpke, S., Imhof, H., Piehl, S. et al. (2017). Environmental chemistry microplastic in the environment. *Chemie in Unserer Zeit* 51 (6): 402–412.

150 Primpke, S., Wirth, M., Lorenz, C. et al. (2018). Reference database design for the automated analysis of microplastic samples based on Fourier transform infrared (FTIR) spectroscopy. *Analytical and Bioanalytical Chemistry* 410 (21): 5131–5141.

151 Koelmans, A.A., Mohamed Nor, N.H., Hermsen, E. et al. (2019). Microplastics in freshwaters and drinking water: Critical review and assessment of data quality. *Water Research* 155: 410–422.

152 Wang, F., Wong, C.S., Chen, D. et al. (2018). Interaction of toxic chemicals with microplastics: A critical review. *Water Research* 139: 208–219.

153 Peñalver, R., Arroyo-Manzanares, N., López-García, I. et al. (2020). An overview of microplastics characterization by thermal analysis. *Chemosphere* 242: 125170.
154 Fischer, M. and Scholz-Böttcher, B.M. (2017). Simultaneous trace identification and quantification of common types of microplastics in environmental samples by pyrolysis-gas chromatography–mass spectrometry. *Environmental Science & Technology* 51 (9): 5052–5060.
155 Parsi, Z., Hartog, N., Górecki, T. et al. (2007). Analytical pyrolysis as a tool for the characterization of natural organic matter—A comparison of different approaches. *Journal of Analytical and Applied Pyrolysis* 79 (1–2): 9–15.
156 Steinmetz, Z., Kintzi, A., Muñoz, K. et al. (2020). A simple method for the selective quantification of polyethylene, polypropylene, and polystyrene plastic debris in soil by pyrolysis-gas chromatography/mass spectrometry. *Journal of Analytical and Applied Pyrolysis* 147: 104803.
157 Fries, E., Dekiff, J.H., Willmeyer, J. et al. (2013). Identification of polymer types and additives in marine microplastic particles using pyrolysis-GC/MS and scanning electron microscopy. *Environmental Science: Processes & Impacts* 15 (10): 1949–1956.
158 Funck, M., Yildirim, A., Nickel, C. et al. (2020). Identification of microplastics in wastewater after cascade filtration using Pyrolysis-GC–MS. *MethodsX* 7: 100778.
159 Kucerik, J., Demyan, M.S., and Siewert, C. (2016). Practical application of thermogravimetry in soil science. *Journal of Thermal Analysis and Calorimetry* 123 (3): 2441–2450.
160 David, J., Weissmannová, H.D., Steinmetz, Z. et al. (2019). Introducing a soil universal model method (SUMM) and its application for qualitative and quantitative determination of poly (ethylene), poly (styrene), poly (vinyl chloride) and poly (ethylene terephthalate) microplastics in a model soil. *Chemosphere* 225: 810–819.
161 Majewsky, M., Bitter, H., Eiche, E. et al. (2016). Determination of microplastic polyethylene (PE) and polypropylene (PP) in environmental samples using thermal analysis (TGA-DSC). *Science of the Total Environment* 568: 507–511.
162 Dubrawski, J. (1987). The effect of particle size on the determination of quartz by differential scanning calorimetry. *Thermochimica Acta* 120: 257–260.
163 Chialanza, M.R., Sierra, I., Parada, A.P. et al. (2018). Identification and quantitation of semi-crystalline microplastics using image analysis and differential scanning calorimetry. *Environmental Science and Pollution Research* 25 (17): 16767–16775.
164 Statheropoulos, M., Kyriakou, S., and Pappa, A. (1999). Repetitive pulsed sampling interface for combined thermogravimetry/mass spectrometry. *Thermochimica Acta* 329 (1): 83–88.

165 Dümichen, E., Braun, U., Senz, R. et al. (2014). Assessment of a new method for the analysis of decomposition gases of polymers by a combining thermogravimetric solid-phase extraction and thermal desorption gas chromatography mass spectrometry. *Journal of Chromatography A* 1354: 117–128.

166 Dümichen, E., Braun, U., Kraemer, R. et al. (2015). Thermal extraction combined with thermal desorption: A powerful tool to investigate the thermo-oxidative degradation of polyamide 66 materials. *Journal of Analytical and Applied Pyrolysis* 115: 288–298.

167 Dümichen, E., Eisentraut, P., Celina, M. et al. (2019). Automated thermal extraction-desorption gas chromatography mass spectrometry: A multifunctional tool for comprehensive characterization of polymers and their degradation products. *Journal of Chromatography A* 1592: 133–142.

168 Dümichen, E., Eisentraut, P., Bannick, C.G. et al. (2017). Fast identification of microplastics in complex environmental samples by a thermal degradation method. *Chemosphere* 174: 572–584.

169 Lee, J. and Chae, K.-J. (2020). A systematic protocol of microplastics analysis from their identification to quantification in water environment: A comprehensive review. *Journal of Hazardous Materials* 403: 124049.

170 Colognato, R., Bonelli, A., Ponti, J. et al. (2008). Comparative genotoxicity of cobalt nanoparticles and ions on human peripheral leukocytes in vitro. *Mutagenesis* 23 (5): 377–382.

171 Kittler, S., Greulich, C., Gebauer, J. et al. (2010). The influence of proteins on the dispersability and cell-biological activity of silver nanoparticles. *Journal of Materials Chemistry* 20 (3): 512–518.

172 Linsinger, T.P., Gerganova, T., Kestens, V. et al. (2019). Preparation and characterisation of two polydisperse, non-spherical materials as certified reference materials for particle size distribution by static image analysis and laser diffraction. *Powder Technology* 343: 652–661.

173 Frisken, B.J. (2001). Revisiting the method of cumulants for the analysis of dynamic light-scattering data. *Applied Optics* 40 (24): 4087–4091.

174 Hernandez, L.M., Yousefi, N., and Tufenkji, N. (2017). Are there nanoplastics in your personal care products? *Environmental Science & Technology Letters* 4 (7): 280–285.

175 Ter Halle, A., Jeanneau, L., Martignac, M. et al. (2017). Nanoplastic in the North Atlantic subtropical gyre. *Environmental Science & Technology* 51 (23): 13689–13697.

176 Gigault, J., Pedrono, B., Maxit, B. et al. (2016). Marine plastic litter: The unanalyzed nano-fraction. *Environmental Science: Nano* 3 (2): 346–350.

177 Besseling, E., Quik, J.T., Sun, M. et al. (2017). Fate of nano-and microplastic in freshwater systems: A modeling study. *Environmental Pollution* 220: 540–548.

178 Correia, M. and Loeschner, K. (2018). Detection of nanoplastics in food by asymmetric flow field-flow fractionation coupled to multi-angle light scattering: Possibilities, challenges and analytical limitations. *Analytical and Bioanalytical Chemistry* 410 (22): 5603–5615.

179 Gallego-Urrea, J.A., Tuoriniemi, J., and Hassellöv, M. (2011). Applications of particle-tracking analysis to the determination of size distributions and concentrations of nanoparticles in environmental, biological and food samples. *Trac-trends in Analytical Chemistry* 30 (3): 473–483.

180 Filipe, V., Hawe, A., and Jiskoot, W. (2010). Critical evaluation of Nanoparticle Tracking Analysis (NTA) by NanoSight for the measurement of nanoparticles and protein aggregates. *Pharmaceutical Research* 27 (5): 796–810.

181 Singh, P., Bodycomb, J., Travers, B. et al. (2019). Particle size analyses of polydisperse liposome formulations with a novel multispectral advanced nanoparticle tracking technology. *International Journal of Pharmaceutics* 566: 680–686.

182 Lambert, S. and Wagner, M. (2016). Characterisation of nanoplastics during the degradation of polystyrene. *Chemosphere* 145: 265–268.

183 Lambert, S. and Wagner, M. (2016). Formation of microscopic particles during the degradation of different polymers. *Chemosphere* 161: 510–517.

184 Lambert, S., Sinclair, C.J., Bradley, E.L. et al. (2013). Effects of environmental conditions on latex degradation in aquatic systems. *Science of the Total Environment* 447: 225–234.

185 Bootz, A., Vogel, V., Schubert, D. et al. (2004). Comparison of scanning electron microscopy, dynamic light scattering and analytical ultracentrifugation for the sizing of poly (butyl cyanoacrylate) nanoparticles. *European Journal of Pharmaceutics and Biopharmaceutics* 57 (2): 369–375.

186 Domingos, R.F., Baalousha, M.A., Ju-Nam, Y. et al. (2009). Characterizing manufactured nanoparticles in the environment: Multimethod determination of particle sizes. *Environmental Science & Technology* 43 (19): 7277–7284.

187 Hassan, P.A., Rana, S., and Verma, G. (2015). Making sense of Brownian motion: Colloid characterization by dynamic light scattering. *Langmuir* 31 (1): 3–12.

188 Kang, P., Ji, B., Zhao, Y. et al. (2020). How can we trace microplastics in wastewater treatment plants: A review of the current knowledge on their analysis approaches. *Science of the Total Environment* 745: 140943.

189 Torre, M., Digka, N., Anastasopoulou, A. et al. (2016). Anthropogenic microfibres pollution in marine biota. A new and simple methodology to minimize airborne contamination. *Marine Pollution Bulletin* 113 (1–2): 55–61.

190 Xu, Q., Gao, Y., Xu, L. et al. (2020). Investigation of the microplastics profile in sludge from China's largest water reclamation plant using a feasible isolation device. *Journal of Hazardous Materials* 388: 122067.

191 Li, J., Song, Y., and Cai, Y. (2020). Focus topics on microplastics in soil: Analytical methods, occurrence, transport, and ecological risks. *Environmental Pollution* 257: 113570.

192 Primpke, S., Lorenz, C., Rascher-Friesenhausen, R. et al. (2017). An automated approach for microplastics analysis using focal plane array (FPA) FTIR microscopy and image analysis. *Analytical Methods* 9 (9): 1499–1511.

2

Occurrence and Removal of Microplastics in Drinking Water Systems

Junyeol Kim[1], Yongli Z. Wager[1,], Carol Miller[1], and John Norton[2]*

[1] *Department of Civil and Environmental Engineering, Wayne State University, Detroit, MI, USA*
[2] *Energy, Research, & Innovation, Great Lakes Water Authority, Detroit, MI, USA*
* *Corresponding author*

2.1 Introduction

Plastic materials are widely used in daily life, and their annual production is exponentially increasing worldwide due to increasing demand. The growth of plastic production over the past 70 years exceeds that of any other manufactured material. Compared to the first large-scale plastic production in the 1950s, an increase of two orders of magnitude in the annual production was recorded in 2019 [1, 2]. Global plastic production recorded as 1.7 million tons (MTs) in the 1950s surpassed 100 MTs in the 1990s and reached 368 MTs in 2019 (Figure 2.1) [1–3]. This increase is expected to continue, and the annual production is predicted to double (600 MTs) by 2025 and triple (> 1,000 MTs) by 2050, considering the current population growth and consumption rate of plastics [4]. Another prediction estimates that 8,300 MTs have been produced so far and about 6,300 MTs of plastic waste were generated in 2015 alone [5]. It was also reported that only 9% of the plastic produced was recycled, 12% was burned, and 79% was accumulated in landfills or the natural environment. Lastly, by 2050, approximately 12,000 MTs of plastic waste will exist in landfills and the natural environment.

The wide range of applications and high production volume result in the release of many plastic materials into the environment. As a result, large amounts of discarded plastics are causing multiple environmental problems such as marine pollution and landfill saturation [6]. Contamination by microplastics is a representative example of pollution caused by plastic waste. Small

Microplastics in Urban Water Management, First Edition. Edited by Bing-Jie Ni, Qiuxiang Xu, and Wei Wei.
© 2023 John Wiley & Sons, Inc. Published 2023 by John Wiley & Sons, Inc.

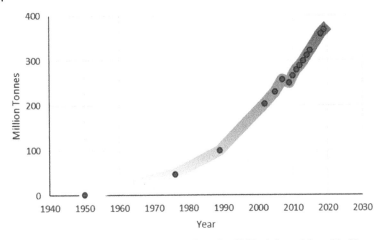

Figure 2.1 Global plastic production since the 1950s. Adapted from [1–3].

fragments of plastic that are not easily biodegradable and have weathered for many years are now being found around the world. These tiny particles, called microplastics, are found in a variety of aquatic environments including both freshwater and seawater [7, 8]. Microplastics have also been found in tap and bottled water. A 2018 survey of 159 drinking water samples (156 tap water and 3 bottled water samples) in 14 countries revealed that 81% of water samples contained anthropogenic particles, including microplastics [9, 10]. Microplastics are detected not only in the aquatic environment, but also in food, soil, and air.

Currently, there is insufficient information to draw conclusions on the toxicity of microplastics. However, many recent studies have found that microplastics could cause adverse health impacts, such as the accumulation of microplastics in organs, respiratory bursts, oxidative stresses, inflammation, and alteration in the expression of stress-related genes in various organisms [11–17]. Moreover, microplastics have been reported as carriers for chemicals, accelerating the transport of organic pollutants and heavy metals in the environment [12, 18–20]. Therefore, microplastic pollution, which is prevalent around the world, is rapidly emerging as one of the greatest environmental and public health challenges. As a result, it is critical to fully understand the pathway through which microplastics are generated and the compartments of the environment into which the microplastics flow. In addition, research on how to mitigate microplastics and minimize contamination is urgently needed.

2.2 What Are Microplastics?

Plastic particles smaller than 5 mm typically fall into the range of microplastics. Since microplastics have a wide range of materials, shapes, colors, and sizes, the definition is not completely uniform among researchers. Considering these complex and diverse microplastics, many researchers described microplastics in marine environments and drinking water as particles smaller than 5 mm, while others focused on microplastics smaller than 5 mm with the lower bound as 1 μm [6, 21–24]. Also, the definition of microplastic size varies by government agencies and research institutes. The United Nations Environment Programme (UNEP) defines solid plastic particles less than 5 mm that cannot be decomposed in water as "microplastic" [25]. The international standardization ISO/TC 61 (Plastics)/SC 14 (Environmental Aspect) defines microplastics as solid and water-insoluble plastic particles of any dimension 1–1000 μm in "terms and definitions" of the ISO/TR 21960:2020 standard [26]. Recently, the State Water Resources Control Board of California also defined microplastics as a particulate polymer material with chemical additives or other substances added thereto, with at least three dimensions larger than 1 nm and smaller than 5 mm [27]. In this definition, polymers that are of natural origin and have not undergone chemical modification (except for hydrolysis) are excluded. The size range of microplastics and their size comparison with other particulate matter are shown in Figure 2.2.

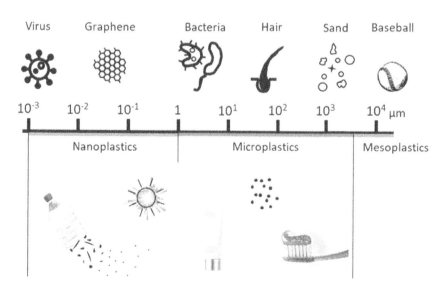

Figure 2.2 Size comparison of plastics with other particulate matter.

2.2.1 Primary and Secondary Microplastics

Microplastics are subdivided into *primary* microplastics and *secondary* microplastics. Primary microplastics are manufactured synthetic particles that fall within the microplastic size category, including microbeads in liquid soaps and cosmetics, abrasives in cleaning agents, and nano-sized latex plastics in paints and coatings, as shown in Figure 2.3 [6, 28]. Owing to the mass production and use of consumer products containing microplastics, primary microplastics are found in natural and urban environments and are becoming a worldwide issue because of slow degradation.

When plastics enter the natural environment, they may break down into small pieces from weathering (Figure 2.3). Small plastic pieces fragmented in this way are called secondary microplastics [24, 29]. Sources of secondary microplastics include any plastics entering into the environment, such as plastic water bottles, fishing nets, plastic bags, and plastic containers. The occurrence rate of secondary microplastics is likely determined by the physicochemical properties of the plastic product, the strength and time of weathering (e.g., photodegradation), and local environmental factors [24, 29]. For example, when plastic surfaces are exposed to strong UV radiation and high temperature for a long time, physicochemical changes that occur on the plastic surfaces can make the plastic material brittle and thus more susceptible to fragmentation and decomposition [30, 31]. Data on the rate of fragmentation and degradation of microplastics is limited in drinking water environments.

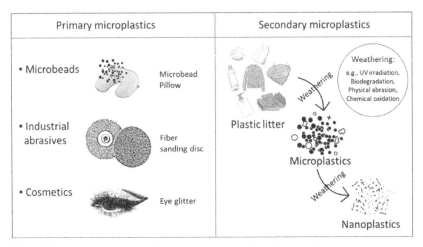

Figure 2.3 Primary and secondary microplastics and their examples.

2.3 The Emergence of Microplastics

According to the production trend in Figure 2.1, an estimated 300 MTs of plastic materials were produced worldwide in 2021. An additional concern is that the degradation rate of plastics in the environment is too slow compared to the rate at which plastic materials are used and discarded. This allows discarded plastic materials to accumulate in ecosystems, including soil, groundwater, and surface water [32–34]. The harm caused by the ingestion of microplastics is also a concern. When ingested, microplastics could cause a number of adverse impacts (e.g., accumulation in organs, inflammation, oxidative stresses, etc.) on various organisms. In addition, microplastics are well known for carrying other harmful chemical compounds, such as persistent organic pollutants and heavy metals [11, 20, 35]. Despite the harmful effects caused by microplastics, only 9% of the plastic produced by 2015 has been recycled. With this recycling rate, it is expected that 12,000 MTs of waste plastics will be thrown into landfills or the environment by 2050 [5]. If the release of waste plastic into the environment continues at the current rate, the risks posed by microplastics are likely to become more extensive. Therefore, it is imperative to understand the primary sources of microplastics and their fate and transport once released into the environment.

2.3.1 Sources

The sources of microplastics can come from numerous ways in everyday life. This is because items made of plastics are used everywhere. Representative consumer products that potentially cause microplastic pollution include cosmetics, personal care products, clothing and textiles, bottled water, automobile tires, abrasives, nets, disposable food containers, packaging materials, and plastic bags. In addition, all industries that use plastics as raw materials can be potential sources of contamination for microplastics, because microplastics can be released once plastic litter is exposed to long-term weathering. Typical microplastic sources and their introduction into the freshwater environment are illustrated in Figure 2.4.

One study identified plastic bottles and bottle caps as suspicious sources of microplastics [36]. Dust from automobile tires has been found to contribute more than 50% of total primary microplastic emissions in mainland China [37]. Even everyday clothes release tens or hundreds of mg of microplastics per kg of washed fabric during washing [38]. Fragmented microplastics released from the degradation of fishing gears are also well-known sources of microplastic pollution in the environment [22, 39, 40]. There is also research showing that the act of opening plastic packaging itself can generate microplastics. About 0.46–250 microplastics (MPs)/cm are generated whether using scissors, hand, or knife when

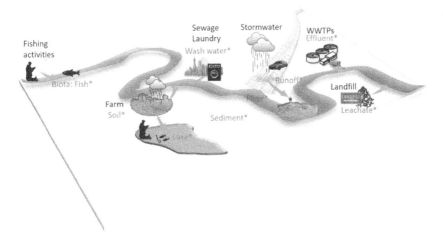

Figure 2.4 Representative sources of microplastics and their introduction to the freshwater environment. WWTPs: wastewater treatment plants. Environmental samples in which microplastics were detected were marked with *.

opening plastic packaging, and the amount of microplastics generated depends on the stiffness, thickness, anisotropy, and density of the packaging material [19]. This means that the use of plastic material itself produces microplastics.

In some cases, fibrous microplastics discharged from textile industrial parks become a major source of microplastics. Despite the removal of 85–99% of microfibers by the industrial complex's sewage treatment plants, the average concentration of microfibers in the discharged water was still 537.5 MPs/L, which was significantly higher than that of municipal sewage treatment plants. This means that the wastewater treatment plants in the textile industrial complex release 430 billion microfibers per day [41]. It shows that the source of microplastics can be derived from nearby industrial facilities and that current water treatment facilities may not be sufficient to prevent water resource pollution by microplastics. In addition, even if the removal efficiency of microplastics reaches 99%, continuous monitoring is still required because of the total amount of microplastics released daily.

Although landfills are considered one of the final destinations for plastic pollutants, recent research shows that microplastics found in leachate from landfills can contribute to microplastic contamination in drinking water sources. For example, one study detected microplastics ranging from 0.42 MPs/L to 24.58 MPs/L in 12 leachate samples collected from six landfills in four cities (Shanghai, Wuxi, Suzhou, and Changzhou) in China [42]. Agricultural practices, called plasticulture, are also considered a potential source of microplastic pollution. Greenhouses and walk-in tunnel covers and plastic mulch are representative examples of plastics used in plasticulture. PVC and LDPE/HDPE used in irrigation have also been identified as one of the potential sources of microplastic pollution in Lake Ontario [43].

Meanwhile, there are also reports and concerns about microplastics arising from the water treatment process. A greater number of microplastics were found in the effluent of ozone treatment process than that in the influent water [44]. The authors explain that this is because plastic particles are broken into smaller pieces by shearing force, increasing the number of microplastics [44, 45]. In addition, in some cases, higher concentrations of polyacrylamide (PAM) than raw water were found in the effluent of water treatment facilities. This is believed to arise from the use of a coagulant containing PAM in the coagulation process [46, 47]. There are concerns about the occurrence of microplastics not only in the ozonation and coagulation process, but also in the filtering process by membrane. Currently, membrane filtration systems are widely used in water treatment facilities, and the filter process by membrane is recognized as an effective water treatment process to remove microplastics [48–50]. Membrane filters require cleaning through backwashing between operations to prevent clogging and fouling. The high pressure and air from backwashing affect the elasticity of the membrane that has been used for a long time and cause damage to the membrane, leading to a possibility that part of the membrane surface may fragment in the form of small particles. In addition, chemical agents such as NaOH, H_2O_2, and NaClO provide an oxidizing environment that makes the membrane more brittle, allowing materials such as PVC to dissolve from the membrane [51]. Based on these facts, Ding et al. consider that physical flushing, chemical agents, mechanical stress, aging, and wear can cause microplastics to be released from organic membrane filters in water treatment plants. The infrastructure of membrane filter systems (e.g., the filling, sealing, casing, gaskets, etc.) can also release microplastics by prolonged mechanical weathering when old and worn [52].

2.3.2 Transformation

Plastic materials discarded into the environment are eventually broken into microplastics by sunlight, mechanical abrasion, and waves. Because a larger amount of micro- and nano-plastics are generated and transformed by prolonged exposure to sunlight, and the transformed microplastics potentially carry metal and organic pollutants to biota, there are increasing concerns about the transformation of microplastics [53–57].

The transformation of microplastics occurs during the plastic degradation process. For instance, oxidized chemical groups were detected in plastics as a result of long-term exposure to seawater. Physical and morphological transformation (in which the size of the plastic was reduced by fragmentation) was also observed after the internalization of microplastics by aquatic animals [58]. Transformation of microplastics occurs through several pathways such as chemical, bio-, photo-, thermal, and mechanical degradation. Chemical degradation of microplastics occurs mainly by hydrolysis and oxidation. Hydrolysis induces the cleavage of C–O bonds in the backbone of plastic polymers, and oxidation causes the chain scission of plastic polymers with carbon-carbon backbones [31]. Photodegradation begins when discarded plastic materials are exposed to sunlight. UV light is a major factor in the degradation of plastics released into the environment and makes plastic materials weak and brittle by reducing their elasticity. A representative example of biodegradation is microbial degradation. Microbial degradation begins when bacteria adhere to the plastic surface and colonize it. Microbial enzymes convert polymers into low-weight oligomers, dimers, and monomers through a multistep chemical reaction of hydrolytic cleavage, which finally breaks them down into CO_2 and H_2O [31, 59].

Weathering, which causes the degradation of microplastics, results in the physicochemical transformation of microplastic surfaces during the degradation process, affecting the behavior of microplastics. The surface of microplastics chemically transformed by weathering affects their adsorption mechanism. As a result of weathering, a rough surface and oxygen functional groups (e.g., carboxylic and hydroxyl groups) were generated, reducing the hydrophobicity of microplastics, and thus adsorption of hydrophobic organic compounds was reduced [60–62]. Also, the microplastic surface, transformed to have oxygen functional groups and a more negative surface charge by weathering, potentially enhances hydrogen bonding and electrostatic interactions in adsorption with hydrophilic and polar organic compounds and metals [57, 60, 63]. Fragmentation by weathering is a representative example of the physical transformation of microplastics. By fragmentation, the surface of a microplastic has a larger surface area, which leads to more adsorption of contaminants [57, 64]. The biofilm involved in biodegradation of microplastics increases heteroaggregation of microplastics from extracellular polymeric substances, which then affects the sorption of microplastics with other compounds [65].

The transformation of microplastics during weathering affects not only the behavior of microplastics, but also their toxicity. Microplastics that have been reduced in size by fragmentation can be more easily ingested by biota, and the decrease in size can also affect the distribution of ingested microplastics in the organs in vivo [66, 67]. The surface charge of microplastics increased by transformation from weathering affects not only the aggregation and dispersion of microplastics in organisms, but also their binding abilities to cell tissues [68]. In addition, it was found that zooplankton prefers to consume microplastics that have been weathered because of biofilm coatings on microplastics [69].

2.4 Occurrence of Microplastics in Drinking Water Systems

Since the 1950s, when production and use of plastics began to increase exponentially, the discharge of plastic materials to the environment has also increased rapidly. As a result, rivers and lakes around the world are contaminated by waste plastics. Environmental pollution caused by the influx of plastics and microplastics has reached a point where it is impossible to mitigate this pollution only by the self-purification of the nature [30].

Globally, microplastics widely occur in water systems around the world. Considering the various sources and pathways of microplastic pollution (Figure 2.4), it is not surprising that microplastics are widely presented in freshwater systems around the world. According to various reports and studies, microplastics have been detected in many countries in Europe, Asia, America, and Africa with diverse levels of detected concentrations. Some places show high pollution levels with concentrations of more than 12,000 MPs/m^3 in the Los Angeles River, the United States and 6,000 MPs/L in the Yangtze River, China [44, 70–72]. The concentrations of microplastics vary by 10^8 times depending on the region [73]. It was reported that relatively high concentrations of microplastics occurred in stormwater and urban canals, which are related to urban centers with high population density and the wide use of plastics. For example, 48–187 MPs/L were found in samples collected from different urban canal waters in Amsterdam, the Netherlands [74]. Also, 490–22,894 MPs/m^3 were detected in seven stormwater retention ponds serving highway, residential areas, and industries in the North of Jutland, Denmark [75]. On the other hand, in natural environments where human activities are minimized, such as Arctic rivers and lakes, microplastics have a relatively low concentration. For example, an average of <1 anthropogenic particle/L was detected in water samples, and 90 MPs/m^2 were detected in sediment samples of an Arctic freshwater lake [76]. In addition, dozens of microplastics were found per m^3 in rivers located in Siberia [77]. One peculiarity is that high concentrations of microplastics were detected in glaciers and snow, which is interpreted as a result of the continuous concentration of microplastics from limited wash-off during snow deposition [73, 78].

2.4.1 Abundance

Several studies have reported that microplastic contamination has occurred in drinking water sources [44, 46, 79]. For example, depending on the sampling location, 1.6–12.6 MPs/L were found at the Three Gorges Reservoir in China in 2018. Also, 3.4–25.8 MPs/L were observed at several sampling points in China's Taihu Lake in 2016.

There are also reports that microplastics are found in tap water. Microplastics were detected in tap water sampled from 17 locations in Denmark. On average, 15.6 particles with a size of more than 100 μm were found, of which 3% turned out to be microplastics. The microplastics detected were mainly polyethylene terephthalate (PET), polypropylene (PP), and polystyrene (PS). Microplastics in drinking water have also been detected in China. As a result of sampling and analyzing 38 tap water samples from different cities in China, 440 ± 275 MPs/L on average were detected, and microplastics with a size less than 50 μm were the majority [80]. Additionally, fragment-type microplastics accounted for a high proportion, and most of the detected plastics were polyethylene (PE) and PP. In a microplastic investigation conducted in Qingdao, China, 0.3–1.6 MPs/L was found in tap water. Fibers accounted for 99.2% of the detected microplastics, with rayon (48.9%) and PET (29.6%) the mainly detected plastic types. An investigation on 159 tap water samples collected from 14 countries worldwide revealed an average of 5.45 anthropogenic particles per liter in 81% of the samples. Most of the particles found were identified as fibers (98.3%), and the rest were fragments or films [9]. Microplastics have also been detected in water fountains at metro stations and individual houses [80–83]. In Shruti's study, as a result of examining water in drinking fountains at 42 metro stations in Mexico City, microplastics were detected in the range from 5 ± 2 to 91 ± 14 MPs/L (18 ± 7 MP/L on average). Of the detected microplastics, 75% is in the size range of 0.1–1 mm, and polyesters and epoxy resins account for the majority of the detected plastics.

In contrast, there is one study reporting that no microplastic was detected in the tap water. In the study, random samples were collected from three house connections, one transfer station, and five consumption taps (a single-family house, an apartment, an educational facility, a residential building, and a commercial enterprise) in a medium-sized city with 65,000 residents in Germany. Microplastics with a size of 10 μm or larger than 10 μm were analyzed [84]. In the analysis, no experimental group of tap water showed a significant difference of microplastics from the blank. Moreover, no microplastics were detected from the tap water samples of the educational facility.

Although microplastics are frequently found in drinking water sources around the world, there are variations in the abundance of microplastics depending on the region and the water source. Cases of high levels of contamination by microplastics have been increasingly reported in recent years. In studies investigating drinking water treatment plants (DWTPs) using rivers as drinking water sources, 6,614 ± 1,132 MPs/L and 3,605 ± 497 MPs/L have been found in the lower Yangtze River in China and in a river in the Czech Republic [44, 47]. Unlike these two studies, there are also freshwater environments where lesser amounts of microplastics were detected. In the microplastic analysis conducted in the

2.4 Occurrence of Microplastics in Drinking Water Systems

Ganga River, a drinking water source in India, a remarkably small amount of microplastics was found in raw water. As a result of five samplings and analyses at six sites from November 2019 to March 2020, an average of 17.86 ± 2.66 MPs/L was reported in the raw water of the DWTP [85]. This is contrary to the results of other two studies in China (6614 ± 1132 MPs/L) and the Czech Republic (3605 ± 497 MPs/L). The authors (Sarkar et al.) speculate that the difference of two orders of magnitude in the abundance is due to the fact that the Ganga River is relatively less polluted. A lower abundance of microplastics was also observed in the Sinos River in Brazil. An average of 330.2 MPs/L and 105.8 MPs/L were detected in raw water and drinking water samples, respectively [86].

Differences in the abundance of microplastics are known to be determined by a range of factors, such as the type of water body, human activity, and the surrounding environment including local weather conditions. A study conducted in 29 Great Lakes tributaries is a good example of how human activities and the surrounding environment have a profound influence on the abundance of microplastics. The study found an average of 1.9 MPs/m^3 and up to 32 MPs/m^3. In particular, microplastics in the form of fragments, foams, and films were detected at higher concentrations in watersheds with more urbanized regions [87]. In another study conducted in the Great Lakes, the average abundance was about 43,000 MPs/km^2, and more than 466,000 MPs/km^2 were detected in a station located downstream of two major cities. In the lake current convergence with the most abundant concentration, microplastics were thought to be produced near the urban effluent [88]. These studies indicate the more urbanization progresses, the higher the probability that microplastics are detected in the water system [6, 87–89]. On the other hand, some studies show that the abundance of microplastics may be independent of the degree of urbanization in the freshwater system. The Sinos River in Brazil is an important drinking water source for 1.3 million residents around the river. As a result of analyzing river water collected from seven municipalities located along the tributaries of the Sinos River, no correlation was found between the number of microplastic particles and the urbanization gradient [86].

The abundance of microplastics detected depends not only on environmental factors (e.g., the type of water body, the level of urbanization, population density) but also on analytical conditions (e.g., analysis method, pretreatment, and sample collection). For example, the detected microplastic abundance may vary depending on the sampling location, sampling technique such as mesh size, and analysis method [6]. He et al. collected microplastics using two conventional methods (trawling and filtering water) at 10 sampling sites from upstream to downstream of the Yangtze River [90]. The trawling method showed an average abundance of 2.2 ± 0.8 to 56.6 ± 51.6 MPs/m^3, while the filtering method resulted in an average of 800.0 ± 300.0 to 3,088.9 ± 330.6 MPs/m^3. The abundance of microplastics by the trawling method was lower than the filtering method with several orders of magnitude. One of the main causes of this difference is that small microplastics pass through the trawl net and, as a result, are not collected during the sampling.

Collectively, the detected abundance of microplastics can vary greatly depending on the sampling area and surrounding environment (e.g., the degree of urbanization and population density) as well as the sample collection, pretreatment, and analytical methods. With the advancement of analytical technology, microplastics with a size less than 10 μm were analyzed in several recent studies. Although microplastics of 1–5 μm in diameter accounted for the majority of plastic particles in these recent studies, they were not within the scope of analysis in many previous studies [36, 44, 46, 47, 91]. The abundance of microplastics may have been underestimated in many studies that could not measure microplastics smaller than 10 μm because of technical limitations. This potential underestimation can be minimized by improving detection capabilities and limits of analytical instruments.

2.4.2 Distribution

Microplastics are distributed within the drinking water sources through various routes such as effluent discharge from WWTPs, stormwater runoff, and precipitation through the atmosphere as shown in Figure 2.5. Household sewage discharged from homes also contributes to the distribution of microplastics. For example, it is already publicly known that microplastics are detected in sewage discharged from washing machines [37, 38]. In addition, microplastics accumulated on the land are distributed to the freshwater environment through urban and highway stormwater [75]. For instance, analysis of microplastics in stormwater runoffs in a semiarid region (Tijuana, Mexico) revealed that microplastics had the highest abundance with higher precipitation. This study concluded that stormwater could be the major contributor to microplastic distribution to the water body [92].

WWTPs can also contribute to the spatial distribution of microplastics. It has been found that microplastics that have not been completely removed in WWTPs are present in the effluent and are discharged to the freshwater system [41, 44, 47]. Microplastics were also detected in the supply pipeline of DWTPs in Sweden. Most of the microplastics detected in the two pipelines installed in 2000 and 2011 were less than 150 μm and 32% were microplastics less than 20 μm. The total number of microplastics detected was 174 ± 405 MPs/m^3 on average and showed great variability from a minimum of 0 MPs/m^3 to a maximum of 1,219 MPs/m^3 [93].

Microplastics in the water can be transported to the atmosphere through atmospheric entrainment or bubble burst ejection [94]. Also, studies on the transport and

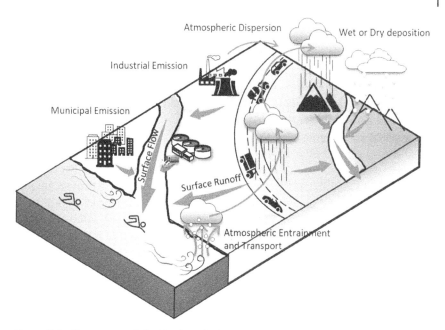

Figure 2.5 Transport and distribution of microplastics in drinking water systems.

deposition of microplastics by the atmosphere have shown that microplastics can move to water bodies or mountainous areas by atmospheric wet or dry deposition, and the transport distance can reach up to 95 km [95]. Microplastics distributed through the atmosphere can accumulate in glaciers at the top of the mountain and redistribute again as glaciers melt and flow, causing potential freshwater pollution [78].

2.4.2.1 Size Distribution

In one recent study that analyzed the size distribution of microplastics in the freshwater environment, the proportion of microplastics with a size of 10 μm or less accounted for the majority. 1–5 μm microplastics accounted for 40–60% of the total microplastics, and 5–10 μm microplastics accounted for 30–40%. Microplastics larger than 10 μm did not exceed 10%, and microplastics larger than 100 μm did not exceed 2% [47]. Comparable size distribution trends have been found in other investigations. In the study of Wang et al., 1–5 μm and 5–10 μm microplasts accounted for 55–58% and 20–28% of the total plastic particles detected. It has also been reported that most microplastics (84–87%) that have passed through a water treatment facility have a size of 1–5 μm [44]. In another report, plastic particles with a size of 1–5 μm accounted for 50% of the total microplastics, and the proportion increased to 65% in the treated water of a water treatment facility [46].

This means that as the size of microplastics increases the proportion of the microplastics in the total detected microplastics may decrease.

Recent studies that lower the detection limit of microplastics to 1 μm show that microplastics with a size of 1–5 μm account for the majority of the detected microplastics. This 1–5 μm size range is, however, rarely covered in previous studies. According to Koelmans et al., most previous studies have only analyzed microplastics larger than 5 μm [91]. This is partly because the size of the detectable microplastics was limited due to the detection limit of the analytical instrument used. The size distribution results of previous studies, not able to analyze smaller microplastics (e.g., 1–5 μm), need to be revisited and reinterpreted. For accurate and detailed size distribution studies, the development of analytical techniques for the detection and quantification of microplastics of 1 μm or less is urgent. If the detection capability of microplastics continues to improve, efforts to reflect such detection capacity in the detection and quantification of microplastics are required to reevaluate the levels of microplastic contamination.

2.4.2.2 Morphological Distribution

Morphological analysis of microplastics shows that fragments and fibers are the mainly observed microplastics. According to a study in the Yangtze River, of the total microplastics detected by the trawling method, 47.9% appeared as fragments, followed by 32.1% as fibers and 17.4% as spheres [90]. The results of using different sampling methods (the filtering method) in the same study showed that 63.4% of microplastics were identified as fibers, 29.4% were fragments, and only 2.6% were spheres. In a study by Pivokonsky et al., 71–76% of the microplastics found in samples from water treatment facilities were fragments. In the sample of another water treatment facility, fragments (42–48%) and fibers (37–61%) accounted for a large part of the microplastic morphology distribution [47]. In addition, in many environmental samples, fibrous microplastics are the dominant species in number. Fibers accounted for 28.6% to 90.5% of microplastics in surface water samples collected from the Yangtze River in China [79]. Also, as a result of analyzing fish living in the estuary, 96% of the detected microplastics were fiber microplastics [96]. In a three-year microplastics study conducted at 12 stations on Lake Winnipeg in Canada, fibrous microplastics accounted for at least 60% at all stations. Following fibrous microplastics, fragments were the second largest proportion, while no microplastics in the form of pellets or beads were identified [97]. Also, in the microplastic studies conducted in the Tapi-Phumduang River and Bandon Bay in Thailand, most (>90%) of the detected microplastics were found to be fibrous microplastics [98]. Microplastics other than fragments or fibers were also found, e.g., film, foam, pellet, sphere, line, bead, flake, sheet, granule, paint, foil, nurdle. However, these types of microplastics generally do not account for the majority [91].

Although fragments and fibers have been found to be major microplastics present in freshwater, much information on size and morphology distribution has not yet been collected as microplastics in many regions still require investigation. Moreover, as the detection capability of microplastics improves, the size and morphology distribution of microplastics must also be continuously updated. Finally, since microplastics with a size of 1 μm or less still have technical difficulties in their qualitative and quantitative analysis, the improvement and development of analytical technology should also be prioritized.

2.4.3 Composition

Plastics are formed by a chemical reaction called polymerization, in which organic molecules have long multiple chains. Plastics are mixed with various additives (e.g., colorants, plasticizers, stabilizers, fillers, and reinforcements) during polymerization, and their chemical composition is determined by the type of additive. Synthetic plastics are generally divided into two types of plastics: thermoplastic and thermoset plastic. Thermoplastics are plastics that have the property of being soft when heated and can be recycled because it becomes hard and reformed during the process of cooling. Thermoplastic types include PE, PP, PET, PS, polyvinylchloride (PVC), polycarbonates (PC), and polyamides (PA). Thermoset plastics, contrary to thermoplastic, do not soften when heated. There are PUR, epoxy resins, acrylic resins, and some polyester for thermoset plastics [6]. Both types of plastics are used in numerous ways in daily life. For example, PE is mainly used for toys, shampoo bottles, food packaging, and pipes, and PP is often used for food and snack packaging, and automobile parts. PET is frequently used in water bottles and beverage bottles, and PS is widely used as food containers, glasses, and insulation. PVC is used to insulate window frames, pipes, and cables, and PUR is widely used as insulation, pillow and mattress, and insulation foam. PE, PP, PVC, PS, and PET account for 90% of the total production of plastics currently in use. 75% of the world's primary plastic production is discarded, and among different uses, packaging materials account for the largest portion while the period of packaging use is the shortest (less than 6 months) [5].

Since packaging containers using PP, PE, and PS as raw materials account for the largest portion, the proportion of PE, PP, and PS is likely to be high for microplastics detected in the environment. According to the analysis of microplastics collected from the Yangtze River, among the microplastics identified, PP was the most abundant (48.7%), followed by PE (32.1%), PS (8.8%), PE-PP (6.7%), PVC (2.1%), PC (1.0%), and Nylon (0.5%) [90]. Among all the microplastics found in 38 tap water samples across China, 26.8% of them were identified as PE, followed by PP (24.4%), PE + PP (22%), PPS (7.3%), PS (6.5%), PET (3.3%), and other polymers

(9.8%) including PMS, PTFE, PC, PMMA, PBT, PB, nylon, and PVC [80]. According to Pivokonsky's study, PET microplastics accounted for the largest proportion (more than 50% on average) of the analyzed plastic particles in the raw water entering the water treatment facility [47]. PP particles (16–26%) was the second-largest proportion after PET. PE also accounted for a large proportion, but only in certain raw waters. Other types of polymers such as PS, PAM, polybutylacrylate (PBA), poly(methyl methacrylate) (PMMA), poly-p-phenylene terephthalamide (PPTA), and polytrimethylene terephthalate (PTT) were detected at less than 10%.

2.5 Removal of Microplastics in Drinking Water Systems

The drinking water system, including drinking water sources, plays a pivotal role in supplying drinking water. The drinking water system prioritizes supplying safe water and has direct impact on the health of the population served. DWTPs are a representative example of minimizing pollutants from drinking water systems for a safe water supply. The processes of DWTPs are primarily designed to provide safe water that meets water quality criteria. In each designed process, organic compounds or heavy metals pollutants are removed, pathogens causing disease are deactivated, and suspended particles increasing turbidity are also treated. Currently, microplastics are pollutants not yet included in the water quality criteria. Therefore, there is no treatment process purposely designed for the treatment of microplastics solely in water treatment facilities. However, conventional water treatment plants have treated particulate matter; therefore, it is expected that DWTPs enables a certain level of microplastic removal. Nevertheless, since microplastics are pollutants that fall outside of water quality criteria, limited studies have examined the removal of microplastics in DWTPs.

Although data on the removal of microplastics in water treatment plants is limited, in general, these water treatment facilities have shown substantial efficiency in removing colloidal particles (e.g., particulate organic matter) smaller than microplastics. Conventional treatment methods such as coagulation, flocculation, sedimentation, and filtration have shown to be effective in treating particles below micrometer size. Advanced water treatment processes such as nanofiltration and ultrafiltration can treat nano-sized particles and particles of 10 nm or more, respectively. However, in many countries, these water treatment facilities are not available and have not been optimized to treat microplastics. According to the WHO, approximately 785 million people lacked access to improved water supplies as of 2017. Most of them live in rural areas where improved water supplies are not easily accessible. Also, considering plastic materials are not easily decomposed with their long half-life, microplastics may not be properly treated via treatment processes in DWTPs. Currently, limited data is available regarding the removal of microplastics by DWTPs.

2.5.1 Water Treatment Plant

Microplastics in the raw water influent are subjected to a number of water treatment processes in DWTPs such as coagulation, flocculation, and filtration (Figure 2.6). They could also undergo advanced water treatment such as ozone treatment in advanced drinking water treatment plants (ADWTPs). Microplastics have similar physical properties to natural particles. These properties (size, density, surface charge, etc.) play a key role in the removal of microplastics as in the removal of natural particles. Mechanisms for microplastic removal include adsorption, sedimentation, filtration, and straining. Furthermore, plastic particles are also removed by membrane processes (microfiltration, ultrafiltration, nanofiltration, or reverse osmosis). Water treatment facilities play a significant role in reducing the concentration of microplastics presented in the raw water. However, because of the lack of research and data, it is still unclear which treatment processes remove most of the introduced microplastics.

2.5.1.1 Removal of Microplastics by Water Treatment

Currently, DWTPs are not able to completely remove microplastics. Analyses performed on the influent (raw water) and effluent water from three DTWPs in the Czech Republic show that the conventional DWTP is not capable of completely treating microplastics [47]. In a 2018 study, $1,473 \pm 34$, $1,812 \pm 35$, and $3,605 \pm 497$ MPs/L microplastics were detected in raw water samples entering into three DWTPs (WTP1, WTP2, and WTP3), and 443 ± 10, 338 ± 76, and 628 ± 28 MPs/L microplastics were found in the treated water from the three DWTPs. Microplastics

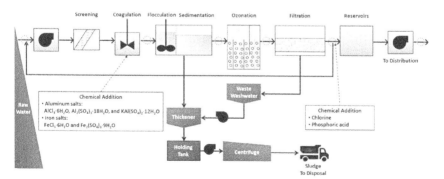

Figure 2.6 Typical water treatment processes in DWTPs. Dashed circles indicate water samples used for microplastic analyses in literature Adapted from [44, 46, 47, 85].

were removed by 70%, 81%, and 83% from WTP1, WTP2, and WTP3, respectively. In the other study conducted by the same researcher group in the Czech Republic, 23 ± 2 and 1,296 ± 35 microplastics detected in the Úhlava River were reduced to 14 ± 1 and 151 ± 4, respectively, in the treated water of DWTPs, achieving a 40% and 88% removal rate [46]. This suggests the necessity of treatment process optimization for higher efficiency of microplastic removal. Notably, DWTPs with a removal rate of more than 80% had an additional granular activated carbon (GAC) filtration process. DWTPs with a conventional sand filtration process only had a 70% removal rate. In a study on the removal of microplastics carried out at a DWTP located in the Yangtze River Delta in China, 6,614 ± 1,132 MPs/L and 930 ± 71 MPs/L were detected in the raw water and treated water, respectively, showing a similar level of removal rate as WTP3 [44]. Meanwhile, in India, a study found that 17.86 ± 2.66 MPs/L microplastics in raw water decreased to 2.75 ± 0.92 MPs/L in the treated water, showing a removal rate of about 85% [85]. Overall, the removal rate of microplastics in DWTPs is currently insufficient to completely treat microplastics and needs to be improved through the development of new treatment processes or optimization of existing processes to achieve high removal efficacy. The removal rates of microplastics by the overall water treatment process of DWTPs are summarized in Table 2.1.

2.5.1.2 Removal Rate of Microplastics Depending on the Size

In one study on the removal of microplastics from DWTPs, it was observed that most microplastics in raw water have a size of less than 10 μm, i.e., 54.6–58% for 1–5 μm and 20.0–27.6% for 5–10 μm [44]. This water treatment facility showed an overall removal rate of 82.1–88.6% for the size of 1–5 μm microplastics with 3,760 ± 726 MPs/L in the influent (raw water) and 793 ± 53 MPs/L in the effluent (treated water). Although the number of microplastics with 1–5 μm has significantly reduced, their proportion in the effluent increased from 54.6–58% to 84.4–86.7%. This suggests that the DWTP is more efficient in removing microplastics with a size bigger than 5 μm. The increased proportion of 1–5 μm MPs was also observed in a study by Pivokonsky et al., showing a 65% proportion in the effluent compared to 50% in the raw water [46]. With a higher removal rate than the size of 1–5 μm, the number of microplastics (5–10 μm) decreased from 1,520 ± 258 in raw water to 136 ± 22 MPs/L in the effluent. The number of microplastics with sizes of 10–50, 50–100, and >100 μm were found to be 731 ± 216, 379 ± 117, and 224 ± 126 MPs/L in raw water, respectively, but 1 ± 1, 0, and 0 in the effluent.

Better removal of larger microplastics was confirmed in another study by Pivokonsky et al. Microplastics larger than 50 μm were almost completely removed and no microplastics bigger than 100 μm were found in the effluent, indicating 100% removal of microplastics (>100 μm) by DWTPs. In contrast, microplastics of

Table 2.1 Overall microplastics removal efficiency of DWTPs.

Water source	Country	Sampling year	Treatment Processes	MPs size (μm) min	MPs size (μm) max	Microplastics/L Raw water	Microplastics/L Treated water	Removal (%)	References
Valley water reservoir	Czech Republic	Nov. 2017–Jan. 2018	Coagulation/flocculation and sand filtration	1	>100	1473 ± 34	443 ± 10	70	[47]
Water reservoir	Czech Republic	Nov. 2017–Jan. 2018	Coagulation/flocculation, sedimentation, and sand and granular activated carbon filtration	1	>100	1812 ± 35	338 ± 76	81	[47]
River water	Czech Republic	Nov. 2017–Jan. 2018	Coagulation-flocculation, flotation, sand filtration, and granular activated carbon filtration	1	>100	3605 ± 497	628 ± 28	83	[47]
Yangtze River	China	Dec. 2018–Jan. 2019	Coagulation/flocculation, sedimentation, sand filtration, and ozonation combined with GAC filtration	1	>100	6614 ± 1132	930 ± 71	82.1–88.6	[44]

(Continued)

Table 2.1 (Continued)

Water source	Country	Sampling year	Treatment Processes	MPs size (μm) min	MPs size (μm) max	Microplastics/L Raw water	Microplastics/L Treated water	Removal (%)	References
Úhlava River, Milence	Czech Republic	Winter 2019–2020	Coagulation, flocculation, sand filtration, hardness stabilization (CO_2 and lime), and Disinfection (ClO_2)	1	>100	23 ± 2	14 ± 1	40	[46]
Úhlava River, Plzeň	Czech Republic	Winter 2019–2020	pH adjustment, coagulation, flocculation, sedimentation, Mn oxidation, sand filtration, ozonation, GAC filtration, UV treatment, hardness stabilization (CO_2 and lime), and disinfection (chlorine)	1	>100	1296 ± 35	151 ± 4	88	[46]
Ganga River	India	Nov. 2019–Mar. 2020	Chlorination, coagulation, pulse clarification and sand filtration	<25	>100	17.86 ± 2.66	2.75 ± 0.92	85.39	[85]

1–5 μm and 5–10 μm accounted for 25–60% and 30–50%, respectively, among all microplastics detected in treated water [47]. This indicates that microplastics larger than 50 μm can be efficiently removed by DWTPs. However, water treatment processes need to be optimized or improved in order to remove microplastics with a size smaller than 10 μm.

2.5.1.3 Removal Rate Depending on the Type of Microplastics

Microplastics detected in DWTPs are largely in the form of fibers, spheres, and fragments [44,46,47]. In the study of Wang et al., microfibers accounted for 53.9–73.9% and 51.6–78.9% of microplastics detected in raw and treated water, respectively. The number of microfibers was found to be 4,295 ± 1,109 MPs/L in raw water and 620 ± 88 MPs/L in treated water, indicating that the removal rate of fibrous microplastics was 82.9–87.5% [44]. In a DWTP of the Czech Republic, fibrous microplastics accounted for 8–13% of microplasts found in raw water [46]. Of the detected microfibers, no fibers less than 5 μm were observed, and microplastics of 5–10 μm size and larger size were always detected in all samples. The most abundant size category was 50–100 μm microplastic, accounting for 46% of the total fiber, followed by 10–50 μm (28%) and over 100 μm (21%). After DWTP's treatment, the number of fibers was reduced from 126 ± 20 MPs/L in raw water to 12 ± 5 MPs/L in treated water. 59% of the fibers found in treated water were smaller than 50 μm, indicating that small-sized microfibers less than 50 μm are more difficult to be removed than large-sized microfibers in DWTPs. In a study conducted at three different Czech DWTPs (WTP1, WTP2, and WTP3), fibers had a high ratio (37–61%) in the raw water of WTP3 whereas accontant for the lowest proportion among three forms of microplastics (fibers, spheres, and fragments) in WTP1 and WTP2 [47]. As a result of water treatment, the number of fibers (168 ± 35 MPs/L) in the raw water of WTP1 decreased to 126 ± 34 MPs/L in treated water. The number of fibrous microplastics (111 ± 50 MPs/L) in the raw water of WTP2 was reduced to 12 ± 6 MPs/L in treated water. In WTP3, the number of fibers was reduced from 1,325 ± 118 in raw water to 294 ± 42 in treated water. Interestingly, it was shown that the proportion of fiber was rather increased in the treated water of WTP1, only 25% of the fiber being removed from WTP1 compared to WTP2 and WTP3 showing 80–90% removal efficiency of fibers. According to the authors, there could be some correlations between the removal of different shapes of microplastics and various water treatment technologies, but additional studies are required to understand the correlation.

In terms of fragment microplastics, according to Pivokonsky et al., fragments accounted for 87–92% of microplasts in the raw water of the DWTP studied. 51% of the detected fragment microplastics were 1–5 μm and 30% were determined as 5–10 μm. The number of fragments (1,170 ± 17 MPs/L) in raw water was reduced to 139 ± 2 MPs/L in treated water. 95% of the fragments in the effluent were found to be smaller than 10 μm, and more than 50% of the effluent fragments were found to be 1–5 μm in size. This shows that the smaller the fragment microplastic is, the more difficult it is to treat. In Pivokonsky's study conducted at three DTWPs (WTP1, WTP2, and WTP3), fragments accounted for 71–76% in the raw water of WTP1 and WTP2. In WTP3, fragments accounted for a 42–48% ratio [47]. As a result of water treatment, the number of fragments was significantly reduced. For example, 1,053 ± 116 fragment MPs/L in raw water of WTP1 was reduced to 270 ± 26 in treated water. In WTP2 and WTP3, 1,377 ± 182 and 1,729 ± 193 MPs/L fragments in raw water decreased to 362 ± 48 and 250 ± 39 MPs/L, respectively, in treated water. Meanwhile, a DWTP in the Yangtze River, China showed that fragments made up 17.6% and 6% of microplastics in raw and treated water, respectively. After water treatment, 1,356 ± 213 MPs/L fragments in raw water was reduced to 228 ± 140 Mps/L, showing a removal rate of 73.1–88.9% for fragments.

Spherical microplastics seemed to have the smallest proportion of the three microplastics types (fibers, spheres, and fragments) in the investigated DWTPs. Spherical MPs were not observed in any samples of raw and treated water of two DWTPs at the Úhlava River (Czech Republic) [46]. Also, in the three DWTPs (WTP1, WTP2, and WTP3), spherical particles were the least abundant, comprising 7–20% of the total particle count in both raw and treated water. The number of spheres in raw and treated water in WTP1 was 215 ± 23 MPs/L and 37 ± 9 MPs/L, respectively. In WTP2, 341 ± 95 MPs/L spheres in raw water decreased to 43 ± 10 MPs/L in treated water. Finally, in WTP3, the number of spheres (251 ± 32 MPs/L) in raw water was reduced to 61 ± 30 MPs/L in treated water [47]. Meanwhile, in a DWTP of Yangtze River, China, 8% (963 ± 365 MPs/L) and 12% (82 ± 22 MPs/L) of spherical microplastics in water and treated water was observed, showing an 89.1–92.7% removal rate.

In summary, studies analyzing samples collected from DWTPs indicate that small-sized microplastics, regardless of their shape, were not efficiently removed by DWTPs. This suggests that the focus should be placed on the treatment of small-sized microplastics in the development or optimization of microplastic treatment processes.

2.5.1.4 Removal Efficiency Depending on the Composition of Microplastics

Research on the treatment and removal rate according to the chemical composition of microplastics showed that PET (55.4–63.1%) accounts for most microplastics in raw water, followed by PE (15.1–23.8%), PP (8.4–18.2%), PS (<5%), and PVC (<5%) [44]. According to Wang et al., microplastics found in raw water were 3,843 ± 598 PETs/L, 1,376 ± 508 PEs/L, 872 ± 294 PPs/L, and 486 ± 118 MPs/L. The amount of microplastics was reduced to 485 ± 53, 125 ± 54, 125 ± 27, and 82 ± 21 in treated water. After the DWTP water treatment PET particles accounted for 47.2–58.8% of the microplastics in treated water, making up the majority of microplastics found in treated water and showing a removal rate of 87.0 ± 3.2%. The removal efficiency of PE, PP, and others were 89.5 ± 7.6%, 85.0 ± 3.1%, and 82.5 ± 5.0%, respectively, and the overall removal of microplastics was 79.7–95.4%. The results suggest that there is no significant difference in the removal rate of different type of plastic polymers. One peculiarity is that the number of PAM increased after treatment by DWTP, i.e., 37 ± 33 PAM plastics in raw water but 112 ± 15 in the effluent (treated water). According to the authors, an increase in PAM in the effluent may be associated with coagulants containing PAM used in the treatment process.

2.5.1.5 Removal of Microplastics by Coagulation, Flocculation, and Sedimentation

Coagulation, flocculation, and sedimentation (CFS) are common processes widely used in water treatment for treating the particulate matter. Pollution and turbidity by particulate matter in raw water are improved by these processes. Chemical coagulants such as aluminum salts and iron salts are extensively used in water treatment facilities for high removal efficiency and cost savings. Coagulants are mixed with raw water through a rapid-mix step for homogenous distribution. During the coagulation process, the electrostatic stability of particulate matter is destabilized and agglomerated by the added coagulant, forming small clumps. Formed clumps go through the flocculation process where the water and flocculants are gently mixed and increase in size to form larger clumps, called *floc*, and then sink or float on the water surface. Large enough flocs are settled in the sedimentation basin (sometimes called clarification) under gravity.

In one study, a 40.5–54.5% removal rate of microplastics was achieved through coagulation with a sedimentation process [44]. It is noteworthy that almost all microplastics larger than 10 μm were removed in this study, whereas microplastics of 5–10 μm showed 44.9–75.0% removal and 1–10 μm showed 28.3–47.5% removal. This indicates that the smaller the particle size, the lower the removal efficiency in CFS. In terms of the shape of microplastics, fiber microplastics showed the highest removal rate of 50.7–60.6% among all types of microplastic shapes detected. The authors speculate that the fiber may have been removed by the agglomeration of the fiber particles into floccules during the coagulation process.

CFS process in other DWTPs showed similar microplastic removal rates. According to Pivokonsky's study [46], after 1,296 ± 35 MPs/L in raw water were treated by CFS, the number was significantly reduced to 497 ± 44 MPs/L. This shows 62% MPs/L of removal, indicating that the conventional water treatment process is somewhat effective in removing microplastics. In the same DWTP, 126 ± 20 fibrous MPs/L were detected in raw water and decreased to 51 ± 10 MPs/L after passing through CFS. In the case of fragments, 1,170 ± 17 MPs/L were found in raw water, and 446 ± 34 MPs/L were found after the CFS process.

In the other study by Pivokonsky et al. [47], although the number of microplastics was reported only in the raw water and effluent of the whole process of DWTP, it was shown that the type of process used for microplastic treatment affected the removal of microplastics. The overall microplastic removal rate was 81–83% in the DTWP, which was subjected to two-step separation of aggregates and granular activated carbon (GAC) filtration including the conventional coagulation-flocculation process, while only 70% of the microplastics were removed in the single-step separation of aggregates and no GAC filtration. On the other hand, it was also observed that the coagulation process did not show any effect on removing microplastics. In DWTP located in the Ganga River in India, microplastics, which were 17.88 MPs/L in raw water, appeared to be 17.11 MPs/L after coagulation and flocculation, showing no significant decrease in number [85]. However, the number of microplastics decreased to 6.99 MPs/L after pulse clarification, showing a microplastic removal rate of about 60%.

2.5.1.6 Removal of Microplastics by Filtration

The filtration process used in water treatment facilities treats flocs and particulates that have not been completely treated in sedimentation by passing the influent water through the pores of filter media. The filtration process uses physicochemical mechanisms such as adsorption, precipitation, and straining to treat pollutants, and several types of media such as sand, anthracite, and gravel are used as filtration media. So far, only a few studies investigated the treatment of microplastics by the DWTP filtration process, and more research must be conducted to understand the efficiency of microplastic treatment in the filtration process.

In a study by Wang et al., sand filtration showed a microplastic removal rate of 29.0–44.4% [44]. Interestingly, no microplastics larger than 50 μm were found after filtration, showing the retention effect of large-sized MPs. The removal rate of fibers, spheres, and fragments was 30.9–49.3%, 23.5–50.9%, and 18.9–27.5%, respectively. These results indicates that sand filtration is not considered the primary method for removing microplastics.

The other study that investigated the microplastic removal rate of sand filtration in DWTP also showed similar or slightly higher levels of microplastic removal [46]. 497 ± 44 MPs/L microplastics were detected in the influent water of the sand filtration process, and the number was reduced to 243 ± 17 MPs/L after filtration, showing a removal rate of 51%. The removal rate of each microplastic type showed that 51 ± 10 MPs/L fibers in the inflow water of the filtration facility decreased to 31 ± 4 MPs/L in the effluent water, resulting in a removal rate of 39%. In terms of fragments, their number decreased from 446 ± 34 MPs/L in filtration inflow to 213 ± 16 MPs/L in filtration outflow, showing a removal rate of 52%. These results showed similar efficiency to the fiber removal rate reported in the study of Wang et al., but different results were shown for fragments.

Meanwhile, in the DWTP of Ganga River, India, an experiment was conducted on the removal rate of microplastics by sand filtration. As a result, 11.17 MPs/L microplastics were reduced to 2.75 MPs/L after filtration, showing 75% removal of microplastics [85]. Compared to the aforementioned studies by Wang et al. and Pivokonsky et al., the microplastic removal rate was significantly higher. However, it is difficult to compare removal rates only with the results given in the paper without detailed sand filtration specifications (e.g., flow rate, pore size, and depth of filter media). Therefore, it is premature to conclude what kind of microplastics are well removed or not by simply comparing the removal rate without detailed specifications of the filtration facility.

A filtration process using granular activated carbon (GAC) as filter media reported 56.8–60.9% removal, showing an improved microplastic removal rate. In particular, the removal rate of 73.7–98.5% was observed for microplastics with a size of 1–5 μm that accounted for a significant part of microplastics in the influent of DWTP [44]. In addition, a removal rates of 38–52.1%, 76.8–86.3%, and 60.3–69.1% was observed for microplastic fibers, spheres, and fragments, respectively. The other study that investigated the microplastic removal in a GAC filtration of DWTP showed that the number of microplastics decreased from 224 ± 3 MPs/L to 149 ± 1 MPs/L after passing through the filtration process, resulting in 34% microplastic removal [46]. This is a significantly lower removal rate compared to the work of Wang et al., and detailed information on various parameters used in the GAC filtration process is also required for the analysis of the discrepancy.

2.5.1.7 Removal of Microplastics by Ozonation

The ozone process using ozone generated by passing electric current to pure oxygen is one of the advanced water treatment processes. This process has the advantage of being more effective in removing viruses than chlorine and not generating harmful residues after ozone treatment. In addition, by oxidizing iron, manganese, sulfides, and organic chemicals, the ozone process not only removes microorganisms, but also removes the color, taste, and odor of water. The efficiency of the ozone process is mainly determined by the contact time and ozone concentration.

Compared to other water treatment processes, the ozone process is sparser in research on the removal rate of microplastics, and the reported results of microplastic removal are also in contrast to those of other processes. According to the result of ozone treatment by Wang et al., the number of microplastics did not show much difference compared to the influent water after passing through the ozone process, or interestingly, the number of microplastics increased, showing a negative removal rate. For example, the number of microplastics showed similar abundance from 243 ± 17 MPs/L to 224 ± 3 MPs/L after passing through the ozone process. And the abundance of 1–5 μm MPs resulted in an increase of 2.8–16.0% in the effluent of the ozone process. According to Wang et al., the shearing force of the water flow may have crushed the microplastic, resulting in a larger number of microplastic. Since these studies show the result that the DWTP could be a source of microplastics, more case studies are required, and detailed research is needed to investigate what mechanisms generate microplastics during ozonation.

2.5.2 Microplastic Removal in Lab-scale Studies

The removal rate of microplastics by coagulation and flocculation in laboratory experiments is markedly different compared to the removal efficiency of microplastics in actual DWTPs. In a study by Ma et al., conventional coagulation and flocculation processes showed poor PE microplastic removal efficiency of less than 10% even though high coagulant concentrations larger than 60 mg/L as $AlCl_3$ were used [99]. Considering that the concentration of the coagulant used in the actual DWTP is typically 20 mg/L, the removal rate of less than 10% is not desirable. $FeCl_3$ coagulant did not show high removal efficiencies either. In experiments with $FeCl_3 \cdot 6H_2O$, 0.5 mM (27.9 mg Fe/L) and 5 mM (279.2 mg Fe/L) of $FeCl_3 \cdot 6H_2O$ showed 8.24 ± 1.22% and 12.65 ± 1.09% removal rates, respectively for PE plastics smaller than 500 μm at pH 7. In addition, interestingly, the microplastics with sizes of <0.5 mm, 0.5 < d < 1 mm, 1 < d < 2 mm, and 2 < d < 5 mm showed 8.28 ± 1.06%, 2.71 ± 0.84%, 1.68 ± 0.63%, and 1.02 ± 0.37% removal rates, respectively. This shows a higher removal rate as the microplastic particle size decreases, which is opposite to the analysis results performed in actual DWTPs. According to Wang et al., microplastics over 10 μm were almost completely removed in the coagulation with sedimentation process, whereas 44.9–75.0% of 5–10 μm microplastics were removed and 28.3–47.5% of 1–10 μm microplastics were removed [44].

In the jar test using aluminum sulfate as a coagulant, it was found that the experimental conditions suitable for removing Kaolin, a conventional colloid material, can be effectively used for removing model plastic spheres (1–5 μm in diameter). According to the results, in tap water containing PE microplastics at concentrations of 5 and 25 mg/L, $Al_2(SO_4)_3 \cdot 18H_2O$ (5 mg Al/L) lowered the turbidity from 16 NTU to less than 1 NTU, reducing the turbidity by 97% and 99%, respectively [100]. In the other jar test by Lapointe et al., spiked 140 μm PE plastics showed 82% removal in the presence of coagulant (2.73 mg Al/L) and flocculant (0.3 mg PAM/L) [101]. The removal rate of more than 80% showed a significant difference when compared to Ma's experiment where 5 mm $AlCl_3 \cdot 6H_2O$ (134.9 mg Al/L) with cationic or anionic PAM (15 mg/L) removed 45.34 ± 3.93% or 61.19 ± 3.67% of PE (d < 0.5 mm), respectively.

On the other hand, it was also found that the removal rate of microplastics did not improve even with the use of a coagulant aid (diallyldimethylammonium chloride). According to Zhang et al., in microplastic removal experiments in which particles of various sizes were used with 20 ppm alum, the removal rates for 1.2 μm, 10–20 μm, 45–53 μm, and 106–125 μm microplastics were <0.1%, 1.8 ± 1.2%, 0.3 ± 0.3%, and 1.4 ± 1.2%, respectively [102]. When 0.5 ppm coagulant aid was additionally used, there was no significant difference in removal rate except for the size of 45–53 μm. The removal rate was significantly improved from 0.3% to 13.6% only for the size of 45–53 μm. Nonetheless, this enhanced removal rate is insufficient to treat microplastics effectively. Within the same study, the removal efficiency of microplastics by filtration was also investigated. The filtration treatment showed removal rates of 94.9 ± 0.4%, 86.9 ± 4.9%, 97.0 ± 3.0%, and 99.9 ± 0.1% for microplastics of 1.2 μm, 10–20 μm, 45–53 μm, and 106–125 μm, respectively. This suggests that the filtration process can be a more efficient process for removing microplastics than CFS processes in DWTPs.

Meanwhile, Lapointe et al. found that the ability to incorporate plastic particles into the flocs plays an important role in plastic removal. In the experiment on the removal mechanism of plastic particles, when the concentration of PAM was reduced from 0.30 to 0.10 mg/L, they found the size of the floc decreased from 977 to 504 mm, as well as the removal efficiency of 140 μm-sized particles was reduced from 82% to 72% [101]. However, the removal of smaller microplastics (15 μm) was not significantly affected by the change in the concentration of PAM. From this result, the ability to incorporate particles into flocs seems to play an important role in the removal of microplastics of 140 μm. Lapointe et al. additionally tested the removal efficiency of pristine PE as well as weathered PE by coagulation and flocculation. As a result, they found that the removal efficiency of weathered PE increased from 64 to 89% compared to pristine PE (140 μm), when 0.45 mg Al/L was used during coagulation. When a higher concentration of coagulant (2.73 mg Al/L) was used, the removal rates of weathered PE and pristine PE were 99% and 82%, respectively, indicating that the weathered PE was more easily removed. The authors concluded that this is because the potential interaction of weathered PE with coagulant and flocculant increased from the increase in physicochemical heterogeneity of the weathered plastic surface via UV photooxidation combined with NOM coating.

In addition, it was found that the removal rate of weathered PE varies depending on the type of coagulant. For example, 89% of the microplastic was removed when alum (0.45 mg Al/L) was used whereas the same amount of ACH (0.45 mg Al/L) increased the removal rate to 93% at pH 7. Under the same experimental conditions where only the pH was different (pH 8), ACH and alum showed 85% and 69% of microplastic removal, respectively, indicating that the microplastic removal rate of ACH was more effective. The type of microplastics was also found to have an effect on the removal rate. At the concentration of coagulant and flocculant (2.73 mg Al/L) added to achieve the target turbidity of less than 1 NTU, PEST fiber showed a 99% removal rate, whereas PE microsphere showed only an 82% removal rate. In the case of PS microplastics, the removal of microspheres (140 μm) was 84%, showing a similar removal rate to PE. This indicates that PS and PE have similar interactions with the coagulant and flocculant used in their experiment [101].

Meanwhile, membrane technology can also be used to remove microplastics. The removal of pollutants using a membrane is a technique frequently used for the advanced treatment of drinking water and has made a great contribution to improving the quality of treated water. With excellent selectivity and separation, membrane technology has been widely used to treat organic contaminants, heavy metals, and disinfection byproducts. Membranes are divided into ultrafiltration, nanofiltration, and reverse osmosis depending on the pore size. Ultrafiltration is widely used in water treatment facilities all over the world because it shows excellent performance in drinking water treatment. However, regarding the treatment of microplastics, few studies have been conducted on the efficiency of ultrafiltration, and there is not much knowledge about the removal rate. In the study of Ma et al., UF was used to remove PE microplastics (d < 0.5 mm), and PE microplastics were completely rejected by UF (30 nm pore size on average). In addition, it was observed that the smaller the size of the microplastic, the worse the fouling on the membrane surface, and the larger the size of the plastic, the more the fouling was alleviated. This suggests that UF has a high potential to be widely used for microplastic removal in DWTPs in the future if the fouling issue is resolved. So far, there are not many studies on the removal of microplastics using membranes yet. Moreover, the number of research on the removal of microplastics by the membrane in DWTPs is scarce. More research is needed in the future to evaluate the effectiveness of microplastic removal using membrane technology in DWTPs.

2.6 Summary and Prospects

Owing to the exponential production of microplastics since the 1950s, water resources around the world are now contaminated with enormous amounts of plastics. In addition to the mass production of plastics, plastics are not easily decomposed in the natural environment because of their hard-to-degrade chemical structure. Therefore, the accumulation of plastics in the environment is becoming a significant concern worldwide. Also, the occurrence, behavior, and toxicity of secondary plastics (i.e., microplastics) caused by the gradual decomposition of discarded plastics are not well known.

2.6 Summary and Prospects

Currently, microplastics are found in most environments of freshwater systems such as rivers, lakes, glaciers, and canals, and the abundance of microplastics varies depending on population density, urbanization, and industrialization. This is because the pollutant source of microplastics is mainly caused by anthropogenic emissions resulting from human activities, such as the use of synthetic fibers, washing clothes, fishing nets, driving cars, and the everyday use and disposal of plastic products. Although some studies have dealt with the pollution sources of microplastics and the current pollution status in the drinking water system, most studies have not been able to cover the entire size range of microplastics. In recent years, a few studies on microplastics down to 1 μm in size have been published, but the number of studies is still insufficient. Furthermore, it is difficult to find studies investigating the abundance of nanoplastics less than 1 μm in the freshwater system. This is mainly related to technical issues such as the qualitative and quantitative limitations of analytical instruments and methodological issues, such as the absence of standard analysis methods. These issues not only pose challenges in measuring and comparing the scales of interregional pollution of microplastics but also become a major obstacle to understanding the distribution, behavior, and transport of microplastics. In order to minimize these obstacles, the establishment of a standard analysis method is required promptly, and the development of analytical instrument that qualitatively and quantitatively analyzes the chemical (i.e., chemical composition) and physical (i.e., size and morphology) properties of micro- and nanoplastics is also warranted.

As the occurrence of microplastics from human activities is inevitable and microplastics are already presented worldwide, research on the removal of microplastics in drinking water systems is also urgent. Currently, only a few limited studies on the microplastic treatment of DWTPs have been conducted. Discrepancies in the results of MP studies in terms of the number, size, shape, and chemical composition of MPs indicate the high variability of MP loads in the drinking water depending on various sources and different countries. In addition to the diversity of microplastics detected in raw water, the removal rates of microplastics differed greatly from study to study, even in the same type of water treatment processes in DWTPs. Although each uses the same removal mechanisms, such as coagulation and filtration, the size of the study pool (i.e., the number of studies) is too small to evaluate and compare the microplastic removal rates of each process. In order to understand the removal of microplastics in DWTPs, some tasks should be prioritized. First, many studies have to be conducted using water samples collected from actual DWTPs. With only a limited number of studies, it is hard to draw conclusions about the removal of microplastics in DWTPs. Second, a standard method of analyzing microplastics in drinking water must be established. This method should include all the procedures including sample collection, sample treatment, microplastics extraction, microplastics quantification, polymer identification, detection limit, etc. If different analysis methods are used or the target size ranges of microplastics are different, the removal rate of microplastics will inevitably vary even in the same water treatment process. Because of the differences in the analytical instrument used and the performance of the instrument, the size range of target microplastics varies from study to study. Moreover, evaluating the accurate removal rate of microplastics is difficult as a consequence of the poor detection limit. Only after improvement of the analytical instrument can more accurate removal rate of microplastics be discussed in each water treatment process. Lastly, detailed specifications of the water treatment process (e.g., operating conditions and parameters, chemical additive concentration and properties, etc.) should be disclosed. Depending on the operating conditions and the type and concentration of additives, the removal rate of microplastics may vary. It is essential to consider these operating conditions and parameters for evaluating the microplastic removal rate of each water treatment process.

In addition, the minimization of microplastics generated by the water treatment facility itself will be an important topic in microplastic research. An increase in the number of microplastics (e.g., PAM) has already been found in the coagulation/flocculation, and ozone processes may increase the number of microplastics by breaking microplastics with shearing forces during ozonation. Furthermore, concerns about microplastics that may occur in the membrane filtration process via the release of micro- and/or nano-particles from old and worn membranes are also being raised. In order to supply safer drinking water, approches to minimizing microplastics generated in DWTPs should be prepared.

References

1 PlasticsEurope (2010). Plastics –The Facts 2011: an analysis of European plastics production, demand and recovery for 2010. *PlasticsEurope*.
2 PlasticEurope (2020). Plastics – The Facts 2020: an analysis of European plastics production, demand and waste data. *PlasticEurope*.
3 PlasticsEurope (2016). Plastics – The Facts 2016: an analysis of European plastics production, demand and waste data. *PlasticsEurope*.
4 Lusher, A., Hollman, P., and Mandoza-Hill, J. (2017). Microplastics in fisheries and aquaculture: Status of knowledge on their occurrence and implications for aquatic organisms and food safety. *FAO Fisheries and Aquaculture Technical Paper*.
5 Geyer, R., Jambeck, J.R., and Law, K.L. (2017). Production, use, and fate of all plastics ever made. *Science Advances* 3 (7): 25–29.

6 Geneva: World Health Organization (2019). Nanoplastics in drinking-water.
7 Cole, M., Lindeque, P., Halsband, C. et al. (2011). Microplastics as contaminants in the marine environment: a review. *Marine Pollution Bulletin* 62 (12): 2588–2597.
8 Novotna, K., Cermakova, L., Pivokonska, L. et al. (2019). Microplastics in drinking water treatment – current knowledge and research needs. *Science of the Total Environment* 667: 730–740.
9 Kosuth, M., Mason, S.A., and Wattenberg, E.V. (2018). Anthropogenic contamination of tap water, beer, and sea salt. *PLOS ONE* 13 (4): e0194970.
10 Mason, S.A., Welch, V.G., and Neratko, J. (2018). Synthetic polymer contamination in bottled water. *Frontiers in Chemistry* 6: 407.
11 Bhagat, J., Nishimura, N., and Shimada, Y. (2020). Toxicological interactions of microplastics/nanoplastics and environmental contaminants: current knowledge and future perspectives. *Journal of Hazardous Materials* 405: 123913.
12 Rainieri, S., Conlledo, N., Larsen, B.K. et al. (2018). Combined effects of microplastics and chemical contaminants on the organ toxicity of zebrafish (Danio rerio). *Environmental Research* 162: 135–143.
13 Wang, W., Ge, J., and Yu, X. (2020). Bioavailability and toxicity of microplastics to fish species: a review. *Ecotoxicology and Environmental Safety* 189: 109913.
14 Alomar, C., Sureda, A., Capó, X. et al. (2017). Microplastic ingestion by Mullus surmuletus Linnaeus, 1758 fish and its potential for causing oxidative stress. *Environmental Research* 159: 135–142.
15 Espinosa, C., Beltran, J.M.G., Esteban, M.A. et al. (2017). In vitro effects of virgin microplastics on fish head-kidney leucocyte activities. *Environmental Pollution* 235: 30–38.
16 Wu, F., Wang, Y., Leung, J.Y.S. et al. (2020). Accumulation of microplastics in typical commercial aquatic species: a case study at a productive aquaculture site in China. *Science of the Total Environment* 708: 135432.
17 Malafaia, G., de Souza, A.M., Pereira, A.C. et al. (2020). Developmental toxicity in zebrafish exposed to polyethylene microplastics under static and semi-static aquatic systems. *Science of the Total Environment* 700: 134867.
18 Browne, M.A., Dissanayake, A., Galloway, T.S. et al. (2008). Ingested microscopic plastic translocates to the circulatory system of the mussel, Mytilus edulis (L.). *Environmental Science and Technology* 42 (13): 5026–5031.
19 Sobhani, Z., Lei, Y., Tang, Y. et al. (2020). Microplastics generated when opening plastic packaging. *Scientific Reports* 10 (1): 1–7.
20 Zhang, W., Ma, X., Zhang, Z. et al. (2015). Persistent organic pollutants carried on plastic resin pellets from two beaches in China. *Marine Pollution Bulletin* 99 (1–2): 28–34.
21 Koelmans, A.A., Besseling, E., and Shim, W.J. (2015). Nanoplastics in the aquatic environment. Critical review. In: *Marine Anthropogenic Litter* (M. Bergmann, L. Gutow, M. Klages eds) 325–340. Springer International Publishing.

22 Eerkes-Medrano, D., Thompson, R.C., and Aldridge, D.C. (2015). Microplastics in freshwater systems: a review of the emerging threats, identification of knowledge gaps and prioritisation of research needs. *Water Research* 75: 63–82.

23 Andrady, A.L. (2011). Microplastics in the marine environment. *Marine Pollution Bulletin* 62 (8): 1596–1605.

24 Arthur, C., Baker, J., and Bamford, H. (2009). Proceedings of the international research workshop on the occurrence, effects, and fate of microplastic marine debris. *NOAA Technical Memorandum NOS-OR&R-30* 530.

25 Leslie, H.A. (2015). Plastic in cosmetic - are we polluting the environment through our personal care? - Plastic ingredients that contribute to marine microplastic litter.

26 ISO/TR 21960:2020(en) (2020). Plastics — environmental aspects — state of knowledge and methodologies.

27 State Water Resources Control Board (2020). Adoption of definition of microplastics in drinking water.

28 van Wezel, A., Caris, I., and Kools, S.A.E. (2016). Release of primary microplastics from consumer products to wastewater in the Netherlands. *Environmental Toxicology and Chemistry* 35 (7): 1627–1631.

29 Conkle, J.L., Báez Del Valle, C.D., and Turner, J.W. (2018). Are we underestimating microplastic contamination in aquatic environments? *Environmental Management* 61 (1): 1–8.

30 Chamas, A., Moon, H., Zheng, J. et al. (2020). Degradation rates of plastics in the environment. *ACS Sustainable Chemistry and Engineering* 8 (9): 3494–3511.

31 Gewert, B., Plassmann, M.M., and Macleod, M. (2015). Pathways for degradation of plastic polymers floating in the marine environment. *Environmental Sciences: Processes and Impacts* 17 (9): 1513–1521.

32 Lebreton, L.C.M., Van Der Zwet, J., Damsteeg, J.W. et al. (2017). River plastic emissions to the world's oceans. *Nature Communications* 8 (1): 1–10.

33 He, D., Luo, Y., Lu, S. et al. (2018). Microplastics in soils: analytical methods, pollution characteristics and ecological risks. *Trends in Analytical Chemistry* 109: 163–172.

34 João, P.D.C., Teresa, R.-S., and Armando, C.D. (2020). The environmental impacts of plastics and micro-plastics use, waste and pollution: EU and national measures. *European Union*.

35 Wright, S.L., Thompson, R.C., and Galloway, T.S. (2013). The physical impacts of microplastics on marine organisms: a review. *Environmental Pollution* 178: 483–492.

36 Oßmann, B.E., Sarau, G., Holtmannspötter, H. et al. (2018). Small-sized microplastics and pigmented particles in bottled mineral water. *Water Research* 141: 307–316.

37 Wang, T., Li, B., Zou, X. et al. (2019). Emission of primary microplastics in mainland China: invisible but not negligible. *Water Research* 162: 214–224.

38 De Falco, F., Di Pace, E., Cocca, M. et al. (2019). The contribution of washing processes of synthetic clothes to microplastic pollution. *Scientific Reports* 9 (1): 1–11.

39 Hidalgo-Ruz, V., Gutow, L., Thompson, R.C. et al. (2012). Microplastics in the marine environment: a review of the methods used for identification and quantification. *Environmental Science and Technology* 46 (6): 3060–3075.

40 Free, C.M., Jensen, O.P., Mason, S.A. et al. (2014). High-levels of microplastic pollution in a large, remote, mountain lake. *Marine Pollution Bulletin* 85 (1): 156–163.

41 Zhou, H., Zhou, L., and Ma, K. (2020). Microfiber from textile dyeing and printing wastewater of a typical industrial park in China: occurrence, removal and release. *Science of the Total Environment* 739: 140329.

42 He, P., Chen, L., Shao, L. et al. (2019). Municipal solid waste (MSW) landfill: a source of microplastics? - evidence of microplastics in landfill leachate. *Water Research* 159: 38–45.

43 Grbić, J., Helm, P., Athey, S. et al. (2020). Microplastics entering northwestern Lake Ontario are diverse and linked to urban sources. *Water Research* 174: 115623.

44 Wang, Z., Lin, T., and Chen, W. (2020). Occurrence and removal of microplastics in an advanced drinking water treatment plant (ADWTP). *Science of the Total Environment* 700: 134520.

45 Horton, A.A., Walton, A., Spurgeon, D.J. et al. (2017). Microplastics in freshwater and terrestrial environments: Evaluating the current understanding to identify the knowledge gaps and future research priorities. *Science of the Total Environment* 586: 127–141.

46 Pivokonský, M., Pivokonská, L., Novotná, K. et al. (2020). Occurrence and fate of microplastics at two different drinking water treatment plants within a river catchment. *Science of the Total Environment* 741: 140236.

47 Pivokonsky, M., Cermakova, L., Novotna, K. et al. (2018). Occurrence of microplastics in raw and treated drinking water. *Science of the Total Environment* 643: 1644–1651.

48 Malankowska, M., Echaide-Gorriz, C., and Coronas, J. (2021). Microplastics in marine environment: a review on sources, classification, and potential remediation by membrane technology. *Environmental Science: Water Research & Technology* 7 (2): 243–258.

49 Shen, M., Song, B., Zhu, Y. et al. (2020). Removal of microplastics via drinking water treatment: current knowledge and future directions. *Chemosphere* 251: 126612.

50 Poerio, T., Piacentini, E., and Mazzei, R. (2019). Membrane processes for microplastic removal. *Molecules* 24 (22): 4148.

51 Robinson, S., Abdullah, S.Z., Bérubé, P. et al. (2016). Ageing of membranes for water treatment: linking changes to performance. *Journal of Membrane Science* 503: 177–187.

52 Ding, H., Zhang, J., He, H. et al. (2021). Do membrane filtration systems in drinking water treatment plants release nano/microplastics? *Science of the Total Environment* 755: 142658.

53 Wang, T., Wang, L., Chen, Q. et al. (2020). Interactions between microplastics and organic pollutants: effects on toxicity, bioaccumulation, degradation, and transport. *Science of the Total Environment* 748: 142427.

54 Godoy, V., Blázquez, G., Calero, M. et al. (2019). The potential of microplastics as carriers of metals. *Environmental Pollution* 255: 113363.

55 Koelmans, A.A., Bakir, A., Burton, G.A. et al. (2016). Microplastic as a vector for chemicals in the aquatic environment: critical review and model-supported reinterpretation of empirical studies. *Environmental Science and Technology* 50 (7): 3315–3326.

56 Zhu, K., Jia, H., Sun, Y. et al. (2020). Long-term phototransformation of microplastics under simulated sunlight irradiation in aquatic environments: roles of reactive oxygen species. *Water Research* 173: 115564.

57 Zhang, H., Wang, J., Zhou, B. et al. (2018). Enhanced adsorption of oxytetracycline to weathered microplastic polystyrene: kinetics, isotherms and influencing factors. *Environmental Pollution* 243: 1550–1557.

58 Padervand, M., Lichtfouse, E., Robert, D. et al. (2020). Removal of microplastics from the environment. A review. *Environmental Chemistry Letters* 18 (3): 807–828.

59 Takada, H. and Karapanagioti, H.K. (ed.) (2019). *Hazardous Chemicals Associated with Plastics in the Marine Environment*. Cham: Springer.

60 Liu, G., Zhu, Z., Yang, Y. et al. (2019). Sorption behavior and mechanism of hydrophilic organic chemicals to virgin and aged microplastics in freshwater and seawater. *Environmental Pollution* 246: 26–33.

61 Müller, A., Becker, R., Dorgerloh, U. et al. (2018). The effect of polymer aging on the uptake of fuel aromatics and ethers by microplastics. *Environmental Pollution* 240: 639–646.

62 Hüffer, T., Weniger, A.K., and Hofmann, T. (2018). Sorption of organic compounds by aged polystyrene microplastic particles. *Environmental Pollution* 236: 218–225.

63 Wang, Q., Zhang, Y., Wangjin, X. et al. (2020). The adsorption behavior of metals in aqueous solution by microplastics effected by UV radiation. *Journal of Environmental Sciences (China)* 87: 272–280.

64 Liu, P., Qian, L., Wang, H. et al. (2019). New insights into the aging behavior of microplastics accelerated by advanced oxidation processes. *Environmental Science and Technology* 53 (7): 3579–3588.

65 Rummel, C.D., Jahnke, A., Gorokhova, E. et al. (2017). Impacts of biofilm formation on the fate and potential effects of microplastic in the aquatic environment. *Environmental Science and Technology Letters* 4 (7): 258–267.

66 Lu, Y., Zhang, Y., Deng, Y. et al. (2016). Uptake and accumulation of polystyrene microplastics in zebrafish (danio rerio) and toxic effects in liver. *Environmental Science and Technology* 50 (7): 4054–4060.

67 Cole, M. and Galloway, T.S. (2015). Ingestion of nanoplastics and microplastics by pacific oyster larvae. *Environmental Science and Technology* 49 (24): 14625–14632.

68 Kim, D., Chae, Y., and An, Y.J. (2017). Mixture toxicity of nickel and microplastics with different functional groups on Daphnia magna. *Environmental Science and Technology* 51 (21): 12852–12858.

69 Vroom, R.J.E., Koelmans, A.A., Besseling, E. et al. (2017). Aging of microplastics promotes their ingestion by marine zooplankton. *Environmental Pollution* 231: 987–996.

70 Shahul Hamid, F., Bhatti, M.S., Anuar, N. et al. (2018). Worldwide distribution and abundance of microplastic: how dire is the situation? *Waste Management and Research* 36 (10): 873–897.

71 Yu, Q., Hu, X., Yang, B. et al. (2020). Distribution, abundance and risks of microplastics in the environment. *Chemosphere* 249: 126059.

72 Auta, H.S., Emenike, C.U., and Fauziah, S.H. (2017). Distribution and importance of microplastics in the marine environment: a review of the sources, fate, effects, and potential solutions. *Environment International* 102: 165–176.

73 Koutnik, V.S., Leonard, J., Alkidim, S. et al. (2021). Distribution of microplastics in soil and freshwater environments: global analysis and framework for transport modeling. *Environmental Pollution* 274: 116552.

74 Leslie, H.A., Brandsma, S.H., van Velzen, M.J.M. et al. (2017). Microplastics en route: field measurements in the Dutch river delta and Amsterdam canals, wastewater treatment plants, North Sea sediments and biota. *Environment International* 101: 133–142.

75 Liu, F., Olesen, K.B., Borregaard, A.R. et al. (2019). Microplastics in urban and highway stormwater retention ponds. *Science of the Total Environment* 671: 992–1000.

76 González-Pleiter, M., Velázquez, D., Edo, C. et al. (2020). Fibers spreading worldwide: Microplastics and other anthropogenic litter in an Arctic freshwater lake. *Science of the Total Environment* 722: 137904.

77 Frank, Y.A., Vorobiev, E.D., Vorobiev, D.S. et al. (2020). Preliminary screening for microplastic concentrations in the surface water of the Ob and Tom Rivers in Siberia, Russia. *Sustainability* 13 (1): 80.

78 Ambrosini, R., Azzoni, R.S., Pittino, F. et al. (2019). First evidence of microplastic contamination in the supraglacial debris of an alpine glacier. *Environmental Pollution* 253: 297–301.

79 Di, M. and Wang, J. (2018). Microplastics in surface waters and sediments of the Three Gorges Reservoir, China. *Science of the Total Environment* 616–617: 1620–1627.

80 Tong, H., Jiang, Q., Hu, X. et al. (2020). Occurrence and identification of microplastics in tap water from China. *Chemosphere* 252: 126493.

81 Zhang, M., Li, J., Ding, H. et al. (2020). Distribution characteristics and influencing factors of microplastics in urban tap water and water sources in Qingdao, China. *Analytical Letters* 53 (8): 1312–1327.

82 Strand, J., Feld, L., Murphy, F. et al. (2018). Analysis of microplastic particles in Danish drinking water.

83 Shruti, V.C., Pérez-Guevara, F., and Kutralam-Muniasamy, G. (2020). Metro station free drinking water fountain- a potential "microplastics hotspot" for human consumption. *Environmental Pollution* 261: 114227.

84 Weber, F., Kerpen, J., Wolff, S. et al. (2021). Investigation of microplastics contamination in drinking water of a German city. *Science of the Total Environment* 755: 143421.

85 Sarkar, D.J., Das Sarkar, S., Das, B.K. et al. (2021). Microplastics removal efficiency of drinking water treatment plant with pulse clarifier. *Journal of Hazardous Materials* 413: 125347.

86 Ferraz, M., Bauer, A.L., Valiati, V.H. et al. (2020). Microplastic concentrations in raw and drinking water in the sinos river, southern Brazil. *Water* 12 (11): 1–10.

87 Baldwin, A.K., Corsi, S.R., and Mason, S.A. (2016). Plastic debris in 29 Great Lakes tributaries: relations to watershed attributes and hydrology. *Environmental Science and Technology* 50 (19): 10377–10385.

88 Eriksen, M., Mason, S., Wilson, S. et al. (2013). Microplastic pollution in the surface waters of the Laurentian Great Lakes. *Marine Pollution Bulletin* 77 (1–2): 177–182.

89 Zhao, S., Zhu, L., and Li, D. (2015). Microplastic in three urban estuaries, China. *Environmental Pollution* 206: 597–604.

90 He, D., Chen, X., Zhao, W. et al. (2021). Microplastics contamination in the surface water of the Yangtze River from upstream to estuary based on different sampling methods. *Environmental Research* 196: 110908.

91 Koelmans, A.A., Mohamed Nor, N.H., Hermsen, E. et al. (2019). Microplastics in freshwaters and drinking water: critical review and assessment of data quality. *Water Research* 155: 410–422.

92 Piñon-Colin, T.D.J., Rodriguez-Jimenez, R., Rogel-Hernandez, E. et al. (2020). Microplastics in stormwater runoff in a semiarid region, Tijuana, Mexico. *Science of the Total Environment* 704: 135411.

93 Kirstein, I.V., Hensel, F., Gomiero, A. et al. (2021). Drinking plastics? – quantification and qualification of microplastics in drinking water distribution systems by µFTIR and Py-GCMS. *Water Research* 188: 116519.

94 Allen, S., Allen, D., Moss, K. et al. (2020). Examination of the ocean as a source for atmospheric microplastics. *PLOS ONE* 15 (5): e0232746.

95 Allen, S., Allen, D., Phoenix, V.R. et al. (2019). Atmospheric transport and deposition of microplastics in a remote mountain catchment. *Nature Geoscience* 12 (5): 339–344.

96 Bessa, F., Barría, P., Neto, J.M. et al. (2018). Occurrence of microplastics in commercial fish from a natural estuarine environment. *Marine Pollution Bulletin* 128: 575–584.

97 Anderson, P.J., Warrack, S., Langen, V. et al. (2017). Microplastic contamination in Lake Winnipeg, Canada. *Environmental Pollution* 225: 223–231.

98 Chinfak, N., Sompongchaiyakul, P., Charoenpong, C. et al. (2021). Abundance, composition, and fate of microplastics in water, sediment, and shellfish in the Tapi-Phumduang River system and Bandon Bay, Thailand. *Science of the Total Environment* 781: 146700.

99 Ma, B., Xue, W., Hu, C. et al. (2019). Characteristics of microplastic removal via coagulation and ultrafiltration during drinking water treatment. *Chemical Engineering Journal* 359: 159–167.

100 Skaf, D.W., Punzi, V.L., Rolle, J.T. et al. (2020). Removal of micron-sized microplastic particles from simulated drinking water via alum coagulation. *Chemical Engineering Journal* 386: 123807.

101 Lapointe, M., Farner, J.M., Hernandez, L.M. et al. (2020). Understanding and improving microplastic removal during water treatment: impact of coagulation and flocculation. *Environmental Science and Technology* 54 (14): 8719–8727.

102 Zhang, Y., Diehl, A., Lewandowski, A. et al. (2020). Removal efficiency of micro- and nanoplastics (180 nm–125 μm) during drinking water treatment. *Science of the Total Environment* 720: 137383.

3

Occurrence of Microplastics in Wastewater Treatment Plants

Kang Song* and Lu Li

State Key Laboratory of Freshwater Ecology and Biotechnology, Institute of Hydrobiology, Chinese Academy of Sciences, Wuhan, China
* Corresponding author

3.1 Introduction

In addition to drinking water systems, wastewater systems are also closely related to human life. Meanwhile, occurrence of microplastics in wastewater treatment plants (WWTPs) has also aroused wide concern. WWTPs are the final step of the human activity water cycle, receiving contaminants from industry, domestic wastewater, storm water and so forth. WWTPs are an important link between anthropogenic and natural water cycles, and act as a barrier against undesired contaminants before discharge into the natural aquatic system [1]. It has been reported that WWTPs are a main source that releases microplastics to the environment [2–6]. This is because most of the microplastics generated by human activities enter the WWTPs through the sewer system. Meanwhile, WWTPs play a large role in reducing the abundance of microplastics during the treatment process, and mitigate the amount of microplastics released into the water system [7–9]. Even though plastic contamination has been studied since last century, research on microplastics has only emerged in recent decades. Microplastic contamination of the aquatic system has been widely reported. Microplastics in wastewater treatment plants are generally divided into two types, namely primary microplastics and secondary microplastics. Primary microplastics are mainly (1) rubber granules, (2) personal care products, (3) raw materials for plastics production, (4) paints, and (5) blasting abrasives. Secondary microplastics are generally generated from degraded large plastic materials, mainly (1) plastic building materials, (2) cooking utensils, scouring sponges, and cloths, (3) textiles, (4) paints, (5) road markings, (6) ship paints, (7) footwear, (8) other uses, and (9) tires [10–12]. Most personal care products from daily life, such as facial cleanser and toothpaste,

Microplastics in Urban Water Management, First Edition. Edited by Bing-Jie Ni, Qiuxiang Xu, and Wei Wei.
© 2023 John Wiley & Sons, Inc. Published 2023 by John Wiley & Sons, Inc.

include plastic microbeads. It is reported that facial cleanser has a large amount of microplastics, with 100 ml of scrub including 0.42–11.12 g plastic microbeads with average particle size of 37.66–95.95 μm. It is assumed that every single person will release 15.2 mg of these types of plastic microbeads to the sewage system [13–15]. Carr et al. found that 1.6 g of toothpaste could include over 4000 polyethylene particles, with size of 100–600 μm [16]. Laundry is a big source of fiber-type microplastics in domestic sewage. Over 1900 pieces of fiber microplastics can be released when washing one piece of clothing. Secondary microplastics have always been regarded as sourced from degradation, such as fragmentation, weathering, biodegradation, or photodegradation. Secondary microplastics are also claimed to be the main source of microplastics in oceans [17, 18].

Most plastic microbeads enter the sewage system with domestic wastewater or rainwater, and then go to the wastewater treatment plants. As a consequence of their size, conventional wastewater treatment process cannot effectively reject them, and most finally end up in sludge or discharged to rivers with effluent. According to earlier estimates, over 90% of microplastics entering wastewater treatment plants generally goes to sludge and the environmental aquatic system [19–22]. The discharge of microplastics in the aquatic system has led to a high accumulation of microplastics bringing potential risks and pressure. The wastewater treatment plant effluent is a main source of microplastic contamination of the environment. WWTPs could also be a main treatment terminal in mitigating microplastic contamination of the aquatic system. Thus, it is important to summarize the occurrence and fate of microplastics in wastewater treatment plants. A better understanding of microplastic removal performance in conventional wastewater treatment systems could provide suggestions for improving the microplastic removal and mitigate its discharge into natural waters. In this chapter, the occurrence, abundance, size, color, shape, and material of those microplastics are summarized and discussed. The characteristics of microplastics from domestic and industrial WWTPs were all summarized. The contribution of different treatment units in the removal of microplastics from WWTPs was compared.

3.2 The Abundance and Removal Performance of Microplastics in WWTPs

The microplastic abundance in WWTPs is varied in different regions or countries. Many review articles have focused on the discussion and analysis of the microplastics occurrence, fate, removal performance, microplastics size, material, shape information and so forth. In general, most of the WWTPs summarized are domestic WWTPs, as very limited information about the microplastic contamination in industrial WWTPs has been reported. Magnusson and Norén has investigated in the influent, effluent, and sludge of a wastewater treatment plant in Sweden [26]. It was reported that the microplastics abundance in the influent was 3 200 000 particles/h and 1770 particles/h in the effluent, with over 99% of the microplastics separated to the sludge. Only particle sizes over 300 μm were considered in this study. It has been reported that wastewater treatment plants are effective in rejecting microplastics of larger size. Thus, microplastics smaller than 300 μm could be substantial and have largely been ignored. In general, a tertiary treatment process might improve the rejection efficiency of smaller size microplastics. Talvitie et al. investigated microplastics removal from four typical municipal wastewater treatment plants using different post-treatment process, and implied that a membrane bioreactor could remove 99.9% of microplastics [2]. Li et al. also reported that a membrane bioreactor could remove microplastics from polluted surface water in drinking water treatment, mainly attributed to the rejection of microplastics by the membrane [63]. Wang et al. investigated the effect of microplastic accumulation in a membrane bioreactor system for microplastic removal during wastewater treatment. Polypropylene (PP) microplastics inhibited microbial growth in a membrane bioreactor with PP at range 0.14–0.30 g/L [64]. The PP microplastics mitigated membrane fouling by a continuous scouring effect and improved microbial community diversity. High concentration of PP also improved the clostridia abundance and reduced the proteobacteria. Maliwan et al. also obtained similar results demonstrating accumulation of microplastics in a membrane bioreactor system could both mitigate membrane fouling in long-term operation and affect microbial diversity [65]. Mishra et al. [66] reviewed membrane bioreactors in treating microplastic pollutants, including discharges, materials, and dimensions. They pointed out that low-cost membrane and technology improvement are important factors in membrane bioreactor application. Vuori and Ollikainen conducted cost effectiveness analysis of technologies in treating microplastics from a wastewater treatment system [67]. It is suggested that the membrane bioreactor process was the most cost-saving technology in both private and social cost, reducing the annual aquatic release of microplastics by billions of particles. The combining of MBR with incineration of sludge could mitigate the potential microplastic contamination to the soils from sludge utilization, and thus reduce the impact to the environment. Xu et al. summarized the removal and generation of microplastics in 17 WWTPs spanning France, the United States, the Netherlands, China, Italy, Korea, Spain, and Australia. The microplastic abundance in the influent of those WWTPs ranged from 1 particle/L to 597.9–675.5 particles/L. The analyses were mostly visual combined with FTIR or Raman [68]. Ngo et al. [69] summarized the pathway, classification, and removal efficiency of microplastics in WWTPs, reporting that the microplastics in the influent of WWTPs ranged 2.06–206 particles/L. The microplastic abundance in the influent of WWTPs in four different city of China

Table 3.1 Microplastics occurrence, removal in different types of wastewater treatment plants.

No.	Country/location	Treatment capacity (m^3/d)	Equivalent population	Treatment process	Influent abundance (MPs/L)	Effluent (MPs/L)	Removal rate	Material	Size (mm)	MP shape	color	References
1	Central Italy	18 000	80 000	Full scale CAS	3.6	–	86%	PES, PE, PP	0.1–0.5	fiber, film	–	[23]
				UASB + AnMBR			94%					
2	Hong Kong, China	84 000	300 000	–	7.1 ± 6.0–12.8 ± 5.8	–	–	PE, PES, PP, PU	1.0–5.0	fiber, sheet, fragment	transparent, blue, black, red, white, yellow	[24]
3	Italy	–	–	Tertiary WWTPs	–	–	84%	–	–	–	–	[25]
4	Sweden	–	–	Secondary WWTPs	–	–	99.90%	–	–	–	–	[26]
5	Israel	~30 000	210 000	Tertiary WWTPs	17 ± 7.49 (≥20 μm); 39.56 ± 12.49 (≥0.45 μm)	–	97.0% (≥20 μm); 94.4% (≥0.45 μm)	PI fibers, PE, others,	0.2–0.85	fragments, films, pellets, beads, foam	black, blue,	[27]
6	Canada	80 000	–	Secondary WWTPs + UV disinfection	–	1.76	–	–	–	–	–	[28]

7	Zhengzhou, China	30 0000	100 000	Tertiary WWTPs	16	2.9	81.90%	PP, PE, PA, PET	0.08–1.7	fiber	black, red	[29]
8	Beijing, China	1 000 000	2 400 000	water reclamation plant	12.03 ± 1.29	0.59 ± 0.22	95%	PET, PE, PP	0.68 ± 0.53	spherical, granule, fragment, film,	black, transparent, blue	[30]
9	Sweden	369 000	~790 000	–	202.2 kg/d	0.7 kg/d	99.60%	PE, PP, PES, PVC, PS, PU	–	–	–	[31]
10	Hangzhou, China	–	36–481	A^2/O + constructed wetland	0.43–2.154	–	11.8%–100%	PP, PS, PET	–	fragments, fiber	white, clear, red, gray	[32]
11	Finland	~30 000	160 000	Tertiary WWTPs	61 ± 26	–	–	–	–	–	–	[33]
12	–	40–90 L/h	test for treating WWTPs effluent	Membrane bioreactor	6.9	0.005	99.90%	PES, PE, PSC, PVC etc. (13 types)	0.02–0.3	fiber, fragment, film	–	[34]
13	–	–	by advanced oxidation processes	Rapid sand filter	0.7	0.02	97%					
14	–	–		Dissolved air flotation,	2.0	0.1	95%					
15	–	~480		Disc filter	0.5–2.0	0.03–0.3	40–98.5%					

(Continued)

Table 3.1 (Continued)

No.	Country/location	Treatment capacity (m³/d)	Equivalent population	Treatment process	Influent abundance (MPs/L)	Effluent (MPs/L)	Removal rate	Material	Size (mm)	MP shape	color	References
16	United states	–	–	Gravity filter	–	–	99.90%	PE	0.045–0.4	fragments	blue, white	[16]
17	Spain	28 400	–	Primary + A²/O	171 ± 43	–	93.70%	–	–	–	–	[35]
18	review	–	–	–	0.28–31 400	0.01–297	–	–	–	–	–	[36]
19	Netherlands	–	–	–	68–91	51–81	–	–	–	–	–	[37]
20	Guilin, China	4–10×10⁴	120 000–380 000	A/O, A2/O	0.7–8.72	0.07–0.78	89.2–93.6%	PP, PE, PET, PAN	0.5–5	fiber, fragment, film	–	[38]
21	USA	11.4–83 300	32 000–180 000	Secondary WWTPs	90–250	1.0–30.0	85.2 ± 6.0 ~ 97.6 ± 1.2%	–	–	–	–	[4]
22	Adana, Turkey		1 500 000	Secondary WWTPs	26 ± 3 ~ 23 ± 4	7 ± 0.7 ~ 4 ± 0.3	73–79%	–	–	–	–	[39]
23	Glasgow, Scotland	26 0954	650 000	Secondary WWTPs	15.7 ± 5.23	0.25 ± 0.04	98.41%	PE, PP, PS	–	–	Red, blue, green, clear (12)	[5]

24	Australia	48000	190 000	Secondary WWTPs	11.8 ± 1.10	2.76 ± 0.11	~76.6%	PP, PES, PA	0.0015–1	fiber, fragments, glitter, foams, beads, films	–	[40]
25	UK	111 496–184 703	184 500	Tertiary WWTPs	8.1×10^8 MPs/day	2.2×10^7 MPs/day	96%	PP	–	fiber, films, fragments	–	[41]
26	Review	–	–	–	1.01–31 400	0.004–447	10.2 ~ 99.9%	–	–	–	–	[42]
27	Iran	22 000	105 800	Secondary WWTPs	–	–	–	–	–	–	–	[43]
28	Changzhou, China	10 000–30 0 000	–	Secondary WWTPs	196 ± 11.89	9.04 ± 1.12	89.17–97.15%	PET, Rayon, PP, PE, PS	0.1–0.5	fiber	white, transparent, black, blue, brown	[44]
29	Canada	49 3 000	1 300 000	Secondary WWTPs	31.1 ± 6.7	0.5 ± 0.2	99%	–	–	–	–	[45]
30	Jiangsu, China	120 000	–	tertiary WWTPs	0–4	–	99.50%	PET, PS, PE, PP	>0.5	fragment, fiber	–	[46]
31	Turkey	12 000–150 000	120 000–1 010 000	Secondary and Tertiary WWTPs	1.5–3.1	0.6–1.6	48–73%	–	–	fiber	–	[47]
32	Spain	35 000	210 000	Secondary WWTPs	12.43 ± 2.70	1.23 ± 0.15	90.30%	PE (17)	0.4–0.6	fragment, film, bad, fiber, foam	–	[48]

(Continued)

Table 3.1 (Continued)

No.	Country/ location	Treatment capacity (m^3/d)	Equivalent population	Treatment process	Influent abundance (MPs/L)	Effluent (MPs/L)	Removal rate	Material	Size (mm)	MP shape	color	References
33	Harbin, China	600 000	3 100 000	Secondary WWTPs	126 ± 14	30.6 ± 7.8	75.70%	PES, PA, PET, PE,	0.02–0.1, 0.1–0.5	fiber, fragment	transparent, gray, white, black	[49]
34	Wuhan, China	20 000	–	Secondary WWTPs	79.9 ± 9.3	28.4 ± 7.0	64.40%	PA, PE, PP, PVC	0.02–0.3	fiber, fragment	transparent, black, brown	[50]
35	Changzhou, China	15 000–100 000	–	Secondary and Tertiary WWTPs	538.67 ± 22.05 – 1290 ± 65.26	20.44 ± 1.19 – 40.67 ± 1.12	66.7–99.9%	PES, PET, PP, rayon, PE (14)	<1.0	fiber, non-fiber	pink, transparent, blue, dark blue	[51]
36	Guangzhou, China	50 000–550 000	130 000–1 427 000	Secondary and Tertiary WWTPs	–	1.72 ± 1.04	–	PET, PP, PE, HDPE (37)	0.1–0.55	fiber, fragment, pellets	–	[52]
37	South Korea	250 000	–	A2O	–	–	−78%	PP, PE	–	–	–	[53]
38	South Korea	20 840 ~ 469 249	–	Secondary and Tertiary WWTPs	4200 ~ 31 400	33–297	>98%	–	–	microbead, fiber, fragment, sheet	–	[54]
39	Shanghai, China	–	2 930 000; 3 557 600;	Tertiary WWTPs	226.27 ± 83; 171.89 ± 62.98	83.16 ± 17.22; 69.03 ± 12.21	63.25%; 59.84%	PET, PA, PE, PP	0.08–1.0	fiber, fragment, pellet	transparent, blue, black, red	[8]

	Location			Treatment				Polymer		Shape	Color	Ref.
40	Spain	12 000	29 777	Tertiary WWTPs	4.40 ± 1.01	0.92 ± 0.21 (MBR); 1.08 ± 0.28 (RSF);	79.01% (MBR); 75.49% (RSF)	LDPE (14)	–	fiber, film, fragment, bead	–	[55]
41	France	80 000	415 000	Tertiary WWTPs	244	2.84	98.83%	PS, PE (10)	0.02–0.2	fiber, film, fragment	–	[7]
42	Nanjing, China	80 000–150 000	250 000–600 000	Tertiary WWTPs	10.3	0.24	97.67–98.46%	PE, PP, phenolic resin, PS, PET	<1.1	fragment, granules, film, fibers	transparent, white (6)	[56]
43	USA	2.35–382 000	3500–56 000 000	Secondary and Tertiary WWTPs	–	0.05 ± 0.024	–	–	–	fiber, fragment	–	[57]
44	Italy	400 000	1 200 000	Secondary and Tertiary WWTPs	2.50 ± 0.3	0.4 ± 0.1	84%	ABS (14)	0.1–5	line, film, fragment, microfiber	–	[25]
45	Xiamen, China	–	3 500 000	Secondary WWTPs	1.57–13.69	0.20–1.73	79.3–97.8%	PP, PE, PS, PEC, PET (8)	–	granules, fragments, fibers, pellets	White, clear	[58]
46	Wuhan, China	70 000–300 000	–	Tertiary WWTPs	23.3–80.5	7.9–30.3	66.1–62.7%	PVC, PA (7)	~0.8	fiber, fragment, microbead	transparent, white (9)	[59]

(Continued)

Table 3.1 (Continued)

No.	Country / location	Treatment capacity (m³/d)	Equivalent population	Treatment process	Influent abundance (MPs/L)	Effluent (MPs/L)	Removal rate	Material	Size (mm)	MP shape	color	References
47	Australia	65 000–150 000	234 000–700 000	Secondary WWTPs	55–98	0.7	69–79%	PET, PE, PP	>0.025	fibers, fragments	–	[60]
48	Thailand	24 000–200 000	51 000–580 000	Secondary WWTPs	12.2	2	84%	PES, PE, PA, PP (13)	–	fiber, sheet, fragment, sphere	–	[61]
49	South Korea (50 WWTPs)	21 000–1 584 000	51 000–3 339 000	Secondary and Tertiary WWTPs	10–470	0.004–0.51	98.7–99.99%	PP, PE, PET (8)	–	fiber, fragment	–	[62]

Marks: Abbreviations: Polyester: PES; Polyethylene: PE; Polypropylene: PP; polyurethane: PU; polyamide: PA; polyethylene terephthalate: PET; polyvinyl chloride: PVC; polystyrene: PS; polyacrylates: PAL; Polyacrylonitrile: PAN; acrylonitrile butadiene styrene: ABS; The number in parentheses in the material list indicates the types of microplastics material reported in corresponding literature, only most frequently detected material was listed in Table 3.1. The same applies for the shape and color, only the most frequently detected shape and color was listed in Table 3.1.

was highly varied. For example, the microplastic abundance was 206, 6.5, 12, and 2.06 particles/L in Wuhan, Xiamen, Beijing, and Hong Kong, respectively. The variation could be related to the wastewater source, difference in sampling, pretreatment, and analysis methods used in the different countries and studies. Cao et al. [24] estimated the daily load of microplastics in nine different WWTPs with microplastics calculated as ranging 1.55×10^6–1.51×10^{10} particles/day. It is suggested that 24-h sequential sampling should be conducted for multiple days to improve the calculation accuracy of daily microplastic loads.

Here we also summarized the most recent microplastic abundance, characteristics, and removal performance in WWTPs around the world (Table 3.1). The information covers Italy, China, Sweden, Israel, Canada, Finland, the United States, Spain, Turkey, UK, Iran, South Korea, France, Australia, and Thailand. The WWTPs treatment capacity can be as high as 1 584 000 m^3/d in South Korea, where the highest equivalent capacity is 56 000 000 in USA. Among all the WWTPs, the influent microplastic abundance ranged 0–31 400 particles/L, and the effluent was in the range 0.004–447 particles/L. The highest influent and effluent microplastic abundances were a WWTP in South Korea [54]. Park et al. investigated 50 typical WWTPs in all of South Korea, where the treatment capacity ranges 21 000–1 584 000 m^3/d, the influent and effluent microplastic abundance are 10–470 and 0.004–0.51 particles/L, respectively [62]. The removal performance for all those 50 WWTPs, including secondary and tertiary WWTPs, ranged 98.7–99.99%. The removal performance reported by Park et al. was relatively higher than other WWTPs summarized [62]. As shown in Table 3.1, the influent microplastic abundance in WWTPs generally range 0–100 particles/L, where only a few WWTPs had extremely high microplastic abundance in the influent.

Xu et al. investigated the microplastic abundance in WWTPs in Changzhou, China. The influent microplastics are 538.67 ± 22.05 – 1290 ± 65.26 particles/L, the effluent is 20.44 ± 1.19 – 40.67 ± 1.12 particles/L, and the removal rate in both secondary and tertiary WWTPs are 66.7–99.9% [51]. As Changzhou is a famous "clothing center" of China, the clothing production process produces many fibers and other corelated microplastics. This could be the reason that the microplastic abundance detected in Changzhou city is far higher than other WWTPs surveyed. Another example of high influent microplastics was reported by Jia et al. of two WWTPs in Shanghai. These two large scale WWTPs in Shanghai have a treatment capacity at 2 930 000 and 3 557 600 equivalent population [8]. The influent microplastic abundances were 226.27 ± 83 and 171.89 ± 62.98 particles/L, and the effluent were 83.16 ± 17.22 and 69.03 ± 12.21 particles/L. The removal rate was relatively low compared to other reported WWTPs at only 59.84% and 63.52%. It is concluded that these two WWTPs have low treatment performance for microplastics, and the large amount of microplastics remaining in the effluent could pose high risks to the receiving water ecosystems.

In general, the average influent microplastic abundances of the summarized WWTPs range 21.27 ± 29.45 particles/L, excluding the extremely high values (only less than 100 particles/L were included). Under this circumstance, the corresponding effluent abundances were at 1.86 ± 4.04 particles/L, and the removal performances were at 93.87 ± 6.05%. This shows that the microplastic abundances in the influent of WWTPs around the world is relatively variable. Meanwhile, except for several abnormally high sites, most of the WWTPs have an average influent microplastic abundance around 21 particles/L. The effluent microplastic abundance can be as low as 0.004 particles/L, and the average effluent abundance for Table 3.1 (excluding those extremely high values) was 1.86 particles/L. The average removal rate as calculated from Table 3.1 is ca. 93.87% (those with typically low value < 80% were excluded). As shown in Table 3.1, among the 49 pieces of data collected (most of which include several WWTPs in one dataset), only 10 had a microplastics removal lower than 80%. At the same time, those WWTPs with a removal rate lower than 80% are generally ca. 60–70%, and only one system with an extremely low removal rate of 11.8%. This outlier is not a typical WWTP, but a constructed wetland combined treatment system [32]. This showed that WWTPs generally had a microplastic removal rate over 90% around the world, and at worst, over 60% removal can be achieved. Those WWTPs with a removal rate lower than 80% are from Turkey [39, 47], Australia [40], China (Harbin city) [49], Wuhan city [50, 59]; Shanghai city [8], South Korea [53], Spain [55], and Australia [60]. Considering the huge amount of wastewater consumption and discharge, the microplastics entering aquatic systems through effluent should in no be way ignored.

3.3 The Microplastics Composition in WWTPs

In general, the microplastic characteristics in WWTPs are summarized in the microplastics-related investigations. These characteristics are mostly including not only the microplastic abundance, but also the size distribution, materials, color, and shape. That information could benefit the identification of microplastics, potential removal performance, potential ecological risks, possible source, and so forth [36, 70]. This study is also summarized in Table 3.1.

3.3.1 Microplastics Size Distribution

Microplastic size is an important factor that could affect the performance and transformation in WWTPs. Smaller size microplastics are more prone to be ingested by plankton, filter feeding organisms, fish, and finally might bring toxicological effects to these organisms [71]. Microplastic sizes are commonly identified by two methods, one is based on the size of sieves used during the extraction of microplastics, and the other is the identity of microplastics under microscope. In Table 3.1, only the most frequently observed microplastic sizes were listed. The size distribution in various WWTPs is quite different, some of which had a size larger than 1 mm and less than 5 mm, while most of had sizes less than 1 mm [7, 16, 23, 27, 30, 34, 44, 48]. Many other papers did not provide size information for the microplastics [53, 57, 58, 61, 62]. In Sun et al. [22] review, the dimensions of 0.025, 0.1, 0.5 mm were the most widely used in size classification, while from our up-to-date view, the size distribution in different studies were largely varied based on their research methodology and need (Table 3.1). Cao et al. investigated the size distribution in all processes of one WWTP in Hong Kong, China [24]. The size distribution in all sampling sites was 12.24–25.37%, 68.63–86%, and 1.48–10.76% for sizes of 0.5–1 mm, 1–5 mm, and 0.1–0.5 mm, respectively. Talvitie et al. found that the dominant microplastic size distribution in one of the WWTPs in Finland is 0.02–0.1 mm, other than 0.1–0.3 mm and > 0.3 mm fractions [34]. Tang et al. investigated two WWTPs in Wuhan, China, where the microplastic size distribution used was 0.02–0.3 mm, 0.3–1 mm, and 1–5 mm [59]. Among which, the most dominant size distribution was 0.3–1 mm followed by 0.02–0.3 mm. The percentage of 1–5 mm is around 10% in all the sampling sites in the WWTPs. Liu et al. found that microplastics with size less than 1 mm ranged 65–86.9% in the influent and 81–91% in the effluent [50]. It is suggested that with the treatment process, the primary microplastics are decomposed to secondary microplastics by physical, chemical, or biological processes. Thus, sometimes the effluent microplastic abundance in small size was increased as compared with the influent of WWTPs. In general, the microplastics observed have a size less than 1 mm, and this can be used as a guidance for the design of biological toxicity tests and environmental transformation tests. He et al. have investigated the effect of microplastic particle size on the nitrification and denitrification efficiency in the wastewater treatment process and reported that particle size has close correlation with microbe activities [21]. In earlier publications, research focused on the effect of suspended solids in natural system nitrogen cycling, and emphasized the importance of particle sizes [72–74]. The effect of microplastic size to the activated sludge process in wastewater treatment plants must be considered in the future.

3.3.2 Microplastic Shapes

The analysis of microplastics is generally classified by physical and chemical characteristics, where the physical characteristics refer to the size, shape, and color. The morphology of microplastics is generally characterized by using stereomicroscope. As this process depends strongly on the operator, the visual identification is open to bias. It was estimated that observation error can be up to 70%, and this error could increase with a decrease in particle size [18, 22]. Magnusson and Norén reported that it is difficult to distinguish synthetic and natural fibers, for example, the cotton made textile fiber. Shape is a very important factor applied in microplastics classification. This factor could reflect the microplastic removal efficiency in wastewater treatment processes. The microplastic shape is also an important factor in interaction with microbiological process in the activated sludge process, and could couple with other contaminants in the wastewater [22, 75]. The microplastic shapes in WWTPs are generally categorized as six different types: fiber, granular, pellet, film, foam, and fragment [27, 69, 76]. The microplastics in WWTPs are largely varied by different sources in different regions or countries. Among these, fiber and fragments are the most dominant shapes in most WWTPs (Table 3.1). Fibers in WWTPs generally come from domestic or industrial laundry discharge. Pellets (beads in spherical shape) and granular (hard small pieces) shape microplastics generally come from personal care products, like body scrubs, toothpaste, and so forth. The film (thin particle), foam (sponge like), and fragments (broken small pieces) come from the erosion process of different types of plastic products in daily life. In general, fiber is the most dominant in the influent and effluent of WWTPs, especially in some industrial WWTPs. It is reported by earlier literature [58] that WWTPs are more effective in removing fragments (91.3%) and granules (91.4%) than fibers (82.8%) and pellets (78.9%). In this study, the fiber removal was improved from 17.7 to 30.4% from influent to effluent. This also highlighted a lower removal of fibers from WWTPs. The high abundance of fiber in WWTPs also could be attributed to the difficulty in separating synthetic fiber from natural fibers. Effective differentiation methods and detection processes for natural and synthetic fibers is very important in quantifying the amount and shape of microplastics in WWTPs, especially for fibers.

3.3.3 Microplastic Materials

The chemical makeup of microplastics is an important factor that affects the flotation and sinking properties, then affects the removal performance of WWTPs. It is reported by Sun et al. that over 30 types of microplastic materials have been observed in WWTPs, where the most commonly found are polyester (PES), polyethylene (PE), polyethylene terephthalate (PET) and polyamide (PA) [22]. The microplastic materials that occurred in WWTPs are summarized in Table 3.1. Note, only the dominant microplastics are listed in the table. It is clear that PE, PES, and PET are generally the most abundant materials in most WWTPs. The abundance and material composition in different WWTPs are varied. Among those materials, PES, PET, and PA are widely used in synthetic clothes, textiles, carpets, etc. PE is widely used in personal care products, plastic bottles, packing films, and so forth. Polypropylene (PP), polyurethane (PU), polystyrene (PS), polyvinyl chloride (PVC), polyacrylates (PAL), polyacrylonitrile (PAN), and acrylonitrile butadiene styrene (ABS) are all common plastic product materials in daily life. Zou et al. [52] investigated six WWTPs in Guangzhou, China. A total of 37 types of polymer microplastics were observed in the six WWTPs. The most abundant polymers were PET (31.86%), PP (26.55%), PE (9.73%), and high-density PE (5.16%). Liu et al. [50] investigated a WWTP in Wuhan, China, and reported that the most common polymer found is PA (54.8%), followed by PE (9.0%), PP (9.6%), and PVC (2.5%). A WWTP in Harbin reports 89.5% of the influent polymers are PES, PA, PET, and PE [49]. Yang et al. [30] investigated China's largest reclamation plants, where PET (42.25%), PES (19.09%), and PP (13.05%) are the most abundant microplastics, and where PET and PES are detected in the form of fibers possibly originating from textiles from domestic sources. The level of PE (1.64%) in this plant is much lower as compared with Europe (~14%) [34] or US (>90%) [16]. The variation of majority polymer material distribution shows that the microplastic material in WWTPs has regional features. The dominant microplastics in a typical region could have correlation with various aspects of the region, including the resident lifestyle, culture, commercial and industrial activities, economic developing level, and so forth.

3.3.4 Microplastic Color

Microplastic color also generally investigated as an important factor in related studies. Color is important as it can potentially be used to trace the source of some microplastics. In nature, the microplastic color sometimes puzzles living organisms. Seabirds, turtles, seals, and other aquatic vertebrate creatures can be confused by different colored microplastics, consume them as food, or be wrapped and suffocated by larger plastics. Thus, researchers generally regard color as one of the important parameters in microplastics-related studies. Color is also summarized here, and the most frequently detected colors are transparent, black, blue, red, and white in most of the investigated WWTPs (Table 3.1). This is probably because that these are the most frequently used colors in human daily life, thus the detection frequency of them was high in WWTPs. Meanwhile, some microplastic color could change during the treatment process in WWTPs. As reported by Jia et al. [8] the microplastics detected in two WWTPs in Shanghai showed microplastics in transparent color occupied 77.73% and 75.70%, respectively. It was suggested that the high percentage of transparent color could result from the effect of erosion and UV irradiation in the environment, where the initial color was changed to transparent. In aquatic systems, transparent color was generally not the dominant color. The in-depth reason for this difference requires further investigation [19].

3.4 Removal of Microplastics in WWTPs and Contribution of Each Process

WWTPs as a receptor must concentrate large amount of microplastics from sewer systems. Generally, WWTPs include physical, chemical, and biological treatment distributed in three stages: primary, secondary, and tertiary treatment. The primary and secondary treatment are designed for removal of most organic matter and nitrogen to meet the discharging standards. Tertiary treatments or extra advanced oxidation processes are aimed at improving treatment performance for water reclamation or reuse. Even though WWTPs are not designed for microplastic removal, the removal performance of microplastics is considerable, at over 80% in most WWTPs. Nonetheless, the microplastics discharged from WWTPs to the environment are significant because of the large volumes of treated wastewater discharged. For example, Saskatoon WWTPs discharge 1.76 particles/L from effluent, equaling to 141 million particles/day [28]. While the microplastics mitigated from WWTPs are generally transferred to the sludge or discharged with the effluent to the aquatic system, a limited amount can be degraded or removed by the treatment process. The microplastics that are discharged into the aquatic system enter the lakes, rivers, or sea, and finally can cause harm to aquatic organisms or ecosystem. McCormick et al. found that microplastics in the rivers downstream of WWTPs were significantly higher than upstream [77]. Talvite et al. [2] reported that the WWTPs effluent has 25× fibers and 3× fragments compared to its receiving water. The other microplastics that go into the sludge are commonly further treated in the sludge treatment process, then enter the soil system (for unrestricted irrigation or safe disposal to land) and affect corresponding ecosystems [78]. Li et al. [79] analyzed the microplastics abundance in sewage sludge of 28 WWTPs and reported that the microplastic abundances were $1.6–56.4\times10^3$ particles/kg dry sludge. Liu et al. [36] looked at the microplastics removal in global WWTPs and reported that the microplastic abundance in sludge were in the range $0.044–2.4\times10^5$ particles/kg. Accompanying this is an estimated daily microplastics discharge into aquatic environment by WWTPs of $5\times10^5–1.39\times10^{10}$ particles. An earlier publication reported that 97–99% of the microplastics are transferred to the sludge, mainly attributed to the primary and secondary treatment [5, 9, 16, 27, 34]. The contribution of different treatment stages in WWTPs in removing microplastics are the focus of this section.

3.4.1 Primary Treatment

Most current WWTPs are not designed for removing microplastics. Primary treatment in WWTPs is the main contributor in removing microplastics, since primary processes are designed for removing suspended solids through filtration, sedimentation, and aeration processes (Figure 3.1). Many microplastics also can be regarded as suspended solids, to be effectively removed during the primary treatment process and settled into sludge [16, 56]. At the same time, preliminary treatment is commonly designed to remove materials that may inhibit followup biological processes. Those materials include sticks, rags, grease, grits, and so forth [80]. Preliminary and primary treatment generally include coarse and fine screening, grease and grit removal, skimming, and primary sedimentation. Thus, microplastics larger than the screen size can all be removed. Liu et al. [50] reported that after screening (mesh size at 6 mm), the largest size of plastics remained was 1.6 mm. Sun et al. summarized the microplastic removal from different process of WWTPs, and reported that the preliminary and primary process removed 35–59% of microplastics [22]. The skimming process contributes to floating microplastic removal, the settling process contributes to the high density microplastic removal, and the flocculation process enhances the trapping of microplastic floc. Ren et al. [29] investigated the microplastic removal from WWTPs in Zhengzhou, China. It is reported that the primary treatment has better performance in removing large microplastics, where secondary process is more efficient in reducing fiber and small microplastics [59, 81]. Talvitie et al. [2] reported that fiber removal was over 90% after primary sedimentation. Michielssen et al. found that small litter is removed 84–88% after primary treatment [82]. In summary, the primary treatment process is a main contributor in microplastic removal in WWTPs.

Microplastic removal in the primary process is significantly affected by density and gravity, microplastics with higher density are prone to be removed by settling process [36, 69]. The microplastics with low or moderate density are easily removed by the air flotation process, where microbubbles can act as carriers adhering to the surface of suspended matter, and rising against gravity to form floating foam removed by the skimming processes [36, 51, 83]. With the combination of screening, air flotation, skimming, and settlement, a large portion of microplastics can be removed during the preliminary and primary treatment in WWTPs.

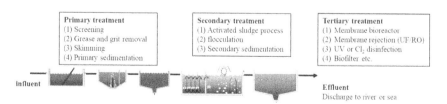

Figure 3.1 Microplastic removal-related process in WWTPs.

3.4.2 Secondary Treatment

Secondary treatment in WWTPs is typically a biological process, designed to further reduce residual suspended and dissolved solids after primary treatment (Figure 3.1). In almost all the 49 data sets collected (Table 3.1), a variety of activated sludge processes were involved in the secondary treatment. The secondary treatment contribution to microplastic removal mostly relies on the secondary sedimentation tank. This process has much lower removal rate for microplastics compared to the primary process. In this stage, the sludge floc in the aeration tank can assist in capturing microplastic debris, then settle to sludge in the clarification tank [22, 69]. Microplastics also might be ingested by protozoa or metazoa and then be trapped by sludge floc [56]. Chemicals in wastewater treatment plants could react as flocculants and aid microplastic removal by forming floc. The flocculation process in secondary treatment could increase the adsorption effect between microplastics and flocculants, increase the settlement ability, and assist the precipitation of microplastics. Meanwhile, the interaction and mechanism between microplastics, flocs, flocculants are rarely reported. The effect and mechanism of the flocculation process on microplastics removal in secondary sedimentation should be further investigated.

Most research has focused on investigating the effect of biofilms on microplastics in recent years [84–89]. In fact, microplastic occurrence in activated sludge could easily form heteroaggregates, with biofilms and other pollutants aggregated surrounding the microplastic particles. This is also can be regarded as a floc, increasing the microplastics' density, and allowing settling during the sedimentation process. This is also the reason that a secondary treatment process has a lower removal performance in treating microplastics. Whether these heteroaggregates show positive or negative effects in wastewater treatment plants, followup microplastic treatment processes, and problems caused after microplastics are released into the environment are not well demonstrated. Meanwhile, aging microplastics are a perfect vector for microbes, heavy metals, viruses, antibiotic resistant genes, and other contaminants to attach. The occurrence and fate of this type of heteroaggregates in both the activated sludge process and the aquatic systems should be investigated in the near future.

3.4.3 Tertiary Treatment

The tertiary treatment is designed for improving biodegradable organic matters (BOD), nitrogen, phosphorus, bacteria etc. pollutants after secondary treatment. In general, a tertiary process has very limited contribution to improving organic matter removal. Tertiary treatment performance for removal of microplastics also depends on the process used, so a membrane-based technology could show high improvement in microplastic rejection [34, 44, 90, 91]. Pittura et al. [23] reported that the overall microplastic removal by pilot scale UAS + AnMBR was greater than full scale CAS system. Ultrafiltration was the main contributor for the rejection of microplastics in their study. Iyare et al. [80] implied that the membrane bioreactors have higher potential in microplastic removal (82.1–99.9%) as compared with other tertiary treatment processes. As can be seen from Table 3.1, WWTPs with a tertiary treatment process have higher microplastic removal as compared with those only have secondary treatment, but do not show dramatic improvement. After tertiary treatment, microbeads and fibers in the smallest size fraction (0.02–0.19 mm) were found to be the most abundant microplastics [90]. Tertiary treatment is the last barrier before microplastics discharging into natural water bodies. More effort should be expended if we want less microplastics release into natural systems through WWTPs effluent.

Tertiary treatment commonly includes membrane bioreactors, membrane filtration, sand filtration, coagulation, advanced oxidation processes (AOPs), disinfection process, and so forth. Chlorination, ozonation, and UV irradiation are widely used AOPs in tertiary WWTPs. Chlorination might increase microplastics abundance by cracking of microplastics during the treatment process [46, 92]. Chlorination also has the potential to change the chemical structure of microplastics, such as forming carbon-chlorine bonds and increasing the toxicity or hydrophobicity [93, 94]. UV oxidation also could cause microplastic characteristics changes [36, 95]. The consequence of using AOPs as tertiary treatment for microplastic removal, and the treated water system should be considered and investigated. Tertiary treatment could contribute both positively and negatively to microplastic removal in WWTPs [42, 96].

3.5 Summary and Future Outlooks

WWTPs receive a large amount of microplastics. Most of those microplastics are removed in the primary stage through skimming or sedimentation process. Even though microplastics in the effluent of WWTPs are largely reduced as compared with the influent, a large amount of microplastics could still enter the environment through effluent discharge or sludge discharge. Thus, it is vital to reduce the microplastics from the source. For example, the cosmetic and beauty products industry are a large production source and discharge point for microplastics and should begin to phase-out and replace the additives with environmentally friendly products to reduce the release of microplastics to the environment. WWTPs also should update treatment technology to improve microplastic removal and reduce emission from the effluent into the aquatic system. Considering the potential of toxic contaminants carried by microplastics, disinfection of wastewater treatment plant effluent should be carried out to mitigate those risks. Currently, it is difficult to reject microplastics through membrane-related technology in wastewater treatment plants because of cost. Additional methods to reduce microplastics from domestic sources, improve rejection from sewer systems, and update primary treatment processes should be investigated. Technologies for improving microplastics from waste sludge should also be investigated in the future, such as incineration during sludge treatment and resource recovery processes. By improving these processes, the release of microplastics through sludge discharge or wastewater treatment plant effluent, and potential contamination to ecological system can be further mitigated.

References

1 Ou, H., and Zeng, E.Y. (eds.) (2018). Occurrence and fate of microplastics in wastewater treatment plants. In: *Microplastic Contamination in Aquatic Environments* (ed. T. Page), 317–338. UK: Elsevier.
2 Talvitie, J., Heinonen, M., Pääkkönen, J.P. et al. (2015). Do wastewater treatment plants act as a potential point source of microplastics? Preliminary study in the coastal Gulf of Finland, Baltic Sea. *Water Science and Technology* 72 (9): 1495–1504.
3 Kay, P., Hiscoe, R., Moberley, I. et al. (2018). Wastewater treatment plants as a source of microplastics in river catchments. *Environmental Science and Pollution Research* 25 (20): 20264–20267.
4 Conley, K., Clum, A., Deepe, J. et al. (2019). Wastewater treatment plants as a source of microplastics to an urban estuary: removal efficiencies and loading per capita over one year. *Water Research X* 3: 100030.
5 Murphy, F., Ewins, C., Carbonnier, F. et al. (2016). Wastewater treatment works (WwTW) as a source of microplastics in the aquatic environment. *Environmental Science & Technology* 50 (11): 5800–5808.
6 Okoffo, E.D., O'Brien, S., O'Brien, J.W. et al. (2019). Wastewater treatment plants as a source of plastics in the environment: a review of occurrence, methods for identification, quantification and fate. *Environmental Science: Water Research & Technology* 5 (11): 1908–1931.
7 Kazour, M., Terki, S., Rabhi, K. et al. (2019). Sources of microplastics pollution in the marine environment: importance of wastewater treatment plant and coastal landfill. *Marine Pollution Bulletin* 146: 608–618.
8 Jia, Q.L., Chen, H., Zhao, X. et al. (2019). Removal of microplastics by different treatment processes in Shanghai large municipal wastewater treatment plants. *Huanjing Kexue* 40 (9): 4105–4112. in chinese.

9 Freeman, S., Booth, A.M., Sabbah, I. et al. (2020). Between source and sea: the role of wastewater treatment in reducing marine microplastics. *Journal of Environmental Management* 266: 110642.

10 Vo, H.C. and Pham, M.H. (2021). Ecotoxicological effects of microplastics on aquatic organisms: a review. *Environmental Science and Pollution Research* 28 (33): 44716–44725.

11 Andrady, A.L. (2017). The plastic in microplastics: a review. *Marine Pollution Bulletin* 119 (1): 12–22.

12 Lassen, C., Hansen, S.F., Magnusson, K. et al. (2015). Microplastics: Occurrence, effects and sources of releases to the environment in Denmark. Danish Environmental Protection AgencyD Danish Environmetal Protection Agency. http://mst.dk/service/publikationer/publikationsarkiv/2015/nov/rapport-om-mikroplast (accessed June 2022)

13 Fendall, L.S. and Sewell, M.A. (2009). Contributing to marine pollution by washing your face: microplastics in facial cleansers. *Marine Pollution Bulletin* 58 (8): 1225–1228.

14 Zitko, V. and Hanlon, M.J.M.P.B. (1991). Another source of pollution by plastics: skin cleaners with plastic scrubbers. *Marine Pollution Bulletin* 22 (1): 41–42.

15 Kalčíková, G., Alič, B., Skalar, T. et al. (2017). Wastewater treatment plant effluents as source of cosmetic polyethylene microbeads to freshwater. *Chemosphere* 188: 25–31.

16 Carr, S.A., Liu, J., and Tesoro, A.G. (2016). Transport and fate of microplastic particles in wastewater treatment plants. *Water Research* 91: 174–182.

17 Gouin, T., Roche, N., Lohmann, R. et al. (2011). A thermodynamic approach for assessing the environmental exposure of chemicals absorbed to microplastic. *Environmental Science & Technology* 45 (4): 1466–1472.

18 Hidalgo-Ruz, V., Gutow, L., Thompson, R.C. et al. (2012). Microplastics in the marine environment: a review of the methods used for identification and quantification. *Environmental Science & Technology* 46 (6): 3060–3075.

19 Li, L., Geng, S., Wu, C. et al. (2019). Microplastics contamination in different trophic state lakes along the middle and lower reaches of Yangtze River Basin. *Environmental Pollution* 254: 112951.

20 Li, L., Song, K., Yeerken, S. et al. (2020). Effect evaluation of microplastics on activated sludge nitrification and denitrification. *Science of the Total Environment* 707: 135953.

21 He, Y., Li, L., Song, K. et al. (2021). Effect of microplastic particle size to the nutrients removal in activated sludge system. *Marine Pollution Bulletin* 163: 111972.

22 Sun, J., Dai, X., Wang, Q. et al. (2019). Microplastics in wastewater treatment plants: detection, occurrence and removal. *Water Research* 152: 21–37.

23 Pittura, L., Foglia, A., Akyol, Ç. et al. (2021). Microplastics in real wastewater treatment schemes: comparative assessment and relevant inhibition effects on anaerobic processes. *Chemosphere* 262: 128415.

24 Cao, Y., Wang, Q., Ruan, Y. et al. (2020). Intra-day microplastic variations in wastewater: a case study of a sewage treatment plant in Hong Kong. *Marine Pollution Bulletin* 160: 111535.

25 Magni, S., Binelli, A., Pittura, L. et al. (2019). The fate of microplastics in an Italian wastewater treatment plant. *Science of the Total Environment* 652: 602–610.

26 Magnusson, K. and Norén, F. (2014). Screening of microplastic particles in and down-stream a wastewater treatment plant.

27 Ben-David, E.A., Habibi, M., Haddad, E. et al. (2021). Microplastic distributions in a domestic wastewater treatment plant: removal efficiency, seasonal variation and influence of sampling technique. *Science of the Total Environment* 752: 141880.

28 Prajapati, S., Beal, M., Maley, J. et al. (2021). Qualitative and quantitative analysis of microplastics and microfiber contamination in effluents of the City of Saskatoon wastewater treatment plant. *Environmental Science and Pollution Research* 28: 32545-32553.

29 Ren, P., Dou, M., Wang, C. et al. (2020). Abundance and removal characteristics of microplastics at a wastewater treatment plant in Zhengzhou. *Environmental Science and Pollution Research* 27 (29): 36295–36305.

30 Yang, L., Li, K., Cui, S. et al. (2019). Removal of microplastics in municipal sewage from China's largest water reclamation plant. *Water Research* 155: 175–181.

31 Rasmussen, L.A., Iordachescu, L., Tumlin, S. et al. (2021). A complete mass balance for plastics in a wastewater treatment plant macroplastics contributes more than microplastics. *Water Research* 201: 117307.

32 Wei, S., Luo, H., Zou, J. et al. (2020). Characteristics and removal of microplastics in rural domestic wastewater treatment facilities of China. *Science of the Total Environment* 739: 139935.

33 Salmi, P., Ryymin, K., Karjalainen, A.K. et al. (2020). Particle balance and return loops for microplastics in a tertiary-level wastewater treatment plant. *Water Science and Technology* 84 (1): 89–100.

34 Talvitie, J., Mikola, A., Koistinen, A. et al. (2017). Solutions to microplastic pollution–removal of microplastics from wastewater effluent with advanced wastewater treatment technologies. *Water Research* 123: 401–407.

35 Edo, C., González-Pleiter, M., Leganés, F. et al. (2020). Fate of microplastics in wastewater treatment plants and their environmental dispersion with effluent and sludge. *Environmental Pollution* 259: 113837.

36 Liu, W., Zhang, J., Liu, H. et al. (2021). A review of the removal of microplastics in global wastewater treatment plants: characteristics and mechanisms. *Environment International* 146: 106277.

37 Leslie, H.A., Brandsma, S.H., Van Velzen, M.J.M. et al. (2017). Microplastics en route: field measurements in the Dutch river delta and Amsterdam canals, wastewater treatment plants, North Sea sediments and biota. *Environment International* 101: 133–142.

38 Zhang, L., Liu, J., Xie, Y. et al. (2021). Occurrence and removal of microplastics from wastewater treatment plants in a typical tourist city in China. *Journal of Cleaner Production* 291: 125968.

39 Gündoğdu, S., Çevik, C., Güzel, E. et al. (2018). Microplastics in municipal wastewater treatment plants in Turkey: a comparison of the influent and secondary effluent concentrations. *Environmental Monitoring and Assessment* 190 (11): 1–10.

40 Raju, S., Carbery, M., Kuttykattil, A. et al. (2020). Improved methodology to determine the fate and transport of microplastics in a secondary wastewater treatment plant. *Water Research* 173: 115549.

41 Blair, R.M., Waldron, S., and Gauchotte-Lindsay, C. (2019). Average daily flow of microplastics through a tertiary wastewater treatment plant over a ten-month period. *Water Research* 163: 114909.

42 Cheng, Y.L., Kim, J.G., Kim, H.B. et al. (2021). Occurrence and removal of microplastics in wastewater treatment plants and drinking water purification facilities: a review. *Chemical Engineering Journal* 410: 128381.

43 Petroody, S.S.A., Hashemi, S.H., and van Gestel, C.A. (2021). Transport and accumulation of microplastics through wastewater treatment sludge processes. *Chemosphere* 278: 130471.

44 Xu, X., Jian, Y., Xue, Y. et al. (2019). Microplastics in the wastewater treatment plants (WWTPs): occurrence and removal. *Chemosphere* 235: 1089–1096.

45 Gies, E.A., LeNoble, J.L., Noël, M. et al. (2018). Retention of microplastics in a major secondary wastewater treatment plant in Vancouver, Canada. *Marine Pollution Bulletin* 133: 553–561.

46 Lv, X., Dong, Q., Zuo, Z. et al. (2019). Microplastics in a municipal wastewater treatment plant: fate, dynamic distribution, removal efficiencies, and control strategies. *Journal of Cleaner Production* 225: 579–586.

47 Akarsu, C., Kumbur, H., Gökdağ, K. et al. (2020). Microplastics composition and load from three wastewater treatment plants discharging into Mersin Bay, north eastern Mediterranean Sea. *Marine Pollution Bulletin* 150: 110776.

48 Bayo, J., Olmos, S., and López-Castellanos, J. (2020). Microplastics in an urban wastewater treatment plant: the influence of physicochemical parameters and environmental factors. *Chemosphere* 238: 124593.

49 Jiang, J., Wang, X., Ren, H. et al. (2020). Investigation and fate of microplastics in wastewater and sludge filter cake from a wastewater treatment plant in China. *Science of the Total Environment* 746: 141378.

50 Liu, X., Yuan, W., Di, M. et al. (2019). Transfer and fate of microplastics during the conventional activated sludge process in one wastewater treatment plant of China. *Chemical Engineering Journal* 362: 176–182.

51 Xu, X., Zhang, L., Jian, Y. et al. (2021). Influence of wastewater treatment process on pollution characteristics and fate of microplastics. *Marine Pollution Bulletin* 169: 112448.

52 Zou, Y., Ye, C., and Pan, Y. (2021). Abundance and characteristics of microplastics in municipal wastewater treatment plant effluent: a case study of Guangzhou, China. *Environmental Science and Pollution Research* 28 (9): 11572–11585.

53 Nguyen, N.B., Kim, M.K., Le, Q.T. et al. (2021). Spectroscopic analysis of microplastic contaminants in an urban wastewater treatment plant from Seoul, South Korea. *Chemosphere* 263: 127812.

54 Hidayaturrahman, H. and Lee, T.G. (2019). A study on characteristics of microplastic in wastewater of South Korea: identification, quantification, and fate of microplastics during treatment process. *Marine Pollution Bulletin* 146: 696–702.

55 Bayo, J., López-Castellanos, J., and Olmos, S. (2020). Membrane bioreactor and rapid sand filtration for the removal of microplastics in an urban wastewater treatment plant. *Marine Pollution Bulletin* 156: 111211.

56 Yuan, F., Zhao, H., Sun, H. et al. (2021). Abundance, morphology, and removal efficiency of microplastics in two wastewater treatment plants in Nanjing, China. *Environmental Science and Pollution Research* 28 (8): 9327–9337.

57 Mason, S.A., Garneau, D., Sutton, R. et al. (2016). Microplastic pollution is widely detected in US municipal wastewater treatment plant effluent. *Environmental Pollution* 218: 1045–1054.

58 Long, Z., Pan, Z., Wang, W. et al. (2019). Microplastic abundance, characteristics, and removal in wastewater treatment plants in a coastal city of China. *Water Research* 155: 255–265.

59 Tang, N., Liu, X., and Xing, W. (2020). Microplastics in wastewater treatment plants of Wuhan, Central China: abundance, removal, and potential source in household wastewater. *Science of the Total Environment* 745: 141026.

60 Ziajahromi, S., Neale, P.A., Silveira, I.T. et al. (2021). An audit of microplastic abundance throughout three Australian wastewater treatment plants. *Chemosphere* 263: 128294.

61 Hongprasith, N., Kittimethawong, C., Lertluksanaporn, R. et al. (2020). IR micro-spectroscopic identification of microplastics in municipal wastewater treatment plants. *Environmental Science and Pollution Research* 27 (15): 18557–18564.

62 Park, H.J., Oh, M.J., Kim, P.G. et al. (2020). National reconnaissance survey of microplastics in municipal wastewater treatment plants in Korea. *Environmental Science & Technology* 54 (3): 1503–1512.

63 Li, L., Liu, D., Song, K. et al. (2020). Performance evaluation of MBR in treating microplastics polyvinylchloride contaminated polluted surface water. *Marine Pollution Bulletin* 150: 110724.

64 Wang, Q., Li, Y., Liu, Y. et al. (2022). Effects of microplastics accumulation on performance of membrane bioreactor for wastewater treatment. *Chemosphere* 287: 131968.

65 Maliwan, T., Pungrasmi, W., and Lohwacharin, J. (2021). Effects of microplastic accumulation on floc characteristics and fouling behavior in a membrane bioreactor. *Journal of Hazardous Materials* 411: 124991.

66 Mishra, S., Singh, R.P., Rout, P.K. et al. (2022). Membrane bioreactor (MBR) as an advanced wastewater treatment technology for removal of synthetic microplastics. *Development in Wastewater Treatment Research and Processes* 2022: 45–60.

67 Vuori, L. and Ollikainen, M. (2022). How to remove microplastics in wastewater? A cost-effectiveness analysis. *Ecological Economics* 192: 107246.

68 Xu, Z., Bai, X., and Ye, Z. (2021). Removal and generation of microplastics in wastewater treatment plants: a review. *Journal of Cleaner Production* 291: 125982.

69 Ngo, P.L., Pramanik, B.K., Shah, K. et al. (2019). Pathway, classification and removal efficiency of microplastics in wastewater treatment plants. *Environmental Pollution* 255: 113326.

70 Lehtiniemi, M., Hartikainen, S., Näkki, P. et al. (2018). Size matters more than shape: ingestion of primary and secondary microplastics by small predators. *Food Webs* 17: e00097.

71 Qiao, R., Deng, Y., Zhang, S. et al. (2019). Accumulation of different shapes of microplastics initiates intestinal injury and gut microbiota dysbiosis in the gut of zebrafish. *Chemosphere* 236: 124334.

72 Xia, X., Jia, Z., Liu, T. et al. (2017). Coupled nitrification-denitrification caused by suspended sediment (SPS) in rivers: importance of SPS size and composition. *Environmental Science & Technology* 51 (1): 212–221.

73 Jia, Z., Liu, T., Xia, X. et al. (2016). Effect of particle size and composition of suspended sediment on denitrification in river water. *Science of the Total Environment* 541: 934–940.

74 Zhang, X., Xia, X., Li, H. et al. (2015). Bioavailability of pyrene associated with suspended sediment of different grain sizes to Daphnia magna as investigated by passive dosing devices. *Environmental Science & Technology* 49 (16): 10127–10135.

75 Li, L., Li, Z.Y., Liu, D. et al. (2020). Evaluation of partial nitrification efficiency as a response to cadmium concentration and microplastic polyvinylchloride abundance during landfill leachate treatment. *Chemosphere* 247: 125903.

76 Hamidian, A.H., Ozumchelouei, E.J., Feizi, F. et al. (2021). A review on the characteristics of microplastics in wastewater treatment plants: a source for toxic chemicals. *Journal of Cleaner Production* 295: 126480.

77 McCormick, A., Hoellein, T.J., Mason, S.A. et al. (2014). Microplastic is an abundant and distinct microbial habitat in an urban river. *Environmental Science & Technology* 48 (20): 11863–11871.

78 Nizzetto, L., Futter, M., and Langaas, S. (2016). Are agricultural soils dumps for microplastics of urban origin? 10777–10779.

79 Li, X., Chen, L., Mei, Q. et al. (2018). Microplastics in sewage sludge from the wastewater treatment plants in China. *Water Research* 142: 75–85.

80 Iyare, P.U., Ouki, S.K., and Bond, T. (2020). Microplastics removal in wastewater treatment plants: a critical review. *Environmental Science: Water Research & Technology* 6 (10): 2664–2675.

81 Dris, R., Imhof, H., Sanchez, W. et al. (2015). Beyond the ocean: contamination of freshwater ecosystems with (micro-) plastic particles. *Environmental Chemistry* 12 (5): 539–550.

82 Michielssen, M.R., Michielssen, E.R., Ni, J. et al. (2016). Fate of microplastics and other small anthropogenic litter (SAL) in wastewater treatment plants depends on unit processes employed. *Environmental Science: Water Research & Technology* 2 (6): 1064–1073.

83 Bui, X.T., Nguyen, P.T., Nguyen, V.T. et al. (2020). Microplastics pollution in wastewater: characteristics, occurrence and removal technologies. *Environmental Technology & Innovation* 19: 101013.

84 Michels, J., Stippkugel, A., Lenz, M. et al. (2018). Rapid aggregation of biofilm-covered microplastics with marine biogenic particles. *Proceedings of the Royal Society B* 285 (1885): 20181203.

85 Miao, L., Wang, P., Hou, J. et al. (2019). Distinct community structure and microbial functions of biofilms colonizing microplastics. *Science of the Total Environment* 650: 2395–2402.

86 Oberbeckmann, S., Löder, M.G., and Labrenz, M. (2015). Marine microplastic-associated biofilms–a review. *Environmental Chemistry* 12 (5): 551–562.

87 Tu, C., Chen, T., Zhou, Q. et al. (2020). Biofilm formation and its influences on the properties of microplastics as affected by exposure time and depth in the seawater. *Science of the Total Environment* 734: 139237.

88 Rummel, C.D., Jahnke, A., Gorokhova, E. et al. (2017). Impacts of biofilm formation on the fate and potential effects of microplastic in the aquatic environment. *Environmental Science & Technology Letters* 4 (7): 258–267.

89 Wang, J., Guo, X., and Xue, J. (2021). Biofilm-developed microplastics as vectors of pollutants in aquatic environments. *Environmental Science & Technology* 55 (19): 12780–12790.

90 Ziajahromi, S., Neale, P.A., Rintoul, L. et al. (2017). Wastewater treatment plants as a pathway for microplastics: development of a new approach to sample wastewater-based microplastics. *Water Research* 112: 93–99.

91 Mahon, A.M., O'Connell, B., Healy, M.G. et al. (2017). Microplastics in sewage sludge: effects of treatment. *Environmental Science & Technology* 51 (2): 810–818.

92 Ruan, Y., Zhang, K., Wu, C. et al. (2019). A preliminary screening of HBCD enantiomers transported by microplastics in wastewater treatment plants. *Science of the Total Environment* 674: 171–178.

93 Kelkar, V.P., Rolsky, C.B., Pant, A. et al. (2019). Chemical and physical changes of microplastics during sterilization by chlorination. *Water Research* 163: 114871.

94 Wang, F., Wong, C.S., Chen, D. et al. (2018). Interaction of toxic chemicals with microplastics: a critical review. *Water Research* 139: 208–219.

95 Cai, L., Wang, J., Peng, J. et al. (2018). Observation of the degradation of three types of plastic pellets exposed to UV irradiation in three different environments. *Science of the Total Environment* 628: 740–747.

96 Bayo, J., Olmos, S., and López-Castellanos, J. (2021). Assessment of microplastics in a municipal wastewater treatment plant with tertiary treatment: removal efficiencies and loading per day into the environment. *Water* 13 (10): 1339.

4

Effects of Microplastics on Wastewater Treatment Processes

Yan Laam Cheng[1,#], Tsz Ching Tse[1,#], Ziying Li[1], Yuguang Wang[2], and Yiu Fai Tsang[1,*]

[1] Department of Science and Environmental Studies and State Key Laboratory in Marine Pollution, The Education University of Hong Kong, Tai Po, New Territories, Hong Kong SAR, China
[2] Department of Civil Engineering, University of Nottingham Ningbo China, Ningbo, China
[#] Co-first author
[*] Corresponding author

4.1 Biological Treatment Processes

4.1.1 Conventional Unit Operations and Processes

As introduced in the prior chapter, microplastics have been frequently detected in raw sewage of wastewater treatment plants [1, 2]. Microplastics are expected to pose a considerable challenge to the existing design of wastewater treatment systems, especially biological processes, as a result of their sizes [3, 4]. It has been found that a significant proportion of microplastics can be removed via adsorption to solid surfaces or sludge during the initial stages of primary and secondary treatments [5, 6]. Iyare et al. (2020) suggested that the removal rate of microplastics through the preliminary and primary treatment processes has reached 72% (ranging 32–93%) [7]. Gatidou et al. (2019) have also suggested that relatively high removal efficiency could be achieved during primary treatment (i.e., ca. 72%) compared with that during secondary treatment (i.e., ca. 7–20%) [8]. The presence of tiny microplastics may deteriorate the normal biodegradation of organic pollutants and inhibit the growth of sludge bacteria in the activated sludge process (ASP). The potential adverse effects of microplastics on biological treatment processes include alteration of microbial community structure, inhibition of microbial growth and reproduction, and induction of contaminant adsorption [4, 9]. In this section, the fates of microplastics in biological treatment processes, including the ASP, oxidation ditch, anaerobic-anoxic-aerobic (A^2O) process, sequencing batch reactor (SBR), biological filter, moving bed biofilm reactor (MBBR), and membrane

Microplastics in Urban Water Management, First Edition. Edited by Bing-Jie Ni, Qiuxiang Xu, and Wei Wei.
© 2023 John Wiley & Sons, Inc. Published 2023 by John Wiley & Sons, Inc.

bioreactor (MBR), were examined. The effects of microplastics on the corresponding treatment processes are also discussed.

4.1.1.1 Suspended-Growth Processes

In ASP, microorganisms grow under aerobic conditions in the aeration tank to consume the organic pollutants in wastewater. The microorganisms secrete extracellular polymeric substances (EPS) to adsorb organic matters and nutrients, such as nitrogen and phosphorus. Some contaminants that microorganisms can easily access are also degraded in this process [10]. Microplastics might be prone to settle in ASP sludge because the biological flocs entrap some of them, forming a biofilm on their surfaces [11–14]. Bayo et al. (2020) suggested that the removal rate of microplastics in the shapes of fiber and fragment after ASP was 82.0% and 81.1%, respectively [11]. The biological flocs entrapped microplastics regardless of shape, but significant amounts of microplastics in fiber and fragment shapes were accumulated in ASP sludge. Most of the microplastics observed in secondary sludge appeared in fibers (50–100%), followed by fragments (5.7–25%) [15–18]. Since aeration facilitates microorganism growth, fibrous microplastics might be entrapped by biological flocs or adhered to the surface of biological flocs and eventually settle in the sedimentation tank [12]. A high proportion of fibrous microplastics was found in ASP sludge, and relatively high removal efficiency was achieved in ASP [10].

The effects of microplastics on treatment performance in different conventional and advanced biological wastewater treatment systems are summarized in Table 4.1. The presence of microplastics in wastewater might increase the concentration of emerging organic pollutants, such as bisphenol A (BPA) and phthalates (PAE) and deteriorate the treatment performance in ASP (in terms of organic removal and solid reduction) [3]. The increase in microplastic concentration in the ASP may alter the interaction between EPS and polystyrene (PS) nanoplastics, leading to the inhibition of microbial growth. Microplastics adhering to microbial cells may lead to cell wall pitting and cell malfunctioning [22]. The microbial community structure may be significantly shifted with the exposure to microplastics. Another study found that polypropylene (PP), polyethylene (PE), polyester (PES), PS, and polyvinyl chloride (PVC) microplastics at a concentration of 1000–10 000 particles/L impacted the treatment performance of the system [23]. A high concentration of microplastics posed exhibitory effects on denitrification but slight inhibitory effects on nitrification [23]. However, the exhibitory or inhibitory effects on the treatment performance were highly dependent on polymer types of microplastics. Because of the accumulation of toxic chemicals with a high concentration of PVC microplastics in the tank, denitrification was inhibited, and nitrous oxide emission was inhibited during nitrification [19, 24].

Table 4.1 Effects of microplastics on treatment performance in different conventional and advanced biological treatment systems.

Unit operation and process	MP polymer type: dosage	Parameter(s) studied	Effects of MPs on treatment performance	References
ASP	PE, polyester, PP, PS: 1000–10 000 particles/L	Ammonia oxidation efficiency	Promoted denitrification	[19]
ASP	PVC: 1000–10 000 particles/L	N_2O emission	Inhibited N_2O emission during nitrification	
SBR	PA, PE, PS, PVC: 1 mg/L	Ammonia oxidation efficiency	Promoted EPS secretion	[20]
MBR	PP: 5 g/L	Ammonia oxidation efficiency	Promoted the removal of ammonia nitrogen	[21]
MBR	PP: 2.34–5 g/L	Microbial richness and activities	Promoted relative abundance of microbial community and enzyme activities of microorganisms related to nitrification and denitrification	

Notes: ASP: Activated sludge process; MBR: Membrane bioreactor; MP: Microplastic; N_2O: Nitrous oxide; PA: Polyamide; PE: Polyethylene; PP: Polypropylene; PS: Polystyrene; PVC: Polyvinyl chloride; SBR: Sequencing batch reactor

In other suspended-growth processes, such as oxidation ditch and A^2O process, the main goals are to remove biochemical oxygen demand (BOD) to acceptable concentrations and convert organic matters into stable solids, carbon dioxide, or other stable nontoxic substances [25]. The oxidation ditch has been regarded as a relatively effective unit operation for removing microplastics up to 95.6%, in which most of the microplastics were in fiber form (i.e., 92.3%) [17]. This result might also be due to the adherence of fibrous microplastics on the surface of biological flocs. A relatively low removal of microplastics was found in the A^2O process (i.e., ca. 15%), because microplastics might be accumulated when the sludge is recirculated [17]. Besides, the abundance of fibrous microplastics was increased by 22.1–257.1%, and those with sizes 0.1–1 mm were increased by 115.8% [17, 26]. Both the stirring and aeration in tanks could increase the friction force in the wastewater, causing microplastic fragmentation and thereby an increase in small fibrous microplastic abundance [12].

SBR is another typical suspended-growth process adopted in wastewater treatment plants, and operates as a cycle, including fill, react, settle, decant, and idle periods in a batch reactor. The manipulation of the five periods mentioned allows the removal of biological nutrients [27]. With SBR, more than 99% of the microplastics were removed from the wastewater [28]. Some studies indicated the effects of microplastics on SBR, but no significant effect was observed. Kalčíková et al. (2017) found no effects of microplastics on SBR, while Wang et al. (2021) found that different dosages of polyamide (PA), PE, PS, and PVC microplastics did not affect the treatment performance of SBR [20, 29]. The microplastics at a concentration of 1 mg/L could promote the secretion of EPS regardless of polymer types, and the presence of microplastics did not affect ammonia oxidation performance [20].

4.1.1.2 Attached-Growth Processes

Biological active filter (BAF), rotating biological contactor, submerged aerated filter (SAF), and trickling filter are the typical techniques of attached-growth processes in biological wastewater treatment around the globe. Because of potential microplastic fragmentation in BAF, the abundance of fibrous and fragment microplastics was increased by 4.7% and 210.9%, respectively [14, 30]. In other attached-growth processes, microplastics might be adsorbed on the surfaces of the filters and thereby retained in the sludge. However, not many studies have investigated the fate of microplastics in attached-growth processes. More research is needed to examine the effects of microplastics on the treatment performance of different biological filter systems.

4.1.1.3 Advanced Wastewater Treatment Processes

MBR and MBBR are advanced wastewater treatment technologies that can further remove BOD, suspended solids, and nutrients in wastewater [25]. In the MBR process, organic matters and nutrients in wastewater are removed by ASP, and the employment of membrane effectively reduces the abundance of suspended solids and microbes. In general, microfiltration membranes with 0.1–50 μm and ultrafiltration membranes with 0.001–0.1 μm are applied in the MBR process [10]. With membranes of tiny pore size, microplastics are directly captured in the process [12]. It is also proved that 99.9% of the microplastics in wastewater could be trapped by the membranes, while only a small proportion of fibrous microplastics could pass through the membrane [31]. Lares et al. (2018) indicated that 88.3% and 11.7% of microplastic in MBR sludge were fibers and fragments, respectively [16]. Lv et al. (2019) could not find any fibrous microplastics in MBR sludge, but microplastics in the shapes of fragment (75%) and film (25%) [17]. The variation in shape composition of microplastics in MBR sludge implies that the removal in MBR is a function of size but not shape [12]. Moreover, when an ultrafiltration membrane with a nominal size of 0.2 μm was adopted, complete removal of microplastics can be achieved [32]. MBR has also been regarded as the most effective technique for microplastic removal when the membrane pore size is smaller than 0.1 μm [17]. In the MBBR process, microorganisms grow on carriers moving freely in the reactor. Those carriers are kept inside the reactor by a sieve installed at the outlet of the reactor [33]. However, the fates of microplastics in MBBR have not been clearly described, urging more research on this topic.

The research on the effects of microplastics in advanced wastewater treatment processes is limited. One MBR served as a control without the addition of microplastics, while another two MBRs were dosed with 0.3 and 5.0 g/L of PP microplastics, respectively [21]. The removal efficiency of ammonia nitrogen in the MBR with a high dosage was higher than the control and that with a low dosage [21]. The presence of microplastics significantly affected the treatment performance of biological wastewater treatment and the efficiency of nitrification and denitrification. The effects of microplastics on the MBR process were studied, including floc sizes, surface properties, and chemical compositions [34]. The results indicated that the abundance of microplastics accumulated in the MBRs at a concentration of 7, 15, and 75 particles/L of feed per day were determined to be 18, 52, and 359 particles/g, respectively [34]. Results also suggested that the performances of MBRs can be maintained at acceptable levels with the presence of microplastics. The removal efficiency of Total Kjeldahl Nitrogen in all four MBRs was above 99% through the nitrification process. When the microplastic accumulation increased from 0 to 359 particles/g, the floc size and the relative hydrophobicity were decreased by 33.0% and 21.2%, respectively. The decrease of flocs' hydrophobicity is beneficial in reducing the reactor fouling propensity caused by bacteria adherence to the membrane surface. With the increase in microplastic accumulation, the relative abundance of Actinobacteria decreased significantly from 53.5 to 39.3%, Planctomycetes decreased from 14.1 to 9.4%, while Proteobacteria increased from 17.7 to 22.1%, and Verrucomicrobia increased from 1.4 to 4.0%. The microbial diversity in reactors with a high dosage of microplastics was significantly different from the control. Therefore, the accumulation of microplastics considerably affects the microbial structure and diversity within sludge flocs.

4.2 Interactions Between Sludge and Microplastics

Microplastics in different shapes may have different impacts on sludge bacteria. Fibrous microplastics may induce adverse effects through adherence and the pitting of cell membranes. However, microplastics in fragment form may enter the cell to exhibit an inhibitory effect on microbial growth. EPS consisting of proteins, polysaccharides, humic acid, and lipids can affect sludge aggregation in biological treatment systems [35]. The presence of microplastics may promote sludge aggregation, but the leached additives may eventually deteriorate the treatment performance. Although microplastics may facilitate the functional groups in EPS, such as carbonyl and amide groups, to enhance sludge aggregation, the changes in the proteins-to-polysaccharides ratio in EPS may deteriorate bioflocculation [3, 36]. The leached additives from microplastics ruptured the connection between EPS and cell wall and further affected the sludge solubilization [24]. However, short exposure to PVC microplastics promoted EPS secretion and further enhanced phosphate removal in an aerobic bioreactor [37]. EPS could function as a shield to prevent microplastics from penetrating the cells by covering the surface of microplastics and protecting the cells against antimicrobial contaminants [36, 38]. The cell envelope of microorganisms is a complex, multilayered structure, which comprises the outer membrane, peptidoglycan cell wall, and cytoplasmic membrane. Fu et al. (2018) found that high microplastic abundance during anaerobic digestion caused a reduction in hydrogen and methane generation and delayed achieving 90% cumulative biogas production [22]. When microplastics adhere to microbial cells, cell wall pitting, and cell malfunctioning may occur. The adsorbed antimicrobial contaminants on the surface of microplastics may damage the organic materials of the cell envelope, causing cell inactivation and cell lysis [39]. In this section, the interactions between sludge and microplastics will be discussed. The effects of microplastics on different types of sludge properties are summarized in Table 4.2.

Table 4.2 Effects of MPs on the properties of different types of sludge.

Type of sludge	MP polymer type: dosage	Parameter(s) studied	Effects of MPs on sludge properties	References
Activated sludge	PA, PVC: 1 mg/L	Ammonia nitrogen removal efficiency	Inhibited nitrification	[40]
		Microbial morphology	Altered microbial morphology	
	PE and PS: 0.24 mg/mL	Antibiotic-resistant bacteria and pathogens	Facilitated the proliferation of ARB and dissemination of antibiotic-resistant genes via horizontal gene transfer	[41]
	PVC: 10, 20, 40, and 60 particles/g-TS	Methane production	Significant increased methane production but inhibited methane production at high concentrations	[42]

Table 4.2 (Continued)

Type of sludge	MP polymer type: dosage	Parameter(s) studied	Effects of MPs on sludge properties	References
Aerobic granular sludge	PA66: 0.1 g/L	Microbial diversity	Enhanced microbial richness, reduced microbial diversity	[43]
	PA66: 0.1 g/L, 0.2 g/L, 0.5 g/L	COD removal efficiency	Promoted the removal of COD	
		Ammonia nitrogen removal efficiency	Slightly promoted the removal of ammonia	
	Polyester: 0, 0.1, 0.2, and 0.5 g/L	Nitrifying process, microbial community structure	Accumulation of nitrite nitrogen and affect the process of nitrogen metabolism	[44]
	PVC: 30 mg/L	Nitrite oxidizing bacteria activity	Promoted nitrite oxidizing bacteria activity	[37]
	PVC: 50 mg/L	EPS abundance	Enhanced EPS abundance	
	PA, PS, PVC: 10 mg/L	AGS morphology	Damaged AGS surface	[20]
	PE: 10 mg/L	EPS abundance	Unaffected EPS abundance	
Anaerobic granular sludge	PET: 75–300 particles/L	COD removal efficiency, methane production, SCFA abundance	Inhibited the removal of COD and production of methane, promoted the accumulation of SCFA	[35]
	PET: 300 particles/L	EPS abundance	Reduced EPS abundance	
	PET: 15 particles/L	Contents of proteins and DNA in EPS	Increased abundance of polysaccharides, lipids, and humic acids	

Notes: AGS: Aerobic granular sludge; COD: Chemical oxygen demand; EPS: Extracellular polymeric substances; PA: Polyamide; PE: Polyethylene; PET: Polyethylene terephthalate; PP: Polypropylene; PS: Polystyrene; PVC: Polyvinyl chloride; SCFA: Short chain fatty acids

4.2.1 Activated Sludge

Activated sludge (AS) in wastewater treatment processes is abundant in microorganisms, including bacteria, protozoa, and fungi, which remove organic matters in wastewater. Nitrification and denitrification are the key functions of AS with the support of ammonia-oxidizing archaea, ammonia-oxidizing bacteria, and nitrite-oxidizing bacteria [40]. The nitrification of AS was inhibited with the exposure of PA and PVC microplastics in the concentration of 1 mg/L for 14 days [40]. As suggested, the densities of PA and PVC microplastics are relatively higher than the wastewater inside the reactor, which corresponded to settling in AS. The accumulation of these microplastics increased their concentration in the tanks, providing an opportunity to contact nitrifying sludge [40]. In the same experiment, the microbial morphology of sludge changed after adding PA and PVC microplastics at 1 mg/L concentration. The shape of the nitrifying sludge is round, which is believed to be related to the friction with PA and PVC microplastics [40].

The presence of PVC microplastics of 10 particles/g can promote methane production, while higher concentrations of PVC microplastics (i.e., 20, 40, and 60 particles/g) inhibit methane production [24]. Mechanistic studies showed that BPA leaching from PVC microplastics was the primary reason for the decreased methane production, causing significant inhibitory effects on the hydrolysis-acidification process. The long-term effects of PVC microplastics revealed that the microbial community was shifted in the direction against hydrolysis-acidification and methanation [24]. Moreover, PE and PS microplastics in 6 mg/25 mL could selectively promote antibiotic-resistant bacteria and pathogens [41]. High cell density can serve as a protected breeding ground facilitating the proliferation of antibiotic-resistant bacteria and dissemination of antibiotic-resistant genes via horizontal gene transfer in the long term once they enter the environment [41].

4.2.2 Aerobic Granular Sludge

Aerobic granular sludge (AGS) is broadly applied to treat various types of wastewater [45]. The main components of AGS are microorganisms and the EPS secreted by microorganisms [10]. It is usually applied in the treatment of wastewater containing a high concentration of organic matter, as the organic removal efficiency is high in AGS [10]. The high-density structure of the microbial community contributes to AGS's ability to treat high-strength organic wastewater [46]. It is also regarded as a wastewater bio-technique with assurance [20]. AGS is composed of numerous microorganisms that form microbial aggregates, which are basically of a large particle size, which is why the depth of diffusion of some oxygen molecules is limited and microbial aggregates favor the formation of anoxic/anaerobic zones in the granular sludge [43]. The numerous pores on the surface of AGS benefit the adsorption and degradation of organic matter [47, 48]. Moreover, AGS could provide wastewater with a suitable environment for removing nutrients, such as nitrogen and phosphorous [49].

PA66 microplastics in 0.1 g/L concentration can promote the microbial richness of AGS, while the microbial diversity was reduced [43]. The same study showed that when PA66 microplastics at concentrations of 0.1, 0.2, and 0.5 g/L were dosed in AGS, the removal efficiency of chemical oxygen demand (COD) was enhanced. After 20 days of microbial adaptation in the system, the removal efficiency of ammonia nitrogen was also enhanced slightly [43]. Microplastics' effects on AGS depend significantly on the polymer type and concentration. PVC microplastics promoted nitrite-oxidizing bacteria and promoted EPS secretion by microorganisms [37]. When PVC microplastics were dosed at a concentration of 30 mg/L, ammonia and nitrite were not detected in the effluent of the system, but nitrate was detected. These results imply that PVC microplastics at a concentration of 30 mg/L promote the growth of nitrite-oxidizing bacteria [37]. When PVC microplastics at a concentration of 50 mg/L were dosed, the abundance of EPS increased. Microorganisms usually secrete EPS to protect themselves against toxic substances. In the study, the average diameter of PVC microplastics was 10 times smaller than that of AGS, which is favorable for the penetration of PVC microplastics into AGS, stimulating the secretion of EPS [37].

PA, PS, and PVC microplastics at a concentration of 10 mg/L inhibited the nitrification of AGS [20]. Reactive oxygen species (ROS) are induced to protect AGS against the intracellular oxidative stress caused by polymer types of microplastics. However, this induction causes inhibitory effects on the nitrification of AGS [20]. The presence of PE microplastics had no significant effects on EPS secretion [20], indicating toxicity of PE microplastics to AGS might be lower than that of other polymer types. The morphology of AGS was observed in advance, and the findings show that PS, PS, and PVC microplastics deteriorated the surface of AGS. Damages and cracks were observed on the AGS surface, thereby proving the deterioration effects on settlement performance and thus the bulking of AGS [20]. Besides, adding 0.5 g/L PES microplastics could inhibit the specific nitrate reduction [44]. The increase of nitrate reductase will further reduce nitrate nitrogen to nitrite nitrogen. Therefore, the increase of nitrite nitrogen cannot be converted timely, which will lead to the accumulation of nitrite nitrogen and affect the process of nitrogen metabolism [44].

4.2.3 Anaerobic Granular Sludge

Reactors containing anaerobic granular sludge (AnGS) have been widely used to treat wastewater with high concentrations of organic matter. AnGS has a more complicated hierarchical structure than AGS, including acidogens, syntrophic microcolonies, and aceticlastic methanogens [23]. As in AGS, EPS are essential components in AnGS, thus endowing AnGS with high tolerance to toxic substances [50]. Moreover, EPS are essential in maintaining granules' structure [51]. The wastewater treatment capacity for organic matter and microbial concentration could also be enhanced in AnGS [52]. Research on the effects of polyethylene terephthalate (PET) microplastics on AnGS found that PET microplastics at a concentration of 15, 75, 150, and 300 particles/L were dosed in a separate AnGS system. PET microplastics at a concentration of 75–300 particles/L have inhibitory effects on AnGS as well as reduced COD removal efficiency and methane production. Moreover, more short-chain fatty acids (SCFA) accumulated in the system [35].

The sizes of sludge decreased with the exposure to PET microplastics [35]. When PET microplastics were exposed to AnGS for a long time, the abundance of EPS, especially center proteins, decreased remarkably, posing adverse impacts on the structure of granules [53]. When AnGS exposed to lower PET microplastics (i.e., 15 particles/L), the performance of AnGS was not affected. Instead, more polysaccharides, lipids, and humic acids were observed in EPS, implying that PET microplastics dose would not pose inhibitory effects on AnGS at low concentrations but enhances EPS secretion to protect microorganisms against toxic substances [53].

Another study on the effects of PES microplastics on AnGS suggested that the COD removal efficiency would be decreased when PES appeared at 0.5 g/L. The interaction between PES microplastics and BPA caused the inhibition of the anaerobic granular sludge. BPA inhibited the hydrolysis-acidification process, high level of BPA concentration was toxic for the bacterial activity and resulted in low COD removal efficiency. The soft ether chain in PES makes it flexible and relatively hydrophilic; therefore, it would adsorb various pollutants, including BPA, and lower the COD removal efficiency [54].

4.3 Effects of Microplastics on Microorganisms and Key Enzymes

4.3.1 Heterotrophic Bacteria

Heterotrophic bacteria utilize organic compounds as sources of carbon and energy for synthesis. Microbes utilize the organic carbon from microplastics as their source through the colonization on the surface of microplastics. During early colonization, biofilm formation is the approach taken by microbes for survival in unfavorable conditions, and it is initiated by planktonic bacteria through cohesion with each other [55]. After the physical forces are applied for a while, the attachment becomes immobilized and permanently adsorbed on the surface, in which microcolonies formed, and EPS produced. Martínez-Campos et al. (2021) identified phyla Proteobacteria, including 24.2% of Betaproteobacteria, 21.4% of Alphaproteobacteria, and 12.0% of Gammaproteobacteria. Bacteirodetes and Actinobacteria are the dominant heterotrophic bacteria on the surface of microplastics in the form of biofilm [55]. The predominance of Alphaproteobacteria confirmed the colonization ability among other bacteria in the initial stage. Meanwhile, Actinobacteria contributed to the processing of organic matter, and Bacteirodetes functioned as biofilm conditioning through the degradation of dissolved organic matter [55]. The possible inhibition mechanisms of microplastics on the nitrifying process, methane production, and microplastic-promoted secretion of EPS in activated sludge, aerobic granular sludge, and anaerobic granular sludge are shown in Figure 4.1.

4.3.2 Ammonia-Oxidizing Bacteria

Ammonia-oxidizing bacteria is one of the major bacteria groups assisting in nitrification and denitrification. Microplastics, in terms of abundance, size, and polymer type, inhibit ammonia-oxidizing bacteria that enhance or deteriorate the performance of nitrification and denitrification. Cui et al. (2021) indicated that a low abundance of PA microplastics (i.e., 50–100 mg/L) completely oxidized ammonia with a lower rate. Therefore, the influence on ammonia-oxidizing bacteria was negligible [56]. However, when the PA microplastic dosage reached 200 mg/L, the activity of ammonia-oxidizing bacteria was inhibited, and thus incomplete oxidation was found with the residual ammonia at 7.6 mg/L [56]. PP and PVC slightly enhanced ammonia oxidation rate under the dosage of 1000 particles/L of microplastics (Table 4.3). However, the performance deteriorated with the increase in the dosage to 10 000 particles/L [19]. Although the presence of PE, PS, and polyester inhibited the performance of ammonia oxidation, the increase in dosage from 1000 to 10 000 particles/L slightly eliminated the adverse effects on the performance (Table 4.3) [56]. Song et al. found that the addition of PVC microplastics heavily suppressed the activity of ammonia-oxidizing bacteria. The baseline ammonia oxidation rate decreased by 28.2 to 50.0% with the increase in dosage (i.e., 1000–10 000 particles/L of PVC microplastics) [57]. The toxic compounds released from PVC microplastics are attributed to ammonia-oxidizing bacteria inhibition.

The abundance and polymer types of microplastics inhibited the performance of nitrification by altering the nitrification genes. With a fixed dosage of 1 mg/L of microplastics, the abundance of ammonia-oxidizing bacteria amoA was reduced significantly by the presence of PA (49.7%), followed by PE (45.7%), PS (24.4%), and PVC (17.9%) [20]. The degree of inhibition was significantly increased with the dosed microplastics. The abundance of ammonia-oxidizing bacteria amoA

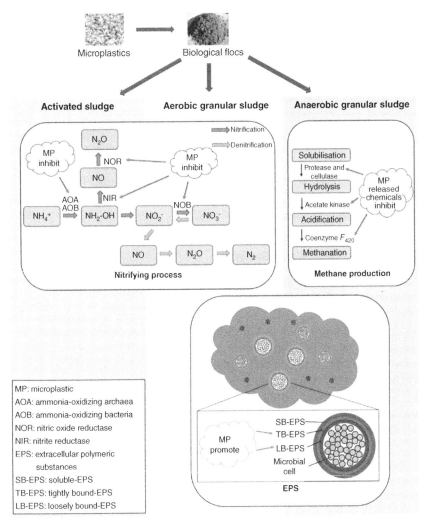

Figure 4.1 Inhibition mechanisms of microplastics on nitrifying process (aerobic condition) and methane production (anaerobic condition) and microplastic-promoted secretion of extracellular polymeric substances in activated sludge, aerobic granular sludge, and anaerobic granular sludge.

further declined by 61.2%, 57.5%, 51.7%, and 44.9% with the addition of PA, PE, PS, and PVC microplastics at 10 mg/L, respectively [20]. Under the addition of 100 mg/L of PA, PE, PS, and PVC microplastics, the abundance decreased by 92.1%, 47.3%, 49.2%, and 52.6%, correspondingly [20]. Winkler et al. (2012) indicated that the location of ammonia-oxidizing bacteria was closer to the outside of aerobic granular sludge. It might be more affected by microplastics when they adhered to

Table 4.3 Effects of microplastics on ammonia oxidation, nitrate production, and denitrification rate in biological wastewater treatment.

Polymer types	Dosage (particles/L)	Effects on ammonia oxidation (Compared with blank)	Effects on nitrate production (Compared with blank)	Effects on denitrification (Compared with blank)	References
PA	10	−8.4%	/	/	[56]
	50	−28.1%	/	/	
	100	−44.7%	/	/	
	200	−56.5%	/	/	
PE	1000	−3.7%	−3.2%	−5.4%	[19]
	5000	−0.9%	−4.5%	−21.4%	
	10 000	0%	−7.8%	−14.3%	
Polyester	1000	−7.3%	−14.3%	−25.0%	
	5000	+0.9%	−3.9%	+21.4%	
	10 000	+0.9%	0%	+3.6%	
PP	1000	+5.5%	−1.9%	−21.4%	
	5000	+1.8%	−11.7%	−14.3%	
	10 000	−0.9%	−3.9%	+7.1%	
PS	1000	−6.4%	−7.8%	−17.9%	
	5000	−4.6%	−7.7%	+3.6%	
	10 000	−4.6%	−11.0%	−1.8%	
PVC	1000	+0.9%	−6.5%	+3.6%	
	5000	−1.8%	−6.5%	+21.4%	
	10 000	−1.6%	−10.4%	−7.1%	
PVC	1000	−28.2%	0%	/	[57]
	5000	−43.6%	−25.0%	/	
	10 000	−50.0%	−40.0%	/	

Notes: PA: Polyamide; PE: Polyethylene; PP: Polypropylene; PS: Polystyrene; PVC: Polyvinyl chloride

the surface of aerobic granular sludge [58]. Despite the reduction of the abundance of ammonia-oxidizing bacteria in the early stage, they remained at a high level of activity in terms of the completion of ammonia oxidation. At the same time, after a period of acclimation, the microorganisms resumed their growth, and the abundance of ammonia-oxidizing bacteria showed an increasing trend from adaption to the presence of PA microplastics at the concentration of 50 mg/L [56].

4.3.3 Nitrite-Oxidizing Bacteria

Like ammonia-oxidizing bacteria, nitrite-oxidizing bacteria is another bacteria group leading to nitrification and denitrification. The influence of PA microplastics on the activity of nitrite-oxidizing bacteria was less significant than that of ammonia-oxidizing bacteria, in which the specific nitrite oxidation rate remained below 2.0 mg N/g MLSS·h [56]. PE, polyester, PP, PS, and PVC microplastics inhibited the nitrate production rate (Table 4.3). The nitrite-oxidizing activity during nitrification was inhibited by microplastics regardless of the abundance of microplastics [19]. The inhibition of nitrite-oxidizing bacteria was eliminated with 1000 particles/L of PVC microplastics because the growth rate of ammonia-oxidizing bacteria was 2.6 times faster than that of nitrite-oxidizing bacteria, and ammonia-oxidizing bacteria were the dominant strains in the partial nitrification process [56]. However, with the increase in dosage of PVC microplastics to 10 000 particles/L, the inhibition of nitrite-oxidizing bacteria occurred, and the nitrite production reduced 25% to 40% of control [57].

Like ammonia-oxidizing bacteria, the shift of nitrification genes was influenced by the polymer types and the abundance of microplastics. With the fixed dosage of microplastics at 1 mg/L, the abundance of nitrite-oxidizing bacteria (Nitrospira) increased significantly with the addition of PS (127.2%), followed by PE (84.2%) and PA (49.1%) [20]. The presence of PVC slightly impacted the increase in abundance of nitrite-oxidizing bacteria by 4.7%. The toxic compounds released from PVC microplastics might inhibit the growth of nitrite-oxidizing bacteria [20]. When the PS, PE, PA, and PVC microplastic dosages reached 100 mg/L, the abundance of nitrite-oxidizing bacteria grew by 286.7%, 397.5%, 187.3%, and 60.1%, respectively [20]. These results indicated that the abundance of nitrite-oxidizing bacteria was negatively correlated with ammonia-oxidizing bacteria when the dosed microplastics inhibited the growth of ammonia-oxidizing bacteria but favored the abundance of nitrite-oxidizing bacteria. Interestingly, PA microplastics did not influence the abundance of nitrite-oxidizing bacteria (Nitrospira). However, they posed slight impacts on the reduction of the abundance of the total bacteria, in which the effects on partial nitrification were mainly from alteration of ammonia-oxidizing bacteria. The presence of microplastics might not threaten to the abundance and activity of nitrite-oxidizing bacteria [56].

4.3.4 Key Enzymes

To date, little research has been done to examine the impacts of microplastics on the abundance of key enzymes and enzymatic activity during biological wastewater treatment. The presence of microplastics might reduce or inactivate the enzymatic activity, affecting the performance of nitrification and denitrification.

The adherence of microplastics on the cell membrane surface increased the contact between microplastics and the key enzymes and reduced protective barriers of EPS [35]. Meanwhile, tiny microplastics might be more prone to penetrate the gaps between the biopolymer chains. This penetration could induce the risk of contacts among microplastics, proteins, and phospholipids such that cell-associated toxicity might occur, eliminating the enzymatic activities [59].

Another possible reason for enzyme reduction might be the induction of oxidative stress at the cellular level because of the toxicity of leached chemicals from microplastics. Microplastics might generate ROS in cells, causing cell distortion, lipid peroxidation, and damage to cell membrane structure and skeleton. This damage hinders the energy exchange outside and inside the cell and eliminates the electron transfer. When the concentration of ROS exceeded the ability of cell repairment, the cell eventually died as the number of bacterial enzyme molecules was limited, and the cell could not remove the excessive ROS [10]. Although the detailed mechanisms behind the impacts of microplastics on enzymatic activity are limited, microplastics might physically block the transfer of electrons and energy inside and outside the cell or induce ROS generation, and thus adversely affect enzymatic activity.

The characteristics and components of EPS are crucial factors affecting microbial aggregation in biological treatment [60]. EPS are essential to protect the microbial community again emerging organic contaminants and maintain the structure of sludge flocs [35, 61]. However, microplastics and emerging organic contaminants caused inhibitory/synergistic effects on the secretion of EPS in biological treatment [35, 37, 43]. The exposure time of a particular polymer type and the dosage of microplastics can have different impacts on EPS [35, 37, 43, 52]. The short exposure of PVC microplastics can promote the secretion of EPS and enhance phosphate removal in wastewater [37]. However, PVC microplastics eventually break down the structure of sludge flocs and accelerate cell death in the long run [52]. On the other hand, exposure to different polymer types might have opposite effects on microorganisms. Short exposure to PA66 microplastics caused damage to microorganisms in sludge and deteriorated cell growth. In a long run, the microorganisms in sludge adapted to the presence of PA66 microplastics, and PA66 microplastics made the sludge produce more signaling molecules, thereby promoting the secretion of EPS [43]. Meanwhile, exposure to different microplastic dosages can have different effects on EPS. A low level of PET microplastics enhanced EPS secretion, whereas EPS generation was suppressed with exposure to high PET microplastic dosage [35].

On the other hand, microplastics might protect enzymes against the toxic environment through a protective barrier from loosely bound EPS and the generation of extra protein from tightly bound EPS. The tryptophan-like proteins in loosely bound EPS might eliminate the contact between the microbes in anaerobic granular sludge and the toxicity released from polyether sulfone microplastics and BPA [53]. Meanwhile, the introduction of microplastics and BPA might stimulate the production of tightly bound EPS in terms of the protein content's increase, thus protecting against the toxic leachate from microplastics [53].

4.4 Effects on Sludge Stabilization and Dewatering

Research on the fate of microplastics in sludge is still in its early stages. Despite the lack of comprehensive studies on microplastics from raw sludge to dewatered sludge, there is no unit operation and process in sludge treatment that has been designed to facilitate microplastic degradation [12]. Indeed, the existence of microplastic in sludge is acknowledged to cause inhibitory and synergistic effects on the performance of sludge treatment (Table 4.4). The potential adverse effects of microplastics include inhibition of microbial growth, desorption of emerging organic pollutants, and leaching of plastic additives [59, 63].

4.4.1 Aerobic Digestion

To date, little recent research has been done to examine the impacts of microplastics on aerobic digestion. Wei et al. (2021) implied that PET microplastics posed little threat to sludge solubilization, but they seriously inhibited hydrolysis [64]. Therefore, the performance of aerobic digestion was adversely affected. The addition of PET microplastics significantly deteriorated the degradation of waste activated sludge from $29.0 \pm 0.5\%$ to $21.2 \pm 0.5\%$. The fraction in waste activated sludge increased and the stability of aerobic digestion reduced [64]. The authors explained that PET microplastics might improve the solubilization of generated waste activated sludge but hinder the degradation of soluble organics [64]. Because of the induced oxidative stress and the leachate of toxic chemicals from PET microplastics, the populations of key microorganisms were reduced, and thus the performance was inhibited [64].

In contrast, Zhang et al. (2021) found no significant impacts ($p > 0.05$) on volatile solid degradation during aerobic digestion with the addition of different polymer types of microplastics [65]. The rates of volatile solid degradation were $43.8 \pm 4.2\%$, $43.6 \pm 3.3\%$, and $44.3 \pm 4.1\%$ with the presence of PVC, PE, and PET microplastics, respectively, when the degradation rate in control was $46.4 \pm 3.9\%$ [65]. However, the presence of microplastics, regardless of the polymer types, adversely affected the removal efficiency of antibiotic resistance genes in sludge during aerobic digestion. The total absolute abundance of antibiotic resistance genes significantly increased due to the presence of PET microplastics (129.6%),

Table 4.4 Effects of MPs on methane production and model-based analysis in terms of biochemical methane potential and hydrolysis rate under different dosages of different polymer types in anaerobic digestion.

Polymer types	Dosage (particles/g)	Effects on methane production	Biochemical methane potential (B_0)	Hydrolysis rate (k)	References
PA6	5	+4.8%	+	No effect	[38]
	10	+39.5%	+	No effect	
	20	+16.1%	+	No effect	
	50	+12.9%	+	No effect	
PE	10	0%	No effect	No effect	[62]
	30	0%	No effect	No effect	
	60	0%	No effect	No effect	
	100	−12.4%	−	−	
	200	−27.5%	−	−	
Polyester	1–6	−9.−11.5%	−	−	[23]
	10	−4.9%	+	+	
	30	−9.7%	−	−	
	60	−6.8%	−	−	
	100	−7.1%	−	−	
	200	−7.3%	−	−	
PVC	10	+5.9%	+	No effect	[24]
	20	−9.4%	−	−	
	40	−19.5%	−	−	
	60	−24.2%	−	−	

Notes: PA: Polyamide; PE: Polyethylene; PP: Polypropylene; PS: Polystyrene; PVC: Polyvinyl chloride

followed by PE microplastics (137.0%), and PVC microplastics (129.6%) [65]. Based on the shift of microbial structure, the authors concluded that the presence of microplastics might induce the alteration in the structure of the microbial community, the increase in the abundance of potential microbial hosts, and the possible enhancement of horizontal transfer of antibiotic resistance genes [65]. These imply that the presence of microplastics in different polymer types might require further investigation of the impacts on the performance of aerobic digestion.

The research on the potentials of microplastics in nitrous oxide emission and mitigation is rarely investigated. Nitrous oxide is a strong greenhouse gas with a global warming potential of 265 CO_2-equivalents [66]. The concentration of nitrous oxide is influenced by nitrite and ammonia in hydroxylamine oxidation and the abundance of ammonia-oxidizing bacteria [19, 56, 67]. In addition, microplastics alter the biophysical environment via the absorption of organic nitrogen and the structure of sludge (e.g., water content and porosity), thus affecting the availability of oxygen and inducing incomplete denitrification [68]. Microplastics inhibited the anaerobic microenvironment during denitrification and promoted the activity of ammonia-oxidizing bacteria, thereby increasing the production of nitrous oxide [19, 49]. Nevertheless, low emission of nitrous oxide has been found with the presence of microplastics under anoxic conditions [56].

4.4.2 Dewatering

To date, little research has been done to investigate the potential impacts of microplastics on sludge dewaterability. Xu et al. (2021) conducted the test of capillary suction time through static filtration [69]. The results indicated the significant influence of microplastics on sludge dewaterability. A negative correlation was found between the abundance of small PS microplastics and the capillary suction time. The increase in the concentrations of small PS microplastics from 0 to 10 mg/L caused a significant reduction in capillary suction time by 17.0% [69]. This implies that low abundance of tiny PS microplastics facilitated sludge dewaterability. However, when the abundance of tiny PS microplastics reached 100 mg/L, the sludge dewaterability was hindered shown by the rise in capillary suction time. The high concentration of microplastics might inhibit the activity of microorganisms and reduce the abundance of key microorganisms, thus altering the composition and distribution of EPS. Further research needs to be done to clarify the possible mechanisms.

Meanwhile, the increase in the abundance of large PS microplastics resulted in a steady increase in the capillary suction time by 29.9% from 0 to 300 mg/L [69]. Since the large microplastics inhibited sludge dewaterability, the size of microplastics might play a vital role in sludge dewatering. The possible mechanism might be the physical crushing of sludge flocs, deteriorating the ability of sludge dewatering [69]. Furthermore, considering the differences in polymer types of the same size, PE microplastics deteriorated sludge dewaterability by approximately 30.0%, similar to PS microplastics. In contrast, PVC microplastics exhibited adverse effects on sludge dewaterability by 47.7% [69]. Similar to previous discussions, PVC might release toxic compounds, which inhibit the structure of the microbial community. The mechanisms require further investigation on the correlations between the properties of microplastics and sludge dewaterability.

4.5 Perspectives

The major function of biological treatment is to reduce suspended solids and residual organics in primary effluent, and thus microplastics are not the targeted impurities to be removed in the secondary treatment. The pathways of microplastics in secondary treatment are not fully explored, especially the contribution of secondary sedimentation and the redistribution from return flow. Despite the relatively low concentration of microplastics in MBR effluent, the availability of membrane plays a major role in trapping various shapes and sizes of microplastics, instead of the interactions with microorganisms in activated sludge. Meanwhile, the role of secondary sedimentation on microplastic removal is neglected. Most research has determined the removal efficiency of microplastics in secondary treatment by the differences in microplastic abundance between primary effluent and secondary effluent. Secondary sedimentation might enhance the removal by settling the biofilm-formed microplastics or floc-entrapped microplastics, in which microplastics might transfer from wastewater to the sludge stream. The changes in microplastic abundance before and after biological treatment might establish the foundation of microbial responses to the presence of microplastics. Besides, the differences in microplastic abundance before and after secondary sedimentation might illuminate the pathways including retention in secondary effluent or transferring to sludge treatment from floc formation and agglomeration. Furthermore, little research has considered the redistribution of microplastics from the return flow of activated sludge. The actual removal efficiency of microplastics should be reflected by determining the mass balance of microplastics within the biological treatment. The contribution of microplastics from the return flow of activated sludge might miscalculate the actual removal performance of biological treatment. Determining the mass balance of microplastics in biological treatment and investigating the occurrence of microplastics in each unit operation and process of biological treatment should be included in future investigations.

It is still the early phase of laboratory-scale research on microplastic removal by different biological treatments. Either the corresponding effects of microplastics on treatment performance or their removal mechanisms in different unit operations and processes have remained unanswered. The unique design of each unit operation and process in biological treatment contributes to different responses to the presence of microplastics. In the suspended-growth biological treatment, microplastics might have indirect interactions with microorganisms by way of floc formation and agglomeration. Yet, the operating conditions and tank setting of biological treatment are seldom considered in laboratory-scale research. The scale of biofilm formation on microplastics or floc agglomeration with microplastics might depend on the retention time, the flow direction in the tank, and the availability of oxygen. The removal mechanisms or the pathways of microplastics might be explained through understanding the interactions between the activity of microorganisms and microplastics.

On the other hand, in the attached-growth biological treatment, microplastics trended to have direct interactions with microorganisms, which served as the media for their growth. The polymer types and shapes of microplastic might determine the scale of biofilm formation. In general, microplastics with an aromatic group and rougher surfaces are more favorable for the attachment of microorganisms. In the initial stage, these properties determined the attachment of microorganisms on the surface, and microorganisms secreted EPS to further facilitate the attachment. Finally, the proliferation of microorganisms is formed on the surface of microplastics, in which the surface properties of microplastics are changed. The hydrophobicity, specific surface area, density, and functional groups of microplastics were changed according to the extent of biofilm formation. Meanwhile, the leached additives from a particular polymer type of microplastics might have inhibitory effects on the microbial growth on their surfaces. Different polymer types of microplastics contain different additives. The concentrations of leached additives from each polymer type should be evaluated to determine the levels of inhibitory effects on microorganisms. Importantly, the dosage of leached additives should be determined based on the actual concentrations of each polymer type of microplastics in primary effluent, instead of a known-effect dosage. Rather than the inhibitory effects of microplastics on treatment performance, these characteristics should be included to evaluate the roles of microplastics on nitrification and denitrification in attached-growth biological treatment.

In general, most studies have shown insignificant effects of microplastics on nitrification and denitrification, because the activity of corresponding microorganisms and key enzymes are not inhibited by the presence of microplastics. The spherical or powder form of microplastics did not adversely affect the microbial growth and its abundance and diversity, even in high dosage. Considering various shapes, polymer types, and sizes, the complexity of microplastics might inhibit the actual performance of diverse biological treatments and thereby the real removal efficiency. Each characteristic of microplastics might affect the floc formation differently. For instance, PVC microplastics had greater adverse effects on microorganisms and inhibited the treatment performance more severely, while other polymer types of microplastics had no obvious impacts on both microorganism activity and treatment performance. However, the presence of PVC microplastics in primary effluent was in low concentration, in which the actual response of microorganisms to the presence of microplastics was not truly explored. The microbial response to the presence of a variety of microplastics might be different from the response to a single type of microplastics. Apart from polymer type compositions, different shapes and sizes of microplastics might have different pathways affecting the activity of microorganisms. In general, tiny fibrous microplastics are easier to pass through the cell membrane than other shapes or larger sizes of microplastics. Thus, they might have greater inhibitory effects on microorganisms, rather than the adherence to the microbial cells. Rather than considering the single polymer type in sphere or powder in the dose-response experiment, the mixtures of different polymer types, shapes, and sizes in either virgin or aged microplastics should be included in the final stage of the dose-response experiment. In this way, the influence on the treatment performance and functional microorganisms could be truly reflected through investigating the complexity of microplastic existence.

4.6 Conclusion

The fates and impacts of microplastics in different biological treatment processes were examined. Microplastics' size and polymer type might influence the removal mechanisms and treatment performance in wastewater treatment. In the suspended-growth process, microplastics might be entrapped by biological flocs or adhere to the surface of biological flocs. In the attached-growth process, they might be adsorbed on the surface of filters and act as growth media for microorganisms. In advanced wastewater treatment processes, microplastics might be more prone to direct screen by membrane filter, where the removal efficiency is relatively higher. Yet, the dynamic between different microplastic characteristics (i.e., size distribution, shape composition, and polymer type composition) and the formation of biological flocs should be further investigated in laboratory-scale wastewater treatment systems. The inhibitory effects of microplastics on treatment performance could be eliminated, and the syngenetic effects of microplastics on microorganism activity could be enhanced.

In general, the low dosage of microplastics had no obvious effects on microorganisms and key enzymes. Instead, the presence of microplastic-associated pollutants (e.g., BPA) influenced EPS secretion, thereby posing inhibitory effects on microorganisms and key enzymes. The following perspectives are suggested to further examine the inhibitory effects of microplastics on microorganisms and key enzymes: (1) The co-occurrence of different organic contaminants and antibiotics on microplastics should be investigated to examine the toxicity effects on microorganisms and key enzymes; and (2) different characteristics of microplastics, especially small size and fibrous shape, should be explored to understand the pathways of microplastics entering the cell.

Acknowledgments

This work was supported by the Environment and Conservation Fund of the Hong Kong Special Administrative Region Government (No. 2020-76), the State Key Laboratory of Marine Pollution (SKLMP) Seed Collaborative Research Fund (No. SCRF/0026), and the Dean's Research Fund (No. FLASS/DRF/IRS-12/ROP-16/RMP-7) and Internal Research Grant (No. RG30/2021-2022R) of The Education University of Hong Kong.

References

1. Khan, M.T., Cheng, Y.L., Hafeez, S. et al. (2020). Microplastics in wastewater: Environmental and health impacts, detection, and remediation strategies. In: *Handbook of Microplastics in the Environment* (ed. T. Rocha-Santos, M. Costa, and C. Mouneyrac), 1–33. Cham: Springer.
2. Sun, J., Dai, X., Wang, Q. et al. (2019). Microplastics in wastewater treatment plants: Detection, occurrence and removal. *Water Research* 152: 21–37.
3. Feng, L.J., Wang, J.J., Liu, S.C. et al. (2018). Role of extracellular polymeric substances in the acute inhibition of activated sludge by polystyrene nanoparticles. *Environmental Pollution* 238: 859–865.
4. Feng, Y., Feng, L.J., Liu, S.C. et al. (2018). Emerging investigator series: Inhibition and recovery of anaerobic granular sludge performance in response to short-term polystyrene nanoparticle exposure. *Environmental Science: Water Research and Technology* 4 (12): 1902–1911.
5. Carr, S.A., Liu, J., and Tesoro, A.G. (2016). Transport and fate of microplastic particles in wastewater treatment plants. *Water Research* 91: 174–182.
6. Hamidian, A.H., Ozumchelouei, E.J., Feizi, F. et al. (2021). A review on the characteristics of microplastics in wastewater treatment plants: A source for toxic chemicals. *Journal of Cleaner Production* 295: 126480.
7. Iyare, P.U., Ouki, S.K., and Bond, T. (2020). Microplastics removal in wastewater treatment plants: A critical review. *Environmental Science: Water Research & Technology* 6 (10): 2664–2675.
8. Gatidou, G., Arvaniti, O.S., and Stasinakis, A.S. (2019). Review on the occurrence and fate of microplastics in Sewage Treatment Plants. *Journal of Hazardous Materials* 367: 504–512.
9. da Costa, J.P., Santos, P.S., Duarte, A.C. et al. (2016). (Nano)plastics in the environment–Sources, fates and effects. *Science of the Total Environment* 566–567: 15–26.

10 Zhang, X., Chen, J., and Li, J. (2020). The removal of microplastics in the wastewater treatment process and their potential impact on anaerobic digestion due to pollutants association. *Chemosphere* 251: 126360.

11 Bayo, J., Olmos, S., and López-Castellanos, J. (2020). Microplastics in an urban wastewater treatment plant: The influence of physicochemical parameters and environmental factors. *Chemosphere* 238: 124593.

12 Cheng, Y.L., Kim, J.G., Kim, H.B. et al. (2021). Occurrence and removal of microplastics in wastewater treatment plants and drinking water purification facilities: A review. *Chemical Engineering Journal* 410: 128381.

13 Hidayaturrahman, H. and Lee, T.G. (2019). A study on characteristics of microplastic in wastewater of South Korea: Identification, quantification, and fate of microplastics during treatment process. *Marine Pollution Bulletin* 146: 696–702.

14 Talvitie, J., Mikola, A., Setälä, O. et al. (2017). How well is microlitter purified from wastewater? – A detailed study on the stepwise removal of microlitter in a tertiary level wastewater treatment plant. *Water Research* 109: 164–172.

15 Gies, E.A., LeNoble, J.L., Noël, M. et al. (2018). Retention of microplastics in a major secondary wastewater treatment plant in Vancouver, Canada. *Marine Pollution Bulletin* 133: 553–561.

16 Lares, M., Ncibi, M.C., Sillanpää, M. et al. (2018). Occurrence, identification and removal of microplastic particles and fibers in conventional activated sludge process and advanced MBR technology. *Water Research* 133: 236–246.

17 Lv, X., Dong, Q., Zuo, Z. et al. (2019). Microplastics in a municipal wastewater treatment plant: Fate, dynamic distribution, removal efficiencies, and control strategies. *Journal of Cleaner Production* 225: 579–586.

18 Magni, S., Binelli, A., Pittura, L. et al. (2019). The fate of microplastics in an Italian Wastewater Treatment Plant. *Science of the Total Environment* 652: 602–610.

19 Li, L., Song, K., Yeerken, S. et al. (2020). Effect evaluation of microplastics on activated sludge nitrification and denitrification. *Science of the Total Environment* 707: 135953.

20 Wang, Z., Gao, J., Dai, H. et al. (2021). Microplastics affect the ammonia oxidation performance of aerobic granular sludge and enrich the intracellular and extracellular antibiotic resistance genes. *Journal of Hazardous Materials* 409: 124981.

21 Wang, Q., Li, Y., Liu, Y. et al. (2021). Effects of microplastics accumulation on performance of membrane bioreactor for wastewater treatment. *Chemosphere* 278 (Part 1): 131968.

22 Fu, S.F., Ding, J.N., Zhang, Y. et al. (2018). Exposure to polystyrene nanoplastic leads to inhibition of anaerobic digestion system. *Science of the Total Environment* 625: 64–70.

23 Li, L., Geng, S., Li, Z. et al. (2020). Effect of microplastic on anaerobic digestion of wasted activated sludge. *Chemosphere* 247: 125874.

24 Wei, W., Huang, Q.S., Sun, J. et al. (2019). Polyvinyl chloride microplastics affect methane production from the anaerobic digestion of waste activated sludge through leaching toxic bisphenol-A. *Environmental Science & Technology* 53 (5): 2509–2517.

25 Spellman, F.R. (2008). *Handbook of Water and Wastewater Treatment Plant Operations*. CRC press.

26 Liu, X., Yuan, W., Di, M. et al. (2019). Transfer and fate of microplastics during the conventional activated sludge process in one wastewater treatment plant of China. *Chemical Engineering Journal* 362: 176–182.

27 Singh, M. and Srivastava, R.K. (2011). Sequencing batch reactor technology for biological wastewater treatment: A review. *Asia-Pacific Journal of Chemical Engineering* 6 (1): 3–13.

28 Lee, H. and Kim, Y. (2018). Treatment characteristics of microplastics at biological sewage treatment facilities in Korea. *Marine Pollution Bulletin* 137: 1–8.

29 Kalčíková, G., Alič, B., Skalar, T. et al. (2017). Wastewater treatment plant effluents as source of cosmetic polyethylene microbeads to freshwater. *Chemosphere* 188: 25–31.

30 Talvitie, J., Heinonen, M., Pääkkönen, J.P. et al. (2015). Do wastewater treatment plants act as a potential point source of microplastics? Preliminary study in the coastal Gulf of Finland, Baltic Sea. *Water Science & Technology* 72 (9): 1495–1504.

31 Talvitie, J., Mikola, A., Koistinen, A. et al. (2017a). Solutions to microplastic pollution – Removal of microplastics from wastewater effluent with advanced wastewater treatment technologies. *Water Research* 123: 401–407.

32 Baresel, C., Harding, M., and Fång, J. (2019). Ultrafiltration/granulated active carbon-biofilter: Efficient removal of a broad range of micropollutants. *Applied Sciences* 9 (4): 710.

33 Ødegaard, H. (2006). Innovations in wastewater treatment: –The moving bed biofilm process. *Water Science & Technology* 53 (9): 17–33.

34 Maliwan, T., Pungrasmi, W., and Lohwacharin, J. (2021). Effects of microplastic accumulation on floc characteristics and fouling behavior in a membrane bioreactor. *Journal of Hazardous Materials* 411: 124991.

35 Zhang, Y.T., Wei, W., Huang, Q.S. et al. (2020). Insights into the microbial response of anaerobic granular sludge during long-term exposure to polyethylene terephthalate microplastics. *Water Research* 179: 115898.

36 Li, H., Xu, S., Wang, S. et al. (2020). New insight into the effect of short-term exposure to polystyrene nanoparticles on activated sludge performance. *Journal of Water Process Engineering* 38: 101559.

37 Dai, H.H., Gao, J.F., Wang, Z.Q. et al. (2020). Behavior of nitrogen, phosphorus and antibiotic resistance genes under polyvinyl chloride microplastics pressures in an aerobic granular sludge system. *Journal of Cleaner Production* 256: 120402.
38 Chen, H., Tang, M., Yang, X. et al. (2021). Polyamide 6 microplastics facilitate methane production during anaerobic digestion of waste activated sludge. *Chemical Engineering Journal* 408: 127251.
39 Chislett, M., Guo, J., Bond, P.L. et al. (2020). Structural changes in cell-wall and cell-membrane organic materials following exposure to free nitrous acid. *Environmental Science & Technology* 54 (16): 10301–10312.
40 Wang, Z., Gao, J., Li, D. et al. (2020). Co-occurrence of microplastics and triclosan inhibited nitrification function and enriched antibiotic resistance genes in nitrifying sludge. *Journal of Hazardous Materials* 399: 123049.
41 Pham, D.N., Clark, L., and Li, M. (2021). Microplastics as hubs enriching antibiotic-resistant bacteria and pathogens in municipal activated sludge. *Journal of Hazardous Materials Letters* 2: 100014.
42 Wei, W., Zhang, Y.T., Huang, Q.S. et al. (2019). Polyethylene terephthalate microplastics affect hydrogen production from alkaline anaerobic fermentation of waste activated sludge through altering viability and activity of anaerobic microorganisms. *Water Research* 163: 114881.
43 Zhao, L., Su, C., Liu, W. et al. (2020). Exposure to polyamide 66 microplastic leads to effects performance and microbial community structure of aerobic granular sludge. *Ecotoxicology and Environmental Safety* 190 (1): 110070.
44 Qin, R., Su, C., Liu, W. et al. (2020). Effects of exposure to polyether sulfone microplastic on the nitrifying process and microbial community structure in aerobic granular sludge. *Bioresource Technology* 302: 122827.
45 Carrera, P., Campo, R., Méndez, R. et al. (2019). Does the feeding strategy enhance the aerobic granular sludge stability treating saline effluents? *Chemosphere* 226: 865–873.
46 Wang, X., Chen, Z., Kang, J. et al. (2018). Removal of tetracycline by aerobic granular sludge and its bacterial community dynamics in SBR. *RSC Advances* 8: 18284–18293.
47 Wang, X., Chen, Z., Kang, J. et al. (2019). The key role of inoculated sludge in fast start-up of sequencing batch reactor for the domestication of aerobic granular sludge. *Journal of Environmental Sciences* 78: 127–136.
48 Wang, X., Chen, Z., Shen, J. et al. (2019). Impact of carbon to nitrogen ratio on the performance of aerobic granular reactor and microbial population dynamics during aerobic sludge granulation. *Bioresource Technology* 271: 258–265.
49 Wu, D., Zhang, Z., Yu, Z. et al. (2018). Optimization of F/M ratio for stability of aerobic granular process via quantitative sludge discharge. *Bioresource Technology* 252: 150–156.

50 Ma, J., Quan, X., Si, X. et al. (2013). Responses of anaerobic granule and flocculent sludge to ceria nanoparticles and toxic mechanisms. *Bioresource Technology* 149: 346–352.

51 Mu, H., Zheng, X., Chen, Y. et al. (2012). Response of anaerobic granular sludge to a shock load of zinc oxide nanoparticles during biological wastewater treatment. *Environmental Science & Technology* 46 (11): 5997–6003.

52 He, C.S., He, P.P., Yang, H.Y. et al. (2017). Impact of zero-valent iron nanoparticles on the activity of anaerobic granular sludge: From macroscopic to microcosmic investigation. *Water Research* 127: 32–40.

53 Zhang, Y.T., Wei, W., Sun, J. et al. (2020). Long-term effects of polyvinyl chloride microplastics on anaerobic granular sludge for recovering methane from wastewater. *Environmental Science & Technology* 54 (15): 9662–9671.

54 Lin, X., Su, C., Deng, X. et al. (2020). Influence of polyether sulfone microplastics and bisphenol A on anaerobic granular sludge: Performance evaluation and microbial community characterization. *Ecotoxicology and Environmental Safety* 205: 111318.

55 Martínez-Campos, S., González-Pleiter, M., Fernández-Piñas, F. et al. (2021). Early and differential bacterial colonization on microplastics deployed into the effluents of wastewater treatment plants. *Science of the Total Environment* 757: 143832.

56 Cui, Y., Gao, J., Zhang, D. et al. (2021). Responses of performance, antibiotic resistance genes and bacterial communities of partial nitrification system to polyamide microplastics. *Bioresource Technology* 341: 125767.

57 Song, K., Li, Z., Liu, D. et al. (2020). Analysis of the partial nitrification process affected by polyvinylchloride microplastics in treating high-ammonia anaerobic digestates. *ACS Omega* 5 (37): 23836–23842.

58 Winkler, M.K., Kleerebezem, R., Khunjar, W.O. et al. (2012). Evaluating the solid retention time of bacteria in flocculent and granular sludge. *Water Research* 46 (16): 4973–4980.

59 Zhang, Z. and Chen, Y. (2020). Effects of microplastics on wastewater and sewage sludge treatment and their removal: A review. *Chemical Engineering Journal* 382 (15): 122955.

60 Flemming, H.C. and Wingender, J. (2010). The biofilm matrix. *Nature Reviews Microbiology* 8: 623–633.

61 Sheng, G.P., Yu, H.Q., and Li, X.Y. (2010). Extracellular polymeric substances of microbial aggregates in biological wastewater treatment systems: A review. *Biotechnology Advances* 28 (6): 882–894.

62 Wei, W., Huang, Q.S., Sun, J. et al. (2019). Revealing the mechanisms of polyethylene microplastics affecting anaerobic digestion of waste activated sludge. *Environmental Science & Technology* 53 (16): 9604–9613.

63 Pittura, L., Foglia, A., Akyol, Ç. et al. (2021). Microplastics in real wastewater treatment schemes: Comparative assessment and relevant inhibition effects on anaerobic processes. *Chemosphere* 262: 128415.

64 Wei, W., Chen, X., Peng, L. et al. (2021). The entering of polyethylene terephthalate microplastics into biological wastewater treatment system affects aerobic sludge digestion differently from their direct entering into sludge treatment system. *Water Research* 190 (15): 116731.

65 Zhang, Z., Liu, H., Wen, H. et al. (2021). Microplastics deteriorate the removal efficiency of antibiotic resistance genes during aerobic sludge digestion. *Science of the Total Environment* 798 (1): 149344.

66 Oshita, K., Okumura, T., Takaoka, M. et al. (2014). Methane and nitrous oxide emissions following anaerobic digestion of sludge in Japanese sewage treatment facilities. *Bioresource Technology* 171: 175–181.

67 Ren, X., Tang, J., Liu, X. et al. (2020). Effects of microplastics on greenhouse gas emissions and the microbial community in fertilized soil. *Environmental Pollution* 256: 113347.

68 Zhou, J., Wen, Y., Marshall, M.R. et al. (2021). Microplastics as an emerging threat to plant and soil health in agroecosystems. *Science of the Total Environment* 787: 147444.

69 Xu, J., Wang, X., Zhang, Z. et al. (2021). Effects of chronic exposure to different sizes and polymers of microplastics on the characteristics of activated sludge. *Science of the Total Environment* 783: 146954.

5

Microplastics in Sewage Sludge of Wastewater Treatment

Wei Wei[1], Xingdong Shi[1], Yu-Ting Zhang[2], Chen Wang[2], Yun Wang[2], and Bing-Jie Ni[1],*

[1] School of Civil and Environmental Engineering, Centre for Technology in Water and Wastewater, University of Technology Sydney, Sydney, NSW, Australia
[2] State Key Laboratory of Pollution Control and Resources Reuse, College of Environmental Science and Engineering, Tongji University, Shanghai, PR China
* Corresponding author

5.1 Introduction

Plastic, a high-molecular weight polymer, brings enormous benefits to human beings for daily life and industry production. It is reported that the yield of plastics has exponentially risen to 359 million tons in 2018, while their amount was only 2 million tons in 1950 [1]. Over 40% of plastics are utilized as single-use packaging, but only 9–40% are recycled, inducing an inappropriate discharge of substantial plastics debris every year [2]. A considerable portion of waste plastic enters the aquatic environment, with an estimated accumulative potential of 250 million tons by 2025 [3]. Plastic pollution is thus recognized as a top environmental problem and emerging threat in the near to medium-term because of its inertness to biodegradation [4]. Generally, plastic undergoes photooxidation under ultraviolet radiation becoming brittle. Wind and waves further degrade this friable debris into microparticles (100 nm to 5 mm) [5] or even nanoparticles (<100 nm) [6], herein called microplastics and nanoplastics [7].

Microplastics are widely utilized in various personal care products such as exfoliant additives (also called microbeads) in facial cleansers [8]. These microbeads, together with microfibers from machine-washed clothing, are discharged to the municipal wastewater system and ultimately enter a wastewater treatment plant (WWTP) [7, 9]. WWTPs are thus regarded as both the most vital sinks and sources of microplastics among various environments. According to reports, domestic sewage contains over 30 types of microplastics, including polyethylene, polyester, polyamide, etc. [10–12].

Current research indicates that 98% of microplastics can be removed in the WWTPs, whereas over 65 million microplastics still enter the receiving environment daily. According to one report, the aquatic environment could receive up to 8 trillion microplastics from WWTPs in the United States [13]. Except for untreated microplastics, the microplastics in raw sewage are transformed and retained in sewage sludge up to 99% [14, 15]. This phenomenon has been proven by analyzing the content changes of microplastics in individual treatment facilities [16]. The concentration of microplastics in such sludge ranges 1.5×10^3–2.4×10^3 items/kg-dry sludge [17]. These microplastics exist in the sewage and sludge of WWTP in the form of spheres, films, and fibers [18, 19]. Notably, distribution characteristics of microplastics in sludge coming from different treatment procedures (i.e., primary sludge, waste-activated sludge, and dewatered sludge) will exhibit discrepancy along the wastewater treatment process. Therefore, this discrepancy should be explored further to understand the fate of microplastics in the sludge treatment process.

Microplastics present in sludge cause enormous challenges to sludge reduction and disposal because of their potential risks to the environment. Remarkably, waste-activated sludge (WAS), the main byproduct in WWTPs, is reported to be increasingly produced by 13% on annual average [20]. Currently, anaerobic digestion and fermentation, combining resource recovery and biowaste reduction, is a preferred choice for sludge treatment in WWTPs [21]. Anaerobic digestion and fermentation are sensitive to exogenous contaminants, i.e., microplastics, significantly decreasing its ability to produce various valuable chemicals. Numerous studies have evaluated the influence of microplastics on the sludge anaerobic treatment process, finding that it depends on their physical and chemical properties [22]. Therefore, it is critical to comprehensively summarize the effects of microplastics in sewage sludge on sludge treatment process in WWTPs.

More importantly, the massive discharge of microplastics from WWTPs has raised concerns about the toxicity of microplastics. Although microplastics are inert materials, they could still induce harm for creatures through various pathways. As early as 1975, researchers verified that oral administration of polyvinyl chloride particles long-term led to embolization of vessels in animals [23]. The cytotoxicity of microplastics is determined by their physical characteristics, including size, shape, and solubility [24]. Exposure to microplastics for an extended time could further trigger a series of physical effects (e.g., inflammation, oxidative stress, etc.) from its persistence, ultimately resulting in tissue damage or carcinogenesis [2]. Moreover, the manufacture of plastics can introduce various chemicals such as dibutyl phthalate, di-2-ethylhexyl phthalate and other plasticizers. These substances are susceptible to release into the outer environment and then migrate to surrounding cells or organisms because they did not tightly knit the matrix of plastic polymer [25]. The leaching of these chemicals or the polymer itself could directly influence human health as cellular uptake of such microplastics ensures a pathway for these nocuous contaminants to enter the cell [26]. Therefore, developing microplastic removal technologies for both sewage and sludge is particularly critical to reduce the environmental risk of microplastics.

Collectively, WWTPs are both essential sinks and sources of microplastics, and these plastic particles further influence sludge treatment, ultimately discharging to the environment. A relatively comprehensive evaluation of the fate and effects of microplastics on the sludge treatment process is important to intensify the risk management of microplastics. Therefore, this chapter will handle these issues from the aspects of 1) investigating the occurrence and concentration of microplastics in various sewage sludge; 2) analyzing the effects of microplastics on sewage sludge treatment, especially the production performance of various valuable chemicals such as methane, short-chain fatty acids, and hydrogen; 3) discussing the potential risk of transporting microplastics from sludge to soil and landfills; 4) summarizing the available technologies for removing microplastics from sludge. Finally, the future directions concerning microplastics in sewage sludge are proposed.

5.2 Occurrence

Microplastics have been detected in natural environments such as freshwater, terrestrial environments, and even the polar regions. With increased processing and application of various manufactured personal care products, a mass of microplastics will inevitably be released to enter the wastewater treatment plant, now regarded as an important source of microplastics. Wastewater treatment facilities can effectively separate microplastics from raw wastewater, and thus most plastic fragments are retained in sewage sludge. To date, many investigations have confirmed the existence of microplastics in sludge. Figure 5.1 shows a stereomicrograph of representative microplastics extracted from sewage sludge. Generally, the concentration of microplastics occurring in the influent of wastewater treatment plants is in the range 0.28–31 400 items/L, while in the effluent the value decreases to 0.01–297 items/L. As a result, the concentration of microplastics accumulated in sewage sludge is 4.40–240 items/g-dry weight. It is known that raw wastewater in the wastewater treatment system will go through a series of treatment facilities to meet environmental standards. Different facilities are operated according to various principles, resulting in the generation of several types of sludge (i.e., primary sludge, waste-activated sludge, and dewatered sludge). Therefore, the distribution characteristics of microplastics in various kinds of sludge will vary greatly, as shown in Table 5.1.

150 | *5 Microplastics in Sewage Sludge of Wastewater Treatment*

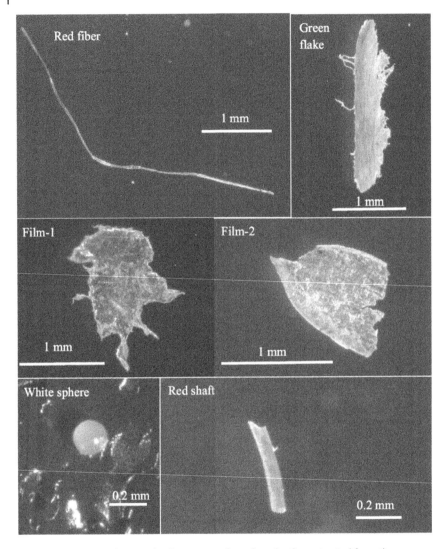

Figure 5.1 Stereomicrograph of representative microplastics extracted from the sewage sludge [39] / with permission of Elsevier.

5.2.1 Primary Sludge

The original sewage entering the wastewater treatment plant will first undergo preliminary and primary treatment prior to entering the subsequent biological treatment system. The preliminary treatment process includes screening, sand and grease removal, and skimming to remove some floating substances, then

Table 5.1 The characteristics of microplastics appeared in different sludge.

Sewage sludge	Source	Microplastic content (items/g-dry sludge)	Chemical composition	Color	Shape
Primary sludge	Primary sedimentation	14.9 ± 6.3–214 ± 16	PET, PA, PP, PES	Black (fiber) Blue (particulate)	Mainly fiber
Waste-activated sludge	Secondary treatment	4.4–113	PET, PA, PEST, PP	Black (fiber) Blue (particulate)	Mainly particulate
Dewatered sludge	Sludge dehydration	129 ± 17	PO, acrylic, PET, PA, alkyd resin, PS	Mainly white and black	Mainly fiber

the primary treatment allows settleable pollutants to form sediment (i.e., primary sludge). The separation of microfibers in sewage occurs during primary sedimentation, while most of the microplastics in particulate form are first adsorbed on the solids then removed in the subsequent treatment step. In detail, the removal of fibrous microplastics in the primary sedimentation process is more significant than that of synthetic particles, representing removal percentages of 92 and 32% respectively [14]. Besides, suspended debris with larger size are more easily retained by preliminary facilities, which can be well removed by gravity-driven settlement while smaller particles not capable of entering primary sedimentation will become the influent composition of subsequent treatment facilities.

The removal extent of microplastics in primary treatment is dependent on several factors. The density of the microplastics appearing in raw wastewater is a key factor related to sedimentation performance, and the higher density implies a higher degree of settling. The overflow speed that determines the residence time of sewage in the treatment facility also affects the settling of microplastics. Besides, the technologies applied in preliminary treatment systems have certain effects on the distribution of microplastics in the primary treatment system. For example, the combination of grid and aerated grit chamber is beneficial to the removal of microplastics, thereby changing the number of microplastics imported into the primary settling tank.

Monitoring efforts have revealed the occurrence of microplastics in the byproduct of the primary system (i.e., primary sludge) [27], aligning with the removal of microplastics in primary sedimentation process. It is reported that microplastics at a wide range of values (from 14.9 ± 6.3 to 214 ± 16 items/ g dry weight) were detected that show unique characteristics. Among them, fibers occupy a higher proportion than other forms of plastics. Most microplastics are larger than 500 μm. The major chemical compositions of the detected microplastics are polyethylene terephthalate (PET), polyamide (PA), polypropylene (PP), acrylic, and polyether sulfones (PES). In addition, the predominant color of these fiber plastics appearing in primary sludge is black, accounting for 72.2%. Although the percentage of particulate plastics in this type of sludge is not high, they are also colorful, with blue being the most common (67.9%). Actually, different forms of microplastics originate from different manufacturing and application processes, which is the main reason they show different colors [28].

5.2.2 Waste-Activated Sludge

The effluent of a primary settling tank subsequently enters the secondary treatment system, a biological treatment step to degrade pollutants together with the process of sedimentation. Thus, the remaining microplastics can further be removed, but the removal extent is lower in comparison with primary treatment [29]. In detail, microbial cells and extracellular polymers in the biological system have strong adsorption capacity on microplastics. Some bacteria (e.g., protozoa and metazoa) may ingest the debris, resulting in the existence of microplastics in sludge floc [30, 31]. Additionally, some flocculants can aggregate microplastic fragments in the form of flocs, then remove them through precipitation [32]. Notably, the unstable flocs of microplastics can disperse in the mixed liquid and be separated in the following secondary settling stage [33]. Contrary to primary treatment, the smaller microplastics in the influent (106–300 μm) can be removed more efficiently than larger fragments (>300 μm), because they are easily adsorbed on the flocs and biofilms. It is also found that the removal efficiency of particle microplastics is higher than that of fibers during secondary treatment [14]. The driving force behind this phenomenon is that the majority of fibers with higher density and larger volume have already been separated from the liquid in the primary treatment system, while remaining fibers may have inertness such as neutral buoyancy, making them difficult to be further removed [34].

It is noteworthy that the actual contents of microplastics in the influent and effluent of the secondary treatment system can assess their occurrence in waste-activated sludge. For example, the removal rates of microplastics in Canada and USA were above 95%, higher than that (70%) reported in Turkey. This indicates that waste-activated sludge in Canada and USA is adsorbing more microplastics. The retention of microplastics in waste-activated sludge may also depend on sewage treatment operations. Specially, different operational conditions of secondary treatment systems lead to assorted sizes of microplastics in the effluent, indicating that the specific microplastics accumulated in waste-activated sludge have different size distributions. Besides, contact time between microplastics and wastewater also affects the removal of microplastics in secondary treatment and their occurrence in waste-activated sludge. It is documented that longer contact time could contribute to the formation of a biomass film on the surface of microplastics [33]. As a result, the microplastics could be denser, making them easier to gravity separate.

A considerable number of studies have detected the amount of microplastics in waste-activated sludge, ranging 4.4–113 items/g-dry sludge [35]. Waste-activated sludge mainly contains plastic particles less dense than water that are difficult to separate during primary treatment. Thus, waste-activated sludge is a collection for plastic particles smaller than 500 μm, and plastics with sizes 125–250 μm account for the largest proportion of all microplastics [36]. Microplastic content in the primary sludge and waste-activated sludge is also compared. The results suggest that primary sludge receives more microplastics from raw sewage than waste-activated sludge, confirming the statement that the primary treatment has a greater role in the removal of microplastics. Moreover, the microplastics existing in waste-activated sludge are mainly composed of PET, PA, polyesters (PEST), and PP, similar to those in primary sludge. Thus, it can be deduced that these particles constitute the main microplastic content in the raw sewage. Like primary sludge, waste-activated sludge received blacker fibrous microplastics and colored particulate plastics [28].

5.2.3 Dewatered Sludge

To achieve sludge reduction and stabilization after wastewater treatment, the generated sludge (including primary and waste-activated sludge) is transferred to a thickener tank to remove part of the water in the sludge. Eventually, the mixed sludge, containing self-precipitating microplastics and flocs with adsorbed microplastics, is dehydrated by physical forces and the supernatant is separated and returned to the inflow. Researchers also investigated the changes in the amount of microplastics in the inflow (642 million items) and outflow (599 million items) of the thickening tank, and results showed a decrease of 6% [37]. The reduction can be caused by the transfer of lighter microplastics from the sludge phase to the water phase, then re-entry into the upstream sewage treatment facilities. In addition, the size of the microplastics existing in dewatered sludge was observed to be smaller than the inlet sludge mixture. This suggests that the technologies involved in sludge dewatering, including mechanical mixing and elevating temperature, might result in the shearing of plastic particles. Correspondingly, changes such as foaming and melting occur in the microplastics from dewatered sludge [34].

It is reported that high numbers of microplastics appear in the dewatered sludge, with the amount of 129 ± 17 items/g-dry sludge [37] and the mass concentration of 240.3 ± 31.4 ng/g-dry sludge [38]. Mass balance analysis indicates that microplastics in dewatered sludge accounted for approximately 0.7% of the total sludge. Microplastics in dewatered sludge are composed of various low-density plastics, including polyolefin (PO), acrylic, PET, PA, alkyd resin, and PS. Furthermore, the surface morphology analysis indicates that the microplastics in dewatered sludge appear to be hackly and brittle [18, 39]. Sludge is a collection of heavy metals, persistent organic pollutants, and pathogenic microbes, and thus a microplastic surface would provide a habitat for these contaminants. As such, the microplastics enriched in dewatered sludge may bring about certain risks to the subsequent disposal of sludge, requiring further analysis. There are many colorful microplastics in dewatered sludge, among which white and blank are the main colors. However, some colored particles are also involved. It is known that colored plastics are more attractive to organisms. The harmful substances adsorbed on the microplastic surface will be inhaled by organisms when ingesting the particles. Meanwhile, fibers occupy a predominant part of microplastics in dewatered sludge, showing a similar trend to the primary sludge. This finding suggests that fibrous microplastics are just the main components of the plastics in the raw wastewater that most likely originated from the release of fiber plastic in manufacturing industry and the washing of various clothes.

Notably, various sludge dehydrating technologies with different principles also have effects on the existence of microplastics in dewatered sludge. Based on the mechanical forces used to remove the water in sludge, the sludge dewatering processes can be divided into centrifugal dewatering and filtering dewatering. Methods such as filter-pressure, belt-type, and plate-frame apply extrusion to achieve water removal from sludge, while centrifugal dehydration is the technology based on the density differences between water phase and sludge phase. During centrifugation, some lighter microplastics will be transferred from the sludge to the water phase, and then enter the front-end treatment process of the wastewater treatment system along with the raw sewage. As such, dewatered sludge derived from centrifugal dehydration contains lower numbers of plastics compared with other commonly used methods. Regarding the regional and climate differences, forest cover, economic status, and urbanization are main factors affecting the spatial distribution of microplastics in dewatered sludge, while the temporal variability of microplastics will also change with temperature and precipitation [39]. The explanation behind these phenomena is that the higher asset and population density lead to higher volume of laundry wastewater, higher consumption of commercial supplies, and thus more discharge of microplastics. Moreover, climate and rainfall may affect surface runoff and thus influence the occurrence of microplastics in water bodies. In conclusion, the strategies for sludge treatment and disposal should be selected according to different regions and climatic conditions. Besides, the microplastic concentrations in sludge can be controlled by optimizing the operating parameters of sewage treatment systems (such as dewatering methods).

5.3 Effects of Microplastics on Sludge Anaerobic Treatment

According to investigations in recent years, the majority of microplastics in municipal wastewater treatment plant (MWTP) are captured and trapped in the sludge because of their special properties such as hydrophobicity and small size [18]. It is reported that the retention rate of microplastics in raw wastewater by sludge is more than 90% [40], and the concentrations of microplastics in sludge were estimated in the range of 1.5×10^3–2.4×10^4 items/kg [18, 41], potentially having certain impact on the subsequent sludge treatment process. Anaerobic digestion and fermentation are frequently used methods for sludge disposal, which can achieve both resource recovery and sludge reduction. As a biochemical process controlled by multiple microbes, contaminants in the sludge could alter the treatment performance, related enzyme activity, as well as microbial community composition in an anaerobic digestion or fermentation system. In last few years, some studies have been focused on the potential influences of microplastics on sludge treatment, and the details are described below.

5.3.1 Methane

In an anaerobic digestion system, methane, as the product converted from sludge organics degradation, often serves as a key index to evaluate the performance of sludge treatment. Effects of microplastics on methane production from anaerobic digestion of sludge have been explored based on amount, size, and type of microplastics. Wei et al. investigated the impacts of PVC and PE microplastics on cumulative methane production after complete digestion of waste-activated sludge [42, 43]. The results demonstrated that a high amount of PE microplastics (i.e., 100 and 200 particles/g TS) and PVC microplastics (i.e., 20, 40, and 60 particles/g TS) inhibit methane production by 12.4–27.5% and 9.4–24.2% respectively. The lower methane potential and hydrolysis coefficient based on one-substrate model analysis were also determined. Increased amounts of PVC or PE microplastics result in an increased degree of inhibition. Similar to the PE and PVC microplastics, different abundances of PES microplastics also exhibit suppression of methane production, methane potential, and hydrolysis coefficient in an anaerobic digestion system [40]. However, a correlation between the inhibition degree of methane production and the amount of PES microplastics is not observed. In contrast, polyamide 6 (PA6) microplastics at 10 particles/g TS is found to be effective in facilitating methane production by 39.5% from sludge anaerobic digestion [44]. The main reason for this phenomenon is stated that leaching of caprolactam (CPL) from PA6 microplastics might promote the activities of key enzymes related to acidification and methanogenesis. Besides, Wei et al. also demonstrated the slight improvement of methane production from sludge anaerobic digestion at low concentration (10 particles/g TS) of PVC microplastics, possibly attributed to the enhancement of solubilization [43]. Thus, the influences of microplastics on methane production from sludge anaerobic digestion depend on microplastic amount and type.

Microplastic size is another key factor regarding their influences on methane production from sludge anaerobic digestion. Zhang et al. [11] evaluated methane generation from sludge exposed to two sizes (80 nm and 5 μm) of PS microplastics. The cumulative methane production was not significantly affected by 0.2 g/L of PS microplastics at these two sizes, while 0.25 g/L of PS microplastics at either size reduced the cumulative methane production. The 80 nm PS nanoplastics show more severe inhibition than 5 μm PS microplastics. Microplastics with smaller size are more likely to penetrate a cytomembrane, causing deeper damage to microorganisms. In addition, some toxicity studies have been performed at the nanoscale. Feng et al. and Fu et al. reported that nanoplastics exposure leads to suppression of methane production and maximum daily methane yield, and the inhibitory extent increased as the nanoplastics concentration increased [45, 46]. In view of the suppression on methane production from sludge, the mechanism is summarized in Figure 5.2, including toxic chemicals leached from microplastics, the excess reactive oxygen species (ROS) induced by microplastics, as well as physical damage through interaction between microbes and microplastics [47].

5.3.2 Short-Chain Fatty Acid

Apart from anaerobic digestion, anaerobic fermentation is also a common method used for sludge treatment, with the high value-added products of short-chain fatty acids (SCFAs) including acetate, propionate, butyrate, and valerate, which has been identified as the appropriate carbon source of wastewater treatment.

At present, many investigators are focused on the study of potential impacts of microplastics on SCFAs production from sludge anaerobic fermentation. Different types of microplastics distinctively affect SCFAs generation. For example, the SCFAs production is enhanced at all tested levels of PA6 microplastics compared to the control without microplastics [44], mainly attributed to the facilitation of

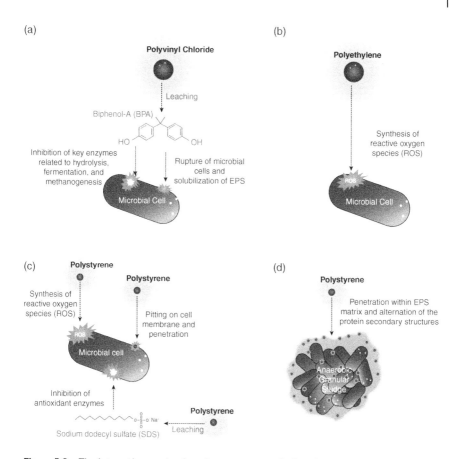

Figure 5.2 The interactive mechanisms between types of microplastics a) polyvinyl chloride, b) polyethylene, c) and d) polystyrene and microbial cell in anaerobic digestion [47] / with permission of Elsevier.

the enzyme activities related to acidification. In contrast, the addition of mixed microplastics, i.e., PE, PET, PS, and PP, in the sludge anaerobic fermenter exhibits inhibition of the maximal SCFAs production [48]. With microplastics addition, the hydrolysis and acidification are both clearly suppressed, thereby deteriorating performance. Similarly, PS microplastics also have been proven to antagonize the SCFAs production by affecting the process of hydrolysis and acidification [49]. Especially, 10 μg/L of PS microplastics exhibits a stronger inhibitory effect than higher levels of PS microplastics at 50, 100, and 1000 μg/L, likely because of less aggregation and greater attachment at lower levels of PS microplastics. However, in the investigation on the response of hydrogen-producing thermophilic bacteria to the acute exposure of PS nanoplastics, Chen et al. documented that the presence of PS nanoplastics stimulate SCFAs production with the increase of acetate and propionate but have a negative effect on the hydrogen formation of system. This may be attributed to the high concentration of propionate [50].

5.3.3 Hydrogen

It is well known that hydrogen is another main product of sludge anaerobic treatment. However, the potential effect of microplastics on hydrogen production in sludge treatment is still in its infancy. A study published by Wei et al. reveals that the presence of PET microplastics restrains hydrogen generation from WAS alkaline anaerobic fermentation at all experimented levels (10, 30, and 60 particles/g-TS), and this deterioration is exacerbated with the increase of PET microplastics concentration [51]. A further mechanistic study disclosed that the toxic di-n-butyl phthalate (DBP) leached from PET microplastics, and excess oxidative stress induced by PET microplastics might be the main factors for causing the inhibitory effects on the steps of hydrolysis, acidogenesis, and acetogenesis involved in WAS alkaline anaerobic fermentation. Besides, Chen et al. explores the influence of PS nanoplastics on hydrogen bioenergy recovered from sludge through acetic acid-type hydrogen fermentation [50]. The results show that the presence of 0.2 g/L PS nanoplastics weakens the protective function of EPS through destroying its structure, composition, and surface charge, resulting in decreasing cumulative hydrogen production at 53.9% of control. Meanwhile, the exposure to PS nanoplastics increases cell permeability and intensifies the imbalance of the antioxidant system, decreasing the cell tolerance to external stimulus (e.g., microplastics and nanoplastics).

5.3.4 Enzyme Activity

Anaerobic treatment of sludge is a complex metabolic process requiring the participation of various enzymes, and the performance of sludge treatment is closely related to enzyme activity. Protease, cellulase, α-amylase, and α-glucosidase are common hydrolases that play a key part in the conversion of proteins and polysaccharides to amino acids and monosaccharides. Acetate kinase (AK) and butyrate kinase (BK) are the key enzymes with respect to the bioprocess of acidogenesis. The BK catalyzes the transformation of amino acids to SCFAs, and AK participates the bioconversion of acetyl-CoA to acetic acids. Moreover, coenzyme F_{420} is essential for the stage of methanogenesis [47]. It is reported that the presence of PVC microplastics and polyether sulfone microplastics can affect the activities of protease, AK, and F_{420}. Among them, F_{420} was more sensitive [43, 52]. However, PA6 microplastics with moderate concentration play a positive role in the activities of key enzymes including protease, α-glucosidase, AK, BK, and F_{420} [44].

Improvement of F_{420} activity is twice as much as that without PA6 microplastics presence. The CPL leached from PA6 microplastics plays a critical role in this facilitation. The catalytic activity of enzymes is enhanced by the combination of CPL and enzyme molecules, which transforms the active site of the enzyme. For nanoplastics, Chen et al. found that in the system of hydrogen recovery by thermophilic bacteria, the activity of α-amylase decreases to 61.8% with the presence of 0.2 g/L PS nanoplastics compared to that without nanoplastics, inhibiting the hydrolysis of sludge and the hydrogen-producing performance [50]. Except for the enzymes controlling each bioprocess involved in anaerobic digestion, the exposure of microplastics can also affect the enzymes responsible for antioxidant defense. Normally, microorganisms protect themselves through a personal antioxidant system against moderate oxidative stress. Superoxide dismutase (SOD) and catalase (CAT) are familiar enzymes that scavenge a certain amount of ROS. However, this neutralization cannot defend against ROS in high concentration on account of limited number of bacterial enzyme molecules [15]. Wei et al. demonstrated that PS nanoplastics exposure restrained the activities of SOD and CAT, thereby impairing the resistance of microorganism to excessive oxidative stress induced by PS nanoplastics [53].

5.3.5 Microbial Community

The performance of sludge anaerobic treatment is closely related to the microbial community of the reaction system, and the distinctions among the structure and metabolism of microbiomes with and without microplastics have been discussed extensively in published literature.

Although microplastics have insignificant impacts on microbial community structure and diversity, their presence changes the relative abundance within the microbial community. Different microbes have disparate tolerance to microplastics. Euryarchaeota, a well-known archaeal phylum containing methanogens, is the dominant phylum in methane-producing anaerobic reactors, and its abundance reduces after microplastics exposure [12, 46, 53]. Aceticlastic Methanosaeta sp. is a type of methanogen that uses acetate for methane generation, and its relative abundance declines in the presence of microplastics, corresponding to the deterioration of methane production performance from sludge [42, 43, 54]. Methanoregula sp. and Methanomassiliicoccus sp. are anaerobic methanogens utilizing H_2, CO_2, and methanol or methylamine to produce methane, and their relative abundances decrease after metal-doped nanoplastics addition, resulting in suppression of methane production in anaerobic digestion [45]. Along with Euryarchaeota, Proteobacteria, Bacteroidetes, Chloroflexi, Firmicutes, and Actinobacteria are the predominant bacteria phyla in sludge anaerobic treatment systems, and many anaerobes in these phyla can convert organic compounds into SCFAs. The abundances of most of these bacteria usually decrease under microplastics stress. In addition, microplastics exposure also can reduce the abundance of many key microbes related to organics degradation, such as hydrolytic microorganisms (Bacteroides sp.) and acidogens (Longilinea sp. and Clostridium_sensu_stricto_12).

In addition to the shift in the relative abundances of key microbes, microplastics can also cause varying degrees of physical damage to microorganisms. The interaction (adsorption, adherence, and pitting) between microplastics and microbial cells could weaken the defense of EPS, even destroy the integrity of cell wall and cell membrane, thereby decreasing cell viability [15, 46].

5.4 Transport of Microplastics from Sludge to Soil and Landfills

Wastewater treatment plants generate a large amount of by-product sludge. In China, approximately 34 million tons of sludge were generated in 2015 [55]. Since most of the microplastics (70–99%) in the wastewater treatment system can be transferred from wastewater to sludge, a large number of microplastics are detected in the sludge [56]. The abundance of microplastics in sludge ranged 10^3–10^5 items/kg [14]. Sludge is usually stabilized by anaerobic digestion and dewatered for reduction in the wastewater treatment plant prior to final disposal [57]. It is reported that high numbers of microplastics appear in the dewatered sludge, with a mass concentration of 240.3 ± 31.4 ng/g-dry sludge [38]. Land application and landfill of the treated sludge from wastewater treatment plant is a common way to properly dispose of the sludge. Therefore, there is a risk of transporting microplastics in the sludge to soil and landfills.

5.4.1 Transport of Microplastics from Sludge to Soil

Land application of sludge refers to the use of treated sludge in the soil to improve fertility [58]. The sludge contains a large amount of organic matter and mineral elements such as nitrogen, phosphorus, potassium, etc., which can meet the nutrient requirements of the soil, improve the soil to a certain extent, and increase the soil fertility [59]. Since this sludge reuse technology is a positive response to the circular economy concept, it has been widely used in soil and brings economic benefits [29, 60]. In Spain, the agricultural sector encourages the application of sludge, and 65% of sludge is recycled through the soil [61]. Unfortunately, the application of sludge to soil also opens up a new path for microplastics to enter the soil, and improper dumping of sludge can also aggravate the accumulation of microplastics in the soil (Figure 5.3). In China, 45% of sludge is used in agriculture [62]. Approximately 55% of the sludge in the United States is used as agricultural

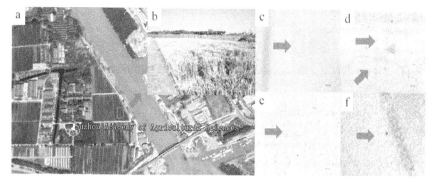

Figure 5.3 The land application of sludge (a, b) and photos of microplastics in the soil ((c) fiber; (d) fragment (e) granule; and (f) sphere) [62] with permission of Elsevier.

fertilizers, with up to 1030 trillion microplastics entering the soil every year [63]. In Europe, 50% of the sludge is applied to the soil as fertilizer, and it is estimated that 63 000–430 000 tons of microplastics are released into the soil every year [29]. Similarly, 75% of sludge in Australia is used in agriculture, and the level of microplastics entering the soil is 2800–19 000 tons of particles per year [36].

Corradini et al. [64] evaluated 31 farmlands in Chile where sludge was repeatedly applied over a 10-year period, and the results shows proved that microplastics in the sludge retain in the soil, and microplastics abundance in the soil is related to the applied sludge. Through long-term application of microplastics-containing sludge products, the total amount of microplastics accumulated in the soil increases over time and is positively correlated with the frequency of sludge application [59]. In Denmark, the number of microplastics in farmland with sludge reuse is twice as high as that in farmland without sludge application [65]. As a result of the low degradability of microplastics, microplastics in the soil can be detected for 15 years after sludge application.

When microplastics enter the soil, they bring many potential risks. For example, they have a potential impact on the physical and chemical properties of the soil including soil structure, bulk density, and water holding capacity [29]. Microplastics will accumulate in earthworms and plants, thereby affecting the soil biota [59]. In addition, microplastics in the sludge have obvious aging characteristics, and most show rough surfaces. Compared with fresh microplastics, microplastics in the sludge have 10 times higher adsorption potential for metal pollutants [66]. With the land application of sludge, the environmental risk brought by microplastics in the sludge may be higher than that of fresh microplastics. This issue requires considerable further attention. More importantly, less than half (16–38%) of microplastics that enter the soil through agricultural soil amendments will remain in the soil, and most of them will migrate from the soil to the aquatic environment (surface water and groundwater) to cause further environmental risks [29].

5.4.2 Transport of Microplastics from Sludge to Landfills

Landfills are the most traditional and simplest way of sludge disposal. It is estimated that the microplastics transferred from sludge to landfills accounts for 79% of all plastic production [67]. Therefore, a landfill, as an important disposal site for sludge, will accumulate a large number of microplastics. In China, 34.5% of the sludge is landfilled, while in Australia only 2% is eventually landfilled [28]. Data from the US Environmental Protection Agency shows that 22% of the sludge is landfilled, that is, up to 464 trillion microplastics enter the landfill [63].

Microplastics transferred to the landfill through sludge will spread from the landfill to the surrounding area through the wind, and will have a negative impact on the surrounding environment [68]. During the operation of the landfill site, various kinds of leachate are generated by the decomposition of garbage. The landfill leachate is rich in pollutants, including heavy metals and organic pollutants [69]. In recent years, emerging pollutant microplastics have been found to be able to carry other pollutants, thereby exacerbating the negative impact of leachate discharge on the surrounding environment [70]. If they are not managed properly, potential risks will increase.

5.5 Enhanced Removal of Microplastics from Sludge

At present, there are no specific microplastic removal units in wastewater treatment plants. However, many studies have shown that various sludge treatment processes including thickening and dewatering, anaerobic digestion, composting, and dry incineration can cause changes in the content of microplastics in the sludge.

5.5.1 Thickening and Dehydration

Thickening and dewatering are physical means to remove water for reducing the volume of sludge. The moisture content of the original sludge can usually reach up to 99.5%, so the sludge must be thickened for subsequent treatment. Alavian Petroody et al. [37] investigated a sewage treatment plant in northern Iran showing that the daily microplastics entering the sludge thickening tank was reduced from ca. 642 million to ca. 599 million, that is, 6% of microplastics were removed from the sludge solids through sludge thickening. Talvitie et al. compared the microplastics abundance in the excess raw sludge and the centrifuged sludge mixture in a Finnish sewage treatment plant, and the results show that the amount of microplastics in the sludge after centrifugation was reduced by ca. 20%, attributed to the separation of microplastics from the sludge through the dehydration process into the discharged water [71]. The sludge digestion process prior to the dehydration process also benefits microplastics removal during the dehydration. Microplastics in the dewatered sludge are 54% less than the digested sludge [37]. The destruction of biological flocs in the digestion stage releases the attached microplastics, making microplastics easier to separate from the sludge and discharge through the dehydration process. However, it is worth noting that microplastics do not disappear once removed from the sludge phase and transferred to the water phase. Therefore, the wastewater needs to be effectively managed through advanced treatment processes such as filtration to further reduce the microplastics in the wastewater, which may also help reduce the amount of microplastics in the final sludge and effluent of sewage treatment plants.

5.5.2 Anaerobic Digestion

Narancic et al. showed that anaerobic digestion of sludge can lead to the biodegradation of plastics such as polylactic acid and polycaprolactone compared to the natural environment. Because of the biodegradability of polymers, the abundance of microplastics may decrease after anaerobic digestion [72]. Mahon et al. [18] collected sludge after anaerobic digestion, thermal drying and lime stabilization from seven sewage plants in Ireland and demonstrated that the anaerobic digestion sludge contained less abundant microplastics particles. This finding indicates that plastic degrading bacteria exist in the anaerobic digestion system, but further evidence is required. Various sludge pretreatment methods are commonly applied in promoting biodegradation by anaerobic digestion. However, different sludge pretreatment methods have different effects on the microplastics abundance. For example, Hurley et al. revealed that sludge pretreatment with 1 and 10 mol/L NaOH can reduce the quality and size of microplastics in the sludge by 59.9% and 27.8%, respectively [73]. Similarly, Li et al. [74] demonstrated that using 1, 5, and 10 mol/L NaOH pretreatments can reduce the mass and size of microplastics in the sludge by up to 53.5% and 16.7%, respectively. In contrast, mechanical mixing increased the abundance of microplastics in the sludge by crushing microplastics into smaller sizes [75].

5.5.3 High Temperature Composting

Composting is one of the most important sludge treatment and recycling technologies. Increased evidence shows that microorganisms are important decomposers of plastics in the environment. Some polyethylene plastic-degrading bacteria have been discovered, such as Arthrobacter and bacillus strains isolated from the plastic surface of coastal debris [76]. At the same time, some researchers successfully obtained polyethylene-degrading bacteria from the biofilm generated on the surface of the sediment microplastics [77]. Compared with the natural environment, the composting process can lead to the biodegradation of plastics [72]. Zhang et al. [78] revealed that microplastics are mainly in flake shape after sludge composting, attributed to the degradation and fragmentation of microplastics through aerobic composting at a high temperature of 70°C. Chen et al. also observed that both ultra-high temperature composting and traditional composting have an impact on microplastics in sludge (Figure 5.4) [75]. The microplastics content through full-size ultra-high temperature (70°C) composting decreased by 43.70% after 45 days of treatment, 10 times those through traditional composting. The analysis of the composting bacterial population reveals that after 56 days of degradation in the ultra-high temperature composting group, the hyperthermophilic mixed bacteria causes 7.3% weight loss of PS microplastics, six times that of the conventional high-temperature (40°C) group. The breaking of C–C bonds under high temperature conditions accelerates the biodegradation of plastics.

5.5.4 Incineration

Incineration is a common reduction method for sludge, by which the sludge will become a valuable and economical mineral product. Especially after incineration, the plastics are completely removed and can no longer enter the environment

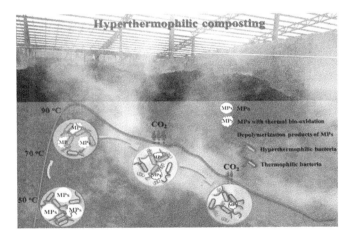

Figure 5.4 The enhanced removal of microplastics by hyperthermophilic composting technology [75] / with permission of Elsevier.

[59]. However, because of the high water content and the large amount of microplastics in the sludge, the incineration process requires high energy consumption and also produces toxic and harmful gases such as dioxins.

5.6 Summary and Outlook

Microplastics have been detected in the water phase and soil phase (sludge system) in WWTPs, regarded as the most critical sinks and sources of microplastics. Over 98% of microplastics in sewage can be removed in WWTPs and are most persistent in the sewage sludge, with 4.4–214 items/g-dry sludge. The content of microplastics in the diverse types of sewage sludge varies with different treatment procedures. Specifically, primary sludge harbors the highest content of microplastics with a diameter of mostly over 500 μm, while remaining microplastics with a particle size smaller than 500 μm are transported to the WAS because of superior adsorption mechanisms. In addition, particles, debris, and fibers are the most common shapes of microplastics in the sewage sludge, with white and black colors. However, current studies investigated microplastics over 20 μm. Smaller microplastics, specifically nanoscaled ones, may have more biotoxicity since they easily across the cell membrane into the intracellular organelle. Therefore, it is necessary to develop further novel detection technologies to reveal the abundance of these tiny particles in the sewage sludge system.

Microplastics have different effects on anaerobic digestion and fermentation, depending on microplastic type, amount, and size. To date, some critical mechanisms regarding the inhibition of microplastics on the yield of methane, short-chain fatty acid and hydrogen have been revealed, including 1) limiting the activity of some key enzymes, 2) altering the microbial structure by reducing the abundance of functional bacterial, 3) leaching some toxic chemicals, especially additives, and 4) producing ROS to damage cell structure. Although the effects of one type of microplastics on anaerobic digestion and fermentation are well documented, several knowledge gaps related to underlying interaction mechanisms and environmental consequences should be resolved in the future. Specifically, the interaction effects among distinct types of microplastics on anaerobic digestion and fermentation should be further investigated. The synergistic or antagonistic interactions should be revealed to understand the role of various kinds of microplastics fully. In addition, the long-term effects of mixed microplastics on continuous reactors should be further discussed.

The final destiny of treated sludge from WWTPs includes landfill or fertilizing crops. In this case, these remaining microplastics can migrate to the terrestrial environment, increasing the potential risk of land-spread microplastics pollution. The soil organisms can be influenced by these microplastics because these tiny particles can be easily ingested and thus accumulate in the food chain. Therefore, knowledge gaps regarding the mobilization and transport of microplastics that are likely to affect the pathway of land-spread sewage sludge microplastics pollution, should be addressed to evaluate and prevent the associated risks.

Although WWTPs have not developed specific microplastic removal processes, current sludge treatment procedures can also change the content of microplastics. Seeking efficient and cost-effective technologies for removing microplastics in the sewage sludge is the top priority issue to be addressed in future research. For example, designing specific treatment units for reducing microplastics is suggested to prohibit the harmful effects of microplastics on sludge and wastewater treatment. In addition, source control is still the most effective method in the current stage. Advocating green packaging and reducing plastic usage from personal care products and garbage bags should be put on the agenda.

References

1 Wang, J., Guo, X., and Xue, J. (2021). Biofilm-developed microplastics as vectors of pollutants in aquatic environments. *Environmental Science & Technology* 55 (19): 12780–12790. http://dx.doi.org/10.1021/acs.est.1c04466.

2 Wright, S.L. and Kelly, F.J. (2017). Plastic and human health: a micro issue? *Environmental Science & Technology* 51 (12): 6634–6647. http://dx.doi.org/10.1021/acs.est.7b00423.

3 Jambeck Jenna, R., Geyer, R., Wilcox, C. et al. (2015). Plastic waste inputs from land into the ocean. *Science* 347 (6223): 768–771. http://dx.doi.org/10.1126/science.1260352.

4 Sutherland, W.J., Clout, M., Côté, I.M. et al. (2010). A horizon scan of global conservation issues for 2010. *Trends in Ecology & Evolution* 25 (1): 1–7. http://dx.doi.org/10.1016/j.tree.2009.10.003.

5 Cózar, A., Echevarría, F., González-Gordillo, J.I. et al. (2014). Plastic debris in the open ocean. *Proceedings of the National Academy of Sciences* 111 (28): 10239. http://dx.doi.org/10.1073/pnas.1314705111.

6 Lambert, S. and Wagner, M. (2016). Characterisation of nanoplastics during the degradation of polystyrene. *Chemosphere* 145: 265–268. http://dx.doi.org/10.1016/j.chemosphere.2015.11.078.

7 Browne, M.A., Crump, P., Niven, S.J. et al. (2011). Accumulation of microplastic on shorelines woldwide: Sources and sinks. *Environmental Science & Technology* 45 (21): 9175–9179. http://dx.doi.org/10.1021/es201811s.

8 Napper, I.E., Bakir, A., Rowland, S.J. et al. (2015). Characterisation, quantity and sorptive properties of microplastics extracted from cosmetics. *Marine*

Pollution Bulletin 99 (1): 178–185. http://dx.doi.org/10.1016/j.marpolbul.2015.07.029.

9   Fendall, L.S. and Sewell, M.A. (2009). Contributing to marine pollution by washing your face: microplastics in facial cleansers. *Marine Pollution Bulletin* 58 (8): 1225–1228. http://doi.org/10.1016/j.marpolbul.2009.04.025.

10 Liu, H., Zhou, X., Ding, W. et al. (2019). Do microplastics affect biological wastewater treatment performance? Implications from bacterial activity experiments. *ACS Sustainable Chemistry & Engineering* 7 (24): 20097–20101. http://dx.doi.org/10.1021/acssuschemeng.9b05960.

11 Zhang, J., Zhao, M., Li, C. et al. (2020a). Evaluation the impact of polystyrene micro and nanoplastics on the methane generation by anaerobic digestion. *Ecotoxicology and Environmental Safety* 205: 111095. http://dx.doi.org/10.1016/j.ecoenv.2020.111095.

12 Zhang, X., Chen, J., and Li, J. (2020c). The removal of microplastics in the wastewater treatment process and their potential impact on anaerobic digestion due to pollutants association. *Chemosphere* 251: 126360. http://dx.doi.org/10.1016/j.chemosphere.2020.126360.

13 Rochman, C.M., Kross, S.M., Armstrong, J.B. et al. (2015). Scientific evidence supports a ban on microbeads. *Environmental Science & Technology* 49 (18): 10759–10761. http://dx.doi.org/10.1021/acs.est.5b03909.

14 Iyare, P.U., Ouki, S.K., and Bond, T. (2020). Microplastics removal in wastewater treatment plants: a critical review. *Environmental Science: Water Research & Technology* 6 (10): 2664–2675. http://dx.doi.org/10.1039/D0EW00397B.

15 Zhang, Z. and Chen, Y. (2020). Effects of microplastics on wastewater and sewage sludge treatment and their removal: a review. *Chemical Engineering Journal* 382: 122955. http://dx.doi.org/10.1016/j.cej.2019.122955.

16 Lv, X., Dong, Q., Zuo, Z. et al. (2019). Microplastics in a municipal wastewater treatment plant: fate, dynamic distribution, removal efficiencies, and control strategies. *Journal of Cleaner Production* 225: 579–586. http://doi.org/10.1016/j.jclepro.2019.03.321.

17 Luo, J., Zhang, Q., Zhao, J. et al. (2020). Potential influences of exogenous pollutants occurred in waste activated sludge on anaerobic digestion: a review. *Journal of Hazardous Materials* 383: 121176. http://dx.doi.org/10.1016/j.jhazmat.2019.121176.

18 Mahon, A.M., O'Connell, B., Healy, M.G. et al. (2017). Microplastics in sewage sludge: effects of treatment. *Environmental Science & Technology* 51 (2): 810–818. http://doi.org/10.1021/acs.est.6b04048.

19 Rolsky, C., Kelkar, V., Driver, E. et al. (2020). Municipal sewage sludge as a source of microplastics in the environment. *Current Opinion in Environmental Science & Health* 14: 16–22. http://dx.doi.org/10.1016/j.coesh.2019.12.001.

20 He, Z.-W., Tang, -C.-C., Liu, W.-Z. et al. (2019). Enhanced short-chain fatty acids production from waste activated sludge with alkaline followed by potassium ferrate treatment. *Bioresource Technology* 289: 121642. http://doi.org/10.1016/j.biortech.2019.121642.

21 Wang, B., Liu, W., Zhang, Y. et al. (2020). Bioenergy recovery from wastewater accelerated by solar power: intermittent electro-driving regulation and capacitive storage in biomass. *Water Research* 175: 115696. http://dx.doi.org/10.1016/j.watres.2020.115696.

22 Hatinoğlu, M.D. and Sanin, F.D. (2021). Sewage sludge as a source of microplastics in the environment: a review of occurrence and fate during sludge treatment. *Journal of Environmental Management* 295: 113028. http://doi.org/10.1016/j.jenvman.2021.113028.

23 Volkheimer, G. (1975). Hematogenous dissemination of ingested polyvinyl chloride particles. *Annals of the New York Academy of Sciences* 246 (1): 164–171. http://dx.doi.org/10.1111/j.1749-6632.1975.tb51092.x.

24 Nel, A., Xia, T., Mädler, L. et al. (2006). Toxic potential of materials at the nanolevel. *Science* 311 (5761): 622–627. http://dx.doi.org/10.1126/science.1114397.

25 Browne, M.A., Niven, S.J., Galloway, T.S. et al. (2013). Microplastic moves pollutants and additives to worms, reducing functions linked to health and biodiversity. *Current Biology* 23 (23): 2388–2392. http://dx.doi.org/10.1016/j.cub.2013.10.012.

26 Khan, F.R., Syberg, K., Shashoua, Y. et al. (2015). Influence of polyethylene microplastic beads on the uptake and localization of silver in zebrafish (Danio rerio). *Environmental Pollution* 206: 73–79. https://doi.org/10.1016/j.envpol.2015.06.009.

27 Gies, E.A., LeNoble, J.L., Noel, M. et al. (2018). Retention of microplastics in a major secondary wastewater treatment plant in Vancouver, Canada. *Marine Pollution Bulletin* 133: 553–561. http://doi.org/10.1016/j.marpolbul.2018.06.006.

28 Raju, S., Carbery, M., Kuttykattil, A. et al. (2018). Transport and fate of microplastics in wastewater treatment plants: implications to environmental health. *Reviews in Environmental Science and Bio/Technology* 17 (4): 637–653. https://doi.org/10.1007/s11157-018-9480-3.

29 Gao, D., Li, X.Y., and Liu, H.T. (2020). Source, occurrence, migration and potential environmental risk of microplastics in sewage sludge and during sludge amendment to soil. *Science of the Total Environment* 742. http://doi.org/10.1016/j.scitotenv.2020.140355.

30 Jeong, C., Won, E., Kang, H. et al. (2016). Microplastic size-dependent toxicity, oxidative stress induction, and p-JNK and p-p38 activation in the monogonont rotifer (Brachionus koreanus). *Environmental Science & Technology* 50 (16): 8849–8857. http://dx.doi.org/10.1021/acs.est.6b01441.

31 Naqash, N., Prakash, S., Kapoor, D. et al. (2020). Interaction of freshwater microplastics with biota and heavy metals: a review. *Environmental Chemistry Letters* 18 (6): 1813–1824. https://doi.org/10.1007/s10311-020-01044-3.

32 Murphy, F., Ewins, C., Carbonnier, F. et al. (2016). Wastewater treatment works (WwTW) as a source of microplastics in the aquatic environment. *Environmental Science & Technology* 50 (11): 5800–5808. http://doi.org/10.1021/acs.est.6b04048.

33 Carr, S.A., Liu, J., and Tesoro, A.G. (2016). Transport and fate of microplastic particles in wastewater treatment plants. *Water Research* 91: 174–182. https://doi.org/10.1016/j.watres.2016.01.002.

34 Sun, J., Dai, X.H., Wang, Q.L. et al. (2019). Microplastics in wastewater treatment plants: detection, occurrence and removal. *Water Research* 152: 21–37. http://doi.org/10.1016/j.watres.2018.12.050.

35 Cheng, Y.L., Kim, J., Kim, H. et al. (2021b). Occurrence and removal of microplastics in wastewater treatment plants and drinking water purification facilities: a review. *Chemical Engineering Journal* 410. https://doi.org/10.1016/j.cej.2020.128381.

36 Raju, S., Carbery, M., Kuttykattil, A. et al. (2020). Improved methodology to determine the fate and transport of microplastics in a secondary wastewater treatment plant. *Water Research* 173: 115549. https://doi.org/10.1016/j.watres.2020.115549.

37 Alavian Petroody, S.S., Hashemi, S.H., and van Gestel, C.A.M. (2021). Transport and accumulation of microplastics through wastewater treatment sludge processes. *Chemosphere* 278: 130471. http://doi.org/10.1016/j.chemosphere.2021.130471.

38 Hamidian, A.H., Ozumcheloui, E.J., Feizi, F. et al. (2021). A review on the characteristics of microplastics in wastewater treatment plants: a source for toxic chemicals. *Journal of Cleaner Production* 295: 126480. http://doi.org/10.1016/j.jclepro.2021.126480.

39 Li, X., Chen, L., Mei, Q. et al. (2018). Microplastics in sewage sludge from the wastewater treatment plants in China. *Water Research* 142: 75–85. http://doi.org/10.1016/j.watres.2018.05.034.

40 Li, L., Geng, S., Li, Z. et al. (2020a). Effect of microplastic on anaerobic digestion of wasted activated sludge. *Chemosphere* 247: 125874. http://doi.org/10.1016/j.chemosphere.2020.125874.

41 Mintenig, S.M., Int-Veen, I., Loeder, M.G.J. et al. (2017). Identification of microplastic in effluents of waste water treatment plants using focal plane array-based micro-Fourier-transform infrared imaging. *Water Research* 108: 365–372. https://doi.org/10.1016/j.watres.2016.11.015.

42 Wei, W., Huang, Q.S., Sun, J. et al. (2019a). Revealing the mechanisms of polyethylene microplastics affecting anaerobic digestion of waste activated

43 Wei, W., Huang, Q.S., Sun, J. et al. (2019b). Polyvinyl chloride microplastics affect methane production from the anaerobic digestion of waste activated sludge through leaching toxic bisphenol-A. *Environmental Science & Technology* 53 (5): 2509–2517. https://doi.org/10.1021/acs.est.8b07069.

44 Chen, H., Tang, M., Yang, X. et al. (2021a). Polyamide 6 microplastics facilitate methane production during anaerobic digestion of waste activated sludge. *Chemical Engineering Journal* 408: 127251. http://doi.org/10.1016/j.cej.2020.127251.

45 Feng, Y., Duan, J.L., Sun, X.D. et al. (2021). Insights on the inhibition of anaerobic digestion performances under short-term exposure of metal-doped nanoplastics via Methanosarcina acetivorans. *Environmental Pollution* 275: 115755. https://doi.org/10.1016/j.envpol.2020.115755.

46 Fu, S.F., Ding, J.N., Zhang, Y. et al. (2018). Exposure to polystyrene nanoplastic leads to inhibition of anaerobic digestion system. *Science of the Total Environment* 625: 64–70. http://doi.org/10.1016/j.scitotenv.2017.12.158.

47 Azizi, S.M.M., Hai, F.I., Lu, W. et al. (2021). A review of mechanisms underlying the impacts of (nano)microplastics on anaerobic digestion. *Bioresource Technology* 329: 124894. http://dx.doi.org/10.1016/j.biortech.2021.124894.

48 Wei, W., Chen, X., and Ni, B.J. (2021a). Different pathways of microplastics entering the sludge treatment system distinctively affect anaerobic sludge fermentation processes. *Environmental Science & Technology* 55 (16): 11274–11283. http://dx.doi.org/10.1021/acs.est.1c02300.

49 Liu, Q., Li, L., Zhao, X. et al. (2021). An evaluation of the effects of nanoplastics on the removal of activated-sludge nutrients and production of short chain fatty acid. *Process Safety and Environmental Protection* 148: 1070–1076. http://doi.org/10.1016/j.psep.2021.02.029.

50 Chen, W., Yuan, D., Shan, M. et al. (2020a). Single and combined effects of amino polystyrene and perfluorooctane sulfonate on hydrogen-producing thermophilic bacteria and the interaction mechanisms. *Science of the Total Environment* 703: 135015. http://doi.org/10.1016/j.scitotenv.2019.135015.

51 Wei, W., Zhang, Y.T., Huang, Q.S., and Ni, B.J. (2019c). Polyethylene terephthalate microplastics affect hydrogen production from alkaline anaerobic fermentation of waste activated sludge through altering viability and activity of anaerobic microorganisms. *Water Research* 163: 114881. https://doi.org/10.1016/j.watres.2019.114881.

52 Lin, X., Su, C., Deng, X. et al. (2020). Influence of polyether sulfone microplastics and bisphenol A on anaerobic granular sludge: performance evaluation and microbial community characterization. *Ecotoxicology and Environmental Safety* 205: 111318. http://doi.org/10.1016/j.ecoenv.2020.111318.

53 Wei, W., Hao, Q., Chen, Z. et al. (2020). Polystyrene nanoplastics reshape the anaerobic granular sludge for recovering methane from wastewater. *Water Research* 182: 116041. https://doi.org/10.1016/j.watres.2020.116041.

54 Zhang, Y.T., Wei, W., Huang, Q.S. et al. (2020d). Insights into the microbial response of anaerobic granular sludge during long-term exposure to polyethylene terephthalate microplastics. *Water Research* 179: 115898. http://dx.doi.org/10.1016/j.watres.2020.115898.

55 Yang, G.J., Xu, Q.X., Wang, D.B. et al. (2018). Free ammonia-based sludge treatment reduces sludge production in the wastewater treatment process. *Chemosphere* 205: 484–492. http://dx.doi.org/10.1016/j.chemosphere.2018.04.140.

56 Petersen, F. and Hubbart, J.A. (2021). The occurrence and transport of microplastics: the state of the science. *Science of the Total Environment* 758: 143936. https://doi.org/10.1016/j.scitotenv.2020.143936.

57 Golwala, H., Zhang, X.Y., Iskander, S.M. et al. (2021). Solid waste: an overlooked source of microplastics to the environment. *Science of the Total Environment* 769.

58 Ziajahromi, S., Neale, P.A., Telles Silveira, I. et al. (2021). An audit of microplastic abundance throughout three Australian wastewater treatment plants. *Chemosphere* 263: 128294. http://dx.doi.org/10.1016/j.chemosphere.2020.128294.

59 Koyuncuoglu, P. and Erden, G. (2021). Sampling, pre-treatment, and identification methods of microplastics in sewage sludge and their effects in agricultural soils: a review. *Environmental Monitoring and Assessment* 193 (4): 175. http://doi.org/10.1016/10.1007/s10661-021-08943-0.

60 Edo, C., González-Pleiter, M., Leganés, F. et al. (2020). Fate of microplastics in wastewater treatment plants and their environmental dispersion with effluent and sludge. *Environmental Pollution* 259: 113837. http://doi.org/10.1016/j.envpol.2019.113837.

61 van den Berg, P., Huerta-Lwanga, E., Corradini, F. et al. (2020). Sewage sludge application as a vehicle for microplastics in eastern Spanish agricultural soils. *Environmental Pollution* 261: 114198. http://doi.org/10.1016/j.envpol.2020.114198.

62 Yang, J., Li, L., Li, R. et al. (2021). Microplastics in an agricultural soil following repeated application of three types of sewage sludge: a field study. *Environmental Pollution* 289: 117943. http://dx.doi.org/10.1016/j.envpol.2021.117943.

63 Koutnik, V.S., Alkidim, S., Leonard, J. et al. (2021). Unaccounted microplastics in wastewater sludge: where do they go? *ACS ES&T Water* 1 (5): 1086–1097. http://doi.org/10.1016/10.1021/acsestwater.0c00267.

64 Corradini, F., Meza, P., Eguiluz, R. et al. (2019). Evidence of microplastic accumulation in agricultural soils from sewage sludge disposal. *Science of the Total Environment* 671: 411–420. https://doi.org/10.1016/j.scitotenv.2019.03.368.

65 Vollertsen, J. and Hansen, A.A. (eds) (2017). Microplastic in Danish wastewater: sources, occurrences and fate. *Danish Environmental Protection Agency* 1906: 55.

66 Zubris, K.A. and Richards, B.K. (2005). Synthetic fibers as an indicator of land application of sludge. *Environmental Pollution* 138 (2): 201–211. http://dx.doi.org/10.1016/j.envpol.2005.04.013.

67 Hou, L.Y., Kumarb, D., Yoob, C.G. et al. (2021). Conversion and removal strategies for microplastics in wastewater treatment plants and landfills. *Chemical Engineering Journal* 406. http://doi.org/10.1016/j.cej.2020.126715.

68 Su, Y., Zhang, Z., Wu, D. et al. (2019). Occurrence of microplastics in landfill systems and their fate with landfill age. *Water Research* 164: 114968. https://doi.org/10.1016/j.watres.2019.114968.

69 Sui, Q., Zhao, W., Cao, X. et al. (2017). Pharmaceuticals and personal care products in the leachates from a typical landfill reservoir of municipal solid waste in Shanghai, China: occurrence and removal by a full-scale membrane bioreactor. *Journal of Hazardous Materials* 323 (Pt A): 99–108. https://doi.org/10.1016/10.1016/j.jhazmat.2016.03.047.

70 Sun, J., Zhu, Z.-R., Li, W.-H. et al. (2021). Revisiting microplastics in landfill leachate: Unnoticed tiny microplastics and their fate in treatment works. *Water Research* 190: 116784. http://doi.org/10.1016/j.watres.2020.116784.

71 Talvitie, J., Mikola, A., Setala, O. et al. (2017). How well is microlitter purified from wastewater? - a detailed study on the stepwise removal of microlitter in a tertiary level wastewater treatment plant. *Water Research* 109: 164–172. http://doi.org/10.1016/j.watres.2016.11.046.

72 Narancic, T., Verstichel, S., Reddy Chaganti, S. et al. (2018). Biodegradable plastic blends create new possibilities for end-of-life management of plastics but they are not a Panacea for plastic pollution. *Environmental Science & Technology* 52 (18): 10441–10452. https://doi.org/10.1021/acs.est.8b02963.

73 Hurley, R.R., Lusher, A.L., Olsen, M. et al. (2018). Validation of a method for extracting microplastics from complex, organic-rich, environmental matrices. *Environmental Science & Technology* 52 (13): 7409–7417. http://doi.org/10.1021/acs.est.8b01517.

74 Li, X.W., Chen, L.B., Ji, Y.Y. et al. (2020b). Effects of chemical pretreatments on microplastic extraction in sewage sludge and their physicochemical characteristics. *Water Research* 171. http://doi.org/10.1016/j.watres.2019.115379.

75 Chen, Z., Zhao, W., Xing, R. et al., Wang, S. and Zhou, S. (2020b). Enhanced in situ biodegradation of microplastics in sewage sludge using hyperthermophilic composting technology. *Journal of Hazardous Materials* 384: 121271. http://doi.org/10.1016/j.jhazmat.2019.121271.

76 Harshvardhan, K. and Jha, B. (2013). Biodegradation of low-density polyethylene by marine bacteria from pelagic waters, Arabian Sea, India. *Marine Pollution Bulletin* 77 (1–2): 100–106. http://doi.org/10.1016/j.marpolbul.2013.10.025.

77 De Tender, C., Devriese, L.I., Haegeman, A. et al. (2017). Temporal dynamics of bacterial and fungal colonization on plastic debris in the North Sea. *Environmental Science & Technology* 51 (13): 7350–7360. http://doi.org/10.1021/acs.est.7b00697.

78 Zhang, L., Xie, Y., Liu, J. et al. (2020b). An overlooked entry pathway of microplastics into agricultural soils from application of sludge-based fertilizers. *Environmental Science & Technology* 54 (7): 4248–4255. http://dx.doi.org/10.1021/acs.est.9b07905.

6

Discharge of Microplastics from Wastewater Treatment Plants

Hongbo Chen and Yi Wu

College of Environment and Resources, Xiangtan University, Xiangtan, China

6.1 Introduction

Wastewater treatment plants (WWTPs) are an important part of the urban drainage system and are considered a potential source of microplastics in the environment [1]. Studies have shown that most of the microplastics are transferred to the sludge after treatment in WWTPs, with a removal efficiency of 70–99% [2, 3]. However, the remaining microplastics are still continuously discharged from WWTPs into the receiving water body [2, 4, 5].

The microplastics in the effluent of WWTPs are discharged directly into rivers, then into lakes, and finally collected in the ocean. This means that the ocean is a huge plastic warehouse. In general, small and light plastic particles float on the water surface, and large and heavy plastic particles are deposited on the bottom [6]. The deposited microplastics are stored in sediments, and the suspended microplastics continue to flow on the surface and are ingested by aquatic organisms. PlasticsEurope predicted that by 2025, a total of 250 million tons of plastic will enter environmental water bodies [7].

Microplastics continuously ingested by aquatic organisms accumulate along the food chain between organisms. Since freshwater and marine resources are important sources of drinking water and seafood, understanding the content and types of microplastics in water bodies is of paramount importance for removing and degrading microplastics in water bodies. This chapter summarizes the fate of microplastics and the current composition of microplastics in terms of abundance and polymer types from WWTPs effluent, freshwater systems, marine systems, and aquatic organisms, aiming to lay the foundation for the removal and degradation of microplastics.

6.2 Microplastics Concentrations in Effluent of WWTPs

Recent studies have shown that WWTPs effluent is an important source of microplastics [3, 8–10]. Therefore, it is necessary to fully understand the concentration of microplastics discharged from WWTPs.

6.2.1 Concentration of Microplastics in Effluent

The concentration of microplastics in the effluent of WWTPs from different countries is shown in Table 6.1. Obviously, the concentration of microplastics in the effluent discharged by WWTPs in developed regions is lower than that in developing regions. For example, the average concentration of microplastics in the effluent of 12 WWTPs from the USA is 0.009 particles/L [8], and the abundance of microplastics in the effluent of 12 WWTPs in Germany ranges 0–0.05 particles/L [11]. In contrast, the average discharge concentration of microplastics in 11 WWTPs in China is as high as 9.04 ± 1.12 particles/L [5].

The treatment process of WWTPs affects the concentration of microplastics in the effluent. In general, the concentration of microplastics in the effluent of WWTPs using the tertiary treatment process (0.006–0.651 particles/L) is lower than that of the WWTPs using only the primary or secondary treatment process (1.0 ± 0.4 particles/L) (Table 6.1). However, studies have also shown that the tertiary treatment of some WWTPs did not further reduce the concentration of microplastics in the effluent [8, 20]. This may arise from the different removal efficiencies of different treatment procedures for microplastics [23].

Although the abundance of microplastics in WWTPs effluent is relatively low, the total amount of microplastics discharged from WWTPs is still quite high because most of these facilities treat millions of liters of wastewater every day. The amount of microplastics in the effluent discharged by WWTPs every day is staggering. According to incomplete statistics, the daily discharge of microplastics from WWTPs is about is ca. 2.4×10^9 particles (Table 6.1). It is estimated that in Europe alone, as many as 520 000 tons of microplastics are discharged from WWTPs every year. In some WWTPs in the Netherlands, the total daily discharge of microplastics even exceeds 1×10^{10} particles/d. Once these microplastics enter a water body, there will be ecological risks. Therefore, WWTPs with high discharge of microplastics urgently need treatment technologies for microplastics control to avoid substantial amounts of them being discharged into the ecosystem.

6.2.2 Types of Microplastics in Effluent

Up to now, more than 30 types of microplastics have been detected in the effluent of WWTPs. Common microplastics include polypropylene (PP), polyethylene (PE), polyester (PES), Polyvinyl chloride (PVC), rayon, polyethylene and

Table 6.1 Microplastics discharged from WWTPs in different regions.

Countries	Cities	WWTPs quantities	Microplastic abundance in effluent	Microplastic emissions	Dominant polymer types	References
China	Changzhou	11	9.04 ± 1.12 particles/L		Rayon, PES	[5]
China	Xiamen	7	0.59 particles/L (Avg.)	6.5×10^8 particles/d	PP, PE, PS	[9]
Finland		1	1.0 ± 0.4 particles/L		PET	[4]
Finland		1	6–651 particles/m^3	1.7×10^6–1.4×10^8 particles/d		[12]
Germany		1	3–5.9 particles/L		PET, PE, PP, PS	[13]
Germany		12	0–50 particles/m^3		PE, PP	[11]
Italy		1	0.4 particles/L (Avg.)	1.6×10^8 particles/d	PES, PA, PE	[14]
UK		1	< 1–3 particles/L	2.2×10^7 particles/d	PP, PE	[15]
UK	Glasgow	1	0.25 ± 0.04 particles/L	6.5×10^7 particles/d	PES, PA, PP	[16]
Netherlands		7	55–81 particles/L	7.48×10^8–4.32×10^{10} particles/d		[17]
France	Le Havre	1	0.099–2.844 particles/L	2.27×10^8 particles/d	PE, PS	[18]
Turkey	Mersin Bay	3	0.9 particles/L (Avg.)	1.8×10^8 particles/d		[19]
Canada	Vancouver	1	0.5 ± 0.2 particles/L	1×10^8–3×10^8 particles/d		[3]
USA		1	2.6 particles/L (Avg.)	4.43×10^6 particles/d		[20]

(Continued)

Table 6.1 (Continued)

Countries	Cities	WWTPs quantities	Microplastic abundance in effluent	Microplastic emissions	Dominant polymer types	References
USA		12	0.004–0.195 particles/L	5.28×10^4–1.49×10^7 particles/d		[21]
USA		1	0.009 particles/L (Avg.)	$\sim 0.93 \times 10^6$ particles/d	PE	[8]
Australia	Sydney	3	0.2–1.5 particles/L	4.72×10^8 particles/d	PET, PE	[22]
Australia		3	0.7 particle/L (Avg.)	2.21×10^7–1.33×10^8 particles/d	PET, PE, PP	[2]

polypropylene copolymer (PE-PP), polyethylene terephthalate (PET), and polyamide (PA) (Table 6.2). Each type of polymer has physicochemical and decomposition properties that are key factors for the formation, distribution, and aggregation of microplastics in the aquatic environment [24].

Table 6.2 summarizes the proportion of certain microplastic polymers in WWTPs effluent in some regions. Comparing the types of microplastics in the effluents of WWTPs in different regions, it is clear that PP and PE are common in the effluents in Asia (mainly China) and Europe (Table 6.2). PP and PE are

Table 6.2 Percentage of different types of microplastics emitted by WWTPs in different countries.

| Countries | Cities | Proportion of microplastics in effluent (%) | | | | | | | | References |
		Rayon	PET	PP	PE	PS	PE-PP	PP-PE	PES	PA	
China	Changzhou	43.45	29.22	14.46	6.28	2.12	1.51				[5]
China	Xiamen		7.5	34.8	17.9	9.6	4.7	13.9			[9]
UK	Glasgow		4	12	4	4			28	20	[16]
Italy					10				35	17	[14]
France	Le Havre				39.7						[18]
Turkey	Mersin Bay			35	51						[19]

plastic types that are widely produced and used worldwide. Both are commonly used in the manufacture of films, packaging bags, and containers. These plastic products are prone to break and enter the sewage pipe network during daily use. For example, a large amount of fiber from the washing of clothes enters the environment. Up to 1900 fibers are released from washing a piece of clothing.

In addition, other common types of microplastics in Asia are PS, PET, and rayon, while in Europe they are PES and PA. In Asia, China for example, there are approximately 3340 WWTPs in operation in mainland China. In all these WWTPs, PP, PE, PS, and PET dominate the effluent. This is because PP, PE, and PS are the main types of polymers consumed in China [9]. In Changzhou, China, the main polymer types of the 11 WWTPs are rayon (43.45%) and PET (29.22%), followed by PP (14.46%), PE (6.28%), PS (2.12%), and PE-PP (1.51%) [5]. The main sources of rayon and PET or PA are laundry and industrial applications, such as packaging and textiles [23]. But in Europe, the PES and PA occupy a certain dominant position. The microplastic polymers discharged from a WWTP in the UK include PES, PA, PP, PE, PET, and PS. Among them, the dominant PES and PA accounted for 28% and 20% of microplastic polymers respectively [16]. The contents of PES and PA in Italian WWTPs are 35% and 17%, respectively, and PE is 10% [14]. PA materials are widely used in various fields of life, including textiles, electrical and electronic components, automotive parts, films, etc. Therefore, PA may come from washing clothes and vehicles in daily life. PES is widely used in the packaging industry, electronic appliances, medical and health, construction, automobiles, and other fields. Packaging is the largest non-fiber application market for polyester. This indicates that a large amount of PES may come from various packaging used in daily life.

In summary, the large amount of microplastics in the effluent of WWTPs and the proportion of several types of microplastic polymers in the effluent are closely related to the mass production and use of plastics worldwide.

6.3 Important Source of the Receiving Waters

WWTPs are an important way for microplastics to enter environmental water bodies [2]. There are two main destinations for microplastics after entering environmental water bodies with the effluent of WWTPs. One of them is deposited at the bottom of rivers and lakes and mixed with the silt at the bottom of rivers and lakes; the other is eventually into the sea with the flow of water. Specifically, they may be deposited on the seabed or suspended after entering the sea (Figure 6.1).

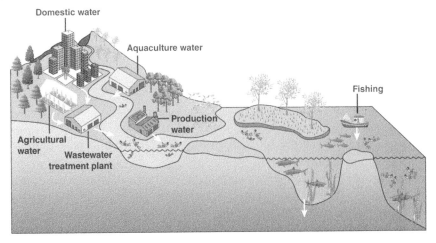

Figure 6.1 The fate of microplastics in urban water systems.

In general, inland rivers and other tributaries that flow into the coast are the main transport channels for microplastics from land to marine waters [25]. Most WWTPs discharge effluent into rivers [14] and eventually the ocean. Therefore, inland water bodies are not only the transmission channel for microplastics from the inland to the ocean, but also a tank for a considerable amount of microplastics. In other words, WWTPs effluent is an important source of daily microplastics in the aquatic environment. In addition, microplastics are easily deposited in estuaries and bays by physical settlement. Their content in sediments will accumulate with the sediments over time and will migrate between the sediments and water. In short, the migration of microplastics in water makes it important to study the quantity and types of microplastics in freshwater, oceans, and sediments.

6.3.1 River

A river is a linear flow that spans multiple regions. The water from many urban pipeline networks is discharged into nearby rivers through WWTPs. Industrial and human activities will affect the concentration, distribution, and type of microplastics on the surface of nearby discharge water bodies. The concentrations and types of microplastics in different rivers around the world are summarized in Table 6.3. In Asia, the surface water of several rivers in China contains elevated levels of microplastics, followed by Pakistan and Indonesia. In particular, the abundance of microplastics detected in the Manas River in China is as high as 10–22 particles/L [26]. The concentration of microplastics in the Ravi River in Pakistan and the Ciwalengke River in Indonesia also reached 16.15 ± 0.08 MPs/L

Table 6.3 Microplastic abundance and types in rivers in different regions.

Regions	Countries	Rivers	Microplastic abundance	Dominant polymer types	References
Asia	China	Beijiang River	0.56 ± 45 particles/m^3	PE, PP, PS	[29]
Asia	China	Suzhou and Huangpu River	1.8–2.4 particles/L	PES, rayon, PP	[30]
Asia	China	Pearl River	0.57 ± 0.71 particles/L	PE, PP, PS	[31]
Asia	Indonesia	Ciwalengke River	5.85 ± 3.28 particles/L		[28]
Asia	Pakistan	Ravi River	16 150 ± 80 particles/m^3	PE, PP, PS	[27]
Asia	China	Manas River	10–22 particles/L	PP, PET, PS	[26]
Europe	Hungary	River Zala	23.85 particles/m^3 (Avg.)	PP, PS	[32]
Europe	Germany	Rhine River	0.05–8.3 particles/m^3		[33]
Europe	Spain	Ebro River	3.5 ± 1.4 particles/m^3	PA, PE, PMMA, PES	[34]
Europe	Italy	Ofanto River	0.9–18 particles/m^3	PE, PS, PP	[35]
Europe	UK	Thames River	8–36.7 particles/m^3	PE, PP	[36]
Europe	Poland	Vistula River	1.6–2.55 particles/L	PS, PP, PA	[37]
Africa		Orange-Vaal River	0.21 ± 0.27 particles/L		[38]
Latin America	Ecuador	Tropical Andean Rivers	0.72–1186 particles/m^3		[39]
North America	USA	Oregon Rivers	0–0.4 particles/m^3		[40]
South America	Colombia	Magdalena River	0–0.14 particles/L	PP, PE	[41]

[27] and 5.85 ± 3.28 particles/L [28], respectively. In addition to Asian countries, Hungary (23.85 particles/m^3 on average), Ecuador (0.72–1186 particle/m^3), Poland (1.6–2.55 particles/L), and the UK (8–36.7 particle/m^3) also have high concentrations of microplastics.

Because of urban areas and factories along the coast, the content of microplastics in the river changes dramatically in different sections [33, 34]. The abundance of microplastics in river sections close to urban areas tends to be higher. The concentration of microbeads within a specified range of 20 kilometers from the Rhine River increased by 166 times, indicating the possibility of microbeads influx. The investigation revealed that there are three WWTPs near 20 kilometers of the river section, two of which receive industrial wastewater [33], which may be the main reason for the surge in microplastics in the Rhine.

Various polymer microplastics such as PP, PE, PS, and PET have been detected in rivers from different regions [27, 29, 31, 34, 35, 38, 41]. Nevertheless, the main types of microplastics in rivers are PP and PE (Table 6.3). As a common raw material for plastic products (e.g., fishing gear, toys, bags, bottles), PP is widely used in modern daily life, while PE is usually used in manufacturing and packaging materials. Therefore, these PP and PE fragments may originate from the degradation of plastic products for daily use. This is similar to the type of microplastics in wastewater discharged from WWTPs.

As an exception, the main type of microplastics in the Ebro River in Spain is PA (24%), followed by PE (16%), poly(methyl methacrylate) (PMMA) (acrylic, 12%), and PES (12%) [34]. In the area where the Ebro River flows, there are two WWTPs directly discharged into it. Not coincidentally, the main fiber types in these two WWTPs wastewater are PES, acrylic and PA [34]. Therefore, the effluent discharge of WWTPs is an important source of microplastics in rivers.

6.3.2 Lake

Lake systems have distinct characteristics compared to other water bodies such as oceans or rivers. First, lakes are widely distributed in various countries and regions. Secondly, the water quality, hydrological environment, and other characteristics of the water body and sediment in the lake system are relatively stable, which makes the deposition of microplastics easier. Lakes have relatively low flow rates and can store water longer than rivers that can stably deposit more microplastics [42]. The concentration of microplastics in lakes is higher than in rivers, which may be the result of multiple rivers merging into the lake.

Microplastic in lakes from different countries show explicit regional differences in abundance. The abundance of microplastics in lakes in China ranges 1.81–34 000 particles/m^3 [42]. Among these lakes, Poyang Lake has the highest microplastic content, reaching 4.5–35.4 particles/L [43]. For the abundance of microplastics in other Asian countries, the average abundance of Red Hills Lake in India is 5.9 particles/L, while that of Al-Hubail Lake and Al-Asfar Lake in Saudi Arabia is 3.7 ± 3.1 particles/L and 2.7 ± 2.9 particles/L, respectively (Table 6.4). In Europe, the content of microplastics in lakes is relatively low. Studies in Hungary

Table 6.4 Microplastic abundance and types in lakes in different regions.

Regions	Countries	Lakes	Microplastic abundance	Dominant polymer types	References
Asia	China	Lake Hovsgol	997–44 435 particles/km^2		[48]
Asia	China	Qinghai Lake	5×10^3–7.58×10^5 particles/km^2	PE, PP	[49]
Asia	India	Red Hills Lake	5.9 particles/L (Avg.)	PP, PE	[50]
Asia	China	Wuliangsuhai Lake	3.12–11.25 particles/L	PS, PP, PE	[51]
Asia	Saudi Arabia	Al-Hubail Lake	3.7 ± 3.1 particles/L		[52]
Asia	Saudi Arabia	Al-Asfar Lake	2.7 ± 2.9 particles/L		[52]
Asia	China	Poyang Lake	4.5–35.4 particles/L	PP, PE	[43]
Europe	Germany	Lake Tollense	0.14 particles/m^3 (Avg.)	PE, PET	[44]
Europe	Hungary	Lake Tisza-tó	23.12 particles/m^3 (Avg.)	PP, PS	[32]
Africa	Kenya	Lake Naivasha	0.407 ± 0.135 particles/m^2 (Avg.)	PP, PE, PES	[46]
Africa	Nigeria	OX-Bow Lake	Dry season: 1004–8329 particles/m^3 Raining season: 201–8369 particles/m^3	PET, PVC	[45]
North America	Canada	Lake Simcoe	0.4–1.3 particles/L	PE, PP	[53]
North America	USA	Lakes Mead and Mohave	0.44–9.7 particles/m^3		[54]
South America	Argentina	La Salada Lake	100–180 particles/m^3		[47]
South America	Brazil	Lake Guaíba	11.3–67.3 particles/m^3	PP, PE	[55]

and Germany show that the average concentrations of microplastics in Lake Tisza-tó and Lake Tollense are 23.12 particles/m^3 and 0.14 particles/m^3, respectively [32,44]. In Africa, the concentration of microplastics in OX-Bow Lake in Nigeria is surprisingly high, with an average concentration of 3701.6 particles/m^3 [45]. But in Kenya, the average concentration is only 0.407 ± 0.135 particles/m^2 [46]. According to Table 6.4, it is obvious that the concentration of microplastics in lakes in North America is lower than that in Asian countries. In South America, the concentration of microplastics in La Salada Lake is as high as 100–180 particles/m^3, significantly higher than other developed countries [47].

According to the above analysis, it can be clearly observed that the pollution of microplastics in the surface water of lakes in China and Saudi Arabia is much more serious than other countries in Europe, North America, and Africa. This shows that the problem of microplastics in developing countries is more serious. Therefore, the difference in development level and economic structure is likely to be the main reason for the regional differences in the distribution of microplastics.

In addition to the differences in the abundance of microplastics, the differences in the types of microplastics in lakes from different countries and regions are also summarized in Table 6.4. The types of microplastics in lakes from different regions include PP, PE, PS, PES, PET, PVC, etc. In Asia, North America, and Africa, the most common polymers are PE and PP. In Africa, PES and PVC account for a large proportion. The world's largest non-fiber plastic production categories are PE (36%) and PP (21%), followed by PVC, PES, and PE [42]. Therefore, the polymer composition and type of microplastics in lakes are consistent with global plastic production.

6.3.3 Sea

As mentioned above, the microplastics in the effluent of WWTPs enter the river system that connects most of the world's surface with the marine environment. Therefore, the microplastics entering the river system will be transported to the ocean. Microplastics are more abundant in the marine environment than in the freshwater environment [56]. Since about half of the world's population lives within 50 miles of the coast, a large amount of plastic produced by people's daily life and production will enter the ocean through a network of rivers [57, 58]. In other words, many microplastics in the ocean come from freshwater systems [59]. According to statistics, the amount of plastic waste flowing into the ocean from rivers in the world is 0.41–4 × 10^6 tons [58]. From coastal areas to high seas, from tropical oceans to polar oceans, and from surface waters to abyssal region, microplastics are everywhere (Table 6.5).

Table 6.5 Microplastic abundance and types in the ocean in different regions.

Regions	Countries	Oceans	Sites	Microplastic abundance	Dominant polymer types	References
Africa	Egypt	Red Sea	Offshore	500.9–680 particles/L		[60]
Africa	Egypt	Mediterranean Sea	Offshore	443.9–703.8 particles/L		[60]
Asia	Malaysia	Offshore waters	Offshore	1900 particles/m^3 (Avg.)	PA, PE, PP	[61]
Asia	China	Chinese coastal and marginal seas	Offshore	0.13–545 particles/m^3		[62]
Asia	China	East China Sea		0.17 ± 0.14 particles/m^3		[63]
Asia	China	Bohai Sea		0.33 ± 0.34 particles/m^3	PE, PP, PS	[64]
Asia	China	Yellow Sea		0.13 ± 0.20 pieces/m^3	PP, PE	[65]
Asia	China	South China Sea		0.469 ± 0.219 particles/m^3	PET	[66]
Europe	Scotland			0–91 128 particles/km^2	PP, PS, PVC	[67]
Europe	Finland	Baltic Sea		16.2 ± 11.2 particles/m^3		[68]
North America	USA	Tampa Bay		1.2–18.1 particles/m^3		[69]
Atlantic		Rockall Trough	Deep Sea	70.8 particles/m^3 (Avg.)	PES	[70]
Antarctic		Antarctic Peninsula	Polar region	755–3524 particles/km^2	PU, PA, PE	[71]
Antarctic		Antarctic continent	Polar region	5.7 particles/L (Avg.)	PE, PP, PS	[72]
Arctic		Arctic (South and Southwest of Svalbard)	Polar region	0–1.31 particles/m^3	PET, PA, PE	[73]

(Continued)

Table 6.5 (Continued)

Regions	Countries	Oceans	Sites	Microplastic abundance	Dominant polymer types	References
Arctic		Seven selected stations during the TUNU-VI Expedition	Polar region	2.4 (± 0.8 SD) particles/m^3	PE, PVC, PP, PS	[74]
		Caspian Sea	Offshore	0.246 ± 0.020 particles/m^3 (Avg.)	PE, PP, PET	[75]
		Northwestern Pacific Ocean		6.4 × 10^2– 4.2 × 10^4 particles/km^2	PE, PP, PA	[76]

6.3.3.1 Microplastics on Beaches and Coastal Areas

Beaches and coastal areas are usually densely populated, close to industrial facilities and river entrances [77], all of which led to accumulation of microplastics. Therefore, the concentration of plastic is higher in coastal areas, including estuaries, deltas, and coastal lagoons. Plastics and microplastics are more abundant in densely populated areas, ranging from small fibers (micrometers) to fragments (a few millimeters) [78]. The content of microplastics is higher in offshore areas. In Africa, the abundance of microplastics in the Red Sea and Mediterranean reached 500.9–680 particles/L and 443.9–703.8 particles/L, respectively [60]. In Asia, the average concentration of microplastics in coastal areas of Malaysia is 1900 particles/m^3 [61].

6.3.3.2 Microplastics on the Surface of Ocean Water

In addition to offshore areas, microplastics are also detected in many marine areas. However, the content of microplastics is relatively low compared to offshore areas. In Tampa Bay in the USA for example, the concentration of microplastics is only 1.2–18.1 particles/m^3 [69]. The concentration of microplastics in the northern Baltic Sea in Finland is 16.2 ± 11.2 particles/m^3 [68]. In the Northwest Pacific, the concentration of microplastics reached 6.4 × 10^2–4.2 × 10^4 particles/km^2 [76]. In the deep sea, the concentration of microplastics is similar to that of nearby surface water and may even increase. In the deep-sea area of Rockall Trough in the Atlantic Ocean, the average concentration of microplastics is as high as 70.8 particles/m^3 [70].

6.3.3.3 Microplastic Pollution in Polar Regions

Even remote areas such as the Arctic and Antarctica, plastic has been found. Since the 1980s, plastic pollution in Antarctica and the Southern Ocean has been documented [79]. Recent studies have shown that although the abundance of microplastics in the Antarctic Peninsula is lower than that in the Northwestern Pacific, it is also as high as 755–3524 particles/km² [71]. In addition to the South Pole, microplastics have also been detected in the Arctic. The abundances of microplastics at seven selected sites during the TU-VI Expedition [74] and South and Southwest of Svalbard [73] was 2.4 ± 0.8 particles/m³ and 0–1.31 particles/m³, respectively. In short, although microplastics have become ubiquitous, their concentration varies with the distance from the area where people live.

Various polymers are synthesized and used for domestic and industrial purposes. There are many microplastics with a density less than seawater, including PE (0.91–0.97 g/cm³) and PP (with a density of 0.9–0.91 g/cm³), causing them to float on the sea. The composition of microplastics varies greatly in different regions. Even in different sea areas in the same geographic area, the composition of microplastics varies greatly. For example, the types of microplastics in the Bohai Sea and the Yellow Sea in China are mainly PE, PP, and PS, while the microplastics in the South China Sea are mainly PET (Table 6.5). In the surface waters of the Svalbard and Antarctic Peninsula in the Arctic, PA is the second most abundant polymer [71, 73]. In Scotland, microplastics in the surrounding waters include 26% PP, 17% PS, 12% PVC, and 10% PE [67]. In Europe, the demand for plastics in 2020 is PE (all types) > PP > PVC (Figure 6.2) [80]. In some studies, PE is the most abundant polymer in seawater [81]. Clearly microplastic pollution produced in human daily life is spreading all over the world, even extending to remote and underdeveloped areas. In particular, microplastics that are used in large quantities are more likely to appear in polar regions.

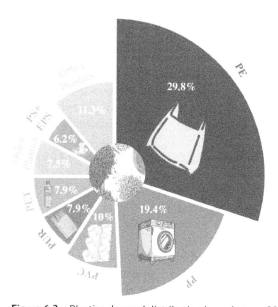

Figure 6.2 Plastics demand distribution by resin type 2019.

6.3.4 Sediments

Microplastics in sediments are the long-term accumulation of microplastics in terrestrial, freshwater, and marine ecosystems [56]. Studies have shown that microplastic contamination is found in 93% of sediment cores [79]. Compared with microplastics in surface water, the concentration of the microplastics in sediments is significantly higher, because some microplastics are denser than water and therefore deposit. There are also low-density microplastics under the action of ocean currents and tides, entraining high-density sand or other substances, thereby sinking to the seafloor. By sedimentation, microplastics accumulate over time, resulting in a relatively high concentration of microplastics in the sediments. In other words, sediments can be considered as the destination of microplastics, although microplastics may also return from sediments to the water [25].

The abundance of microplastics in freshwater system sediments varies greatly from region to region (Table 6.6). In Asia, the abundance of microplastics in the sediments of Red Hills Lake and Veeranam Lake in India are 27 particles/kg (average value) and 92–604 particles/kg, respectively, and that of Vembanad Lake is 252.80 ± 25.76 particles/m^2 [50, 82, 83]. The abundance of microplastics in the sediments of the Ravi River in Pakistan is as high as $40\,536 \pm 202$ particles/m^2 [27]. The abundance of microplastics in Lake Ziway in Africa is unexpectedly high, ranging 400–124 000 particles/kg [84]. Although the abundance of Lake Bizerte is not as high as that of Lake Ziway, it is not as low as 3000–18 000 particles/kg dry sediment [85]. The sediments of Lake Victoria contain less microplastics, ranging from 0 to 108 particles/kg [86]. In addition, the content of microplastics in the sediments of three European countries is relatively high. The microplastic content in the sediments of the Ebro River in Spain, Vesijärvi Lake in Finland, Edgbaston Pool and Hampstead Pond No. 1 in the UK were 2052 ± 746, 395.5 ± 90.7, 250–300, and 539 particles/kg, respectively [34, 87–89]. In North America, the concentration of microplastics in the sediments of Lake Michigan in the USA and Lawrence River in Canada are also alarming, ranging 32.9–6229 particles/kg and 65–7562 particles/kg, respectively [90, 91].

The abundance of microplastics in coastal areas is relatively high. In Africa, the content of microplastics in the sediments of Aero and Masese beaches is 0–1102 particles/kg [86], while in the coast of North Tunisia it is 316.03 ± 123.74 particles/kg [92]. The concentrations of microplastics in the offshore sediments of the Red Sea and Mediterranean Sea are 356–546 and 430–806 particles/kg, respectively [60].

In addition, the abundance of microplastics is positively correlated with the depth of sediments, indicating that microplastics accumulate at the bottom of water bodies [100]. The deeper the riverbed or seabed, the more microplastics

Table 6.6 Microplastic abundance and types in sediments in different regions.

Regions	Countries	Sediments	Microplastic abundance	Dominant polymer types	References
Asia	China	Qinghai Lake	67–1292 particles/m^2	PP, PE	[49]
Asia	China	Poyang Lake	54–506 particles/kg	PP, PE	[43]
Asia	China	Pearl River	685 ± 342 particle/kg	PP, PET	[31]
Asia	India	Red Hills Lake	27 particles/kg (Avg.)	PP, PE	[50]
Asia	India	Veeranam Lake	92–604 particles/kg	PA, PE, PS, PP	[82]
Asia	India	Vembanad Lake	252.80 ± 25.76 particles/m^2	LDPE, PS, PP	[83]
Asia	Indonesia	Ciwalengke River	30.3 ± 15.9 particle/kg		[28]
Asia	Pakistan	Ravi River	40 536 ± 202 particles/m^2	PE, PP	[27]
Africa	Nigeria	OX- Bow Lake	Dry season: 347–4031 particles/kg Raining season: 507–7593 particles/kg	PET, PVC	[45]
Africa	Ethiopia	Lake Ziway	400–124 000 particles/m^3		[84]
Africa	Tunisia	Lake Bizerte	3000–18 000 particles/kg		[85]
Africa		Aero and Masese beaches	0–1102 particles/kg	PE, PP	[86]
Africa		Lake Victoria	0–108 particles/kg	PE, PP	[86]
Africa	Egypt	Red Sea	356–546 particles/kg	PET	[60]
Africa	Egypt	Mediterranean Sea	430–806 particles/kg	PET	[60]

(Continued)

Table 6.6 (Continued)

Regions	Countries	Sediments	Microplastic abundance	Dominant polymer types	References
Africa	Tunisia	North Tunisian coast (Mediterranean Sea)	316.03 ± 123.74 particles/kg	PE, PP, PS	[92]
Europe	UK	Edgbaston Pool	250–300 particles/kg		[89]
Europe	UK	Hampstead Pond No. 1	539 particles/kg (Max)		[88]
Europe	Finland	Vesijärvi Lake	395.5 ± 90.7 particles/kg	PA, PE, PP	[87]
Europe	Hungary	Lake Tisza-tó	0.46–1.62 particles/kg	PP, PS	[32]
Europe	Spain	Ebro River	2052 ± 746 particle/kg	PA, PE, PMMA	[34]
North America	Canada	Lake Simcoe	8–1070 particles/kg	PS, PE	[53]
North America	USA	Lake Michigan	32.9–6229 particles/kg	PET, HDPE, PP	[90]
North America	USA	Lakes Mead and Mohave	Surface: 87.5–1010 particles/kg Core: 220–2040 particles/kg		[54]
North America	Canada	Lawrence River	65–7562 particles/kg		[91]
South America	Colombia	Magdalena River	0–105 particles/kg	PP, PE	[41]
Arctic		Pack ice of Fram Strait	$(1.2 ± 1.4) × 10^7$ particles/m^3	PE, PP	[93]
Arctic		Arctic (Deep Sea)	42–6595 particles/kg	PVC, PA, PP	[94]
Arctic		Arctic Central Basin (Deep Sea)	0–0.2 particles/g		[95]
Arctic		Arctic (Deep Sea)	239–13 331 particles/kg	PA, EPDM	[96]

Table 6.6 (Continued)

Regions	Countries	Sediments	Microplastic abundance	Dominant polymer types	References
Atlantic		North Atlantic Ocean (Deep Sea)	0.197 ± 0.129 particles/g	PES, PP	[97]
Pacific		Western Pacific Ocean (Deep Sea)	240 particles/kg (Avg.)	PP-PE	[98]
Pacific		Pacific Ocean (Deep Sea)	71.1 particles/kg (Avg.)		[99]
		Antarctica and Southern Ocean (Deep Sea)	0–9.52 particles/g	PES	[79]

accumulate in the sediments. In the USA, the concentration of microplastics in the surface sediments of Lake Mead and Lake Mohave is 87.5–1010 particles/kg, while the concentration in the core of the sediments ranges 220–2040 particles/kg, which is significantly higher than that of the surface sediments [54].

A study from the Western Pacific collected sediments in the deep ocean and measured the average abundance of microplastics in the sediments to be 240 particles/kg [79]. The abundance of microplastics in the North Atlantic sediments is 0.197 ± 0.129 particles/g [97]. More importantly, the average concentration of microplastics in Antarctica and Southern Ocean sediments is 0–9.52 particles/g [79]. This value is much higher than in more remote ecosystems, indicating that the amount of microplastics accumulated in the deep ocean of Antarctica is higher than previously expected. The content of microplastics in a research center in the Arctic deep ocean reached 239–13 331 particles/kg [96].

Table 6.6 summarizes the polymer components of microplastics in sediments in freshwater system sediments in different regions. The common polymer components in the sediments of Asian lakes and rivers are PP, PE, PA, PS, and PET. In Europe, PA is the most abundant plastic in sediments, followed by PS and PP. In the USA, PE is the main polymer. In the sediments of African freshwater systems, the main types of microplastics are PET, PVC, PP, and PE. These main types of microplastics are consistent with the common types of microplastics in WWTPs effluent, freshwater, and marine areas. This shows that the types of global microplastic polymers match people's daily consumption, and their main types will also change accordingly with changes in demand.

Although the size and shape of microplastics (such as the aspect ratio of high-density microplastics) and the surface tension of the sea surface may affect their sinking speed, microplastics with a higher density than seawater have a relatively high probability of sinking on the seabed (e.g., PET and PVC) [81]. In other words, the density of plastic material determines its distribution in the water (Figure 6.3). PP can float on the surface of water bodies. On all the continents studied, the proportion of PP in water is greater than that in sediments. In Asia, America, and Africa, the proportion of PE with a density similar to PP in surface water is smaller than that in sediments [42].

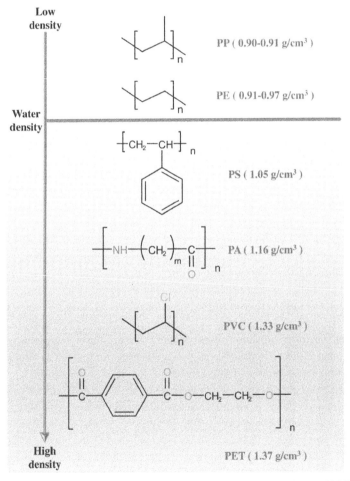

Figure 6.3 Density and structure of different microplastic polymers [101].

6.4 Uptake of Microplastics in Aquatic Organisms

The second destination of microplastics after entering water bodies through WWTPs is to be absorbed and accumulated by organisms, and then enter the food chain, thereby accumulating and spreading in different food chain links. Microplastic pollution poses a threat to marine biota [78]. As a result of their small size, microplastics are easily ingested and absorbed by most aquatic organisms, especially the aquatic organisms that form the basis of the food chain and stay in their bodies for a long time [102]. In addition, microplastics are confused with prey or ingested during passive water filtration, and they may be transferred from prey to predator after being ingested. The ingestion of these microparticles involves a wide range of taxa, from microplankton to macrovertebrates [78]. Microplastics have a biomagnification effect in the food chain. Aquatic organisms with different nutrient levels can easily ingest small plastic particles and enrich them in the food web (Figure 6.4). Therefore, the content of microplastics in aquatic organisms may be much higher than that in water bodies or lake sediments [42].

6.4.1 Freshwater Organisms

The ingestion of microplastics by freshwater fish occurs globally, especially in areas close to or downstream of urbanized area [103]. Fish are the main group that absorbs microplastics in freshwater environments (Table 6.7). As is well known, fish are at a higher trophic level in water bodies and are predators at the top of the food chain. They have a downward control effect on low-trophic organisms in water bodies.

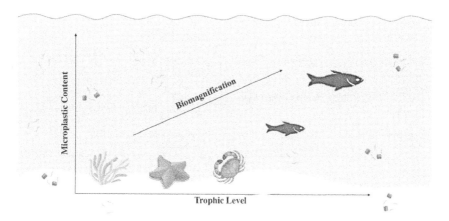

Figure 6.4 Biomagnification of microplastics in food web.

Table 6.7 Microplastic abundance and types in freshwater organisms in different regions.

Regions	Countries	Waters	Organisms	Microplastic abundance	Dominant polymer types	References
Asia	China	Qinghai Lake	Fish	5.4 ± 3.6 particles/fish	PE, PS, PA, PP	[49]
Asia	China	Poyang Lake	Fish	0–18 particles/fish		[43]
Asia	China	Lijiang River	Fish	0.6 ± 0.6 particles/fish	PET	[104]
Asia	Thailand	Chi River	Fish	1.76 ± 0.97 particles/fish		[105]
Europe	Germany	Lakes & rivers	Fish	0.2 ± 0.5 particles/fish		[106]
Europe	UK	River Thames	Fish	0.69 ± 1.25 particles/fish	PE, PP, PES	[103]
Europe	Portugal	Mondego estuary	Fish	1.67 ± 0.27 particles/fish	PES, PP, rayon	[107]
Africa	Ethiopia	Lake Ziway	Fish	4.4 ± 3.6 mg/kg	PE, PP	[84]
North America	USA	Lakes Mead and Mohave	Fish	0–19 particles/fish		[54]
North America	USA	Several rivers	Fish	10 ± 2.3–13 ± 1.6 particles/fish		[108]
South America	Argentina	Rio de La Plata	Fish	18.5 (± 18.9) particles/fish		[109]
South America	Brazil	Pajeú River	Fish	3.6 particles/fish (Avg.)		[110]
South America	Argentina	Paraná River	Fish	9.9–27 particles/fish		[111]
Oceania	Australia	Melbourne Area	Fish	0.6 particles/fish (Avg.)	PES, PA, PP	[112]

The content of microplastics in fish varies in different water bodies. For example, the content of microplastics of fish in Lake Ziway in Africa is 4.4 ± 3.6 mg/kg [84]. The abundance of microplastics of fishes in Qinghai Lake in China is 5.4 ± 3.6 particles/individual [49], while that in fishes in Lijiang River is only 0.6 ± 0.6 particles/individual [104]. In the USA, the abundance of microplastics in fish is 0–19 particles/organism in Lakes (e.g., Lake Mead and Lake Mohave) and 10 ± 2.3–13 ± 1.6 particles/fish in rivers, respectively [54,108]. In South America, the content of microplastics in fish in Pajeú River is 3.6 particles/individual [110], while that in Paraná River is 9.9–27 particles/individual [111]. The content of microplastics in fish may be affected by the dietary habits of fish and the content of microplastics in the environment [113]. Larger fish have an increased demand for energy and a corresponding increase in food intake, thus increasing their intake of microplastics. This also means that different life stages will affect the absorption of microplastics through changes in eating habits.

The microplastics in fish include PP, PE, PES, PS, and PA (Table 6.7). PA can be completely dissolved in fish by digestion, which may result in a lower abundance [42]. In addition, these polymers are basically derived from a large number of commonly used particles in people's daily lives, indicating that WWTPs effluent is an important source of microplastic pollution [103].

6.4.2 Marine Life

As the main fishery resource, marine fish have great economic value. Microplastics were detected in some marine fishes (Table 6.8). Fish living in the ocean absorb lesser amounts of microplastics. For example, the content of microplastics in Mediterranean fish is 1.13 (SD ± 1.99) particles/individual [114]. A consequence of the serious pollution of microplastics in offshore waters, the content of microplastics in offshore fishes is also relatively high. In the North Atlantic Ocean close to the Madeira Islands (Portugal), the average microplastic content of 27 fishes detected was 4.7 ± 4.8 particles/fish [115]. In the Atlantic Ocean, the abundance of microplastics in fish off the coast of the Canary Islands is 2.77 ± 1.91 particles/fish, and the number of microplastics in fishes off the coast of Ghana is as high as 34.0 ± 2.1 particles/fish [116, 117]. In addition, in the Mediterranean area near the coast of Egypt, the microplastic content in fish ranges 0–9.8 particles/fish [60], which is higher than the microplastic abundance in the Mediterranean waters away from the coast (1.13 (SD ± 1.99) particles/individual). These studies indicate that the retention rate of microplastics in fish in coastal areas may be much higher than that in fish in pelagic areas.

PE, PP, PET, and PS are the most produced polymers in the world [80] and are also often found in the digestive tract of fish. Consistent with the results of the

Table 6.8 Microplastic abundance and types in marine organisms in different regions.

Regions	Countries	Oceans	Organisms	Microplastic abundance	Dominant polymer types	References
Asia	China	Bohai Sea	Zooplankton	Rainy season: 2.03 ± 2.87 pieces/m^3 Dry season: 0.41 ± 0.38 pieces/m^3	PET	[118]
Asia	China	Yellow Sea	Zooplankton	12.24 ± 25.70 pieces/m3	Organic oxidation polymers, poly-octenes, PP	[65]
Asia	China	South China Sea	Fish	0–12.0 particles/individual	PET	[119]
Europe	Portugal	North Atlantic Ocean	Fish	4.7 ± 4.8 particles/fish	PP, PE	[115]
Europe		Mediterranean Sea	Fish	1.13 (SD ± 1.99) particles/fish		[114]
Europe	Egypt	Red Sea	Fish	1.18–7.16 particles/fish		[60]
Europe	Egypt	Mediterranean Sea	Fish	0–9.8 particles/fish		[60]
Europe	Finland	Baltic Sea	Fish	1–5 particles/fish		[68]
Africa		the east coast of South Africa	Fish	0.79 ± 1.00 particles/fish	Rayon, PES, PA	[120]
Atlantic		Canary Islands coast	Fish	2.77 ± 1.91 particles/fish		[116]
Atlantic	Ghana	the coast of Ghana	Fish	34.0 ± 2.1 particles/fish		[117]
Oceania	Australia	Southern Australia	Fish	0–17 pieces/fish	Polyolefin	[121]

types of microplastic polymers in the ocean, PP, PE, PA, and PET are also the main types of polymers in marine biological samples. Therefore, the accumulation of microplastics in marine organisms may be affected by their living environment and predation habits.

The density of microplastics determines their vertical position in the water body, which in turn will affect the chance of fishes living in different waters encountering microplastics [122]. For example, pelagic fish are more likely to encounter low-density polymers (e.g., PP and PE), while bottom species may be more susceptible to high-density microplastics (e.g., PVC and PET) [123].

6.4.3 Soil and Crops

Microplastics from inland continue to enter waters and accumulate in sediments for a long time, thereby affecting aquatic life [60]. The abundance of microplastics in benthic organisms is positively correlated with the abundance of microplastics in sediments. Fish that feed on sediment are more likely to be exposed to plastic particles than plankton fish that eat surface foods [84].

In Terra Nova Bay, Ross Sea, Antarctica, and Shelf of Bering and Chukchi Seas in the Arctic, the microplastic content in benthic organisms is 0.01–3.29 particles/mg and 0.04–1.67 particles/individual, respectively [124, 125]. In the Rockel Trough of the North Atlantic, the average microplastic concentration of benthic organisms is 1.582 ± 0.448 particles/g [70]. In Europe, Galway Bay in Ireland and Northwestern Iberian continental shelf in Spain have higher abundances of 0.79 ± 1.14 particles/individual (on average) and 1.03–4.41 particles/individual, respectively [126, 127]. In short, the abundance of microplastics in benthic organisms is relatively higher in areas close to human gathering places.

Benthic organisms also affect the distribution of microplastics in sediments. Once microplastics enter the sediments, they can be ingested by deep-sea creatures and enter the food web. Predatory benthic organisms can excrete microplastics into sediments, and polychaetes that feed on sediments can move microplastics to different depths of sediments through burrowing behavior.

Table 6.9 shows the types of microplastics in benthic organisms in some waters. The main types of microplastics are PP, PE, PA, and PET. This is consistent with the type of microplastic polymers in sediments. The content and types of microplastics in benthic organisms are closely related to their habitat.

In summary, the amount of microplastics ingested by aquatic organisms is higher in the coastal areas than in the ocean, and in deep sea areas is higher than that in shallow layers.

Table 6.9 Microplastic abundance and types in benthic organisms in different regions.

Regions	Countries	Oceans	Organisms	Microplastic abundance	Dominant polymer types	References
Asia	China	South Yellow Sea	Benthic organisms	1.7–47.0 particles/g	PP, PE, PA, PS, PET	[100]
Atlantic		Rockall Trough, North Atlantic Ocean (Deep-Sea)	Benthic invertebrates	1.582 ± 0.448 particles/g (Avg.)	Acrylic	[70]
Europe	Ireland	Galway Bay	Benthic invertebrates	0.79 ± 1.14 particles/individual (Avg.)	PVA, EPDM, PE, PVC	[127]
Europe	Spain	Northwestern Iberian continental shelf	Benthic species	1.03–4.41 particles/individual	PE, PP	[126]
Antarctica		Terra Nova Bay, Ross Sea	Benthic invertebrates	0.01–3.29 particles/mg	PA, PP	[124]
Arctic and sub-Arctic regions		Shelf of Bering and Chukchi Seas	Benthic organisms	0.04–1.67 particles/individual	PA, PE, PET	[125]

6.5 Conclusions and Considerations for Future Work

6.5.1 Conclusions

This chapter summarizes the abundance and types of microplastics discharged from WWTPs, as well as the abundance and types of microplastics in water bodies and aquatic organisms. Overall, microplastics are concentrated in densely populated areas. The closer to people's living areas, the higher the content of microplastics. However, microplastic pollution in remote areas such as the sparsely populated areas of the Arctic and Antarctic cannot be ignored. The types of microplastic polymers commonly found in environmental water bodies and aquatic organisms are PE, PP, PET, and PS, which are microplastics that people use daily. The content and types of microplastics in organisms are closely related to their habitats. Human beings, as the most senior consumers, may also have elevated levels of microplastics in their bodies.

6.5.2 Considerations for Future Work

Existing studies on the distribution of microplastics focus on the concentration and types of microplastics, and little attention is paid to the size of microplastics. Given that the current concentration of microplastics is measured by the number of particles per unit of water or dry weight of sediments, combining the size of microplastics is essential to reflect their true concentration. In addition, the size of microplastics has a direct impact on its influence mechanism. Investigating the size of microplastics is of great significance to further clarify the pollution of microplastics.

Microplastics have been widely detected in rivers, lakes, oceans, and aquatic organisms, but little is known about their migration and transformation rules between WWTPs, freshwater systems, marine systems, and aquatic organisms. Constructing the migration and transformation network of microplastics in the environment and developing corresponding models will help to fully understand the discharge of microplastics from WWTPs and its role in microplastics pollution and control.

Acknowledgments

This work was financially supported by the National Natural Science Foundation of China (NSFC) (51608464) and Hunan Provincial Natural Science Foundation (2020JJ4576).

References

1 Zhang, Z. and Chen, Y. (2020). Effects of microplastics on wastewater and sewage sludge treatment and their removal: A review. *Chemical Engineering Journal* 382: 122955.
2 Ziajahromi, S., Neale, P.A., Telles Silveira, I. et al. (2021). An audit of microplastic abundance throughout three Australian wastewater treatment plants. *Chemosphere* 263: 128294.
3 Gies, E.A., LeNoble, J.L., Noel, M. et al. (2018). Retention of microplastics in a major secondary wastewater treatment plant in Vancouver, Canada. *Marine Pollutution Bulletin* 133: 553–561.
4 Lares, M., Ncibi, M.C., Sillanpaa, M. et al. (2018). Occurrence, identification and removal of microplastic particles and fibers in conventional activated sludge process and advanced MBR technology. *Water Research* 133: 236–246.
5 Xu, X., Jian, Y., Xue, Y. et al. (2019). Microplastics in the wastewater treatment plants (WWTPs): Occurrence and removal. *Chemosphere* 235: 1089–1096.

6 Huang, D., Tao, J., Cheng, M. et al. (2021). Microplastics and nanoplastics in the environment: Macroscopic transport and effects on creatures. *Journal of Hazardous Materials* 407: 124399.
7 Jambeck Jenna, R., Geyer, R., Wilcox, C. et al. (2015). Plastic waste inputs from land into the ocean. *Science* 347 (6223): 768–771.
8 Carr, S.A., Liu, J., and Tesoro, A.G. (2016). Transport and fate of microplastic particles in wastewater treatment plants. *Water Research* 91: 174–182.
9 Long, Z., Pan, Z., Wang, W. et al. (2019). Microplastic abundance, characteristics, and removal in wastewater treatment plants in a coastal city of China. *Water Research* 155: 255–265.
10 Vivekanand, A.C., Mohapatra, S., and Tyagi, V.K. (2021). Microplastics in aquatic environment: Challenges and perspectives. *Chemosphere* 282: 131151.
11 Mintenig, S.M., Int-Veen, I., Loder, M.G.J. et al. (2017). Identification of microplastic in effluents of waste water treatment plants using focal plane array-based micro-Fourier-transform infrared imaging. *Water Research* 108: 365–372.
12 Talvitie, J., Mikola, A., Setala, O. et al. (2017). How well is microlitter purified from wastewater? – A detailed study on the stepwise removal of microlitter in a tertiary level wastewater treatment plant. *Water Research* 109: 164–172.
13 Wolff, S., Kerpen, J., Prediger, J. et al. (2019). Determination of the microplastics emission in the effluent of a municipal waste water treatment plant using Raman microspectroscopy. *Water Research X* 2: 100014.
14 Magni, S., Binelli, A., Pittura, L. et al. (2019). The fate of microplastics in an Italian Wastewater Treatment Plant. *Science of the Total Environment* 652: 602–610.
15 Blair, R.M., Waldron, S., and Gauchotte-Lindsay, C. (2019). Average daily flow of microplastics through a tertiary wastewater treatment plant over a ten-month period. *Water Research* 163: 114909.
16 Murphy, F., Ewins, C., Carbonnier, F. et al. (2016). Wastewater Treatment Works (WwTW) as a source of microplastics in the aquatic environment. *Environmental Science and Technology* 50 (11): 5800–5808.
17 Leslie, H.A., Brandsma, S.H., van Velzen, M.J. et al. (2017). Microplastics en route: Field measurements in the Dutch river delta and Amsterdam canals, wastewater treatment plants, North Sea sediments and biota. *Environment International* 101: 133–142.
18 Kazour, M., Terki, S., Rabhi, K. et al. (2019). Sources of microplastics pollution in the marine environment: Importance of wastewater treatment plant and coastal landfill. *Marine Pollutution Bulletin* 146: 608–618.
19 Akarsu, C., Kumbur, H., Gökdağ, K. et al. (2020). Microplastics composition and load from three wastewater treatment plants discharging into Mersin Bay, north eastern Mediterranean Sea. *Marine Pollution Bulletin* 150: 110776.

20 Michielssen, M.R., Michielssen, E.R., Ni, J. et al. (2016). Fate of microplastics and other small anthropogenic litter (SAL) in wastewater treatment plants depends on unit processes employed. *Environmental Science: Water Research & Technology* 2 (6): 1064–1073.

21 Mason, S.A., Garneau, D., Sutton, R. et al. (2016). Microplastic pollution is widely detected in US municipal wastewater treatment plant effluent. *Environmental Pollution* 218: 1045–1054.

22 Ziajahromi, S., Neale, P.A., Rintoul, L. et al. (2017). Wastewater treatment plants as a pathway for microplastics: Development of a new approach to sample wastewater-based microplastics. *Water Research* 112: 93–99.

23 Hamidian, A.H., Ozumchelouei, E.J., Feizi, F. et al. (2021). A review on the characteristics of microplastics in wastewater treatment plants: A source for toxic chemicals. *Journal of Cleaner Production* 295: 126480.

24 Klein, M. and Fischer, E.K. (2019). Microplastic abundance in atmospheric deposition within the Metropolitan area of Hamburg, Germany. *Science of the Total Environment* 685: 96–103.

25 Peller, J.R., Nelson, C.R., Babu, B.G. et al. (2020). A review of microplastics in freshwater environments: Locations, methods, and pollution loads. In: *Contaminants in Our Water: Identification and Remediation Methods* (ed. S. Ahuja and B.G. Loganathan), 65–90. American Chemical Society.

26 Wang, G., Lu, J., Li, W. et al. (2021). Seasonal variation and risk assessment of microplastics in surface water of the Manas River Basin, China. *Ecotoxicology and Environment Safety* 208: 111477.

27 Irfan, M., Qadir, A., Mumtaz, M. et al. (2020). An unintended challenge of microplastic pollution in the urban surface water system of Lahore, Pakistan. *Environment Science Pollution Research International* 27 (14): 16718–16730.

28 Alam, F.C., Sembiring, E., Muntalif, B.S. et al. (2019). Microplastic distribution in surface water and sediment river around slum and industrial area (case study: Ciwalengke River, Majalaya district, Indonesia). *Chemosphere* 224: 637–645.

29 Tan, X., Yu, X., Cai, L. et al. (2019). Microplastics and associated PAHs in surface water from the Feilaixia Reservoir in the Beijiang River, China. *Chemosphere* 221: 834–840.

30 Luo, W., Su, L., Craig, N.J. et al. (2019). Comparison of microplastic pollution in different water bodies from urban creeks to coastal waters. *Environmental Pollution* 246: 174–182.

31 Fan, Y., Zheng, K., Zhu, Z. et al. (2019). Distribution, sedimentary record, and persistence of microplastics in the Pearl River catchment, China. *Environmental Pollution* 251: 862–870.

32 Bordos, G., Urbanyi, B., Micsinai, A. et al. (2019). Identification of microplastics in fish ponds and natural freshwater environments of the Carpathian basin, Europe. *Chemosphere* 216: 110–116.

33 Mani, T., Blarer, P., Storck, F.R. et al. (2019). Repeated detection of polystyrene microbeads in the Lower Rhine River. *Environmental Pollution* 245: 634–641.
34 Simon-Sánchez, L., Grelaud, M., Garcia-Orellana, J. et al. (2019). River Deltas as hotspots of microplastic accumulation: The case study of the Ebro River (NW Mediterranean). *Science of the Total Environment* 687: 1186–1196.
35 Campanale, C., Stock, F., Massarelli, C. et al. (2020). Microplastics and their possible sources: The example of Ofanto river in southeast Italy. *Environmental Pollution* 258: 113284.
36 Rowley, K.H., Cucknell, A.C., Smith, B.D. et al. (2020). London's river of plastic: High levels of microplastics in the Thames water column. *Science of the Total Environment* 740: 140018.
37 Sekudewicz, I., Dabrowska, A.M., and Syczewski, M.D. (2021). Microplastic pollution in surface water and sediments in the urban section of the Vistula River (Poland). *Science of the Total Environment* 762: 143111.
38 Weideman, E.A., Perold, V., and Ryan, P.G. (2019). Little evidence that dams in the Orange–Vaal River system trap floating microplastics or microfibres. *Marine Pollution Bulletin* 149: 110664.
39 Donoso, J.M. and Rios-Touma, B. (2020). Microplastics in tropical Andean rivers: A perspective from a highly populated Ecuadorian basin without wastewater treatment. *Heliyon* 6 (7).
40 Valine, A.E., Peterson, A.E., Horn, D.A. et al. (2020). Microplastic prevalence in 4 Oregon rivers along a rural to urban gradient applying a cost-effective validation technique. *Environmental Toxicology and Chemistry* 39 (8): 1590–1598.
41 Martínez Silva, P. and Nanny, M.A. (2020). Impact of microplastic fibers from the degradation of nonwoven synthetic textiles to the Magdalena River water column and river sediments by the City of Neiva, Huila (Colombia). *Water* 12 (4): 1210.
42 Yang, S., Zhou, M., Chen, X. et al. (2021). A comparative review of microplastics in lake systems from different countries and regions. *Chemosphere* 286 (Pt 2): 131806.
43 Yuan, W., Liu, X., Wang, W. et al. (2019). Microplastic abundance, distribution and composition in water, sediments, and wild fish from Poyang Lake, China. *Ecotoxicology and Environment Safety* 170: 180–187.
44 Tamminga, M., Stoewer, S.C., and Fischer, E.K. (2019). On the representativeness of pump water samples versus manta sampling in microplastic analysis. *Environmental Pollution* 254 (Pt A): 112970.
45 Oni, B.A., Ayeni, A.O., Agboola, O. et al. (2020). Comparing microplastics contaminants in (dry and raining) seasons for Ox- Bow Lake in Yenagoa, Nigeria. *Ecotoxicology and Environment Safety* 198: 110656.

46 Migwi, F.K., Ogunah, J.A., and Kiratu, J.M. (2020). Occurrence and spatial distribution of microplastics in the surface waters of Lake Naivasha, Kenya. *Environmental Toxicology and Chemistry* 39 (4): 765–774.

47 Alfonso, M.B., Arias, A.H., and Piccolo, M.C. (2020). Microplastics integrating the zooplanktonic fraction in a saline lake of Argentina: Influence of water management. *Environmental Monitoring and Assessment* 192: 2.

48 Free, C.M., Jensen, O.P., Mason, S.A. et al. (2014). High-levels of microplastic pollution in a large, remote, mountain lake. *Marine Pollutution Bulletin* 85 (1): 156–163.

49 Xiong, X., Zhang, K., Chen, X. et al. (2018). Sources and distribution of microplastics in China's largest inland lake – Qinghai Lake. *Environmental Pollution* 235: 899–906.

50 Gopinath, K., Seshachalam, S., Neelavannan, K. et al. (2020). Quantification of microplastic in Red Hills Lake of Chennai city, Tamil Nadu, India. *Environment Science Pollution Research International* 27 (26): 33297–33306.

51 Mao, R., Hu, Y., Zhang, S. et al. (2020). Microplastics in the surface water of Wuliangsuhai Lake, northern China. *Science of the Total Environment* 723: 137820.

52 Pico, Y., Alvarez-Ruiz, R., Alfarhan, A.H. et al. (2020). Pharmaceuticals, pesticides, personal care products and microplastics contamination assessment of Al-Hassa irrigation network (Saudi Arabia) and its shallow lakes. *Science of the Total Environment* 701: 135021.

53 Felismino, M.E.L., Helm, P.A., and Rochman, C.M. (2021). Microplastic and other anthropogenic microparticles in water and sediments of Lake Simcoe. *Journal of Great Lakes Research* 47 (1): 180–189.

54 Baldwin, A.K., Spanjer, A.R., Rosen, M.R. et al. (2020). Microplastics in Lake Mead National Recreation Area, USA: Occurrence and biological uptake. *PLoS One* 15 (5).

55 Bertoldi, C., Lara, L.Z., Mizushima, F.A.L. et al. (2021). First evidence of microplastic contamination in the freshwater of Lake Guaiba, Porto Alegre, Brazil. *Science of the Total Environment* 759: 143503.

56 Kumar, R., Sharma, P., Manna, C. et al. (2021). Abundance, interaction, ingestion, ecological concerns, and mitigation policies of microplastic pollution in riverine ecosystem: A review. *Science of the Total Environment* 782: 146695.

57 Cole, M., Lindeque, P., Halsband, C. et al. (2011). Microplastics as contaminants in the marine environment: A review. *Marine Pollutution Bulletin* 62 (12): 2588–2597.

58 Schmidt, C., Krauth, T., and Wagner, S. (2017). Export of plastic debris by rivers into the sea. *Environmental Science and Technology* 51 (21): 12246–12253.

59 Eerkes-Medrano, D., Thompson, R.C., and Aldridge, D.C. (2015). Microplastics in freshwater systems: A review of the emerging threats, identification of knowledge gaps and prioritisation of research needs. *Water Research* 75: 63–82.

60 Sayed, A.E.H., Hamed, M., Badrey, A.E.A. et al. (2021). Microplastic distribution, abundance, and composition in the sediments, water, and fishes of the Red and Mediterranean seas, Egypt. *Marine Pollutution Bulletin* 173 (Pt A): 112966.

61 Taha, Z.D., Md Amin, R., Anuar, S.T. et al. (2021). Microplastics in seawater and zooplankton: A case study from Terengganu estuary and offshore waters, Malaysia. *Science of the Total Environment* 786: 147466.

62 Jiang, Y., Yang, F., Hassan Kazmi, S.S.U. et al. (2021). A review of microplastic pollution in seawater, sediments and organisms of the Chinese coastal and marginal seas. *Chemosphere* 286 (Pt 1): 131677.

63 Zhao, S., Zhu, L., Wang, T. et al. (2014). Suspended microplastics in the surface water of the Yangtze Estuary System, China: First observations on occurrence, distribution. *Marine Pollutution Bulletin* 86 (1–2): 562–568.

64 Zhang, W., Zhang, S., Wang, J. et al. (2017). Microplastic pollution in the surface waters of the Bohai Sea, China. *Environmental Pollution* 231 (Pt 1): 541–548.

65 Sun, X., Liang, J., Zhu, M. et al. (2018). Microplastics in seawater and zooplankton from the Yellow Sea. *Environmental Pollution* 242 (Pt A): 585–595.

66 Wang, T., Zou, X., Li, B. et al. (2019). Preliminary study of the source apportionment and diversity of microplastics: Taking floating microplastics in the South China Sea as an example. *Environmental Pollution* 245: 965–974.

67 Russell, M. and Webster, L. (2021). Microplastics in sea surface waters around Scotland. *Marine Pollutution Bulletin* 166: 112210.

68 Sainio, E., Lehtiniemi, M., and Setala, O. (2021). Microplastic ingestion by small coastal fish in the northern Baltic Sea, Finland. *Marine Pollutution Bulletin* 172: 112814.

69 McEachern, K., Alegria, H., Kalagher, A.L. et al. (2019). Microplastics in Tampa Bay, Florida: Abundance and variability in estuarine waters and sediments. *Marine Pollutution Bulletin* 148: 97–106.

70 Courtene-Jones, W., Quinn, B., Gary, S.F. et al. (2017). Microplastic pollution identified in deep-sea water and ingested by benthic invertebrates in the Rockall Trough, North Atlantic Ocean. *Environmental Pollution* 231: 271–280.

71 Lacerda, A., Rodrigues, L.D.S., van Sebille, E. et al. (2019). Plastics in sea surface waters around the Antarctic Peninsula. *Scientific Reports* 9 (1): 3977.

72 Suaria, G., Perold, V., Lee, J.R. et al. (2020). Floating macro- and microplastics around the Southern Ocean: Results from the Antarctic Circumnavigation Expedition. *Environment International* 136: 105494.

73 Lusher, A.L., Tirelli, V., O'Connor, I. et al. (2015). Microplastics in Arctic polar waters: The first reported values of particles in surface and sub-surface samples. *Scientific Reports* 5: 1.

74 Morgana, S., Ghigliotti, L., Estevez-Calvar, N. et al. (2018). Microplastics in the Arctic: A case study with sub-surface water and fish samples off Northeast Greenland. *Environmental Pollution* 242 (Pt B): 1078–1086.

75 Manbohi, A., Mehdinia, A., Rahnama, R. et al. (2021). Microplastic pollution in inshore and offshore surface waters of the southern Caspian Sea. *Chemosphere* 281: 130896.

76 Pan, Z., Guo, H., Chen, H. et al. (2019). Microplastics in the Northwestern Pacific: Abundance, distribution, and characteristics. *Science of the Total Environment* 650 (Pt 2): 1913–1922.

77 Bond, A.L., Provencher, J.F., Elliot, R.D. et al. (2013). Ingestion of plastic marine debris by Common and Thick-billed Murres in the northwestern Atlantic from 1985 to 2012. *Marine Pollutution Bulletin* 77 (1–2): 192–195.

78 Barboza, L.G.A., Frias, J.P.G.L., Booth, A.M. et al. (2019). Microplastics pollution in the marine environment. In: *World Seas: An Environmental Evaluation* (ed. C. Sheppard), 329–351. Academic Press.

79 Cunningham, E.M., Ehlers, S.M., Dick, J.T.A. et al. (2020). High abundances of microplastic pollution in deep-sea sediments: Evidence from Antarctica and the Southern Ocean. *Environmental Science and Technology* 54 (21): 13661–13671.

80 PlasticsEurope (2020). Plastics – The facts 2020: An analysis of European plastics production, demand and waste data. Plastics Europe, Belgium.

81 Shim, W.J., Hong, S.H., and Eo, S. (2018). Chapter 1 – Marine microplastics: Abundance, distribution, and composition. In: *Microplastic Contamination in Aquatic Environments* (ed. E.Y. Zeng), 1–26. Elsevier.

82 Bharath, K.M., Srinivasalu, S., Natesan, U. et al. (2021). Microplastics as an emerging threat to the freshwater ecosystems of Veeranam lake in south India: A multidimensional approach. *Chemosphere* 264 (Pt 2): 128502.

83 Sruthy, S. and Ramasamy, E.V. (2017). Microplastic pollution in Vembanad Lake, Kerala, India: The first report of microplastics in lake and estuarine sediments in India. *Environmental Pollution* 222: 315–322.

84 Merga, L.B., Redondo-Hasselerharm, P.E., Van den Brink, P.J. et al. (2020). Distribution of microplastic and small macroplastic particles across four fish species and sediment in an African lake. *Science of the Total Environment* 741: 140527.

85 Abidli, S., Toumi, H., Lahbib, Y. et al. (2017). The first evaluation of microplastics in sediments from the complex lagoon-channel of Bizerte (Northern Tunisia). *Water, Air, & Soil Pollution* 228 (7).

86 Egessa, R., Nankabirwa, A., Basooma, R. et al. (2020). Occurrence, distribution and size relationships of plastic debris along shores and sediment of northern Lake Victoria. *Environmental Pollution* 257: 113442.

87 Scopetani, C., Chelazzi, D., Cincinelli, A. et al. (2019). Assessment of microplastic pollution: Occurrence and characterisation in Vesijarvi lake and Pikku Vesijarvi pond, Finland. *Environmental Monitoring and Assessment* 191 (11): 652.

88 Turner, S., Horton, A.A., Rose, N.L. et al. (2019). A temporal sediment record of microplastics in an urban lake, London, UK. *Journal of Paleolimnology* 61 (4): 449–462.

89 Vaughan, R., Turner, S.D., and Rose, N.L. (2017). Microplastics in the sediments of a UK urban lake. *Environmental Pollution* 229: 10–18.

90 Lenaker, P.L., Baldwin, A.K., Corsi, S.R. et al. (2019). Vertical distribution of microplastics in the water column and surficial sediment from the Milwaukee River Basin to Lake Michigan. *Environmental Science and Technology* 53 (21): 12227–12237.

91 Crew, A., Gregory-Eaves, I., and Ricciardi, A. (2020). Distribution, abundance, and diversity of microplastics in the upper St. Lawrence River. *Environmental Pollution* 260: 113994.

92 Abidli, S., Antunes, J.C., Ferreira, J.L. et al. (2018). Microplastics in sediments from the littoral zone of the north Tunisian coast (Mediterranean Sea). *Estuarine, Coastal and Shelf Science* 205: 1–9.

93 Peeken, I., Primpke, S., Beyer, B. et al. (2018). Arctic sea ice is an important temporal sink and means of transport for microplastic. *Nature Communications* 9 (1): 1505.

94 Bergmann, M., Wirzberger, V., Krumpen, T. et al. (2017). High quantities of microplastic in Arctic deep-sea sediments from the HAUSGARTEN observatory. *Environmental Science and Technology* 51 (19): 11000–11010.

95 Kanhai, L.D.K., Johansson, C., Frias, J.P.G.L. et al. (2019). Deep sea sediments of the Arctic Central Basin: A potential sink for microplastics. *Deep Sea Research Part I: Oceanographic Research Papers* 145: 137–142.

96 Tekman, M.B., Wekerle, C., Lorenz, C. et al. (2020). Tying up loose ends of microplastic pollution in the Arctic: Distribution from the sea surface through the water column to deep-sea sediments at the HAUSGARTEN observatory. *Environmental Science and Technology* 54 (7): 4079–4090.

97 Courtene-Jones, W., Quinn, B., Ewins, C. et al. (2020). Microplastic accumulation in deep-sea sediments from the Rockall Trough. *Marine Pollututation Bulletin* 154: 111092.

98 Zhang, D., Liu, X., Huang, W. et al. (2020). Microplastic pollution in deep-sea sediments and organisms of the Western Pacific Ocean. *Environmental Pollution* 259: 113948.

99   Peng, G., Bellerby, R., Zhang, F. et al. (2020). The ocean's ultimate trashcan: Hadal trenches as major depositories for plastic pollution. *Water Research* 168: 115121.
100 Wang, J., Wang, M., Ru, S. et al. (2019). High levels of microplastic pollution in the sediments and benthic organisms of the South Yellow Sea, China. *Science of the Total Environment* 651 (Pt 2): 1661–1669.
101 Anderson, J.C., Park, B.J., and Palace, V.P. (2016). Microplastics in aquatic environments: Implications for Canadian ecosystems. *Environmental Pollution* 218: 269–280.
102 Cheung, P.K. and Fok, L. (2017). Characterisation of plastic microbeads in facial scrubs and their estimated emissions in Mainland China. *Water Research* 122: 53–61.
103 Horton, A.A., Jurgens, M.D., Lahive, E. et al. (2018). The influence of exposure and physiology on microplastic ingestion by the freshwater fish Rutilus rutilus (roach) in the River Thames, UK. *Environmental Pollution* 236: 188–194.
104 Zhang, L., Xie, Y., Zhong, S. et al. (2021). Microplastics in freshwater and wild fishes from Lijiang River in Guangxi, Southwest China. *Science of the Total Environment* 755 (Pt 1): 142428.
105 Kasamesiri, P. (2020). Microplastics ingestion by freshwater fish in the Chi River, Thailand. *International Journal of GEOMATE* 18: 67.
106 Roch, S., Walter, T., Ittner, L.D. et al. (2019). A systematic study of the microplastic burden in freshwater fishes of south-western Germany – Are we searching at the right scale? *Science of the Total Environment* 689: 1001–1011.
107 Bessa, F., Barria, P., Neto, J.M. et al. (2018). Occurrence of microplastics in commercial fish from a natural estuarine environment. *Marine Pollutution Bulletin* 128: 575–584.
108 McNeish, R.E., Kim, L.H., Barrett, H.A. et al. (2018). Microplastic in riverine fish is connected to species traits. *Scientific Reports* 8 (1): 11639.
109 Pazos, R.S., Maiztegui, T., Colautti, D.C. et al. (2017). Microplastics in gut contents of coastal freshwater fish from Rio de la Plata estuary. *Marine Pollutution Bulletin* 122 (1–2): 85–90.
110 Silva-Cavalcanti, J.S., Silva, J.D.B., Franca, E.J. et al. (2017). Microplastics ingestion by a common tropical freshwater fishing resource. *Environmental Pollution* 221: 218–226.
111 Blettler, M.C.M., Garello, N., Ginon, L. et al. (2019). Massive plastic pollution in a mega-river of a developing country: Sediment deposition and ingestion by fish (Prochilodus lineatus). *Environmental Pollution* 255 (Pt 3): 113348.
112 Su, L., Nan, B., Hassell, K.L. et al. (2019). Microplastics biomonitoring in Australian urban wetlands using a common noxious fish (Gambusia holbrooki). *Chemosphere* 228: 65–74.

113 Collard, F., Gasperi, J., Gabrielsen, G.W. et al. (2019). Plastic particle ingestion by wild freshwater fish: A critical review. *Environmental Science and Technology* 53 (22): 12974–12988.

114 Rios-Fuster, B., Alomar, C., Compa, M. et al. (2019). Anthropogenic particles ingestion in fish species from two areas of the western Mediterranean Sea. *Marine Pollutution Bulletin* 144: 325–333.

115 Gago, J., Portela, S., Filgueiras, A.V. et al. (2020). Ingestion of plastic debris (macro and micro) by longnose lancetfish (Alepisaurus ferox) in the North Atlantic Ocean. *Regional Studies in Marine Science* 33: 100977.

116 Herrera, A., Stindlova, A., Martinez, I. et al. (2019). Microplastic ingestion by Atlantic chub mackerel (Scomber colias) in the Canary Islands coast. *Marine Pollutution Bulletin* 139: 127–135.

117 Adika, S.A., Mahu, E., Crane, R. et al. (2020). Microplastic ingestion by pelagic and demersal fish species from the Eastern Central Atlantic Ocean, off the Coast of Ghana. *Marine Pollutution Bulletin* 153: 110998.

118 Zheng, S., Zhao, Y., Liangwei, W. et al. (2020). Characteristics of microplastics ingested by zooplankton from the Bohai Sea, China. *Science of the Total Environment* 713: 136357.

119 Ding, J., Jiang, F., Li, J. et al. (2019). Microplastics in the coral reef systems from Xisha Islands of South China Sea. *Environmental Science and Technology* 53 (14): 8036–8046.

120 Naidoo, T., Sershen, Thompson, R.C. et al. (2020). Quantification and characterisation of microplastics ingested by selected juvenile fish species associated with mangroves in KwaZulu-Natal, South Africa. *Environmental Pollution* 257: 113635.

121 Wootton, N., Reis-Santos, P., Dowsett, N. et al. (2021). Low abundance of microplastics in commercially caught fish across southern Australia. *Environmental Pollution* 290: 118030.

122 de Sa, L.C., Oliveira, M., Ribeiro, F. et al. (2018). Studies of the effects of microplastics on aquatic organisms: What do we know and where should we focus our efforts in the future? *Science of the Total Environment* 645: 1029–1039.

123 Lusher, A.L., McHugh, M., and Thompson, R.C. (2013). Occurrence of microplastics in the gastrointestinal tract of pelagic and demersal fish from the English Channel. *Marine Pollutution Bulletin* 67 (1–2): 94–99.

124 Sfriso, A.A., Tomio, Y., Rosso, B. et al. (2020). Microplastic accumulation in benthic invertebrates in Terra Nova Bay (Ross Sea, Antarctica). *Environment International* 137: 105587.

125 Fang, C., Zheng, R., Zhang, Y. et al. (2018). Microplastic contamination in benthic organisms from the Arctic and sub-Arctic regions. *Chemosphere* 209: 298–306.

126 Filgueiras, A.V., Preciado, I., Carton, A. et al. (2020). Microplastic ingestion by pelagic and benthic fish and diet composition: A case study in the NW Iberian shelf. *Marine Pollutution Bulletin* 160: 111623.

127 Pagter, E., Nash, R., Frias, J. et al. (2021). Assessing microplastic distribution within infaunal benthic communities in a coastal embayment. *Science of the Total Environment* 791: 148278.

7

Microplastics Removal and Degradation in Urban Water Systems

*Qiuxiang Xu[1] and Bing-Jie Ni[2],**

[1] State Key Laboratory of Pollution Control and Resources Reuse, College of Environmental Science and Engineering, Tongji University, Shanghai, PR China
[2] Centre for Technology in Water and Wastewater, School of Civil and Environmental Engineering, University of Technology Sydney, Sydney, NSW, Australia
* Corresponding author

7.1 Introduction

Although plastic products have brought great convenience to human society, plastics are inevitably discharged into the various environments in large quantities during their continued use [1, 2]. It was reported that up to 13 000 000 tons of plastic were released into the ocean in 2018 [3]. As a consequence of these characteristics of plastic, such as stability and refractory degradation, the discharged plastic has produced serious pollution problems [4], and the environmental problems caused by them have received more and more attention [5]. In addition, under conditions such as sunlight, weathering, erosion, immersion, etc., these plastics discharged into the environment will gradually be decomposed further into small fragments or particles known as microplastics (MPs) (0.1–5 mm in diameter) [6, 7]. Apart from the decomposition of those discharged plastics, small artificial plastic particles added to skin care products and detergents also release substantial amounts of MPs [8].

Currently, a large number of MPs have been found in water environments such as oceans, surface water, groundwater, drinking water, wastewater, and rainwater, because water in open and closed systems can be used as pipes and sinks for MPs migration [9, 10]. Among the above water environments, the urban water environment composed of surface water, groundwater, drinking water, wastewater, and rainwater is highly relevant to human beings. Thus, the environmental behavior of MPs has received extensive attention, and their fate and occurrence have thoroughly been summarized [11–13]. Figure 7.1 presents the main sources and migration processes of MPs in urban water environmental. As shown in Figure 7.1, a large part of MPs generated by humans in their daily life and production activities is released into wastewater and rainwater, then enters the urban drainage system [14–16]. MPs in wastewater are collected at the wastewater treatment plants (WWTPs), while most of those in rainwater go directly into surface water [17, 18]. The main function of WWTPs is to treat degradable pollutants such as degradable organic matter and nitrogen [19]. Given the refractory biodegradability of MPs, most MPs are retained by activated sludge adsorption in sludge, the residual MPs are then released into rivers with the discharge of WWTPs effluent [20]. The migration paths of MPs entering surface water mainly include (1) feeding by aquatic organisms; (2) sedimentation in sediments; (3) migration and transformation in drinking water sources such as rivers, lakes, groundwater, and other water environments, and then entering drinking water treatment plants (DWTPs) [21–23]. At present, it has been widely demonstrated that complete removal of MPs is difficult to achieve in WTPs, thus the MPs generated by humans will inevitably enter the human body through food chains such as drinking water and aquatic food [24, 25]. Therefore, MPs can form an internal cycle in urban waters, resulting in the continuous exposure of humans and aquatic organisms to MPs [26].

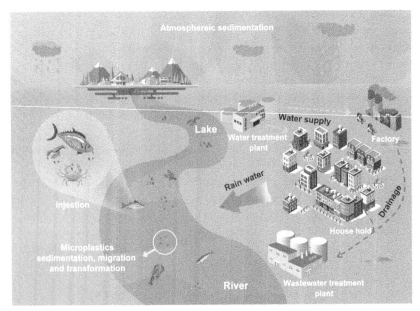

Figure 7.1 The main sources and migration of MPs in urban water systems [26] / with permission of Elsevier.

The persistent exposure risks posed by the widespread presence of MPs have potential impacts on the ecological environment and human health that should be taken seriously. The small size of microplastic particles allows aquatic organisms to ingest MPs through food webs at different nutrient levels [27]. On the one hand, MPs ingestion by aquatic organisms may cause false satiety, digestive difficulties, and intestinal wear [28]; on the other hand, MPs may indirectly induce an oxidative stress response, reducing growth rate, and causing physiological and reproductive diseases [29]. All of the above factors could disturb the normal growth and survival of an aquatic organism [30]. Previous studies have shown that ingestion of polystyrene MPs could alter weight gain and specific growth behaviors of the marine jacopever (Sebastes schlegelii) [31]. Moreover, hazardous chemicals added in plastic production (such as bisphenol A and phthalates) and precursor monomers are easily released from the plastic matrix during decomposition and degradation, and also pose potential hazards to aquatic organisms, and transfer to humans through the food chain, ultimately affecting human health [32, 33]. It has been demonstrated that the above additives and monomers cause endocrine disruption in aquatic organisms, leading to reduced reproduction [34, 35]. As a result of those properties such as hydrophobicity and large specific surface area, MPs have strong adsorption capacity for persistent pollutants and can be used as carriers of harmful microbes (such as pathogens) and chemical contaminants [36, 37]. These adsorbed pollutants could then be desorbed in the organisms during the ingestion of MPs, aggravating the bioaccumulation of pollutants and also causing potential harm to organisms [38, 39].

As mentioned above, MPs ingested by aquatic organisms and MPs in drinking water can be transferred to and ingested by the human body through the food chain, eventually leading to the continuous accumulation of MPs [40]. Previous studies have demonstrated that ingestion is the main way MPs enter the body [41]. Despite limited information on the interaction between MPs and human cells, MPs, as an emerging contaminant, are considered a potential risk to human health [42]. On the one hand, given the high surface area and persistent accumulation properties, MPs could induce oxidative stress and chronic inflammation, resulting in cytotoxicity [43, 44]. For example, it was found that when human brain and epithelial cells were exposed to 0.05–10 mg/L polystyrene and polyethylene MPs, reactive oxygen species were generated, causing cytotoxicity [45]. Moreover, MPs, especially smaller particles, may enter the peripheral tissues and circulatory system through lymphatic aggregation after being ingested into the human body, resulting in systemic exposure [46, 47]. MPs circulating in the body may exacerbate inflammation and risk of pulmonary hypertension [48, 49]. In addition, although unconfirmed thus far, the toxic contaminants (e.g., monomers, harmful additives, heavy metals, and pathogens) released or loaded by MPs have been considered to be potentially harmful to human health [50–53]. In general, with the growing awareness of the huge potential risks of MPs, reducing MPs release through enhancing plastic recycling and enhancing MPs removal to minimize human exposure have received more and more attention [54, 55]. In particular, the urban water environment is closely related to human life, so the related MPs removal methods have been widely studied [56].

In recent years, various techniques and methods have been used to try to reduce the presence of MPs, such as coagulation, flocculation, and sedimentation (CFS) [26]; electrocoagulation [57]; filtration [58]; membrane separation [59]; adsorption [60]; magnetic extraction [61]; microbial degradation [62]; photocatalysis [63]; and chemical oxidation [64]. Among these methods, in the urban water environment, CFS, electrocoagulation, filtration, membrane separation, adsorption, photocatalysis, and chemical oxidation are the research hotspots because each has a large application scale or a good application prospect in the field of water treatment [65–67]. CFS, electrocoagulation, filtration, membrane separation, or adsorption can effectively remove impurities and pollutants in water, and are the mainstream conventional methods of water treatment at present and will remain so in the coming decades [68, 69]. Each method showed good removal performance for MPs from drinking water (including WWTP and bottled water) and WWTP treated wastewater [70, 71]. In comparison with the above-mentioned widely used conventional techniques, photocatalysis, a widely studied pollutant enhancement treatment [72, 73], is regarded as a useful and promising technology for the removal of MPs because it could completely degrade MPs into H_2O and CO_2 using reactive radical species (ROS) produced by the accelerated reaction involving light and a catalyst [74]. Like photocatalysis, chemical oxidation methods such as the Fenton process could also degrade MPs into small chains with lower molecular weight and even further into simple nontoxic molecules (such as water and carbon dioxide) by generating ROS (e.g., hydroxyl and sulfate radicals) [67]. Although previous studies have focused on MPs reduction to minimize the risk of exposure to MPs, each technology has different characteristics, and a single technology may not achieve satisfactory MPs removal performance. Therefore, it is urgent to comprehensively summarize the removal processes and relevant mechanisms of these methods to provide certain technical guidance for the subsequent enhancement of MPs removal through a single technology or multi-technology cooperation.

This chapter is a summary of the main removal and degradation methods of MPs in addition to biotechnology in urban waters, expecting to offer some valuable information and guide future study of strengthening MPs removal to purify urban water bodies and minimize the potential risk of MPs. This chapter firstly introduces the environmental behavior of MPs in the urban water environment, aiming to understand their migration and transformation, potential risks to the ecological environment and human health, and the urgency of reducing their release. Then, the research status of filtration and separation, membrane separation, coagulation, adsorption, photocatalysis, and some other MPs removal methods in an urban water environment is systematically summarized. At the same time, the relevant removal/degradation mechanisms are also comprehensively discussed. Finally, research gaps in MPs removal and degradation and future investigations for enhanced MPs removal are also proposed.

7.2 Use of Separation-based Technology for the Removal of MPs

7.2.1 CFS

CFS, a common process in DWTPs, plays an important role in removing suspended colloidal particles or dissolved pollutants in water [67]. The CFS process has been widely reported to effectively remove MPs both in real WTPs and in jar tests [70, 74]. Figure 7.2 displays the removal of MPs by the CFS process, including the rapid mixing of coagulants and MPs in the coagulation stage, the slow mixing in the flocculation stage, and the sedimentation of most MPs in the final stage. Charge neutralization, sweep flocculation, or their combination are the main mechanisms for the CFS process removal of MPs. During the charge neutralization process, the hydrolyzed products of the coagulants are positively charged and therefore easily adsorbed on the negatively charged surface, thus neutralizing the negatively charged MPs [75]. As for sweep flocculation, after the positively charged hydrolysates of the coagulants adsorb the negatively charged MPs to form flocs, with the increase of the particle size and density of the flocs, they are swept away by the amorphous sediments, forming strong swept flocculation [76].

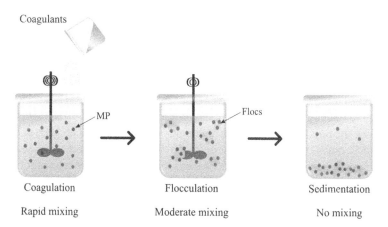

Figure 7.2 Procedures of removing MPs using CFS [74] / with permission of Elsevier.

Currently, MPs in urban waters have been reported to be substantially removed by combined process based on CFS method in DWTPs, and CFS is a major contributor. Pivokonský et al. [77] investigated the fate of MPs ≥ 1 μm at two different DWTPs within a river catchment. It was found that the concentration of MPs decreased from 23 ± 2 in raw water to 14 ± 1 MPs/L in drinking water in DWTP Milence with CFS and sand filtration processes. Much higher removal efficiency of MPs was observed in DWTP Plzeň with higher initial MP concentration and a more complex process. The MPs concentration effectively decreased from 1296 ± 35 in raw water to 497 ± 44 MPs/L after CFS treatment regardless of their size or shape and further decreased to 151 ± 4 MPs/L after other processes including deep-bed filtration, and ozonation and granular activated carbon (GAC) filtration, contributing 62% and 26% removal performance, respectively. The total removal efficiency of 88% was similar to that of several DWTPs reported in the literature that also employed a combined process based on the CFS method. For example, Pivokonský et al. [78] investigated the MPs content of influent and effluent in three different DWTPs supplied by different water bodies, and found that the average abundance of MPs decreased from between 1473 ± 34 and 3605 ± 497 particles/L in raw water to between 338 ± 76 and 628 ± 28 particles/L in drinking water, achieving 70–83% removal efficiency. As reported by Wang et al. [79], the removal efficiency of MPs ≥ 1 μm at a DWTP with CFS and GAC filtration located in the Yangtze River Delta reached 86%, but the removal contribution (approximately 50%) of CFS was lower than that reported by Pivokonský et al. [77] (62%), and mainly for fibrous MPs removal. Besides, similar removal of MPs by CFS was also observed in the effluent of WWTPs. It was demonstrated that the CFS using three inorganic and organic coagulants (i.e., ferric chloride, polyaluminum chloride, and polyamine, respectively) enhanced the removal of MPs of two different sizes: 1 and 6.3 μm [71]. The corresponding maximum removal efficiency reached 99.4%, 98.2%, and 65%. Clearly, ferric chloride and polyaluminum chloride were more effective than polyamines. Further investigation revealed that the removal mechanism caused by ferric chloride might occur from sweep coagulation or a combination of sweep coagulation and charge neutralization, while polyaluminum chloride and polyamine work by charge neutralization [71, 80].

As introduced above, the removal effect of the CFS method on MPs shows certain differences under different water sources, different MPs, different coagulants, etc. Therefore, the removal effect and mechanism of the CFS method on MPs, and the effect of various experimental conditions on the removal effect were also investigated under laboratory conditions. Because of high frequency detection in urban waters and its potential risks to human health [17, 75, 81, 82], the removal characteristics of polyethylene (PE), polystyrene (PS), polyethylene terephthalate (PET), and polyester (PEST) MPs in the CFS method were thus mainly studied. Ma et al. [83] investigated the removal characteristics of PE MPs using Fe-based coagulants, and found although smaller PE particle size showed higher removal rate, all removal rates were low (< 15%) using the CFS method, possibly attributed to the difficulty of sedimentation caused by the low density of PE and the difficulty of adsorption and trapping of PE by Fe-based flocs with relatively small size. However, when polyacrylamide (PAM) was introduced into CFS, the removal performance of PE was largely improved under the condition of high dosage of Fe-based coagulants and PAM. The removal rate of PE (d < 0.5 mm) was over 85% after adding anionic PAM, much higher than the previous 13.27%. This might occur from stronger adsorption bridging effects caused by PAM, increasing the density of flocs, with PE more easily adsorbed/trapped [84, 85]. Similar poor removal performance of PE was also observed by Zhou et al. [75]. It was found that the maximum removal rate of PE was still lower than 20% when the dosage of ferric chloride ($FeCl_3$) reached 180 mg/L, and slightly increased to 29.70% when polyaluminium chloride (PAC) was used for coagulation. However, removal efficiency of PS was high with both $FeCl_3$ and PAC, and its maximum value reached 77.83%. This significant removal difference was mainly from the higher density of PS than PE. Moreover, mechanism investigation revealed that the hydrolysis products of coagulants also played a major role in both MPs removal, in addition to the conventional removal mechanism. The experimental parameter study also found that alkaline conditions and relatively high stirring speed were more conducive to the removal of MPs through promoting the hydrolysis of PAC, increasing the average size of flocs, and increasing the collision probability of coagulant hydrolysate with MPs, respectively [76, 83, 86].

Al-based salts (e.g., $AlCl_3 \cdot 6H_2O$ and $Al_2(SO_4)_3 \cdot 18H_2O$), as another conventional coagulant, were also widely used to investigate the removal of MPs. Ma et al. [87] compared the removal behavior of PE using Al-based salts ($AlCl_3 \cdot 6H_2O$) and Fe-based salts ($FeCl_3 \cdot 6H_2O$), and found that better PE removal efficiency was obtained in the presence of Al-based salts (especially for smaller PE particle size), possibly from the larger specific surface area formed by Al-based salts with the smaller floc particle size and higher zeta potential of flocs [87]. However, the removal efficiency was still low (under 40%) even at high dosage of $AlCl_3 \cdot 6H_2O$ (15 mM). When PAM (especially anionic PAM) was added into the CFS process, the removal performance of PE particles was effectively improved by the densification of loose Al-based flocs [87]. As a result, the removal rate of PE particles with smaller size (d < 0.5 mm) increased from 25.83 ± 2.91 (without PAM) to 61.19 ± 3.67% (15 mg/L PAM). Adding a coagulant aid to improve the removal rate of MPs versus Al-based salts alone was also reported in other studies. Shahi et al. [88] investigated the removal behavior of different MPs with Al-based salts in the presence of cationic polyamine-coated (PC) sand, and found that the removal efficiency of MPs increased to 70.7% as the alum dose increased to 30 mg/L, while a further increase in the dose caused a sharp decrease in removal efficiency. This might be because the aluminum hydroxide precipitate exceeded the adsorption capacity of the soluble originally formed alum material, leading to restabilization of the surface charge of MPs [88]. With addition of 500 mg/L PC sand (cationic polymer), the removal efficiency of MPs respectively increased to 92.7% and 90.2% for 20 and 30 mg/L alum, achieving a 50% reduction in alum dose. This improved removal effect might be attributed to the enhancement of a bridging effect between flocs, particles, and MPs by the cationic polyamine polymer, and improvement of sedimentation of suspended floc MPs by high specific gravity sand [88]. However, because of the poor attachment and difficulty in agglomeration, poor removal of MPs with small-smooth-spherical characteristics was observed for alum alone and alum combined with PC sand [89, 90]. As found by Zhang et al. [91], adding coagulant aid (PolyDADMAC) increased the removal efficiency of MPs in drinking water sources from below 2% with coagulant $Al_2(SO_4)_3$ to 13.6%. This limited promotion induced by PolyDADMAC might result from the low density of MPs and the conventional dosage of coagulant $Al_2(SO_4)_3$ (less than 20 ppm for actual DWTPs) used in the study.

Unlike the above studies, Al-based salts showed good MPs removal in other studies. Zhang et al. [82] reported that the removal rate of PET and weathered PET using the CFS method with $AlCl_3$ reached 100% and 92%. The corresponding removal mechanism was charge neutralization and sweeping flocculation. Moreover, it was also thought that the oxygen-containing functional groups (-OH, -COOH) on the surface of PET and weathered PET effectively reacted with the coagulant, making it easier to form flocs during coagulation compared to PE and PS, thus presenting better removal performance [82]. Similarly, Skaf et al. [92] reported that when the solution containing 5 mg/L microspheres with an initial turbidity of 16 NTU was treated by CFS using alum below 10 mg/L, the highest removal efficiency of 99% and 97% for kaolin and microspheres was achieved at the optimal alum dose of 4.8 mg/L, much higher than the previously reported ca. 20% removal efficiency for PS with size < 0.5 mm using a similar alum concentration. Through the investigation of the floc photos and zeta potential measurements, sweep flocculation was mainly responsible for microspheres removal. In addition to simulated water, the CFS method for MPs removal was also investigated in surface water. Lapointe et al. [93] reported that higher removal efficiency of PEST fibers in surface water (99% removal, less than 5 fibers/L) was achieved by 2.73 mg Al/L and 0.3 mg PAM/L compared to PE (82%) and PS (84%), possibly from the characteristics of PEST. PEST shape facilitated aggregation and the C=O bonds facilitated the attachment of different aluminum hydroxide species [93]. Moreover, the smaller the particles presented the better the removal under the same conditions, similar to that reported in other studies using surface water. Xue et al. [94] found that the MPs size significantly affected its removal performance in the CFS process, with smaller sizes being removed more efficiently. A concentration of 30 mg alum/L, commonly used in DWTPs, could remove over 80% of the smaller (≤ 25 μm) PS microspheres, but only removed ca. 60% of the MPs (e.g., ≥ 45 μm microspheres). Interestingly, changes in turbidity could be a good indicator of MPs removal. It was found that the removal of weathered PE (64 μm) and pristine PEST fiber exhibited a relatively strong correlation with turbidity reduction ($r = 0.96$ and 0.94) [93], in accordance with Xue et al. [94]. For PS microspheres (3-, 6- and 25-μm), the correlation coefficient (r) values between its removal and turbidity removal were respectively 0.984, 0.925, and 0.987, showing a strong positive linear correlation. Therefore, the removal rate of some MPs (e.g., PE, PEST, and PS) might be optimized by enhancing turbidity reduction.

Magnesium hydroxide, an environmentally friendly coagulant with many advantages (such as safety, non-toxicity, strong adsorption capacity and good stability etc.) [95], has also received attention for its potential application in MPs removal. Zhang et al. [96] compared the MPs removal performance of CFS based on magnesium hydroxide coagulant (MHC) and magnetic magnesium hydroxide coagulant (MMHC, the combination of $Mg(OH)_2$ and Fe_3O_4) and found that the removal efficiency of MPs reached 73.4% in the presence of MHC only, and the value further increased to 87.1% under the coagulation system of magnetic magnesium hydroxide with $Mg^{2+}:OH^-$ ratio of 1:1. This increase might be because the bubble structure formed by $Mg(OH)_2$ and Fe_3O_4 increased the collision between the coagulant and MPs in the coagulation process, and formed a uniform and dense floc more conducive to MPs removal [96, 97]. Moreover, PAM once again demonstrated promotion of MPs removal during coagulation. After adding 4 mg/L PAM, the highest removal rate (92.6%) was achieved by MMHC, prepared by 200 mg/L of $Mg(OH)_2$ and 120 mg/L of Fe_3O_4. However, the dosage of MMHC used is much higher than the actual amount of traditional Fe-based salts and Al-based salts coagulant used in actual DTWPs (e.g., 20 g/L and 40 mg/L) [76, 79].

7.2.2 Electrocoagulation

As introduced above, MPs could be removed by the CFS method using coagulants. Electrocoagulation (EC), which uses metal electrodes to electrically generate coagulants, provides an alternative to the CFS method without the addition of chemicals and is a simple process [98, 99]. Generally, from the production of cations to the formation of flocs, EC typically goes through three sequential processes: the generation of electrons and formation of iron or aluminum hydroxide coagulants under action of an electric field, destabilization of particles and colloids in the presence of coagulants, and final formation of flocs via collision between destabilized particles and coagulants [70], as shown in Figure 7.3. Since ions are the only product of the EC process, there is no need to consume oxidant or reductant, and there is no or little pollution to the environment, making EC an environmentally friendly, easy to automate, low-cost technology to be applied to water treatment [100, 101].

Recently, several studies have investigated the performance of EC to remove MPs in synthetic wastewater and actual wastewater under laboratory conditions. Compared with the traditional CFS method, EC showed better removal performance of MPs. For example, Shen et al. [102] found that the removal efficiencies of four microplastics (i.e., PE; polymethylmethacrylate, (PMMA), cellulose acetate (CA), and polypropylene (PP)) by EC in all experiments reached more than 82%, and the highest removal efficiencies reached 93.2%, 91.7%, 98.2%, and 98.4% respectively, indicating that fiber MPs might be easier to remove than granular MPs. Similar removal performances were also shown by the studies of Elkhatib et al. [103] and Perren et al. [57], the corresponding removal efficiencies in their studies were all more than 90%. The corresponding removal mechanisms were relevant to flocculation and charge neutralization [102]. The properties of the wastewater (such as pH, electrolyte concentration, and voltage density) could affect the removal of MPs by EC. Although pH could induce an effect on the removal of MPs, more than 90% of removal efficiency was achieved in the initial pH 6–8, similar to the pH value of municipal wastewater (6–9.2) [104]. This indicated that EC could be effectively applied to the removal of MPs in common municipal wastewater without adding additional chemicals to adjust pH, showing good pH adaptability. Moreover, it was also found that increasing electrolyte concentration and applied voltage density promoted MPs removal, possibly from the increased the amount of Al^{3+} and Fe^{3+} dissolved out [57, 102]. Meanwhile, in addition to the 96.5% removal rate of MPs, it was found in the real wastewater experiment that both high removal rates of COD and thermotolerant coliform colonies (92.2% and 88.8%) were synchronously achieved by EC.

Cost analysis of EC performed by Elkhatib et al. [103] found that the operating costs for the two best operating conditions used in the study were $0.29/m^3

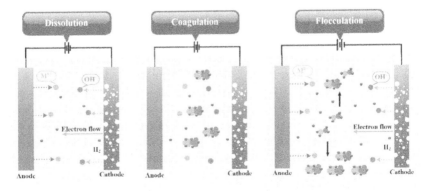

Figure 7.3 Procedures for MPs removal using electrocoagulation [70] / with permission of Elsevier.

wastewater and $1.02/m^3 wastewater, respectively, in the lower range of EC treatment cost values (0.03–3.85 US$/m^3) reported in various studies [104–108]. Similarly, Perren et al. [57] also reported the operating cost of MPs removal by EC was lower than other pollutants (such as domestic wastewater, iron, and bleaching wastewater). As discussed earlier, EC showed good potential in both economic and technical aspects for the removal of MPs from wastewater. However, given the ongoing cost of regular replacement of sacrificial anodes and electricity, large-scale application of EC for MPs removal in urban waters will require additional pilot or large-scale studies and economic evaluation.

7.2.3 Filtration

Filtration, a water treatment process widely used in DWTPs, also plays important role in the removal of MPs. Pivokonský et al. [77] reported that deep-bed filtration and GAC filtration decreased the content of MPs from 497 ± 44 MPs/L after CFS to 151 ± 4 MPs/L in the DWTP Plzeň. Their contribution to total MPs removal amounted to 26%. A higher contribution was observed in the studies of Wang et al. [79], the presence of GAC filtration removed ca. 56.8–60.9% of MPs in a DWTP located in the Yangtze River Delta. GAC filtration was beneficial to the removal of MPs with small particle sizes, such as size category 1–5 μm and 10–50 μm [77, 79]. Taking into account the role of CFS introduced in Section 7.2.1, filtration and CFS combined can remove more than 70% of MPs in most actual DWTPs reported [77–79].

Apart from drinking water, removal of MPs from WWTP effluent by filtration has also been investigated under laboratory conditions [109]. Bayo et al. [58] evaluated role of rapid sand filtration (RSF) in the abatement of MPs from the effluent of a WWTP and found that RSF reduced the average concentration of MPs from 4.40 ± 1.01 to 1.08 ± 0.28 MP/L, achieving 75.49% removal. Among the 14 different MPs identified in the raw effluent, only polyethylene (LDPE), nylon (NYL), and polyvinyl (PV) were not completely removed by RSF, showing good broad removal. The RSF removal capacity could be improved by combination with other methods. As a strategy proposed by Wang et al. [110], RSF integrated with biochar provided significant capacity for the removal of MPs (> 95%) with size of 10 μm, much higher than that of RSF alone (60–80%). A biochar filter showed better retention behavior and higher filter coefficient compared with RSF. Moreover, environmental scanning electron microscope (ESEM) examination and analysis indicated that MPs removal mechanisms of RSF integrated with biochar involved "*stuck*," "*trapped*," and "*entangled*," while RSF was only "*stuck*" [110]. In addition to traditional sand and carbon filters, filtration removal of MPs by other filter media is also gaining attention. Shen et al. [111] used aluminosilicate filter media and cationic surfactant-modified products for the first time to remove MPs from the effluent of WWTPs. This method obtained high removal efficiencies (>96%) and fixation capacity for PE and polyamide (PA), higher than that of RSF (63%).

SEM examination suggested the retention of MPs by this filter involved multiple mechanisms (i.e., *"captured," "trapped,"* and *"entangled"*), resulting in higher removal efficiency of MPs than RSF. Moreover, compared to other filters such as biochar and activated carbon, the filter media (zeolite) used is a kind of natural ore and its price is very low (< 100 vs 564 and 2524 US$) [110], showing good potential application prospects for enhanced removal of MPs in WWTPs.

7.2.4 Membrane Separation

Membrane separation, an advanced treatment technology commonly used in drinking water, has some advantages, such as stable effluent quality, simple operation, and ease of maintenance [112]. Generally, this technique can be divided into ultrafiltration, microfiltration, and reverse osmosis according to different membrane sizes [70]. Because of strong selectivity and separation, various contaminants including MPs, ions, bacteria, and suspended solids in drinking water can be validly removed by membrane separation technology [70, 113], and the corresponding principle of removing contaminants including MPs is shown in Figure 7.4. Membrane separation technology removes MPs through the blocking and entrapment effect of physical barriers. Currently, several studies have investigated the removal of MPs in synthetic water and actual wastewater by membrane separation technology under laboratory conditions and at pilot scale.

Ma et al. [83] found that compared with the lower removal efficiency of CFS for PE, ultrafiltration removed almost all PE. During this process, slight membrane fouling was inevitably induced as a loose cake layer was formed by retaining MPs.

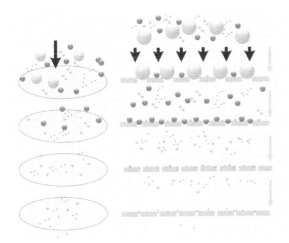

Figure 7.4 Procedures for removal of various contaminants including MPs through membrane separation technology [70] / with permission of Elsevier.

When CFS and ultrafiltration were combined, the membrane fouling was gradually alleviated after coagulation with Fe-based salts. This alleviation could be attributed to a looser cake layer formed after coagulation, causing the larger the porosity. Generally, the larger the MPs, the larger the porosity of the relatively loose cake layer and the less membrane fouling occurred. Therefore, the combination of membrane separation technology and CFS can play a better role, and has been widely used in drinking water treatment with excellent effluent performance [83, 112, 114]. Membrane separation technology also showed good removal performance for MPs in wastewater. As reported by Pramanik et al. [59], ultrafiltration and microfiltration respectively removed up to 96% and 91% of MPs. The difference in removal is because the ultrafiltration membrane with smaller pore size can remove more and smaller MPs. However, this property also caused a decline rate of ultrafiltration significantly higher than that of microfiltration. Tadsuwan et al. [115] further applied an ultrafiltration system to a WWTP, reducing the concentration of MPs in the final effluent from 10.67 ± 3.51 to 2.33 ± 1.53 particles/L. The corresponding overall removal efficiency increased from 86.14 to 96.97%. This suggested that membrane separation technology could be used as a tertiary treatment to further enhance the removal of MPs in WWTPs, reducing their entry into the aquatic environments in the effluent of WWTPs.

7.2.5 Adsorption

Adsorption, as an effective method to alleviate environmental problems, has attracted wide attention for its high efficiency, simple operation, environmental friendliness, low cost, and regenerable adsorbents [116]. Currently, the adsorbents used include activated carbon and biochar prepared from wastes naturally available in the environment, as well as synthetic sponges and some modified materials that can effectively adsorb and remove MPs through electrostatic interactions, chemical bonding interactions, and π-π interactions between MPs and adsorbents [60, 117–121]. For example, Sun et al. [119] used a biodegradable and reusable sponge material synthesized from graphene oxide and oxygen-doped carbon nitride to remove 72–90% of MPs. Compared with other adsorbents, the removal of MP by sponge is very simple, the sponge is placed in MPs-contaminated water and allowed to adsorb the solution for a period of time (Figure 7.5) without stirring or fixing the adsorbent. In addition, given excellent compressibility and mechanical properties, sponges can be reused after desorbing the MPs through a simple press [121]. As for other adsorbents, considerable removal efficiencies were also obtained, and they might be regenerated through ultrasonication and thermal treatment [117, 118].

In addition to removal efficiency, the adsorption capacity, defined as the amount of adsorbed MPs per unit mass of adsorbents, is also an important indicator of the

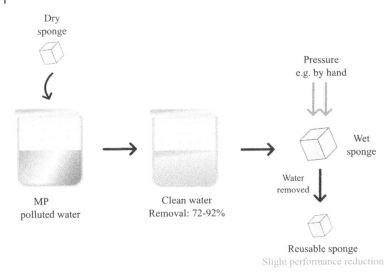

Figure 7.5 Removal of MPs using biodegradable and reusable sponge materials [74] / with permission of Elsevier.

adsorption process. Generally, they are obviously affected by many factors, for example adsorbent characteristics (e.g., porosity and surface area) and experimental conditions (e.g., MPs concentration, pH, and solution ions). The higher porosity and specific surface area made the mechanical interlocking between MPs and adsorbents more stable, thereby improving the adsorption capacity [122]. This observation inspired us to modify the precursor by changing the calcination or pyrolysis temperature to increase the adsorbent porosity and surface area when preparing the adsorbent. Since MPs are negatively charged in solution, when the surface of the adsorbent is modified to make the surface positively charged, its adsorption characteristics are also changed. For example, Shen et al. [111] modified the surface of zeolites and molecular sieves with a cationic surfactant to make the surface hydrophobic and positively charged, significantly improving its adsorption capacity. The correlation between MPs concentration and adsorbent dosage deserves attention. Higher MPs concentration may cause higher adsorption capacity, but the plastic particles are mainly removed by surface binding. If there is not enough adsorbent, the available surface is insufficient, ultimately leading to low removal efficiency. Adsorption technology was also favorable for the removal of MPs under acidic pH conditions, attributed to the increase of H^+ content on protonation of the adsorbent surface increasing its positive charge, resulting in a stronger electrostatic interaction more conducive to the adsorption reaction [60, 123]. The presence of positive and negative ions in solutions containing MPs can interfere with the adsorption process. As mentioned above, the surface of MPs is negatively charged and can be neutralized by some cations, such as

Na^+ and Ca^{2+}. Thus, the electrostatic interaction between the MPs and the adsorbent with positively charged will be weakened [60]. Similarly, the presence of negatively charged ions (e.g., $H_2PO_4^-$) in real water significantly affected the removal of MPs because of competitive adsorption between MPs and anions [118]. Moreover, the coexisting organic matter (e.g., natural organic matter) in MPs solution was also reported to cause adverse effects on the adsorption process from a competitive adsorption effect [118]. The main cost of adsorption technology is the synthesis and regeneration of the adsorbent. Although most of the adsorbents currently in use are low cost, the cost of the regeneration process has not been thoroughly considered [74]. Therefore, when the adsorption method is applied to remove MPs, its overall cost including synthesis and regeneration should be comprehensively evaluated.

7.3 Photocatalysis Degradation of Microplastics

The methods discussed above are mostly based on separation technology to remove MPs, and most cannot achieve complete removal, inevitably resulting in the continued release of large numbers of MPs into bodies of water. Semiconductor photocatalysis technology, which can efficiently degrade various pollutants in air and water into CO_2 and H_2O under the irradiation of light, has received much attention and research because of excellent degradation performance, sustainable reuse, and nontoxic byproducts [124–126]. Recently, several nanoscale metal oxide and metal oxide semiconductor-based photocatalysts have also been used to enhance the degradation of MPs, showing good degradation performance [127–129].

Plastic particles undergo polymer aging, chain crosslinking, and mineralization under photocatalytic reaction conditions, finally forming H_2O and CO_2. This process occurs when a semiconductor, such as titanium dioxide (TiO_2) and bismuth oxychloride, etc., are excited by a light source with energy equal to or greater than its band gap [74]. Charge separation occurs during this process, where electrons (e^-) in the valence band are transferred into the conduction band, while positive holes (h^+) are formed. When H_2O, OH^- and O_2 are present on the semiconductor surface, they react with the electrons and positive holes to generate reactive oxygen species (ROS) (e.g., hydroxyl ($OH\cdot$) and superoxide (O_2^-) radicals) that can break the chains of MPs and fully mineralize them [128, 130]. The conventional process of degrading MPs based on photocatalysis technology is shown in the following Eqs. (7.1)–(7.4) [74].

$$\text{Semiconductor} \xrightarrow{h\nu} h^+ + e^+ \quad (7.1)$$

$$h^+ + H_2O \rightarrow OH^\bullet \tag{7.2}$$

$$e^- + O_2 \rightarrow O_2^- \tag{7.3}$$

$$MPs + ROS(OH^\bullet + O_2^-) \rightarrow CO_2 + H_2O \tag{7.4}$$

7.3.1 Zinc Oxide-based Photocatalysis

Considering its good visible light absorption, high electron mobility, low biotoxicity to humans and aquatic organisms, and low cost, zinc oxide (ZnO) with tailorable defect chemistry has been extensively used for photocatalytic degradation of MPs [130]. Tofa et al. [129] first evaluated heterogeneous photocatalytic degradation of low-density polyethylene (LDPE) MP residues using ZnO nanorods. After 175 h of visible light exposure, many wrinkles, cracks, and spots were visually observed in the surface of LDPE MP. Meanwhile, chain scission of LDPE caused a decrease in elasticity, accompanied by the formation of new functional groups attributed to the formation of low molecular weight compounds during the photocatalytic reaction. For example, it was observed that residues have a 30% increase in the carbonyl index. The proposed mechanism of ZnO photocatalytic degradation of LDPE is shown in Figure 7.6. The generated ROS (such as OH• and O_2^-) by the catalyst attack the weak parts of LDPE (e.g., chromogenic groups and defects) and induce chain cleavage with the generation of polyvinyl

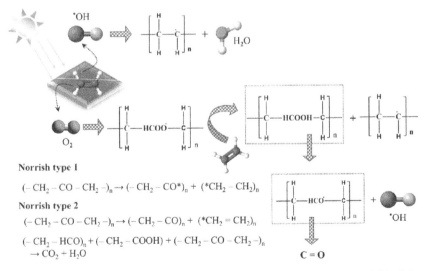

Figure 7.6 Mechanism of photocatalytic degradation of LDPE by ZnO under visible light irradiation [67] / with permission of American Chemcial Society.

alkyl radicals. After LDPE forms hydroperoxide groups by oxidation, it undergoes various degradation reactions such as cross-linking, chain scission, branching, and further oxidation. Then, carbonyl and vinyl group containing species and even CO_2 and H_2O are produced. In order to further enhance the photodegradation of LDPE, Tofa and co-workers modified ZnO nanorods by depositing platinum nanoparticles on the nanorod surface, obtaining ZnO-Pt nanocomposite photocatalysts [131]. It was found that ZnO-Pt have better photocatalytic performance with visible light irradiation compared to bare ZnO nanorods. This improvement was mainly attributed to plasmon absorption from platinum nanoparticles and the diffusion of photogenerated electrons into platinum from the ZnO nanorod interfaces that together resulted in better interfacial exciton separation and improvement of hydroxyl radical activity as well as an increase in visible light absorption.

In addition to LDPE, ZnO NRs also showed a good photocatalytic degradation effect on PP MPs suspended in water. Uheida et al. [130] reported that the average particle volume of PP MPs decreased by 65% after two weeks of visible light irradiation using glass fiber substrates. The corresponding photodegradation mechanism was chain scission. Hydroxyl radicals generated by photoexcitation of ZnO NRs caused polymer chain scission of PP MPs and degradation. More importantly, GC-MS spectra confirmed that most of photocatalytic degradation byproducts were molecules with little or no toxicity to humans and the aquatic environment reported in the literature, such as ethynyloxy/acetyl radicals, hydroxypropyl, butyraldehyde, and acetone.

7.3.2 Titanium Dioxide-based Photocatalysis

To achieve complete decomposition of MPs as much as possible, Nabi et al. [132] introduced a titanium dioxide (TiO_2) film for enhanced mineralization of PS and PE under UV irradiation. It was found that TiO_2 nanoparticle film made with Triton X-100 has a good degradation effect on various sizes of PS, achieving 99.99% degradation rate for 5 μm-sized PS under 254 nm UV irradiation after 24 h reaction and 98.40% for 400 nm-sized PS under 365 nm UV irradiation after 12 h reaction. Figure 7.7 shows the morphology change of PS particles during the photocatalytic process. With the prolonged photoreaction time, the morphology of PS changed significantly, and the sphere size decreased. This change could be attributed to the generation of holes and OH• from TiO_2 under UV light, which could oxidize PS MPs and eventually cause mineralization [132–134] accompanied by generation of carbon dioxide as the main end product. Similar degradation performance was also observed on polyamide 66. As reported by Lee et al. [135], the TiO_2 film showed 97% mass loss for polyamide 66 (PA66) MPs in WWTPs under 254 nm UV irradiation within 48 h.

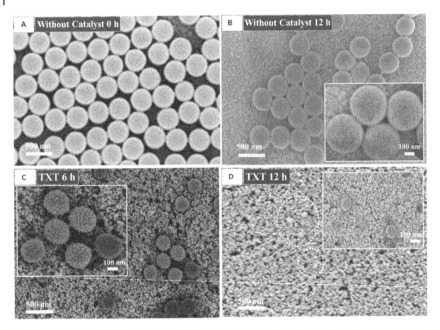

Figure 7.7 SEM images of PS with or without TiO$_2$ films after different irradiation time under 365 nm UV Light [136] / with permission of Elsevier.

Moreover, some work has also been done on the modification of TiO$_2$ for degradation of MPs in an aqueous environment. Ariza-Tarazona et al. [128] synthesized two N-TiO$_2$ using extrapallial fluid of blue mussels and sol-gel respectively for degradation of high-density PE MPs. Mussels-derived N-TiO$_2$ showed excellent capacity to improve the photocatalytic degradation of high-density PE, achieving 6.40% mass loss after 20 h of irradiation, while relatively low mass loss (2.86%) was achieved by sol-gel N-TiO$_2$ after 8 h. In addition, compared with sol-gel N-TiO$_2$, mussels-derived N-TiO$_2$, considered a green synthesis, presented good photocatalytic capacity in both solid and aqueous environments. The findings also show that environmental conditions, interactions between pollutant and catalyst, and catalyst surface area should each be attentively set to avoid photocatalytic stagnation. Similarly, Au@Ni@TiO2-based micromotors also showed excellent capacity to remove extracted MPs and suspended matter in both real and pure water [127]. For example, in just 40 seconds, 12 of 18 MPs were removed under UV light. Meanwhile, the addition of a small percentage of H$_2$O$_2$ can accelerate the degradation process.

The photocatalytic degradation performance of MPs could be affected by some operating variables. Ariza-Tarazona et al. [136] evaluated the impact of pH and temperature on the degradation of high-density PE MPs by a bio-inspired

C,N-TiO$_2$ semiconductor under visible light irradiation, and found that low temperature and low pH favored MPs degradation. Low pH increased the level of H$^+$ during the reaction, favoring the generation of hydrogen peroxide, in turn facilitating the interaction of the colloidal nanoparticles with the high-density PE MPs and eventual MPs degradation. At low temperature, the increase in MPs surface area by fragmentation was suitable for the interaction between C,N-TiO$_2$ and MPs. It was also demonstrated that these two operating variables combined to impact the degradation of PE MPs, resulting in a highest mass loss of 71.77% at pH = 3 at 0°C. The reported findings indicated that the degradation of MPs could be accelerated and enhanced by appropriately changing the operating conditions during the photocatalytic process.

7.3.3 Bismuth-based Photocatalysis

In addition to ZnO and TiO$_2$-based photocatalysis, bismuth oxychloride, an efficient photocatalyst for contaminant pollutant treatment [137], has also been evaluated for its degradation effect on MPs. Jiang et al. [63] reported that hydroxy-rich ultrathin BiOCl (BiOCl-X) presented certain potential for degradation of MPs, reducing the mass of PE-S by 5.38% after 300 min of visible light irradiation. The mass loss was 24 times higher compared to BiOCl nanosheets. This enhancement can be ascribed to BiOCl-X with larger specific surface area and better charge transfer ability, aiding active site utilization and charge separation and transfer [138]. Mechanism exploration suggested that the degradation of PE-S is caused by the attack on C−H bonds by the active OH• generated by BiOCl-X. It was also found that compared with the protection of MPs by Coulombic repulsion under alkaline conditions, acidic conditions were more favorable for their degradation. Toxicity evaluation of PE-S degradation solutions using zebrafish showed that survival and hatchability was not significantly changed in the PE-S degradation solution, indicating the toxicity of degradation byproducts of PE-S may be very weak. However, it must be pointed out that the degradation of PE-S induced by BiOCl-X needs further improvement.

7.4 Chemical Oxidation Degradation of Microplastics

In addition to photocatalytic degradation, some chemical oxidation methods based on traditional advanced oxidation processes (AOPs) proven effective for removal of persistent pollutants by generating reactive radicals [139, 140] have also been applied for the treatment of MPs. For instance, Kang et al. [64] used carbon nanosprings (Mn@NCNTs) to activate PMS for PE MPs degradation under hydrothermal conditions and found that both hydrothermal temperature and

Mn@NCNTs driven PMS-AOPs play an integral role in the degradation. As a result, the highest removal (54 wt %) of PE MPs was achieved by a Mn@NCNTs/PMS system at 160°C. Mechanism exploration indicated radical oxidation rather than nonradical oxidation played a leading role. The generation of the sulfate radical (SO_4^{-}) and OH• by Mn@NCNTs and heat-induced PMS activation oxidized MPs, leading to decomposition and mineralization. Toxicity evaluation tests suggested that the intermediate products generated by MPs degradation were not harmful to microorganisms and could be used as nutrients for aquatic algae. Similarly, Liu et al. [141] also found that heat activated $K_2S_2O_8$ and Fenton treatments accelerated the aging of MPs, but the related degradation has not been investigated and reported.

Additionally, Miao et al. [142] employed an electro-Fenton system with a TiO_2/graphite (TiO_2/C) cathode to degrade polyvinyl chloride (PVC) MPs. This integrated system showed excellent degradation performance for PVC, mainly by oxidation and breakage of the PVC backbone caused by OH• and reductive dichlorination by applied cathodic potential. After reaction at 100°C for 6 h, PVC MPs degradation performance of 75% dechlorination efficiency and 56% weight loss was obtained. Meanwhile, higher temperature could accelerate the reaction, improving the dechlorination of PVC MPs and favoring its decomposition. In the proposed degradation mechanism (Figure 7.8), after receiving electrons from the system cathode, PVC MPs were dechlorinated by cathodic reduction at high temperature. At the same time, the generation of OH• oxidized PVC MPs, forming various intermediate organics and releasing them into solution where the hydrocarbons were further oxidized into small molecular substances (e.g., alcohols and

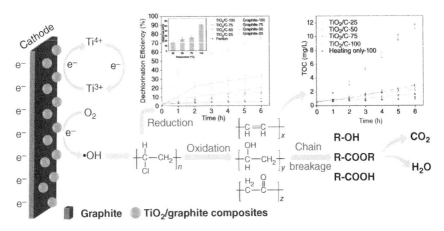

Figure 7.8 Schematic diagram of PVC MPs degradation via electro-Fenton system encompassing TiO_2/graphite cathode [142] / with permission of Elsevier.

esters). Finally, some of the small molecular species were mineralized along with the generation of H_2O and CO_2. The reported technology provided an alternative method to degrade MPs. However, the energy consumption (e.g., high temperature and electricity requirements) of the currently reported methods will bring a certain treatment cost. Therefore future studies should be carried out to strengthen the degradation effect with reduction of energy consumption.

7.5 Future Prospects

MPs pollution, as new global problem, has gained increasing research attention because of potential risks to human health and the ecological environment. This chapter summarized main research on the removal and degradation of MPs, and studied the performances, related mechanisms, and limitations of these removal and degradation technologies. At present, except for membrane separation that can remove almost all MPs, other methods cannot achieve complete removal or degradation of MPs. However, membrane separation is prone to membrane fouling and is therefore not suitable for large-scale and widespread applications [83]. Considering the ever-increasing release of plastics, it's estimated that 2.5 billion tons of plastics will be released by 2025 [3], which will inevitably impose a huge burden on the corresponding control of MPs. Therefore, the first and foremost necessity is reducing the use of plastic products reasonably, especially limiting their abuse while increasing the recycling of plastics, in order to reduce the discharge of plastics to the environment as much as possible. Although many countries have adopted policies to control unnecessary use of plastics, the effect is thusfar unsatisfactory. It is suggested that ecological risks of MPs should be publicized to strengthen the awareness of the need to reduce the use of plastic products and recycle them.

Drinking water is closely related to human daily life. However, current research results show that the removal rate of MPs in DWTPs is only ca. 80–90%, inevitably leading to significant MPs entering the body. On the other hand, the removal of MPs is historically not the primary concern of DWTPs. Considering the large potential risk of MPs to human health and the importance of drinking water, it is recommended to gradually incorporate the removal of MPs as a drinking water assessment indicator, rather than only turbidity, DOM, and other more conventional indicators. On the other hand, because of the limited removal effect of a single process, it is necessary to further the research on optimization of MPs removal by synergistically strengthening multi-process removal. Meanwhile, technologies such as membrane separation can be adopted for the water directly entering the human body to remove MPs as much completely as possible. Moreover, there is also a considerable amount of MPs in the effluent of the

WWTPs [17], but current research on the removal at WWTP-scale is insufficient, leading to continuous entry of MPs into the water environment. Therefore, measures should be taken to remove MPs in wastewater before discharge.

Compared with the removal of MPs based on separation technology, photocatalysis and chemical oxidation technologies have unique advantages in degrading MPs into carbon dioxide in water, providing alternative ideas for the complete removal of MPs. However, the current research in this area has not yet attracted sufficient attention. Photocatalysis and chemical oxidation-related pollutant removal research mainly focuses on the degradation of traditional pollutants such as antibiotics and persistent pollutants. However, it must be pointed out that the potential risks of MPs are more worrying than traditional pollutants. Therefore, more photocatalysis and chemical oxidation research should be carried out for the control of MPs, such as exploring the feasibility of other photocatalysts or chemical oxidation methods to degrade MPs. Considering that the methods of degrading MPs require specific synthetic materials, more severe experimental conditions (such as high temperature, high pressure and additional voltage) will bring increased treatment cost. Therefore, the development of inexpensive photocatalytic materials and chemical oxidation processes, and the reduction of the reaction temperature and energy consumption while maintaining good degradation performance of MPs should be a future research priority. In addition, certain intermediate products will be produced during the process of degrading MPs [63, 64]. However, there are few related toxicity studies, and only respective evaluation tests on zebrafish and microalgae. To comprehensively and objectively assess the impact of photocatalytic and chemical oxidation methods on the control of MPs, more comprehensive toxicity evaluation should be undertaken to evaluate the potential impact of intermediates after degradation.

References

1 Ebrahimbabaie, P., Yousefi, K., and Pichtel, J. (2022). Photocatalytic and biological technologies for elimination of microplastics in water: Current status. *Science of the Total Environment* 806: 150603.

2 Wright, S.L. and Kelly, F.J. (2017). Plastic and human health: A micro issue? *Environmental Science & Technology* 51 (12): 6634–6647.

3 Jambeck, J.R., Geyer, R., Wilcox, C. et al. (2015). Plastic waste inputs from land into the ocean. *Science* 347 (6223): 768–771.

4 MacLeod, M., Arp, H.P.H., Tekman, M.B. et al. (2021). The global threat from plastic pollution. *Science* 373 (6550): 61–65.

5 Schmaltz, E., Melvin, E.C., Diana, Z. et al. (2020). Plastic pollution solutions: Emerging technologies to prevent and collect marine plastic pollution. *Environment International* 144: 106067.

6 Shen, M., Zhang, Y., Zhu, Y. et al. (2019). Recent advances in toxicological research of nanoplastics in the environment: A review. *Environmental Pollution* 252: 511–521.

7 Anderson, P.J., Warrack, S., Langen, V. et al. (2017). Microplastic contamination in lake Winnipeg, Canada. *Environmental Pollution* 225: 223–231.

8 Napper, I.E., Bakir, A., Rowland, S.J. et al. (2015). Characterisation, quantity and sorptive properties of microplastics extracted from cosmetics. *Marine Pollution Bulletin* 99 (1–2): 178–185.

9 Horton, A.A., Walton, A., Spurgeon, D.J. et al. (2017). Microplastics in freshwater and terrestrial environments: Evaluating the current understanding to identify the knowledge gaps and future research priorities. *Science of the Total Environment* 586: 127–141.

10 Jiang, J.Q. (2018). Occurrence of microplastics and its pollution in the environment: A review. *Sustainable Production and Consumption* 13: 16–23.

11 Yi, H., Jiang, M., Huang, D. et al. (2018). Advanced photocatalytic Fenton-like process over biomimetic hemin-Bi_2WO_6 with enhanced pH. *Journal of the Taiwan Institute of Chemical Engineers* 93: 184–192.

12 Duis, K. and Coors, A. (2016). Microplastics in the aquatic and terrestrial environment: Sources (with a specific focus on personal care products), fate and effects. *Environmental Sciences Europe* 28 (1): 1–25.

13 Rodriguez-Narvaez, O.M., Goonetilleke, A., Perez, L., and Bandala, E.R. (2021). Engineered technologies for the separation and degradation of microplastics in water: A review. *Chemical Engineering Journal* 414: 128692.

14 Cheung, P.K. and Fok, L. (2017). Characterisation of plastic microbeads in facial scrubs and their estimated emissions in Mainland China. *Water Research* 122: 53–61.

15 Fendall, L.S. and Sewell, M.A. (2009). Contributing to marine pollution by washing your face: Microplastics in facial cleansers. *Marine Pollution Bulletin* 58 (8): 1225–1228.

16 Zhang, X., Liu, C., Liu, J. et al. (2022). Release of microplastics from typical rainwater facilities during aging process. *Science of the Total Environment* 813: 152674.

17 Sun, J., Dai, X., Wang, Q. et al. (2019). Microplastics in wastewater treatment plants: Detection, occurrence and removal. *Water Research* 152: 21–37.

18 Sang, W., Chen, Z., Mei, L. et al. (2021). The abundance and characteristics of microplastics in rainwater pipelines in Wuhan, China. *Science of the Total Environment* 755: 142606.

19 Sun, Y., Chen, Z., Wu, G. et al. (2016). Characteristics of water quality of municipal wastewater treatment plants in China: Implications for resources utilization and management. *Journal of Cleaner Production* 131: 1–9.

20 Carr, S.A., Liu, J., and Tesoro, A.G. (2016). Transport and fate of microplastic particles in wastewater treatment plants. *Water Research* 91: 174–182.

21 Franzellitti, S., Canesi, L., Auguste, M. et al. (2019). Microplastic exposure and effects in aquatic organisms: A physiological perspective. *Environmental Toxicology and Pharmacology* 68: 37–51.

22 Razeghi, N., Hamidian, A.H., Wu, C. et al. (2021). Microplastic sampling techniques in freshwaters and sediments: A review. *Environmental Chemistry Letters* 19 (6): 4225–4252.

23 Eriksen, M., Mason, S., Wilson, S. et al. (2013). Microplastic pollution in the surface waters of the Laurentian Great Lakes. *Marine Pollution Bulletin* 77 (1–2): 177–182.

24 Li, J., Liu, H., and Chen, J.P. (2018). Microplastics in freshwater systems: A review on occurrence, environmental effects, and methods for microplastics detection. *Water Research* 137: 362–374.

25 Wang, W., Gao, H., Jin, S. et al. (2019). The ecotoxicological effects of microplastics on aquatic food web, from primary producer to human: A review. *Ecotoxicology and Environmental Safety* 173: 110–117.

26 Xu, Q., Huang, Q.S., Luo, T.Y. et al. (2021). Coagulation removal and photocatalytic degradation of microplastics in urban waters. *Chemical Engineering Journal* 416: 129123.

27 Krause, S., Baranov, V., Nel, H.A. et al. (2021). Gathering at the top? Environmental controls of microplastic uptake and biomagnification in freshwater food webs. *Environmental Pollution* 268: 115750.

28 Zhu, J. and Wang, C. (2020). Recent advances in the analysis methodologies for microplastics in aquatic organisms: Current knowledge and research challenges. *Analytical Methods* 12 (23): 2944–2957.

29 Yu, P., Liu, Z., Wu, D. et al. (2018). Accumulation of polystyrene microplastics in juvenile Eriocheir sinensis and oxidative stress effects in the liver. *Aquatic Toxicology* 200: 28–36.

30 Foley, C.J., Feiner, Z.S., Malinich, T.D. et al. (2018). A meta-analysis of the effects of exposure to microplastics on fish and aquatic invertebrates. *Science of the Total Environment* 631: 550–559.

31 Yin, L., Chen, B., Xia, B. et al. (2018). Polystyrene microplastics alter the behavior, energy reserve and nutritional composition of marine jacopever (Sebastes schlegelii). *Journal of Hazardous Materials* 360: 97–105.

32 Liu, W., Zhao, Y., Shi, Z. et al. (2020). Ecotoxicoproteomic assessment of microplastics and plastic additives in aquatic organisms: A review. *Comparative Biochemistry and Physiology Part D: Genomics and Proteomics* 36: 100713.

33 Cormier, B., Gambardella, C., Tato, T. et al. (2021). Chemicals sorbed to environmental microplastics are toxic to early life stages of aquatic organisms. *Ecotoxicology and Environmental Safety* 208: 111665.

34 Iguchi, T., Watanabe, H., and Katsu, Y. (2006). Application of ecotoxicogenomics for studying endocrine disruption in vertebrates and invertebrates. *Environmental Health Perspectives* 114 (Suppl 1): 101–105.

35 Chen, Q., Allgeier, A., Yin, D. et al. (2019). Leaching of endocrine disrupting chemicals from marine microplastics and mesoplastics under common life stress conditions. *Environment International* 130: 104938.

36 Naik, R.K., Naik, M.M., D'Costa, P.M. et al. (2019). Microplastics in ballast water as an emerging source and vector for harmful chemicals, antibiotics, metals, bacterial pathogens and HAB species: A potential risk to the marine environment and human health. *Marine Pollution Bulletin* 149: 110525.

37 Llorca, M., Schirinzi, G., Martínez, M. et al. (2018). Adsorption of perfluoroalkyl substances on microplastics under environmental conditions. *Environmental Pollution* 235: 680–691.

38 Razanajatovo, R.M., Ding, J., Zhang, S. et al. (2018). Sorption and desorption of selected pharmaceuticals by polyethylene microplastics. *Marine Pollution Bulletin* 136: 516–523.

39 Vo, H.C. and Pham, M.H. (2021). Ecotoxicological effects of microplastics on aquatic organisms: A review. *Environmental Science and Pollution Research* 28 (33): 44716–44725.

40 Zhang, Q., Xu, E.G., Li, J. et al. (2020). A review of microplastics in table salt, drinking water, and air: Direct human exposure. *Environmental Science & Technology* 54 (7): 3740–3751.

41 Prata, J.C., da Costa, J.P., Lopes, I. et al. (2020). Environmental exposure to microplastics: An overview on possible human health effects. *Science of the Total Environment* 702: 134455.

42 Triebskorn, R., Braunbeck, T., Grummt, T. et al. (2019). Relevance of nano-and microplastics for freshwater ecosystems: A critical review. *TrAC Trends in Analytical Chemistry* 110: 375–392.

43 Chen, G., Feng, Q., and Wang, J. (2020). Mini-review of microplastics in the atmosphere and their risks to humans. *Science of the Total Environment* 703: 135504.

44 Gasperi, J., Wright, S.L., Dris, R. et al. (2018). Microplastics in air: Are we breathing it in? *Current Opinion in Environmental Science & Health* 1: 1–5.

45 Schirinzi, G.F., Pérez-Pomeda, I., Sanchís, J. et al. (2017). Cytotoxic effects of commonly used nanomaterials and microplastics on cerebral and epithelial human cells. *Environmental Research* 159: 579–587.

46 Yee, M.S.L., Hii, L.W., Looi, C.K. et al. (2021). Impact of microplastics and nanoplastics on human health. *Nanomaterials* 11 (2): 496.

47 Shi, Q., Tang, J., Liu, R. et al. (2021). Toxicity in vitro reveals potential impacts of microplastics and nanoplastics on human health: A review. *Critical Reviews in Environmental Science and Technology*: 1–33.

48 Huang, Z., Weng, Y., Shen, Q. et al. (2021). Microplastic: A potential threat to human and animal health by interfering with the intestinal barrier function and changing the intestinal microenvironment. *Science of the Total Environment* 785: 147365.

49 Zhang, Q., Zhao, Y., Li, J. et al. (2020). Microplastics in food: Health risks. In: *Microplastics in Terrestrial Environments* (D. He and Y. Luo, Eds), 343–356. Cham: Springer.

50 Smith, M., Love, D.C., Rochman, C.M. et al. (2018). Microplastics in seafood and the implications for human health. *Current Environmental Health Reports* 5 (3): 375–386.

51 Campanale, C., Massarelli, C., Savino, I. et al. (2020). A detailed review study on potential effects of microplastics and additives of concern on human health. *International Journal of Environmental Research and Public Health* 17 (4): 1212.

52 Liao, Y.L. and Yang, J.Y. (2020). Microplastic serves as a potential vector for Cr in an in-vitro human digestive model. *Science of the Total Environment* 703: 134805.

53 De-la-torre, G.E. (2020). Microplastics: An emerging threat to food security and human health. *Journal of Food Science and Technology* 57 (5): 1601–1608.

54 Sharma, S. and Chatterjee, S. (2017). Microplastic pollution, a threat to marine ecosystem and human health: A short review. *Environmental Science and Pollution Research* 24 (27): 21530–21547.

55 Padervand, M., Lichtfouse, E., Robert, D. et al. (2020). Removal of microplastics from the environment. A review. *Environmental Chemistry Letters* 18 (3): 807–828.

56 Pal, A., He, Y., Jekel, M. et al. (2014). Emerging contaminants of public health significance as water quality indicator compounds in the urban water cycle. *Environment International* 71: 46–62.

57 Perren, W., Wojtasik, A., and Cai, Q. (2018). Removal of microbeads from wastewater using electrocoagulation. *ACS Omega* 3 (3): 3357–3364.

58 Bayo, J., López-Castellanos, J., and Olmos, S. (2020). Membrane bioreactor and rapid sand filtration for the removal of microplastics in an urban wastewater treatment plant. *Marine Pollution Bulletin* 156: 111211.

59 Pramanik, B.K., Pramanik, S.K., and Monira, S. (2021). Understanding the fragmentation of microplastics into nano-plastics and removal of nano/microplastics from wastewater using membrane, air flotation and nano-ferrofluid processes. *Chemosphere* 282: 131053.

60 Yuan, F., Yue, L., Zhao, H. et al. (2020). Study on the adsorption of polystyrene microplastics by three-dimensional reduced graphene oxide. *Water Science and Technology* 81 (10): 2163–2175.

61 Grbic, J., Nguyen, B., Guo, E. et al. (2019). Magnetic extraction of microplastics from environmental samples. *Environmental Science & Technology Letters* 6 (2): 68–72.

62 Brandon, A.M., Gao, S.H., Tian, R. et al. (2018). Biodegradation of polyethylene and plastic mixtures in mealworms (larvae of Tenebrio molitor) and effects on the gut microbiome. *Environmental Science & Technology* 52 (11): 6526–6533.

63 Jiang, R., Lu, G., Yan, Z. et al. (2021). Microplastic degradation by hydroxy-rich bismuth oxychloride. *Journal of Hazardous Materials* 405: 124247.

64 Kang, J., Zhou, L., Duan, X. et al. (2019). Degradation of cosmetic microplastics via functionalized carbon nanosprings. *Matter* 1 (3): 745–758.

65 Asami, T., Katayama, H., Torrey, J.R. et al. (2016). Evaluation of virus removal efficiency of coagulation-sedimentation and rapid sand filtration processes in a drinking water treatment plant in Bangkok, Thailand. *Water Research* 101: 84–94.

66 Chew, C.M., Aroua, M.K., Hussain, M.A. et al. (2015). Practical performance analysis of an industrial-scale ultrafiltration membrane water treatment plant. *Journal of the Taiwan Institute of Chemical Engineers* 46: 132–139.

67 Bacha, A.U.R., Nabi, I., and Zhang, L. (2021). Mechanisms and the engineering approaches for the degradation of microplastics. *ACS ES&T Engineering* 1 (11): 1481–1501.

68 Touffet, A., Baron, J., Welte, B. et al. (2015). Impact of pretreatment conditions and chemical ageing on ultrafiltration membrane performances. Diagnostic of a coagulation/adsorption/filtration process. *Journal of Membrane Science* 489: 284–291.

69 Bu, F., Gao, B., Yue, Q. et al. (2019). The combination of coagulation and adsorption for controlling ultra-filtration membrane fouling in water treatment. *Water* 11 (1): 90.

70 Shen, M., Song, B., Zhu, Y. et al. (2020). Removal of microplastics via drinking water treatment: Current knowledge and future directions. *Chemosphere* 251: 126612.

71 Rajala, K., Grönfors, O., Hesampour, M. et al. (2020). Removal of microplastics from secondary wastewater treatment plant effluent by coagulation/flocculation with iron, aluminum and polyamine-based chemicals. *Water Research* 183: 116045.

72 Yi, H., Huang, D., Qin, L. et al. (2018). Selective prepared carbon nanomaterials for advanced photocatalytic application in environmental pollutant treatment and hydrogen production. *Applied Catalysis B: Environmental* 239: 408–424.

73 Lee, K.M., Lai, C.W., Ngai, K.S. et al. (2016). Recent developments of zinc oxide based photocatalyst in water treatment technology: A review. *Water Research* 88: 428–448.

74 Hanif, M.A., Ibrahim, N., Dahalan, F.A. et al. (2022). Microplastics and nanoplastics: Recent literature studies and patents on their removal from aqueous environment. *Science of the Total Environment* 810: 152115.

75 Zhou, G., Wang, Q., Li, J. et al. (2021). Removal of polystyrene and polyethylene microplastics using PAC and FeCl3 coagulation: Performance and mechanism. *Science of the Total Environment* 752: 141837.

76 Sillanpää, M., Ncibi, M.C., Matilainen, A. et al. (2018). Removal of natural organic matter in drinking water treatment by coagulation: A comprehensive review. *Chemosphere* 190: 54–71.

77 Pivokonský, M., Pivokonská, L., Novotná, K. et al. (2020). Occurrence and fate of microplastics at two different drinking water treatment plants within a river catchment. *Science of the Total Environment* 741: 140236.

78 Pivokonsky, M., Cermakova, L., Novotna, K. et al. (2018). Occurrence of microplastics in raw and treated drinking water. *Science of the Total Environment* 643: 1644–1651.

79 Wang, Z., Lin, T., and Chen, W. (2020). Occurrence and removal of microplastics in an advanced drinking water treatment plant (ADWTP). *Science of the Total Environment* 700: 134520.

80 Gillberg, L., Hansen, B., Karlsson, I. et al. (2003). About water treatment. *Kemira Kemwater* 171: 195–199, 72–73. 91-631-4344-5.

81 Andrady, A.L. (2011). Microplastics in the marine environment. *Marine Pollution Bulletin* 62 (8): 1596–1605.

82 Lu, S., Liu, L., Yang, Q. et al. (2021). Removal characteristics and mechanism of microplastics and tetracycline composite pollutants by coagulation process. *Science of the Total Environment* 786: 147508.

83 Ma, B., Xue, W., Ding, Y. et al. (2019). Removal characteristics of microplastics by Fe-based coagulants during drinking water treatment. *Journal of Environmental Sciences* 78: 267–275.

84 Aguilar, M.I., Saez, J., Llorens, M. et al. (2003). Microscopic observation of particle reduction in slaughterhouse wastewater by coagulation–flocculation using ferric sulphate as coagulant and different coagulant aids. *Water Research* 37 (9): 2233–2241.

85 Aboulhassan, M.A., Souabi, S., Yaacoubi, A. et al. (2006). Improvement of paint effluents coagulation using natural and synthetic coagulant aids. *Journal of Hazardous Materials* 138 (1): 40–45.

86 Yang, H.M., Gu, J.H., Zhang, D.H. et al. (2017). Analysis of flocculant PAC and PAM in the treatment of pickled cabbage wastewater. *Journal of Civil, Architectural & Environmental Engineering* 39 (04): 95–101.

87 Ma, B., Xue, W., Hu, C. et al. (2019). Characteristics of microplastic removal via coagulation and ultrafiltration during drinking water treatment. *Chemical Engineering Journal* 359: 159–167.

88 Shahi, N.K., Maeng, M., Kim, D. et al. (2020). Removal behavior of microplastics using alum coagulant and its enhancement using polyamine-coated sand. *Process Safety and Environmental Protection* 141: 9–17.

89 Hassas, B.V., Caliskan, H., Guven, O. et al. (2016). Effect of roughness and shape factor on flotation characteristics of glass beads. *Colloids and Surfaces A: Physicochemical and Engineering Aspects* 492: 88–99.

90 Zhanpeng, J. and Yuntao, G. (2006). Flocculation morphology: Effect of particulate shape and coagulant species on flocculation. *Water Science and Technology* 53 (7): 9–16.
91 Zhang, Y., Diehl, A., Lewandowski, A. et al. (2020). Removal efficiency of micro-and nanoplastics (180 nm–125 μm) during drinking water treatment. *Science of the Total Environment* 720: 137383.
92 Skaf, D.W., Punzi, V.L., Rolle, J.T. et al. (2020). Removal of micron-sized microplastic particles from simulated drinking water via alum coagulation. *Chemical Engineering Journal* 386: 123807.
93 Lapointe, M., Farner, J.M., Hernandez, L.M. et al. (2020). Understanding and improving microplastic removal during water treatment: Impact of coagulation and flocculation. *Environmental Science & Technology* 54 (14): 8719–8727.
94 Xue, J., Peldszus, S., Van Dyke, M.I. et al. (2021). Removal of polystyrene microplastic spheres by alum-based coagulation-flocculation-sedimentation (CFS) treatment of surface waters. *Chemical Engineering Journal* 422: 130023.
95 Zhao, J., Liu, S., Chi, Y. et al. (2015). Magnesium hydroxide coagulation performance and floc properties in treating kaolin suspension under high pH. *Desalination and Water Treatment* 53 (3): 579–585.
96 Zhang, Y., Zhao, J., Liu, Z. et al. (2021). Coagulation removal of microplastics from wastewater by magnetic magnesium hydroxide and PAM. *Journal of Water Process Engineering* 43: 102250.
97 Zhang, M., Xiao, F., Wang, D. et al. (2017). Comparison of novel magnetic polyaluminum chlorides involved coagulation with traditional magnetic seeding coagulation: Coagulant characteristics, treating effects, magnetic sedimentation efficiency and floc properties. *Separation and Purification Technology* 182: 118–127.
98 Garcia-Segura, S., Eiband, M.M.S., de Melo, J.V. et al. (2017). Electrocoagulation and advanced electrocoagulation processes: A general review about the fundamentals, emerging applications and its association with other technologies. *Journal of Electroanalytical Chemistry* 801: 267–299.
99 Moussa, D.T., El-Naas, M.H., Nasser, M. et al. (2017). A comprehensive review of electrocoagulation for water treatment: Potentials and challenges. *Journal of Environmental Management* 186: 24–41.
100 Zeboudji, B., Drouiche, N., Lounici, H. et al. (2013). The influence of parameters affecting boron removal by electrocoagulation process. *Separation Science and Technology* 48 (8): 1280–1288.
101 Behbahani, M., Moghaddam, M.A., and Arami, M. (2011). Techno-economical evaluation of fluoride removal by electrocoagulation process: Optimization through response surface methodology. *Desalination* 271 (1–3): 209–218.
102 Shen, M., Zhang, Y., Almatrafi, E. et al. (2022). Efficient removal of microplastics from wastewater by an electrocoagulation process. *Chemical Engineering Journal* 428: 131161.

103 Elkhatib, D., Oyanedel-Craver, V., and Carissimi, E. (2021). Electrocoagulation applied for the removal of microplastics from wastewater treatment facilities. *Separation and Purification Technology* 276: 118877.

104 Ozyonar, F. and Karagozoglu, B. (2011). Operating cost analysis and treatment of domestic wastewater by electrocoagulation using aluminum electrodes. *Polish Journal of Environmental Studies* 20 (1): 173.

105 Kobya, M., Hiz, H., Senturk, E. et al. (2006). Treatment of potato chips manufacturing wastewater by electrocoagulation. *Desalination* 190 (1–3): 201–211.

106 Drogui, P., Asselin, M., Brar, S.K. et al. (2009). Electrochemical removal of organics and oil from sawmill and ship effluents. *Canadian Journal of Civil Engineering* 36 (3): 529–539.

107 Dalvand, A., Gholami, M., Joneidi, A. et al. (2011). Dye removal, energy consumption and operating cost of electrocoagulation of textile wastewater as a clean process. *Clean–Soil, Air, Water* 39 (7): 665–672.

108 Asselin, M., Drogui, P., Brar, S.K. et al. (2008). Organics removal in oily bilgewater by electrocoagulation process. *Journal of Hazardous Materials* 151 (2–3): 446–455.

109 Long, Z., Pan, Z., Wang, W. et al. (2019). Microplastic abundance, characteristics, and removal in wastewater treatment plants in a coastal city of China. *Water Research* 155: 255–265.

110 Wang, Z., Sedighi, M., and Lea-Langton, A. (2020). Filtration of microplastic spheres by biochar: Removal efficiency and immobilisation mechanisms. *Water Research* 184: 116165.

111 Shen, M., Hu, T., Huang, W. et al. (2021). Removal of microplastics from wastewater with aluminosilicate filter media and their surfactant-modified products: Performance, mechanism and utilization. *Chemical Engineering Journal* 421: 129918.

112 Park, H.B., Kamcev, J., Robeson, L.M. et al. (2017). Maximizing the right stuff: The trade-off between membrane permeability and selectivity. *Science* 356 (6343): eaab0530.

113 Wu, M., Liu, W., and Liang, Y. (2019). Probing size characteristics of disinfection by-products precursors during the bioavailability study of soluble microbial products using ultrafiltration fractionation. *Ecotoxicology and Environmental Safety* 175: 1–7.

114 Leiknes, T. (2009). The effect of coupling coagulation and flocculation with membrane filtration in water treatment: A review. *Journal of Environmental Sciences* 21 (1): 8–12.

115 Tadsuwan, K. and Babel, S. (2022). Microplastic abundance and removal via an ultrafiltration system coupled to a conventional municipal wastewater treatment plant in Thailand. *Journal of Environmental Chemical Engineering* 10: 107142.

116 Reineccius, J., Bresien, J., and Waniek, J.J. (2021). Separation of microplastics from mass-limited samples by an effective adsorption technique. *Science of the Total Environment* 788: 147881.

117 Tong, M., He, L., Rong, H. et al. (2020). Transport behaviors of plastic particles in saturated quartz sand without and with biochar/Fe_3O_4-biochar amendment. *Water Research* 169: 115284.

118 Wang, J., Sun, C., Huang, Q.X. et al. (2021). Adsorption and thermal degradation of microplastics from aqueous solutions by Mg/Zn modified magnetic biochars. *Journal of Hazardous Materials* 419: 126486.

119 Sun, C., Wang, Z., Zheng, H. et al. (2021). Biodegradable and re-usable sponge materials made from chitin for efficient removal of microplastics. *Journal of Hazardous Materials* 420: 126599.

120 Chen, Y.J., Chen, Y., Miao, C. et al. (2020). Metal–organic framework-based foams for efficient microplastics removal. *Journal of Materials Chemistry A* 8 (29): 14644–14652.

121 Wang, Z., Sun, C., Li, F. et al. (2021). Fatigue resistance, re-usable and biodegradable sponge materials from plant protein with rapid water adsorption capacity for microplastics removal. *Chemical Engineering Journal* 415: 129006.

122 Ganie, Z.A., Khandelwal, N., Tiwari, E. et al. (2021). Biochar-facilitated remediation of nanoplastic contaminated water: Effect of pyrolysis temperature induced surface modifications. *Journal of Hazardous Materials* 417: 126096.

123 Magid, A.S.I.A., Islam, M.S., Chen, Y. et al. (2021). Enhanced adsorption of polystyrene nanoplastics (PSNPs) onto oxidized corncob biochar with high pyrolysis temperature. *Science of the Total Environment* 784: 147115.

124 Velempini, T., Prabakaran, E., and Pillay, K. (2021). Recent developments in the use of metal oxides for photocatalytic degradation of pharmaceutical pollutants in water—A review. *Materials Today Chemistry* 19: 100380.

125 Li, S., Shan, S., Chen, S. et al. (2021). Photocatalytic degradation of hazardous organic pollutants in water by Fe-MOFs and their composites: A review. *Journal of Environmental Chemical Engineering* 9 (5): 105967.

126 Khasawneh, O.F.S. and Palaniandy, P. (2021). Removal of organic pollutants from water by Fe_2O_3/TiO_2 based photocatalytic degradation: A review. *Environmental Technology & Innovation* 21: 101230.

127 Wang, L., Kaeppler, A., Fischer, D. et al. (2019). Photocatalytic TiO_2 micromotors for removal of microplastics and suspended matter. *ACS Applied Materials & Interfaces* 11 (36): 32937–32944.

128 Ariza-Tarazona, M.C., Villarreal-Chiu, J.F., Barbieri, V. et al. (2019). New strategy for microplastic degradation: Green photocatalysis using a protein-based porous N-TiO_2 semiconductor. *Ceramics International* 45 (7): 9618–9624.

129 Tofa, T.S., Kunjali, K.L., Paul, S. et al. (2019). Visible light photocatalytic degradation of microplastic residues with zinc oxide nanorods. *Environmental Chemistry Letters* 17 (3): 1341–1346.

130 Uheida, A., Mejía, H.G., Abdel-Rehim, M. et al. (2021). Visible light photocatalytic degradation of polypropylene microplastics in a continuous water flow system. *Journal of Hazardous Materials* 406: 124299.

131 Tofa, T.S., Ye, F., Kunjali, K.L. et al. (2019). Enhanced visible light photodegradation of microplastic fragments with plasmonic platinum/zinc oxide nanorod photocatalysts. *Catalysts* 9 (10): 819.

132 Nabi, I., Li, K., Cheng, H. et al. (2020). Complete photocatalytic mineralization of microplastic on TiO2 nanoparticle film. *Iscience* 23 (7): 101326.

133 Kesselman, J.M., Weres, O., Lewis, N.S. et al. (1997). Electrochemical production of hydroxyl radical at polycrystalline Nb-doped TiO2 electrodes and estimation of the partitioning between hydroxyl radical and direct hole oxidation pathways. *The Journal of Physical Chemistry B* 101 (14): 2637–2643.

134 Shang, J., Chai, M., and Zhu, Y. (2003). Photocatalytic degradation of polystyrene plastic under fluorescent light. *Environmental Science & Technology* 37 (19): 4494–4499.

135 Lee, J.M., Busquets, R., Choi, I.C. et al. (2020). Photocatalytic degradation of polyamide 66; evaluating the feasibility of photocatalysis as a microfibre-targeting technology. *Water* 12 (12): 3551.

136 Ariza-Tarazona, M.C., Villarreal-Chiu, J.F., Hernández-López, J.M. et al. (2020). Microplastic pollution reduction by a carbon and nitrogen-doped TiO_2: Effect of pH and temperature in the photocatalytic degradation process. *Journal of Hazardous Materials* 395: 122632.

137 Nabi, I., Fu, Z., Li, K. et al. (2019). A comparative study of bismuth-based photocatalysts with titanium dioxide for perfluorooctanoic acid degradation. *Chinese Chemical Letters* 30 (12): 2225–2230.

138 Hong, Y., Liu, E., Shi, J. et al. (2019). A direct one-step synthesis of ultrathin g-C3N4 nanosheets from thiourea for boosting solar photocatalytic H2 evolution. *International Journal of Hydrogen Energy* 44 (14): 7194–7204.

139 Cheng, H., Han, J., Nabi, I. et al. (2019). Significantly accelerated PEC degradation of organic pollutant with addition of sulfite and mechanism study. *Applied Catalysis B: Environmental* 248: 441–449.

140 Ma, D., Yi, H., Lai, C. et al. (2021). Critical review of advanced oxidation processes in organic wastewater treatment. *Chemosphere* 275: 130104.

141 Liu, P., Qian, L., Wang, H. et al. (2019). New insights into the aging behavior of microplastics accelerated by advanced oxidation processes. *Environmental Science & Technology* 53 (7): 3579–3588.

142 Miao, F., Liu, Y., Gao, M. et al. (2020). Degradation of polyvinyl chloride microplastics via an electro-Fenton-like system with a TiO_2/graphite cathode. *Journal of Hazardous Materials* 399: 123023.

8

Microplastics Contamination in Receiving Water Systems

*Muhammad Junaid and Jun Wang**

Joint Laboratory of Guangdong Province and Hong Kong Region on Marine Bioresource Conservation and Exploitation, College of Marine Sciences, South China Agricultural University, Guangzhou, China
* Corresponding author

8.1 Introduction

Precisely, the current era of human history could be referred to as the Plastic Age [1]. Unregulated consumption and disposal, inadequate collection and recycling, as well as unique physicochemical properties (such as biodegradation-resistant nature) have made plastic pollution hazardous and unaesthetic [2, 3]. Plastics exhibit abundant applications owing to their unique material properties including durability, corrosion resistance, and light weight [4, 5]. Therefore, the plastic production industry showed tremendous growth since 1950 with recent output reaching 360 million metric tons in 2018 (Figure 8.1) [6–8]. Large-sized plastic litter can be partially removed from the environment and subjected to the recycling process. However, miniature plastics (<5 mm), known as microplastics (MPs), are challenging or impractical to remove from the environment [3, 9]. Therefore, MPs pose more serious and widespread ecotoxicological risks [10]. MPs are subdivided into two main categories: primary MPs that enter into the environment predominantly with micrometric size (e.g., from washing and wearing of synthetic textiles or tire wear and tear) or they are manufactured with size less than 5 mm for specific applications (e.g., from hand cleaner, air blast cleaning media, and cosmetics); and secondary MPs produced in the environment as a result of progressive fragmentation, mainly through biological, chemical, and physical weathering [11–14].

MPs can contaminate the terrestrial and aquatic environment through various pathways and sources (Figure 8.2) [10]. The main sources of land-based MPs include runoff from industrial and domestic activities and effluents from wastewater treatment plants (WWTPs) [9, 15]. Studies have shown that a typical

Microplastics in Urban Water Management, First Edition. Edited by Bing-Jie Ni, Qiuxiang Xu, and Wei Wei.
© 2023 John Wiley & Sons, Inc. Published 2023 by John Wiley & Sons, Inc.

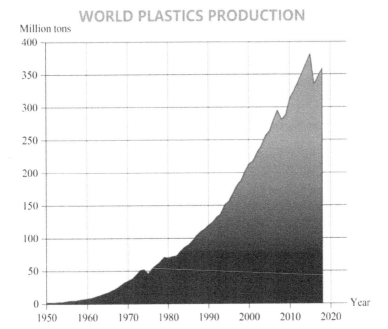

Figure 8.1 Global production of plastics from 1950 to 2018. or, Adapted from [6–8].

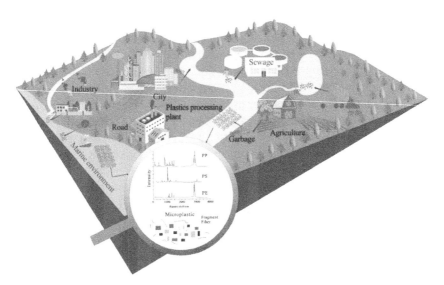

Figure 8.2 Sources and pathways of microplastics contamination in terrestrial and aquatic environments [10] / with permission of Elsevier.

WWTP can remove ca. 88% of MPs, and this percentage can increase to more than 97% if tertiary treatment is also included in the treatment scheme [16]. WWTPs usually trap MPs in oxidation chambers or sewage sludge tanks. However, a large quantity of MPs is yet released into freshwater bodies along with treated effluents through various routes [17]. Moreover, domestic sewage effluents and spillage of plastic powder resin are reported to be among the major sources of primary MPs in the freshwater environment [18, 19]. For the terrestrial environment, the application of domestic sewage sludge laden with synthetic fibers or sedimented MPs from household and personal care products are significant sources [20]. In addition, the occurrence and sources of MPs in the aquatic ecosystem can be influenced by urbanization, proximity to urban centers, dense human population, water retention time, sewage spillage, and waste management types [20, 21]. Moreover, areas adjacent to the plastic production plants are also observed as hotspots for disseminating MPs. For instance, water samples from areas in the proximity of a polyethylene production plant in Sweden were observed with MPs contamination as high as 100 000 particles/m^3 [22]. Sources of secondary MPs contribute a larger proportion of plastics into the environment, mainly through anthropogenic activities such as littering and municipal solid waste collection, transportation, and disposal processes [23]. In the environment, surface runoff, wind dispersal, and soil erosion can drive the transportation of secondary MPs and associated degradation products [20]. Surface runoff from urban and agriculture areas, stormwater runoff from highways (containing tire and road wear particles), and atmospheric fallout from highly urbanized areas (exhibiting a large quantity of fibers), are also considered as the significant sources of secondary MPs in receiving aquatic bodies [24, 25]. Fibers are the most commonly detected form of MPs in the environment, mainly from the the release of washing machine effluents and continuous abrasion of upholstery and clothes manufactured from synthetic textiles [26, 27]. A previous study showed that approximately 1900 fibers/items released during washing ultimately reached aquatic or terrestrial environment through WWTPs [26] Although synthetic fibers (mainly comprised of acrylic, polyester, and polyamide) fall in the category of secondary MPs, they tend to release into the environment with primary MPs [20].

After entering water bodies, MPs can distribute in various water layers such as water column, surface water, and bottom sediments. Distribution of MPs in those water layers can be influenced by water flow conditions, surface biofilms, and polymer characteristics (particle density, polarity, shape, size, etc.), which can then influence the bioavailability and toxicity to aquatic species [28, 29]. In freshwater ecosystems, the dominant polymer types of MPs include polystyrene (PS), polyethylene (PE), polypropylene (PP), polyethylene terephthalate (PET), and polyvinyl chloride (PVC) [5]. Furthermore, fiber, fragment, film, bead, and foam are the most commonly reported shapes of MPs of less than 1 mm size in freshwater bodies, and as size increased the concentration of MPs revealed a decreasing trend [10, 30]. Several other factors also influence the transport and migration of MPs in aquatic bodies, such as the size of the water body, water currents, and wind velocity [31, 32]. In the aquatic environment, microbes quickly colonize plastic fragments, resulting in the formation of biofilm, mainly comprised of bacteria, fungi, and algae [33]. Because of biofilm formation, the physiochemical properties, e.g., surface charge, the buoyant density of MPs can change, ultimately influencing the dissemination and fate of MPs in the aquatic environment [34]. For instance, higher trophic organisms feed on those biofilms, thereby indirectly ingesting MPs along with food, changing the fate of plastics in water bodies [35]. Previously, several studies have been conducted highlighting the impacts of MPs on aquatic organisms at various trophic levels [36–39]. For example, MPs have been detected in fish [40], mussels [41], oysters [42], waterbirds [43], penguins [44], and megafauna [45]. MPs are ingested by aquatic organisms directly or indirectly from various trophic levels, and can have a huge impact on the entire ecosystem, more adverse than that of macroplastics (large fragments), threatening the entire food chain [36, 46]. The distribution of MPs in the aquatic organism may vary based on their sensitivity, feeding activity, regulatory ability, and diverse habitat. Moreover, MPs can transfer along the food chain from low trophic level to high trophic level via prey-predator nexus [38, 47]. For example, the abundance of MPs has been reported in predators on top of the food chain, including waterbirds [48], seals [49], humpbacked dolphins [50], sharks [51], whales [52], and also in humans [53].

In humans, exposure to MPs has been reported through the food chain and direct dietary exposure [54]. For instance, MPs have been detected in drinking water [55–57], table salts [58, 59], and commercial aquatic products [41, 60, 61]. Recently, human exposure to MPs through air inhalation has been highlighted in several studies [62, 63]. MPs have been also detected in human stools at a concentration of 2 particles/g [53]. However, knowledge gaps still exist in precisely monitoring the occurrence, sources, and fate of MPs and nanoplastics (NPs) in the aquatic environment and their associated impact on aquatic life and ultimately on human health. Therefore, this chapter summarizes documented studies on the occurrence, distribution, and compositional profiles of MPs in freshwater resources, including rivers and lakes. Furthermore, factors involved in the aging of MPs have been demonstrated. In addition, the uptake of MPs by organisms (invertebrates, fish, waterbirds, and megafauna), associated ecotoxicological impacts, and their potential interactions with microorganisms (especially bacteria), as well as with humans have been discussed. Last but not the least, research challenges, implications, and future perspectives of MPs contamination in freshwater environments have been highlighted.

8.2 Occurrence of Microplastics in Freshwater Resources

8.2.1 River Surface Waters

Freshwater bodies serve as the primary resource of water for drinking and other human consumption needs; thus, they are considered as the main MPs exposure pathways for humans [64]. MPs enter into the freshwater resources mainly through conventional WWTPs because they cannot remove or fully retain MPs during sewage treatment processes [20, 35]. Besides effluents from WWTPs, atmospheric deposition and surface runoff are also reported as the major pathways for MPs contamination of rivers [65]. Previously, several studies have highlighted the contamination and distribution of MPs in various rivers worldwide. The first evidence of MPs pollution was recorded in 2011 in California, USA [66]. Recently, the abundance of MPs has been reported in various rivers from Shanghai, China [67]; Columbian Rivers [68]; the Thames River, UK [20]; the Ombrone River, Italy [69], the Rhine River, Germany [70]; and multiple Great Lake tributaries, USA [71]. Rivers are considered the major transportation pathway for plastic-related litter. Estimations have shown that a significant proportion of plastic pollution, as high as 70–80%, of marine resources is transported through river efflux [72]. Alarmingly, only two urban rivers San Gabriel and Los Angeles have been releasing ca. 30 metric tons of MP particles over only three days [66]. Similarly, the Danube River in Austria discharges 4.2 tons/day of mesoplastics and MPs using stationary driftnets over two years [73]. Moreover, the urban section of the Seine River in Paris was observed with an annual intercepted quantity of 22–36 tons of floating macroplastic fragments [74].

Previous studies have also investigated the influence of domestic and industrial wastewater on MPs abundance in surface water samples from various rivers, including the Saigon River, Vietnam [75]; the Raritan River, USA [76]; the Seine River, France [65]; the Ottawa River, Canada [77]; and St. Lawrence, North America [78]. Wang et al. studied MPs abundance affected by anthropogenic activities in the urban reaches of the Yangtze River and the Hanjiang River, China (Wuhan section), and compositional profiling revealed fibrous plastics with PET and PP polymers as the most abundant MPs in analyzed samples [79]. Similar studies have been conducted in the Three Gorges Reservoir's tributaries including the Yangtze River [80] and the Xiangxi River [81]. Moreover, anthropogenic activities and MPs are reported to exhibit a spatial relationship at specific sites [82]. For example, a study found MPs contamination in the Danube River associated with industrial sources [83].

Regarding Europe, the abundance of MPs was measured at 58–1265 items/m^3 in the Antua River, Portugal, with PP, PE, PS, and PET as the main polymers with size <5 mm [84]. In comparison, much lower levels of MPs (0.9–13 particles/m^3) were reported in the Ofanto River, Italy, and this study revealed PE, PS, PVC, and PUR as the main polymers, ranged in size 0.3–5 mm [85]. The same-sized PS and

PMMA polymers of MPs were also observed in the Rhine River, Germany with an average concentration of 0.89 particles/m^2 [86]. As far as riverine systems in China are concerned, several recent studies have reported varying contamination levels of MPs. For instance, the abundance of MPs in the Xingxi River was reported at 0.055–34.2 items/m^2 with PS, PP, and PE as the predominant polymers of various sizes (0.112–5 mm) [81]. In comparison, the surface water from the Pearl River was contaminated with significantly higher levels of MPs (379–7924 items/m^3), and compositional profiles showed PP, PE, and PET as the primary polymers with sizes ranging 0.02–5 mm [87]. Similarly, the alarming concentrations of MPs (1597–12 611 particles/m^3) were detected in the Yangtze River (Three Gorges Reservoir) with the abundance of PC, PE, PP, PS, PVC, and VC as the main polymers with sizes <5 mm [88]. A relatively lower concentration of MPs ranging 483–967 items/m^3 was reported in the Tibet Plateau Rivers, comprised of PET, PE, PP, PS, and PA polymers of <5 mm in size [89]. In the Wei River, MPs polymers including PE, PVC, and PS (<5 and >5 mm) were measured at the abundance level of 3.67–10.7 items/L [90]. In conclusion, MPs in rivers from China are comparatively higher than those in other counties, which is attributed to the rapid economic growth of China in recent decades [35].

8.2.2 Lake Surface Waters

Lakes are relatively close water bodies fed with water from surface runoff, precipitation, and groundwater [10]. Therefore, they usually contain plastic litter generated in the catchment area [91, 92]. In the last decade, several studies have investigated MPs occurrence and distribution in lakes located in various regions worldwide. For instance, multiple studies have detected MPs in water samples collected along the shorelines and offshore areas of the US Great Lakes [93–95]. Similarly, MPs were also detected in the Great Lakes of Africa [96]. In China, MPs have been detected in various lakes including Qinghai Lake [97], Taihu Lake [91], Lake Hovsgol, Mongolia [32], and 20 urban lakes of Wuhan [79]. In detail, a relatively lower concentration of MPs was found in the Qinghai Lake, i.e., 0.005–0.758 items/m^2, and the main characterized polymers were PP, PS, PE, and PET with size >0.2–5 mm, implying the significant effect of tourist activities in the lake area [97]. Similarly, the abundance of MPs ranged 3.4–25.8 items/L in Taihu Lake with PET, PES, PP, TA, and CE as the main polymers of various sizes (0.005–5 mm) [91]. A similar contamination level of MPs (5–34 items/L) was reported in Poyang Lake, comprised of PP, PE, PVC, and nylon polymers with sizes <5 and >5 mm [98]. In comparison, the elevated levels of MPs as high as 1760 – 10 120 items/m^3 were reported in Ulansuhai Lake with sizes <5 mm [99]. Similarly, the elevated concentration of MPs ranged 616.7–2316.7 particles/m^3 in Dongting Lake with size <5 mm, and polymers including PP, PS, PE, PVC, and PET [100]. In Hong Lake, the abundance of MPs was 685.5 particles/m^3, comprising of PP, PS, PE, and PVC polymers with sizes ranging 0.05–5 mm [101].

In Europe, Fischer et al. reported the occurrence of MPs at 2.68–3.36 and 0.82–4.42 particles/m^3, respectively in Lake Chiusi and Lake Bolsena, Italy [31]. This study further highlighted that moderate rainfall and heavy winds increased the MPs concentration in the lakes and major contributions were associated with domestic sewage and lateral land-based runoff [31]. Further, the abundance of MPs was reported in Swiss Lakes with adsorbed hydrophobic micropollutants [102]. A relatively lower abundance of MPs (0.27 particles/m^3) was observed in Lake Kallavesi, Finland, consisting of various polymers including PP, PE, PVC, PET, PAN, PET, and PMMA with size >0.33 mm [103]. Similarly, Lake Maggiore, Italy was also observed with lower MP concentration ranging 0.04–0.057 particles/m^3 with PP, PS, and PE as the major polymer types of size <5 mm [104]. In North America, Lake Winnipeg [105], Laurentian Great Lakes [93], Lake Superior [94], and Lake Michigan [95] were observed with lower levels of MPs, respectively as 0.053–0.748, 0.043, 0.037, and 0.0014–0.1 particles/m^3. The PE, PP, and copolymer in Lake Michigan [95] and PE, PP, PS, PR, PET, PVC, PDMS, and CPE in Lake Superior were the main identified MP polymers [94].

8.3 Composition of Microplastics in Freshwater

A hefty amount of literature is available on the composition of MPs by polymer type, size, and shape, reported in freshwater resources worldwide. Contrary to conventional pollutants, the compositional profiles of MPs in the aquatic environment can vary largely owing to their different colors, sizes, shapes, particle densities, polymer types, innate additives, and attached pollutants [82, 106]. Interestingly, the origin of MPs can be determined by characterizing their polymers and concentrations. For instance, the shape, size, color, and chemical composition of MPs found in Great Lakes of North America were similar to those detected in facial cleansers, implying personal care products as the potential sources of MPs in freshwaters [93]. Similarly, the size, shape, and color of MPs observed in the WWTP effluents were similar to those found in toothpaste formulations [107]. Similarly, The abundance of industrial microspheres and resin particles was reported in Lake Erie in the vicinity of the Huron Lake industrial zone [21, 93].

Plastics are synthetic polymers, that mainly contain compounds having diverse chemical properties [108]. Therefore, chemical composition-based categorization is the most common and easy criterion to characterize MPs pollution and it is mainly comprised of PS, PE, PP, PET, and PVC [5]. Considering polymer types, approximately 75% of MPs in freshwater are comprised of PE, PP, PET, and PS [82]. A recent study summarized the composition of MPs in terms of polymer type in global freshwater bodies with the following percentages: PP (24%), PE (24%), PS (13%), PET (11%), PA (6%), PVC (1%), PU (1%), and miscellaneous (20%). Because of the significantly high consumption and production, PP and PE were found with the highest detection frequency in the freshwater aquatic environment [83]. Hence, the conventional sewage treatment plants need advanced modifications to reduce contamination of these abundant MPs in wastewater.

Depending on the primary source, weathering, and retention time in the environment, MPs have various shapes commonly reported as fiber/line, fragment/sheet, sphere/pellet, foam, and film [10]. Regarding the composition of MPs based on their shape/topologies, most fractions were comprised of fibers (59%) and fragment (20%), followed by the film (9%), bead (7%), foam (2%), and miscellaneous (3%) in freshwater [10]. The highest proportion of fiber MPs in the aquatic environment is mainly attributed to their discharge through laundry effluents [109]. Similarly, the high abundance of fragment MPs in the aquatic environment is mainly from their significantly elevated amounts in runoff from crushing large pieces of plastics [110]. Moreover, fragment MPs were the predominant morphology, followed by the fiber, film, and pellet MPs in the Qing River, Beijing, China (Figure 8.3) [111]. The conventional WWTPs are unable to remove such fiber and fragment MPs that ultimately make their way to the freshwater bodies [112].

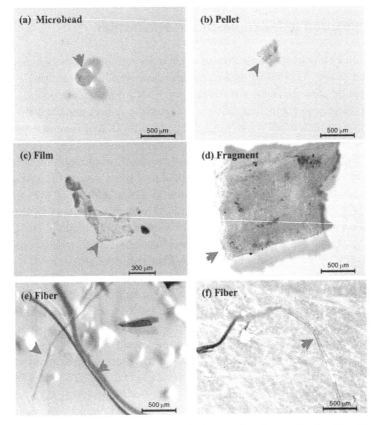

Figure 8.3 Representative images of microplastics observed in the Qing River, Beijing, China [111] / with permission of Elsevier.

8.4 Factors Influencing the Aging of Microplastics

In the natural environment, MPs undergo aging or weathering processes including microbial degradation (biofilm growth and oxidation), UV-irradiation, sunlight photo-oxidation, mechanical wearing and tearing, and heat aging, thus resulting in the generation of new plastic polymers [113]. The aging process can influence the physicochemical properties of MPs such as particle size, shape, color, density, reactivity, crystallinity, hydrophobicity, surface morphology, and surface functionality [39, 113]. After degradation, MPs changed into miniature plastics such as NPs and exhibit rough surfaces, thereby increasing the surface area and exhibiting a higher potential for interactions/sorption for other chemicals and microbial communities [114, 115]. The aging of MPs also occurred as a result of seawater corrosion and sand friction, which also increased their surface area [116]. Perez et al. further reported that the aging process causes a decrease in the molecular weight and alters the surface characteristics of MPs. Moreover, it plays a significant role in the adsorption of hydrophilic pollutants on MP surfaces [117].

Interestingly, the adsorption capacity of PVC- and PS-MPs was significantly enhanced by 20.4% and 123.3% as a result of aging [118]. The aged MPs exhibit oxygenated functional groups on the surface compared with pristine MPs, which also increase the sorption capacity of MPs. Compared to pristine plastics, those oxygen functional groups in aged MPs promote high-level sorption of hydrophilic compounds, e.g., ciprofloxacin [118]. This higher sorption capacity for hydrophilic compounds could be attributed to the decrease in hydrophobicity of MPs as a result of weathering/aging [119]. Similarly, the aged MPs including PS, PP, and PA showed a high sorption capacity for difenoconazole and metformin [120]. The adsorption capacity of PVC for heavy metals (copper and silver) increased significantly after immersion of PVC in the marine aquatic environment for 502 days [121]. A similar study also highlighted the increased sorption capacity of PET and PVC plastics for heavy metals over one year [122]. These studies comprehensively demonstrated the difference between the characteristics of aged and pristine MPs. Therefore, environmental occurrence, distribution, fate, and ecological implication can also differ for aged microplastic; hence, these aspects needed further investigations.

8.5 Uptake and Associated Ecological Impacts of Microplastics in Aquatic Organisms

8.5.1 Invertebrates

The uptake of MPs in the aquatic organism and their trophic transfer along the food chain is portrayed in Figure 8.4. Invertebrates are an important part of the aquatic ecosystem, mainly relying on primary producers as a food source, while

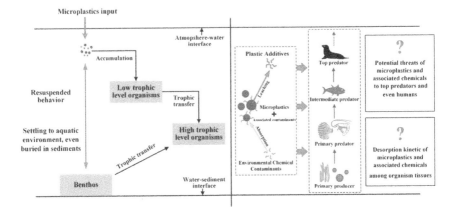

Figure 8.4 The uptake of microplastics and their trophic transfer along aquatic food chains. The uptake of microplastics by aquatic species may enhance the concentration of adsorbed chemicals (plastic additives, heavy metals, and organic pollutants) in organisms, potentially leading to bioaccumulation and biomagnification [39] / with permission of Elsevier.

they serve as a critical dietary source for carnivore species (Figure 8.4). Aquatic invertebrates are more prone to MPs contamination attributed to their feeding behavior and a critical place in the food chain (next to the primary predators). Previously, many studies have highlighted MPs contamination in various species of worm [123], arthropod [124], and mollusk [60]. Importantly, MPs have been detected with an average concentration of 0.14 particles/mg of tissues in 50% of macroinvertebrates such as Hydropsychidae, Heptageniidae, and Baetidae collected from the urban rivers in South Wales [125]. In addition, invertebrate species exhibit varying living properties in the aquatic environment that can affect the uptake, distribution, and metabolism of MPs in invertebrates. For example, the shore crab is a non-filter feeder species; therefore, the uptake of MPs via respiratory exposure could serve as the main pathway [126]. Moreover, a higher bioaccumulation potential of MPs as compared to ingestion was reported in mussels through adherence to soft tissues [127].

Abidli et al. reported various types of MPs with concentrations of 703.9–1482.8 particles/kg in six commercial mollusk species from Bizerte lagoons including one cephalopod *Sepia officinalis*, two gastropods *Bolinus brandaris* and *Hexaplex trunculus*, and three bivalves *Crassostrea gigas*, *Ruditapes decussatus*, and *Mytilus galloprovincialis* [128]. Because of the ubiquitous presence of MPs in bivalve species including clams and mussels both from freshwater and marine water environment, they are referred to as the biological indicators of MPs in the aquatic ecosystem. For instance, the average concentrations of MPs were measured at 0.35 and 0.47 particles/g in *Crassostrea gigas* and *Mytilus edulisi*, mussel species cultured for human consumption [129]. The abundance of MPs (0.9–4.6 particles/g) was also observed in marine mussel *Mytilus edulisi*, collected from 22 coastal sites in China [130]. Similarly, another study on commercial bivalve species from China reported the total MPs concentration ranged 4.3–57.2 particles/individual and 2.1–10.5 particles/g wet weight [131]. In the UK, *Mytilus edulisi* collected from the coastal region also exhibit MPs at concentrations ranging 0.7–2.9 particles/g in tissues [41].

Previous studies showed that exposure to MPs induced multiple toxicological effects in invertebrates, such as disturbance in reproduction, growth, development, survival, and feeding behavior [132–134]. Conversely, some studies highlighted mild toxic effects associated with MPs, e.g., PVC-MPs (0.1–1.0 μm, 0.125 g/L) failed to cause any physiological damages in mussel *Perna perna* after chronic exposure (90 days) [135]. Similarly, PET-MPs (10–150 μm, 0.8–4000 particles/mL) exhibit no significant impact on growth, development, and feeding behavior in Gammarus pulex (a freshwater amphipod) after acute exposure of 24 h [136]. Graham et al. reported that most of the ingested PS-MPs (100–500 μm) were egested in Pacific oyster *Magallana gigas*, implying negligible risks to species at the next trophic level. However, further studies are required to establish concrete evidence for risks associated with MPs through trophic transfer from invertebrates [137].

8.5.2 Waterbirds

Waterbirds, both from freshwater and marine aquatic environment are mainly exposed to MPs through feeding on contaminated food sources [43, 138]. They can act as an important tracer for MPs transportation in the environment, as the migratory birds have been reported with MPs contamination in their excreta and feathers [139]. Several factors can regulate the ingestion of MPs in birds including habitat, species, life stages, and foraging activity as well as the availability of MPs [140, 141]. Like other creatures, a significant portion of ingested MPs in birds also accumulated in the gut [138, 142]. However, studies on the fate and dissemination of MPs in the gastrointestinal tract of birds are still elusive. Besides feces, feathers, and regurgitation behavior (throwing the undigested residues of prey potentially contaminated with MPs) act as additional excretory pathways in birds. Studies have speculated regurgitation in gulls, great skua, and Eurasian waterbirds, as the alternative excretory route for MPs [143–145].

Studies are scarce yet that highlight the contamination of MPs in freshwater birds, compared to those in seabirds. For instance, in the Poyang Lake of China, MPs were found in feces of migratory birds with the mean abundance of 4.93 particles/g and MPs were significantly higher in the active range of birds [146]. Similarly, MPs were detected in the gastrointestinal tract of 86.7% with an average abundance of 5.8

particles/bird in cormorant chicks *Phalacrocorax auritus*, sampled from the Laurentian Great Lakes [48]. This is a clear example of trophic transfer as cormorant parents fed on MPs contaminated prey, then fed that to their chicks. In Canada, MPs (50 μm–5 mm) were measured in 15 individuals (out of 350 in total) belonging to 8 species (out of 15 in total) of freshwater birds such as geese, ducks, and loons [147]. In South Africa, fibrous MPs were reported in 10% of feather and 5% feces samples collected from seven duck species of freshwater wetlands [148].

Studies have also highlighted those waterbirds belonging to different regions uptake varying levels of MPs [140, 149]. Therefore; they are efficiently used for tracing and evaluating the levels and routes of MPs contamination in the oceans, especially northern fulmars Fulmarus glacialis that contain plastic debris in the stomach [150, 151]. In the western Indian Ocean, nine species of seabird were observed with plastic debris contamination in Barau's petrels (6.10 ± 1.29 particles/individual) and the tropical shearwaters (3.84 ± 0.59 particles/individual), respectively with 63% and 79% of plastic debris found in the gut [140]. Similarly, recent investigations revealed 77% and 20% of King and Gentoo penguin scats from the Antarctic regions were contaminated with MPs, mainly fibers [44, 152]. However, the pathways of MPs intake in penguins are still unclear, either directly from surroundings or indirectly feeding on the contaminated prey, hence further relevant studies are recommended.

Although investigations are available on MPs pollution in seabirds, the toxicological implications of MPs exposure in freshwater birds are still elusive. For example, the ingestion of plastic debris caused significant hazardous impacts on Flesh-footed Shearwaters fledglings including morphological disruption. Abundance of plastics was positively correlated with the levels of calcium, cholesterol, uric acid, and amylase in the blood, implying that plastic may induce health risks in birds through disturbance in blood chemistry [153]. However, a study on Japanese quail *Coturnix japonica* feeding experiment revealed that the uptake of PP-MPs (3–4.5 mm) at environmentally relevant concentration caused delay in development and sexual maturation, albeit with no significant toxic impacts in terms of survival or change in population dynamics over two filial generations [154]. Moreover, studies have been also conducted on the mixture toxicity of MPs and associated pollutants such as plastic additives and adsorbed chemicals [155–157]. In conclusion, MPs in the aquatic environment can affect waterbirds because of their predator nature, they can therefore be used as a useful indicator of plastic pollution in water bodies.

8.5.3 Mammals and Megafauna

MPs exhibit deleterious effects on megafauna through various exposure routes, i.e., trophic transfer, filter-feeding, and unintentional ingestion (Figure 8.4) [50, 51].

Moreover, MPs contamination in megafauna (especially in baleen whale and filter-feeding sharks) is a critical issue as they need to filter a large volume of water daily to fulfill their nutrition and food needs. For MP analyses in megafauna, studies are usually performed through dissecting the dead individuals from fishery bycatch or stranding [158]. The gastrointestinal tracts dissected from 50 stranded mammals at the Britain coast were contaminated with 261 MP particles, and those mammals comprised species composition of one whale, two seals, and seven dolphin species [158]. Investigations based on analyses of scat samples showed MPs abundance in various species of seal including northern fur seals *Callorhinus ursinus*, gray seals *Halichoerus grypus atlantica*, harbor seals *Phoca vitulina vitulina*, and fur seals *Arctocephalus australis* [159, 160]. In the Netherlands, the intake of plastic debris was reported in harbor seals Phoca vitulina (11% out of 107 in total) and more MPs were detected in the stomach of young seals [161]. Studies have also reported MPs-mediated parasite enrichment in the intestinal tract of the seals [49]. Interestingly, a feeding study highlighted the trophic transfer of MPs to gray seals *Halichoerus grypus* from Atlantic mackerel, implying MPs potential mechanistic transfer to the top predators from low trophic levels across the food chain [162].

Previously, MPs abundance have been reported in the intestinal tract or gut of multiple dolphin species including *Sousa chinensis* [50], harbor porpoises *Phocoena phocoena* [163], East Asian finless porpoises *Neophocaena asiaeorientalis* sunameri [164], and short-beaked common dolphin *Delphinus delphis* [165]. A recent study reported MPs contamination in the gastrointestinal tract of beluga whales *Delphinapterus leucas* (an omnivorous species) with an average concentration of 97 ± 42 particles/individual [52]. Another study established an efficient approach to detect MPs in marine mammals and reported MPs in the digestive tract and stomach of True's beaked Whales [166]. Similarly, several previous studies have reported the presence of MPs in various species of shark including Porbeagle shark *Lamna nasus*, blackmouth catshark *Galeus melastomus*, whale shark *Rhincodon typus*, and basking shark *Cetorhinus maximus* [51, 167–169]. Another interesting study on whales revealed that feeding strategies and prey presences can affect ingestion of MPs differently in sei whale *Balaenoptera borealis* and minke whale *Balaenoptera acutorostrata* [45]. The indirect skin biopsies of 12 whale sharks from the Gulf of California revealed the presence of plastic additives and CYP1A-like protein in subcutaneous tissues, implying the possibility of negative impacts of MPs contamination on the filter-feeding shark [170]. A string of studies by Fossi et al. highlighted the abundance and associated negative impacts of MPs and other pollutants (phthalate additives and organic pollutants) in Mediterranean fin whales Balaenoptera physalus, speculating the filter-feeding on contaminated prey and direct ingestion from the surrounding environment as the primary pathways of MPs contamination in whales [167, 171, 172]. In addition, Fossi and coworkers also proposed the potential overlap among feeding habitats of whale and MPs contamination hotspot [173]. In conclusion, the top predatory species such as whales, sharks, seals, and dolphins exhibit a great ecosystem value, and such predators serve as the bioindicators for the health of an ecosystem. However, the development of methods to measure accurately the extent of MPs pollution, distribution, and associated toxic implications in those gigantic mammals/megafaunas are still challenging. Similarly, the mechanism involved in clinical pathologies and toxicity mediated by MPs exposure are largely elusive yet; hence, relevant studies are highly recommended.

8.6 Interactions among Microplastics and Microbes (Bacteria)

8.6.1 Microplastic Biofilms: Formation Mechanisms and Characteristics

Previously, a hefty number of studies have highlighted the interaction of MPs with various species of bacteria, especially in terms of attached bacteria on the surface of MPs. Generally, bacteria and other microbes are attached to the surface of MPs through the formation of biofilms [174, 175]. In the last decade, several studies have been conducted regarding biofilm formation on the surface of MPs in the aquatic environment. In microbes, extracellular polymeric substances (EPS) secretion is the main factor involved in their attachment with MPs as well as other surfaces [175]. Previously, researchers speculated that the environmental MPs provide shelter to microbial communities against harsh weather conditions [176]. The formation of biofilms not only provides physical support but also increases bacterial diffusivity and protects against shear forces and mechanical damages [174, 175, 177]. Biofilms also assist in the protection against predators [177]. In addition, MP biofilm formation increases the suspended material deposition in the aquatic environment that can lead to nutrient accumulation on the horizontal surfaces [177]. It also acquires essential metabolites (such as metals) from the surrounding environment to serve as electron receptors and are critical for the regulation of cellular processes in microbes [178, 179]. MP biofilms can affect microbial communities in various ways, e.g., change in cellular behavior and genetic expressions that can influence their motility [177]. Moreover, MP biofilms exhibit unique nature and are less diverse in characteristics compared with the microbial community from the same environmental settings [180–182]. For this reason, the term "plastisphere" is introduced to define the specificity of microbes attached to plastics or MPs [183].

Among bacterial communities, phylum Proteobacteria is the most common and abundant to develop biofilms on MPs in all kinds of aquatic environments [184,

185]. In addition, Cyanobacteria have been detected to frequently formulate biofilm in the freshwater environment [186]. A study by Miao et al. based on MP biofilm alpha diversity speculated that natural substrates as well MPs themselves as the driving factors involved in the selection of microbial communities to attach on the surface [186]. Conversely, a later study reported that the conventional biofilm process is involved in the attachment of microbes on MP surfaces [187]. Results from this study revealed no significant difference among plastic-associated and particle-associated biofilms despite variations concerning the surrounding water environment. However, a clear dichotomy exists among free-living and particle-attached microbes in water from the same sites [182, 188, 189]. Moreover, the community structure of MP biofilms in the aquatic environment has been reported to vary among sites [33, 190]. Although studies are elusive in highlighting the negative impacts of MPs on attached microbial/bacterial biofilms, they can also affect ecological processes of bacterial communities. Therefore, further studies are needed for in-depth understanding the effects on MPs on attached bacterial communities.

8.6.2 Factors Affecting Biofilm Formation

The interaction between MPs and microbial communities could be influenced by several factors including size, polymer type, hydrophobicity, etc. A limited number of studies have highlighted the effects of MPs' size on bacterial/microbial attachment. For instance, the size of MPs revealed no noticeable effects on the type of attached microbial community, albeit this study did not consider the role of various polymer types [32]. However, a study on mesoplatics (PS and PET) reported the enrichment of Proteobacteria (Alpha and Gamma diversities) in comparison to PS-MPs that were dominated by Beta diversity in similar environmental conditions [191]. Conversely, compared to those on macroplastics, different MP polymers such as PS, PE, PP were enriched with Actinobacteria [81]. These studies demonstrate that the size of MPs potentially affects the attachment of a specific microbial community; therefore, further relevant studies should be conducted.

Besides size, the selection of microbes for attachment on MPs is also reported to be influenced by the type of polymer. Various communities of bacteria and eukaryotes were isolated from PP and PE sampled from the North Pacific, North Atlantic, and North Sea [180, 183, 192]. Similarly, compared to other MPs, PVC and PE exhibited a preferred selection of Pseudomonadales and Betaproteobacteriales [193]. Among MP samples collected from the Lake Hovsgol, Mongolia, and the North Pacific Gyre, PE-MPs showed a higher abundance of bacteria compared to that of PP polymer [32, 194]. This varying potential of MP polymers to attach specific microbial communities is also attributed to their chemical properties, e.g., the elevated microbial attachment on PE was observed following the higher adsorption potentially due to relatively large space among its molecules [195].

Hydrophobicity is another significant factor that can affect the attachment of bacteria to MP surfaces [196]. Except for some general findings, no specific information is available to understand the role of hydrophobicity in bacterial attachment on MPs. Among available studies, one investigation highlighted that the hydrophobicity of MPs reduced after 3 weeks of microbial colonization in marine aquatic conditions [197]. Certain other factors can also influence MPs hydrophobicity including chemical interactions, etching, plasma irradiation, and UV irradiation. The microbial biofilm formation is also reported to be affected by surface charges [198]. The ideal surface energy for bacterial colonization was measured at 31–43 mNm^{-1} [199]. Moreover, the selection and attachment of bacteria can also be influenced by the roughness of MPs surface, and it was reported as one of the primary factors [199]. The rough MP surfaces possess a large surface area that can enhance the absorption of nutrients; hence, increasing the attachment of microbial community [174].

Exposure duration can also play a significant role in the formation of biofilms on MP surfaces. For example, the abundance of Epsilonproteobacteria was observed on PP and PET in a six weeks' exposure period, albeit the enrichment of Flavobacteria was relatively higher after five months' exposure under the same conditions in the coastal areas of Germany [200]. Interestingly, the abundance of Sphingobacteria remained unchanged during both short and long exposure durations [200]. This study also reported that the exposure period may change the diversity of the attached microbial communities. In the coastal areas of China, an increase in the number of dominant microbial communities and Shanon diversity index was observed, and the highest diversity was reported for Erythrobacteraceae at a 12-month exposure duration, followed by that at 6 months [201]. Moreover, the spatial and temporal variations are also reported to affect the attachment and composition of microbes on MPs and these factors are more influencing when it comes to the aquatic environment [202]. Apart from the characteristics of MPs and microbial communities, the chemicals/molecules adsorbed on the surface of MPs can also significantly impact the formation of biofilms on MP surfaces [203]. Previous studies have reported a significantly elevated initial microbial colonization in the presence of various adsorbed molecules including lipids, sugars, proteins, fatty acids, and nucleic acids [200].

8.6.3 Role of Microplastic Biofilms in Genetic Material Transfer

Biofilm formation on MPs can also promote evolution and genetic material transfer among microbial communities in the aquatic environment (Figure 8.5). Therefore, because of high cell density and nutrient availability, MPs may serve as

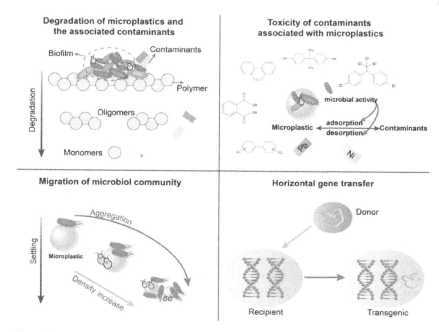

Figure 8.5 Biofilm formation on microplastics and its implications in the aquatic environment [210] / with permission of Elsevier.

hotspots to accumulate antibiotic resistant genes (ARGs) and play a critical in their horizontal gene transfer (HGT) in water bodies [204, 205]. Previously, several studies highlighted the presence of ARGs on MPs. For example, a study found 13 ARGs attached to plastics as compared to only 2 ARGs in the seawater, which implied a critical role of plastics in the generation and transfer of ARGs [206]. Similarly, Laverty et al. isolated the antibiotic resistance Vibrio spp. from MPs. Moreover, class 1 integrons were also found on MPs surface, which play a significant role in ARGs' environmental transport [207]. It is also hypothesized that the HGT of antibiotic resistance elements based on MPs biofilm attachment mainly occurred through conjugation [204]. It is noteworthy that the presence of other pollutants especially antibiotics and metals may enhance the antibiotic resistance characteristic of MPs [181]. Antibiotics usually make complexes with metal cations (zinc, copper, etc.), reducing the activity of the antibiotics [208]. Therefore, bacterial interaction with metals and antibiotics on MPs biofilm may take place through co-selection [209]. There are only preliminary studies available that highlight the role of MPs in ARGs' generation and transfer under different environmental settings; therefore, relevant studies are highly encouraged for in-depth understanding.

8.6.4 Microplastics as Pathogen Carriers

Another important interaction among MPs and microbes is the potential role of plastics as carriers for pathogens, specifically in the aquatic environment. Generally, MPs by their innate properties provide a favorable habitat for both animal and human pathogens. Previously, only a few studies have investigated this critical nexus among microbes and MPs that directly involved ecological and human health implications. For instance, various *E. coli* strains, fish, shrimp, and even plant pathogens have been reported in the biofilms extracted from environmental plastics [185, 211, 212]. Keswani et al. highlighted that MPs are largely responsible for the dispersal and survival of pathogens under various environmental conditions on different environmental media owing to their ubiquitous, persistent, and buoyant characteristics [203]. Moreover, Vibrio species such as *V. parahaemolyticus* were found on plastics collected from the Baltic Sea [213]. A similar study explored the presence of *V. cholera*, *V. vulnificus*, and *V. mimicus* on plastic samples from marine water in Brazil [211]. The plastics recovered from the Western Mediterranean Sea were also observed with pathogenic vibrio species [184]. WWTPs play a significant role in the dissemination of pathogens in the environment because MPs and pathogens co-release from WWTPs, providing an ecological niche to pathogens [214, 215]. For example, Campylobacteraceae was detected on biofilms attached to MPs in effluents from WWTPs [182]. Studies have also speculated that WWTPs might serve as favorable habitat for pathogens attached to MPs (compared to a free-living pathogen) from a lack of nutrient competition and grazing pressure [216]. Several data gaps still exist to understand further the role of MPs in pathogens colonizing, transport, fate, and associated mechanism, especially in soil and sediments.

8.7 Potential Interactions between Microplastics and Humans

8.7.1 Dietary Exposure

Recently, the risks of MPs to human health have become a mainstream concern. In humans, the main exposure pathways to MPs include ingestion, dermal contact, and inhalation [63]. Among them, MPs exposure through diet intake and food sources is the most significant pathway [35]. While evaluating the human health risk associated with dietary intake, MPs contamination level in food items, as well as the amount transferred along the food chain should be investigated. Aquatic products are considered the major source of MPs in the human diet. For example, based on three edible fish species, i.e., Atlantic chub mackerel, Atlantic horse mackerel, and seabass, the estimated intake of MPs was ranged 518–3078 particles/capita/year in Europe [217]. Further, the annual MPs intake/capita varies greatly from country to country based on the quantity of bivalves' consumption and the level of MPs pollution. For instance, the intake of MPs was estimated at 283 and 1800–11 000 particles/capita/year, respectively in Korean and European mussel consumers [60, 129]. However, the estimated intake was relatively lower in bivalve consumers from the UK (123 particles/capita/year), albeit significantly higher (4620 particles/capita/year) in individuals from Belgium, France, and Spain [218]. Therefore, bivalves could be regarded as the global bioindicator of MPs contamination in aquaculture products for human dietary consumption [219].

Previously, many studies have highlighted the contamination of MPs in a vast variety of commercially available aquaculture products and aquatic species such as bivalves, commercial fish, sea urchins, and sea cucumbers [39]. Moreover, MPs have been detected in various daily diets (drinks/food) such as drinking water (tap and bottled), beer, sugar, honey, canned fish, seaweed nori, sea salt, seafood, etc. [35, 39]. Importantly, a recent study reported the contamination of NPs and MPs in vegetables and fruits purchased from the local market in Catania, Italy, and observed an elevated estimated daily intake (EDI) of plastics in children and adults [220]. On a per-year average, an individual intakes 11 000 MP particles from shellfish [129], 4000 MP particles from drinking water [221], and 37–1000 MP particles from edible sea salt [58]. Studies have highlighted that MPs may penetrate leaves, seeds, and fruits of the edible crops, e.g., Li et al. reported MPs (0.2–2 μm) in the roots of lettuce (Lactuca sativa) and wheat (Triticum aestivum) [222]. The plant uptake of MPs can cause adverse human health effects directly and induce a threat to food security indirectly. Interestingly, Schwabl et al. detected the mean levels of MPs in human stool at 2 particles/g, with size ranging 50–500 μm and nine different topologies in the abundance of PET and PP polymers, implying the inevitable uptake of MPs by the body through various sources [53].

Interestingly, MPs have been also detected in commercially available salts (primarily sea salt) processed by more than 120 brands globally [54, 59]. The concentration of MPs in sea salt ranged 0–1674 particles/kg (significantly higher than lake salt and rock salt) obtained from 28 brands belonging to 16 countries, implying that sea salt can also serve as a global indicator of MPs pollution [59]. However, consistent with the bivalves, the concentration of MPs in sea salt greatly differs among various regions worldwide and may range from zero to tens of thousands of particles/kg [54]. This regional varying contamination levels of MPs in salts could be attributed to differences in salt processing methods and MPs analytical quantification techniques. Besides exposure via aquatic products and sea salt, MPs with a concentration range of 0–5.42 × 10^7 [57, 95] and 0–930 particles/L [221, 223, 224] have been detected in bottled or tap water samples collected in various countries. In the US population, the average MPs intake was estimated to be 4000–90 000 particles/capita/year respectively in tap water and bottled water samples [62]. Therefore, it is mandatory to upgrade current drinking water treatment plants to remove MPs [225]. It is important to mention here that the distribution plumbing system for drinking water may be a potential source of plastic additives and NPs/MPs due to the aging of plastic pipes (mainly PE and PVC) mediated by water erosion, disinfectants, temperature, and biofilms [226]. Therefore, this area of research should be explored in-depth to characterize human MPs exposure pathways associated with drinking water to avoid future worst-case scenarios. Because of complex erosion effects, drink packages manufactured using plastic materials can also act as a significant contributing source for MPs contamination [63]. In general, for a better understanding of the human health implications of MPs, the development of more accurate, sophisticated, and standard analytical techniques to analyze MPs in dietary sources is mandatory. Further, to avoid future worst cases, the food safety management practices/guidelines should be devised for the detection, characterization, and quantification of MPs/NPs in food items.

Based on the available data from aquatic products and other food items, it is still challenging to accurately elucidate the actual risks posed by MPs to human health. A recent study highlighted that NPs (500 nm) enhanced the bioaccumulation of two veterinary antibiotics florfenicol and oxytetracycline in edible clams, albeit the estimated human health risk in terms of hazard quotient through the consumption of these contaminated clams was significantly lower than the threshold values [227]. Moreover, the consumption of aquaculture products such as bivalves, fish, and sea cucumber may pose relatively low human health risks because these species are eviscerated before ingestion [228, 229]. Another study revealed a relatively lower MPs level in the flesh of the cooked mussel (-14%) compared to the raw meat, and this decrease was attributed to the thermal degradation and natural variability in MPs [230]. Nonetheless, the trophic transfer of MPs/NPs along the food chain with edible products and other dietary items still need comprehensive investigations. Limitations also exist in the precise characterization and quantification of MPs, especially NPs in the food and dietary items. For instance, because of background contamination and inaccurate identification of MPs in honey and sugar samples, the analytical method was challenged [231]. Conversely, a string of studies also speculated that MPs may not potentially undergo biomagnification along the food chain at the higher trophic levels, and only species at the lower trophic level exhibit the highest risk implications [232, 233]. There is the possibility of negligible risk via food consumption if MPs are rarely detected in dietary items. For example, no evidence of significant MPs contamination in honey samples was reported in a study from Switzerland [234]. Similarly, another study reported negligible human health risk associated with salt consumption from 17 brands, having MPs with size <149 μm and ingestion frequency of 37 particles/capita/year [58]. Therefore, based on the above-mentioned factors, no concrete evidence exists to establish a link between human health impacts of MPs contamination in the dietary products; hence, more relevant studies are highly recommended.

8.7.2 Exposure through Inhalation and Dermal Contact

As mentioned earlier, there are three major uptake pathways of MPs/NPs in humans, i.e., the lungs, the gastrointestinal tract, and the skin (Figure 8.6) [235]. Besides exposure through dietary items and food chain, studies have reported inhalation as one of the major exposure pathways as MPs ubiquitous presence in various atmospheric environment settings including indoor, outdoor, urban, rural, and remote regions, with concentrations ranging from 1–3 fold in magnitude for various sampling sites [5, 236–238]. Additionally, most of the atmospheric MPs are of fibrous morphology, probably exhibiting elevated human health implications and raising serious concerns [62, 239]. After entering the respiratory system, most of the MPs might be trapped in the lining fluid in the lungs or accumulated in the airways, which may cause adverse health implications by averting the clearance process of the lungs and respiratory products. Human exposure to airborne MPs may induce oxidative stress, abnormal gene expressions, immune responses, airway diseases, and interstitial lung inflammation [63, 239]. Therefore, it is immensely important to compare and characterize the difference in human exposure through two major pathways (inhalation and ingestion). Catarino et al. investigated the microfiber intake at household levels through dust inhalation (17 731–68 415 particles/capita/year) at dinnertime, and it was significantly higher than MPs ingestion through consumption of contamination clams (4620 particles/capita/year) from different countries [218]. Considering inhalation as the primary route of exposure in the US population, MPs intake was increased from 3.9–5.2×10^4 to 7.4–12.1×10^4 particles/capita/year [62]. Similarly, a comparative study reported the MPs intake via air inhalation (0–3.0×10^7), drinking water (0–4.7×10^3), and table salt (0–7.3×10^4) [54]. Hence, these results comprehensively portrayed that inhalation might be an underestimated pathway for MPs exposure to MPs compared to dietary exposure.

From indoor air, a human body may inhale approximately 272 particles/day of airborne MPs depending on size [240]. However, NPs could pose a more serious human health risk when it comes to exposure through inhalation and dermal routes. For instance, MP particles with a size less than 2.5 μm can pass the respiratory barriers after accumulation in the lungs [241]. Similarly, dermal exposure is a relatively less important exposure pathway in humans for MPs uptake because particles with a size below 100 nm can absorb through the skin, penetrating the stratum corneum [242]. However, studies on human exposure assessment to NPs are largely elusive because of limitations in relevant quantification methods development. Thus far, no study has comprehensively demonstrated the nexus between NPs' composition, size, and uptake in human exposure scenarios.

8.7.3 Microplastics' Toxicity in Humans

A huge gap exists in the literature concerning toxicity, diseases, and pathologies in humans associated with MPs exposure. This field is still in its infancy and requires extensive studies. Human health risks associated with MPs may be largely dependent on their biological dosage and persistence in the body [239]. There is a high possibility that MPs are resistant to biochemical degradation *in vivo*. These MPs can be referred to as hazardous "micromaterials" as they can induce toxicity in terms of increased oxidative stress, reduced cell viability, cell membrane damage, cellular injury, genetic disruptions, immune responses, and inflammation both at cellular and tissue level [243–247]. Tissues in the human body usually uptake MPs through paracellular persorption and endocytosis (gastrointestinal tract and airways surface) that can be regulated by the size, shape, surface charge, surface functionalization, and hydrophobicity of MPs [239]. The absorption of MPs through endocytosis further comprises micropinocytosis, phagocytosis, caveolae- and clathrin-mediated endocytosis. Previous studies on the animal model showed that MPs translocate from cells to the circulatory or lymphatic system and bioaccumulate in secondary tissues and organs, potentially inducing oxidative stress, immunotoxicity, and cytotoxicity [248, 249]. Further, they can primarily induce imbalanced generation of reactive oxygen species (ROS) that ultimately affect antioxidant enzymes' (e.g., glutathione S-transferase) activity and activate Mitogen-activated Protein Kinase (MAPK) signaling pathway [250, 251]. Importantly, a study based on simulated digestion showed that MPs can make aggregates with oil droplets and inhibit the activities of digestive enzymes, reducing lipid metabolism in the body [252].

Previously, multiple *in vitro* studies using human cell lines and *in vivo* studies employing rodent animal models have reported the pertaining toxic effects of MPs. An *in vitro* study unveiled the interaction among human serum albumin (HSA) and PVC-MPs using a multispectroscopic approach and was attributed to the presence of electrostatic forces, which ultimately induced variations in the microenvironment, caused molecular changes in the secondary structure of HSA, and disseminated to other tissues through the circulatory system [253]. The hazardous impacts of MPs can be influenced by various factors including exposure period, particle properties (e.g., concentration, shape, polymer type, size, surface functional groups, and charges), adsorption kinetics of particles and released additives, and biological responses of cells and tissues [39]. In addition, the smaller-sized MPs can cause a higher degree of toxicity from a relatively higher uptake rate. NPs may exhibit a higher potential to interact with various human cells and can easily penetrate the cell membrane and internalize to the cytoplasm. For instance, Lehner et al. highlighted the uptake of PS-NPs by alveolar epithelial cells in a size-dependent manner (Figure 8.6) [235]. Another *in vitro* investigation on both mono-cell and complex-cell cultures demonstrated the hazardous effects of PS-MPs in terms of cytotoxicity, endocytosis internalization, ROS generation, DNA damage, as well as genetic toxicity [39]. Further, the dose-dependent cellular and gene-level effects of NPs (<100 nm, 1–100 µg/mL) through internalization were reported in human Caco-2 cells (human colorectal adenocarcinoma), although results from this investigation were not statistically significant [254]. Further, an *in vivo* study on Wistar rats reported reproductive toxicity, endocrine disruption, and altered responses of semen biomarkers mediated by virgin PS-MPs (25 and 50 nm) [255].

Although the above-mentioned studies demonstrated the possible hazardous impacts of MPs on humans, there is no direct evidence to support the speculation of MPs exposure-based human health risks or diseases yet. Moreover, most of the *in vitro* or *in vivo* studies used unrealistically high exposure concentrations of MPs that are nearly impossible to be detected in the environment. In addition, the existing studies on MPs and biota interactions are largely focused on exploring the uptake, distribution, and compositional profiles of MPs in various organs or tissues. None of the studies exclusively addressed the human health effects or risks associated with MPs' exposure. Thus, relevant studies are highly recommended to

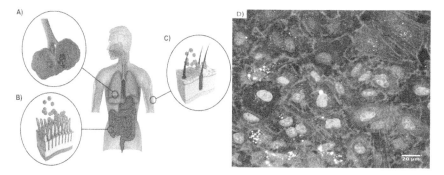

Figure 8.6 The three major uptake pathways of nanoplastics/microplastics in humans, i.e. (A) the lungs, (B) the gastrointestinal (GI) tract, and (C) the skin. (D) The uptake of 200 nm amino-modified polystyrene microplastics labeled with FITC (yellow) by human lung epithelial cancer cells (A549) labeled for F-actin (purple) and the nuclei (blue), after exposure for 24 h [235] / with permission of American Chemical Society.

simulate a realistic scenario for MPs mediated human health implications. So far, the cellular uptake pathways, internalization mechanism, intracellular fate, and tissue level adverse impacts of MPs and NPs have been scarcely investigated. Further, research gaps still exist to obtain robust, precise, and comparable data on human health implications associated with MPs exposure. Extensive human biomonitoring studies should be conducted to elucidate the risks of MPs and NPs, rather than merely focusing on limited types or specific shapes (e.g., fiber, spherical microspheres), as well as targeting only specific tissues and organs (gastrointestinal tracts or lungs). Moreover, ignoring the lack of clinical pathologies at the moment, chronic *in vivo* studies at low levels of NPs/MPs should be conducted to simulate the realistic scenario and associated harmful effects at cellular and tissue levels.

8.8 Implications and Suggestions

Although a hefty amount of literature is available on the abundance of MPs in receiving waters, the relevant studies from South Asia, East Asia, South America, and Africa are elusive yet. Moreover, there are no standard thresholds for reporting of MPs levels in freshwater bodies, therefore; the published results from different studies are difficult to compare and could prove a potential limitation for continuing MPs-related investigations and development of remediation measures [82].

Further field studies are recommended to understand the vertical distribution of MPs in water bodies as well as in rivers downstream to estuaries and their ways to seas. Importantly, loads of MPs in urban freshwater bodies through stormwater and surface runoff should be quantified. Further, elucidating the abundance and variation in air-water and water-sediment depositional fluxes of MPs as well as dry and wet atmospheric fallouts in the aquatic environment is critically needed. In addition, the non-point sources of MPs in receiving waters also need to be investigated. The degradation and fragmentation rate of MPs through the aging process in water bodies should be investigated to elucidate further the associated ecological implications [18].

Sufficient studies have been conducted on the sources and origin of MPs in the environment. Albeit the development of advanced and practical methods is still in infancy and challenging for MPs extraction specifically fibrous fraction from water as well as from complex environmental samples such as sediment. Further, the mechanisms involved in the aging of MPs and those responsible for the conversion of primary MPs to secondary MPs are scarcely reported. Moreover, the controlling factors responsible for preventing the degradation and decomposition of MPs also need further investigations. The physicochemical properties of

MPs make them susceptible to adsorb various environmental pollutants (heavy metals, chemical additives, and organic pollutants), providing a toxic pool for aquatic communities through acting as carriers to assist the internalization of those pollutants. More relevant studies are needed to establish concrete evidence whether MPs in combined exposure with other chemicals induce compound toxic impacts or change the transfer, bioaccumulation, and biomagnification of those chemicals and that of MPs through the food chain. It is now established that MPs <100 μm in size pose the most deleterious effects; however, additional studies are required to understand further the role of various polymer types, shapes, concentrations, and sizes of MPs in posing risk to ecological and human health. Importantly, there is a dire need to develop methods for ecological and human health risk assessment associated with MPs contamination in water bodies [82].

To understand further the interaction of MPs with humans, various *in vivo* (rat, mice, fish) and *in vitro* (cell lines) models could be used. Such experimental investigations are needed to understand the toxic implications of MPs (with varying physicochemical properties) in terms of their intake and excretion through various routes, transport mechanisms, metabolism, and accumulation in the human body. Various exposure routes of MPs in humans such as inhalation and dietary intake through food and beverages need further characterization. Moreover, animal models can also be employed to investigate the interactions among MPs and microbial communities through analyzing the changes in animal or human gastrointestinal microbiota at species and genus level using 16S rRNA high-throughput sequencing. Further, MPs are also involved in the enrichment of antibiotic resistant bacteria, resulting in the accumulation and dissemination of ARGs. Therefore, further studies are needed to investigate the role of MPs in the transformation of resistant bacteria, pathogens, and ARGs in the aquatic environment [35]. The issue of MPs is so critical that joint efforts through public involvement, legislation, health experts, engineering tools, and biotechnological expertise are required to control this gigantic anthropogenic catastrophe [82].

Acknowledgments

This study was funded by the Fund for International Young Scientists, National Natural Science Foundation of China (42150410389), National Key Research and Development Program of China (2018YFD0900604), Guangdong Province Universities and Colleges Pearl River Scholar Funded Scheme (2018), the National Natural Science Foundation of China (42077364), and Key Research Projects of Universities in Guangdong Province (2019KZDXM003 and 2020KZDZX1040).

References

1 Cózar, A., Echevarría, F., González-Gordillo, J.I., Irigoien, X., and Duarte, C.M. (2014). Plastic debris in the open ocean. *Proceedings of the National Academy USA* 111 (28): 10239–10244.

2 Li, J., Liu, H., and Paul Chen, J. (2018). Microplastics in freshwater systems: a review on occurrence, environmental effects, and methods for microplastics detection. *Water Research* 137: 362–374.

3 Strungaru, S.-A., Jijie, R., Nicoara, M., Plavan, G., and Faggio, C. (2019). Micro- (nano) plastics in freshwater ecosystems: abundance, toxicological impact and quantification methodology. *TrAC Trends in Analytical Chemistry* 110: 116–128.

4 Zeng, E.Y. (2018). *Microplastic Contamination in Aquatic Environments: An Emerging Matter of Environmental Urgency*. Elsevier.

5 Zhang, Y., Kang, S., Allen, S., Allen, D., and Sillanp, M. (2020). Atmospheric microplastics: a review on current status and perspectives. *Earth-Science Reviews* 203: 103118.

6 PlasticsEurope. Plastics – The Facts 2018: An Analysis of European Plastics Production, Demand and Waste Data. 2019.

7 Garside, M. (2019). Global production statistics. https://www.statista.com/statistics/282732/global-production-of-plastics-since-1950.

8 Ritchie, H. and Roser, M. (2018). Plastic Pollution. Our Wold in Data. https://ourworldindata.org/plastic-pollution.

9 Abel, D.S.M., Anderson,, Kloas, W., Zarfl, C., Hempel, S., and Rillig, M.C. (2017). Microplastics as an emerging threat to terrestrial ecosystems. *Global Change Biology* 24: 1405.

10 Yang, L., Zhang, Y., Kang, S., Wang, Z., and Wu, C. (2021). Microplastics in freshwater sediment: a review on methods, occurrence, and sources. *Science of the Total Environment* 754: 141948.

11 Cole, M., Lindeque, P., Halsband, C., and Galloway, T.S. (2011). Microplastics as contaminants in the marine environment: a review. *Marine Pollution Bulletin* 62 (12): 2588–2597.

12 Napper, I.E., Bakir, A., Rowland, S.J., and Thompson, R.C. (2015). Characterisation, quantity and sorptive properties of microplastics extracted from cosmetics. *Marine Pollution Bulletin* 99 (1): 178–185.

13 Fendall, L.S. and Sewell, M.A. (2009). Contributing to marine pollution by washing your face: microplastics in facial cleansers. *Marine Pollution Bulletin* 58 (8): 1225–1228.

14 De Falco, F., Cocca, M., Avella, M., and Thompson, R.C. (2020). Microfiber release to water, via laundering, and to air, via everyday use: a comparison between polyester clothing with differing textile parameters. *Environmental Science & Technology* 54 (6): 3288–3296.

15 Andrady, A.L. (2017). The plastic in microplastics: a review. *Marine Pollution Bulletin* 119 (1): 12–22.

16 Sun, J., Dai, X., Wang, Q., van Loosdrecht, M.C.M., and Ni, B.-J. (2019). Microplastics in wastewater treatment plants: detection, occurrence and removal. *Water Research* 152: 21–37.

17 Okoffo, E.D., O'Brien, S., O'Brien, J.W., Tscharke, B.J., and Thomas, K.V. (2019). Wastewater treatment plants as a source of plastics in the environment: a review of occurrence, methods for identification, quantification and fate. *Environmental Science: Water Research & Technology* 5: 1908.

18 Akdogan, Z. and Guven, B. (2019). Microplastics in the environment: a critical review of current understanding and identification of future research needs. *Environmental Pollution* 254: 113011.

19 Ghelardini, C., Bartolini, A., Galeotti, N., Teodori, E., and Gregory, M.R. (1996). Plastic 'scrubbers' in hand cleansers: a further (and minor) source for marine pollution identified. *Marine Pollution Bulletin* 32: 12.

20 Horton, A.A., Walton, A., Spurgeon, D.J., Lahive, E., and Svendsen, C. (2017). Microplastics in freshwater and terrestrial environments: evaluating the current understanding to identify the knowledge gaps and future research priorities. *Science of the Total Environment* 586: 127–141.

21 Zbyszewski, M. and Corcoran, P.L. (2011). Distribution and Degradation of Fresh Water Plastic Particles Along the Beaches of Lake Huron, Canada. *Water Air & Soil Pollution* 220 (1–4): 365–372.

22 Norén, F. and Naustvoll, L.-J. Survey of microscopic anthropogenic particles in Skagerrak. 2010.

23 Duis, K. and Coors, A. (2016). Microplastics in the aquatic and terrestrial environment: sources (with a specific focus on personal care products), fate and effects. *Environmental Sciences Europe* 28 (1): 2.

24 Nizzetto, L., Futter, M., and Langaas, S. (2016). Are Agricultural Soils Dumps for Microplastics of Urban Origin? *Environmental Science & Technology* 50 (20): 10777–10779.

25 Unice, K.M., Weeber, M.P., Abramson, M.M. et al. (2019). Characterizing export of land-based microplastics to the estuary - Part I: application of integrated geospatial microplastic transport models to assess tire and road wear particles in the Seine watershed. *Science of the Total Environment* 646: 1639–1649.

26 Browne, M.A., Crump, P., Niven, S.J. et al. (2011). Accumulation of Microplastic on Shorelines Woldwide: sources and Sinks. *Environmental Science & Technology* 45 (21): 9175–9179.

27 Napper, I.E. and Thompson, R.C. (2016). Release of synthetic microplastic plastic fibres from domestic washing machines: effects of fabric type and washing conditions. *Marine Pollution Bulletin* 112 (1): 39–45.

28 Kane, I.A., Clare, M.A., Miramontes, E. et al. (2020). Seafloor microplastic hotspots controlled by deep-sea circulation. *Science* 368 (6495): 1140–1145.

29 Van Melkebeke, M., Janssen, C., and De Meester, S. (2020). Characteristics and Sinking Behavior of Typical Microplastics Including the Potential Effect of Biofouling: implications for Remediation. *Environmental Science & Technology* 54 (14): 8668–8680.

30 Corcoran, P.L., Norris, T., Ceccanese, T., Walzak, M.J., Helm, P.A., and Marvin, C.H. (2015). Hidden plastics of Lake Ontario, Canada and their potential preservation in the sediment record. *Environmental Pollution* 204: 17–25.

31 Fischer, E.K., Paglialonga, L., Czech, E., and Tamminga, M. (2016). Microplastic pollution in lakes and lake shoreline sediments – A case study on Lake Bolsena and Lake Chiusi (central Italy). *Environmental Pollution* 213: 648–657.

32 Free, C.M., Jensen, O.P., Mason, S.A., Eriksen, M., Williamson, N.J., and Boldgiv, B. (2014). High-levels of microplastic pollution in a large, remote, mountain lake. *Marine Pollution Bulletin* 85 (1): 156–163.

33 Hoellein, T., Rojas, M., Pink, A., Gasior, J., and Kelly, J. (2014). Anthropogenic litter in urban freshwater ecosystems: distribution and microbial interactions. *PLoS ONE* 9 (6): e98485.

34 Hall, R.O., Jr and Meyer, J.L. (1998). The trophic significance of bacteria in a detritus-based stream food web. *Ecology* 79 (6): 1995–2012.

35 Wang, C., Zhao, J., and Xing, B. (2020). Environmental source, fate, and toxicity of microplastics. *Journal of Hazardous Materials* 407: 124357.

36 Carbery, M., O'Connor, W., and Palanisami, T. (2018). Trophic transfer of microplastics and mixed contaminants in the marine food web and implications for human health. *Environment International* 115: 400–409.

37 Shen, M., Zhang, Y., Zhu, Y. et al. (2019). Recent advances in toxicological research of nanoplastics in the environment: a review. *Environmental Pollution* 252 (Pt A): 511–521.

38 Wang, W., Gao, H., Jin, S., Li, R., and Na, G. (2019). The ecotoxicological effects of microplastics on aquatic food web, from primary producer to human: a review. *Ecotoxicology and Environmental Safety* 173: 110–117.

39 Huang, W., Song, B., Liang, J. et al. (2020). Microplastics and associated contaminants in the aquatic environment: a review on their ecotoxicological effects, trophic transfer, and potential impacts to human health. *Journal of Hazardous Materials* 405: 124187.

40 Azevedo-Santos, V.M., Gonçalves, G.R.L., Manoel, P.S., Andrade, M.C., Lima, F.P., and Pelicice, F.M. (2019). Plastic ingestion by fish: a global assessment. *Environmental Pollution* 255 (Pt 1): 112994.

41 Li, J., Green, C., Reynolds, A., Shi, H., and Rotchell, J.M. (2018). Microplastics in mussels sampled from coastal waters and supermarkets in the United Kingdom. *Environmental Pollution* 241: 35–44.

42 Teng, J., Wang, Q., Ran, W. et al. (2019). Microplastic in cultured oysters from different coastal areas of China. *Science of the Total Environment* 653: 1282–1292.
43 Fossi, M.C., Panti, C., Baini, M., and Lavers, J.L. (2018). A Review of Plastic-Associated Pressures: cetaceans of the Mediterranean Sea and Eastern Australian Shearwaters as Case Studies. *Frontiers in Marine Science* 5.
44 Le Guen, C., Suaria, G., Sherley, R.B. et al. (2020). Microplastic study reveals the presence of natural and synthetic fibres in the diet of King Penguins (Aptenodytes patagonicus) foraging from South Georgia. *Environment International* 134: 105303.
45 Burkhardt-Holm, P. and N'Guyen, A. (2019). Ingestion of microplastics by fish and other prey organisms of cetaceans, exemplified for two large baleen whale species. *Marine Pollution Bulletin* 144: 224–234.
46 Wright, S.L., Thompson, R.C., and Galloway, T.S. (2013). The physical impacts of microplastics on marine organisms: a review. *Environmental Pollution* 178: 483–492.
47 Santana, M.F.M., Moreira, F.T., and Turra, A. (2017). Trophic transference of microplastics under a low exposure scenario: insights on the likelihood of particle cascading along marine food-webs. *Marine Pollution Bulletin* 121 (1–2): 154–159.
48 Brookson, C.B., De Solla, S., Fernie, K.J., Cepeda, M.F.E., and Rochman, C.M. (2019). Microplastics in the diet of nestling double-crested cormorants (Phalacrocorax auritus), an obligate piscivore in a freshwater ecosystem. *Canadian Journal of Fisheries and Aquatic Ences* 76: 2156.
49 Hernandez-Milian, G., Lusher, A., MacGabban, S., and Rogan, E. (2019). Microplastics in grey seal (Halichoerus grypus) intestines: are they associated with parasite aggregations? *Marine Pollution Bulletin* 146: 349–354.
50 Zhu, J., Yu, X., Zhang, Q. et al. (2019). Cetaceans and microplastics: first report of microplastic ingestion by a coastal delphinid, Sousa chinensis. *Science of the Total Environment* 659: 649–654.
51 Maes, T., Jel, J.V.D.D., Vethaak, A.D., Desender, M., and Leslie, H.A. (2020). You are what you eat, microplastics in porbeagle sharks from the North East Atlantic: method development and analysis in spiral valve content and tissue. *Frontiers in Marine Science* 7: 273.
52 Moore, R.C., Loseto, L., Noel, M. et al. (2020). Microplastics in beluga whales (Delphinapterus leucas) from the Eastern Beaufort Sea. *Marine Pollution Bulletin* 150: 110723.
53 Schwabl, P., Köppel, S., Königshofer, P. et al. (2019). Detection of various microplastics in human stool: a prospective case series. *Annals of Internal Medicine* 171 (7): 453–457.
54 Zhang, Q., Xu, E.G., Li, J. et al. (2020). A review of microplastics in table salt, drinking water, and air: direct human exposure. *Environmental Science & Technology* 54 (7): 3740–3751.

55 Koelmans, A.A., Mohamed Nor, N.H., Hermsen, E., Kooi, M., Mintenig, S.M., and De France, J. (2019). Microplastics in freshwaters and drinking water: critical review and assessment of data quality. *Water Research* 155: 410–422.

56 Oßmann, B.E., Sarau, G., Holtmannspötter, H., Pischetsrieder, M., Christiansen, S.H., and Dicke, W. (2018). Small-sized microplastics and pigmented particles in bottled mineral water. *Water Research* 141: 307–316.

57 Zuccarello, P., Ferrante, M., Cristaldi, A. et al. (2019). Exposure to microplastics (<10 μm) associated to plastic bottles mineral water consumption: the first quantitative study. *Water Research* 157: 365–371.

58 Karami, A., Golieskardi, A., Keong Choo, C., Larat, V., Galloway, T.S., and Salamatinia, B. (2017). The presence of microplastics in commercial salts from different countries. *Scientific Reports* 7: 46173.

59 Kim, J.S., Lee, H.J., Kim, S.K., and Kim, H.J. (2018). Global pattern of microplastics (MPs) in commercial food-grade salts: sea salt as an indicator of seawater MP pollution. *Environmental Science & Technology* 52 (21): 12819–12828.

60 Cho, Y., Shim, W.J., Jang, M., Han, G.M., and Hong, S.H. (2019). Abundance and characteristics of microplastics in market bivalves from South Korea. *Environmental Pollution* 245: 1107–1116.

61 Feng, Z., Wang, R., Zhang, T. et al. (2020). Microplastics in specific tissues of wild sea urchins along the coastal areas of northern China. *Science of the Total Environment* 728: 138660.

62 Cox, K.D., Covernton, G.A., Davies, H.L., Dower, J.F., Juanes, F., and Dudas, S.E. (2019). Human consumption of microplastics. *Environmental Science & Technology* 53 (12): 7068–7074.

63 Prata, J.C. (2018). Airborne microplastics: consequences to human health? *Environmental Pollution* 234: 115–126.

64 Novotna, K., Cermakova, L., Pivokonska, L., Cajthaml, T., and Pivokonsky, M. (2019). Microplastics in drinking water treatment - Current knowledge and research needs. *Science of the Total Environment* 667: 730–740.

65 Dris, R., Gasperi, J., and Tassin, B. (2018). Sources and fate of microplastics in Urban Areas: a focus on Paris Megacity. In: *Freshwater Microplastics: Emerging Environmental Contaminants?* (ed. M. Wagner and S. Lambert), 69–83. Cham: Springer International Publishing.

66 Moore, C.J., Lattin, G.L., and Zellers, A.F. (2011). Quantity and type of plastic debris flowing from two urban rivers to coastal waters and beaches of Southern California. *Journal of Integrated Coastal Zone Management* 11: 65–73.

67 Peng, G., Xu, P., Zhu, B., Bai, M., and Li, D. (2018). Microplastics in freshwater river sediments in Shanghai, China: a case study of risk assessment in megacities. *Environmental Pollution* 234: 448–456.

68 Kapp, K.J. and Yeatman, E. (2018). Microplastic hotspots in the Snake and Lower Columbia rivers: a journey from the Greater Yellowstone Ecosystem to the Pacific Ocean. *Environmental Pollution* 241: 1082–1090.

69 Guerranti, C., Cannas, S., Scopetani, C., Fastelli, P., Cincinelli, A., and Renzi, M. (2017). Plastic litter in aquatic environments of Maremma Regional Park (Tyrrhenian Sea, Italy): contribution by the Ombrone river and levels in marine sediments. *Marine Pollution Bulletin* 117 (1–2): 366–370.

70 Klein, S., Worch, E., and Knepper, T.P. (2015). Occurrence and Spatial Distribution of Microplastics in River Shore Sediments of the Rhine-Main Area in Germany. *Environmental Science & Technology* 49 (10): 6070–6076.

71 Baldwin, A.K., Corsi, S.R., and Mason, S.A. (2016). Plastic debris in 29 Great Lakes tributaries: relations to watershed attributes and hydrology. *Environmental Science & Technology* 50 (19): 10377–10385.

72 Bowmer, T. (2010). Proceedings of the GESAMP International Workshop on micro-plastic particles as a vector in transporting persistent, bio-accumulating and toxic substances in the oceans. *UNESCO-IOC, Paris*.

73 Lechner, A., Keckeis, H., Lumesberger-Loisl, F. et al. (2014). The Danube so colourful: a potpourri of plastic litter outnumbers fish larvae in Europe's second largest river. *Environmental Pollution* 188 (100): 177–181.

74 Gasperi, J., Dris, R., Bonin, T., Rocher, V., and Tassin, B. (2014). Assessment of floating plastic debris in surface water along the Seine River. *Environmental Pollution* 195: 163–166.

75 Lahens, L., Strady, E., Kieu-Le, T.C. et al. (2018). Macroplastic and microplastic contamination assessment of a tropical river (Saigon River, Vietnam) transversed by a developing megacity. *Environmental Pollution* 236: 661–671.

76 Estahbanati, S. and Fahrenfeld, N.L. (2016). Influence of wastewater treatment plant discharges on microplastic concentrations in surface water. *Chemosphere* 162: 277–284.

77 Vermaire, J.C., Pomeroy, C., Herczegh, S.M., Haggart, O., and Murphy, M. (2017). Microplastic abundance and distribution in the open water and sediment of the Ottawa River, Canada, and its tributaries. *Facets* 2 (1): 301–314.

78 Castaneda, R.A., Avlijas, S., Simard, M.A., and Ricciardi, A. (2014). Microplastic pollution in St. Lawrence River sediments. *Canadian Journal of Fisheries & Aquatic Ences* 71 (1): 21–40.

79 Wang, W., Ndungu, A.W., Li, Z., and Wang, J. (2017). Microplastics pollution in inland freshwaters of China: a case study in urban surface waters of Wuhan, China. *Science of the Total Environment* 575: 1369–1374.

80 Zhang, K., Gong, W., Lv, J., Xiong, X., and Wu, C. (2015). Accumulation of floating microplastics behind the Three Gorges Dam. *Environmental Pollution* 204: 117–123.

81 Zhang, K., Xiong, X., Hu, H. et al. (2017). Occurrence and Characteristics of Microplastic Pollution in Xiangxi Bay of Three Gorges Reservoir, China. *Environmental Science & Technology* 51 (7): 3794–3801.

82 Li, C., Busquets, R., and Campos, L.C. (2020). Assessment of microplastics in freshwater systems: a review. *Science of the Total Environment* 707: 135578.

83 Lechner, A. and Ramler, D. (2015). The discharge of certain amounts of industrial microplastic from a production plant into the River Danube is permitted by the Austrian legislation. *Environmental Pollution* 200: 159–160.

84 Rodrigues, M.O., Abrantes, N., Gonçalves, F.J.M., Nogueira, H., Marques, J.C., and Gonçalves, A.M.M. (2018). Spatial and temporal distribution of microplastics in water and sediments of a freshwater system (Antuã River, Portugal). *Science of the Total Environment* 633: 1549–1559.

85 Campanale, C., Stock, F., Massarelli, C. et al. (2020). Microplastics and their possible sources: the example of Ofanto river in southeast Italy. *Environmental Pollution* 258: 113284.

86 Mani, T., Hauk, A., Walter, U., and Burkhardt-Holm, P. (2015). Microplastics profile along the Rhine River. *Scientific Reports* 5: 17988.

87 Lin, L., Zuo, L.Z., Peng, J.P. et al. (2018). Occurrence and distribution of microplastics in an urban river: a case study in the Pearl River along Guangzhou City, China. *Science of the Total Environment* 644: 375–381.

88 Di, M. and Wang, J. (2018). Microplastics in surface waters and sediments of the Three Gorges Reservoir, China. *Science of the Total Environment* 616-617: 1620–1627.

89 Jiang, C., Yin, L., Li, Z. et al. (2019). Microplastic pollution in the rivers of the Tibet Plateau. *Environmental Pollution* 249: 91–98.

90 Ding, L., Mao, R.F., Guo, X., Yang, X., Zhang, Q., and Yang, C. (2019). Microplastics in surface waters and sediments of the Wei River, in the northwest of China. *Science of the Total Environment* 667: 427–434.

91 Su, L., Xue, Y., Li, L. et al. (2016). Microplastics in Taihu Lake, China. *Environmental Pollution* 216: 711–719.

92 Zhang, Y., Gao, T., Kang, S., and Sillanpää, M. (2019). Importance of atmospheric transport for microplastics deposited in remote areas. *Environmental Pollution* 254 (Pt A): 112953.

93 Eriksen, M., Mason, S., Wilson, S. et al. (2013). Microplastic pollution in the surface waters of the Laurentian Great Lakes. *Marine Pollution Bulletin* 77 (1–2): 177–182.

94 Hendrickson, E., Minor, E.C., and Schreiner, K. (2018). Microplastic Abundance and Composition in Western Lake Superior As Determined via Microscopy, Pyr-GC/MS, and FTIR. *Environmental Science & Technology* 52 (4): 1787–1796.

95 Mason, S.A., Kammin, L., Eriksen, M. et al. (2016). Pelagic plastic pollution within the surface waters of Lake Michigan, USA. *Journal of Great Lakes Research* 42 (4): 753–759.

96 Biginagwa, F.J., Mayoma, B.S., Shashoua, Y., Syberg, K., and Khan, F.R. (2016). First evidence of microplastics in the African Great Lakes: recovery from Lake Victoria Nile perch and Nile tilapia. *Journal of Great Lakes Research* 42 (1): 146–149.

97 Xiong, X., Zhang, K., Chen, X., Shi, H., Luo, Z., and Wu, C. (2018). Sources and distribution of microplastics in China's largest inland lake - Qinghai Lake. *Environmental Pollution* 235: 899–906.

98 Yuan, W., Liu, X., Wang, W., Di, M., and Wang, J. (2019). Microplastic abundance, distribution and composition in water, sediments, and wild fish from Poyang Lake, China. *Ecotoxicology and Environmental Safety* 170: 180–187.

99 Qin, Y., Wang, Z., Li, W., Chang, X., Yang, J., and Yang, F. (2020). Microplastics in the sediment of Lake Ulansuhai of Yellow River Basin, China. *Water Environment Research* 92 (6): 829–839.

100 Jiang, C., Yin, L., Wen, X. et al. (2018). Microplastics in Sediment and Surface Water of West Dongting Lake and South Dongting Lake: abundance, source and composition. *International Journal of Environmental Research and Public Health* 15 (10).

101 Wang, W., Yuan, W., Chen, Y., and Wang, J. (2018). Microplastics in surface waters of Dongting Lake and Hong Lake, China. *Science of the Total Environment* 633: 539–545.

102 Faure, F., Demars, C., Wieser, O., Kunz, M., and Alencastro, L.F.D. (2015). Plastic pollution in Swiss surface waters: nature and concentrations, interaction with pollutants. *Environmental Chemistry* 12: 582–591.

103 Uurasjärvi, E., Hartikainen, S., Setälä, O., Lehtiniemi, M., and Koistinen, A. (2020). Microplastic concentrations, size distribution, and polymer types in the surface waters of a northern European lake. *Water Environment Research* 92 (1): 149–156.

104 Sighicelli, M., Pietrelli, L., Lecce, F. et al. (2018). Microplastic pollution in the surface waters of Italian Subalpine Lakes. *Environmental Pollution* 236: 645–651.

105 Anderson, P.J., Warrack, S., Langen, V., Challis, J.K., Hanson, M.L., and Rennie, M.D. (2017). Microplastic contamination in Lake Winnipeg, Canada. *Environmental Pollution* 225: 223–231.

106 Zhang, K., Shi, H., Peng, J. et al. (2018). Microplastic pollution in China's inland water systems: a review of findings, methods, characteristics, effects, and management. *Science of the Total Environment* 630: 1641–1653.

107 Carr, S.A., Liu, J., and Tesoro, A.G. (2016). Transport and fate of microplastic particles in wastewater treatment plants. *Water Research* 91: 174–182.

108 Hidalgo-Ruz, V., Gutow, L., Thompson, R.C., and Thiel, M. (2012). Microplastics in the marine environment: a review of the methods used for identification and quantification. *Environmental Science & Technology* 46 (6): 3060–3075.

109 Kole, P.J., Löhr, A.J., Van Belleghem, F., and Ragas, A.M.J. (2017). Wear and tear of tyres: a stealthy source of microplastics in the environment. *International Journal of Environmental Research and Public Health* 14: 10.

110 Auta, H.S., Emenike, C.U., and Fauziah, S.H. (2017). Distribution and importance of microplastics in the marine environment: a review of the sources, fate, effects, and potential solutions. *Environment International* 102: 165–176.

111 Wang, C., Xing, R., Sun, M. et al. (2020). Microplastics profile in a typical urban river in Beijing. *Science of the Total Environment* 743: 140708.

112 Browne, M.A. (2015). *Sources and Pathways of Microplastics to Habitats*. Springer International Publishing.

113 Guo, X. and Wang, J. (2019). The chemical behaviors of microplastics in marine environment: a review. *Marine Pollution Bulletin* 142: 1–14.

114 Brennecke, D., Duarte, B., Paiva, F., Caçador, I., and Canning-Clode, J.O. (2016). Microplastics as vector for heavy metal contamination from the marine environment. *Estuarine Coastal & Shelf Science* 178: 189–195.

115 Rios, L.M., Moore, C., and Jones, P.R. (2007). Persistent organic pollutants carried by synthetic polymers in the ocean environment. *Marine Pollution Bulletin* 54 (8): 1230–1237.

116 Eriksson, C. and Burton, H. (2003). Origins and biological accumulation of small plastic particles in fur seals from Macquarie Island. *Ambio* 32 (6): 380–384.

117 Pérez, J.M., Vilas, J.L., Laza, J.M. et al. (2010). Effect of reprocessing and accelerated ageing on thermal and mechanical polycarbonate properties. *Journal of Materials Processing Tech* 210 (5): 727–733.

118 Liu, K., Wang, X., Fang, T., Xu, P., Zhu, L., and Li, D. (2019). Source and potential risk assessment of suspended atmospheric microplastics in Shanghai. *Science of the Total Environment* 675: 462–471.

119 Hüffer, T. and Hofmann, T. (2016). Sorption of non-polar organic compounds by micro-sized plastic particles in aqueous solution. *Environmental Pollution* 214: 194–201.

120 Goedecke, C., Stollin, U.M., Hering, S. et al. (2017). A First Pilot Study on the Sorption of Environmental Pollutants on VariousMicroplastic Materials. *Journal of Environmental Analytical Chemistry* 04 (01).

121 Kedzierski, M., D'Almeida, M., Magueresse, A. et al. (2018). Threat of plastic ageing in marine environment. Adsorption/desorption of micropollutants. *Marine Pollution Bulletin* 127: 684–694.

122 Rochman, C.M., Kurobe, T., Flores, I., and Teh, S.J. (2014). Early warning signs of endocrine disruption in adult fish from the ingestion of polyethylene with and without sorbed chemical pollutants from the marine environment. *Science of the Total Environment* 493: 656–661.

123 Van Cauwenberghe, L., Claessens, M., Vandegehuchte, M.B., and Janssen, C.R. (2015). Microplastics are taken up by mussels (Mytilus edulis) and lugworms (Arenicola marina) living in natural habitats. *Environmental Pollution* 199: 10–17.

124 Desforges, J.P., Galbraith, M., and Ross, P.S. (2015). Ingestion of Microplastics by Zooplankton in the Northeast Pacific Ocean. *Archives of Environmental Contamination and Toxicology* 69 (3): 320–330.

125 Windsor, F.M., Tilley, R.M., Tyler, C.R., and Ormerod, S.J. (2019). Microplastic ingestion by riverine macroinvertebrates. *Science of the Total Environment* 646: 68–74.

126 Watts, A.J., Lewis, C., Goodhead, R.M. et al. (2014). Uptake and retention of microplastics by the shore crab Carcinus maenas. *Environmental Science & Technology* 48 (15): 8823–8830.

127 Kolandhasamy, P., Su, L., Li, J., Qu, X., Jabeen, K., and Shi, H. (2018). Adherence of microplastics to soft tissue of mussels: a novel way to uptake microplastics beyond ingestion. *Science of the Total Environment* 610-611: 635–640.

128 Abidli, S., Lahbib, Y., and Trigui El Menif, N. (2019). Microplastics in commercial molluscs from the lagoon of Bizerte (Northern Tunisia). *Marine Pollution Bulletin* 142: 243–252.

129 Van Cauwenberghe, L. and Janssen, C.R. (2014). Microplastics in bivalves cultured for human consumption. *Environmental Pollution* 193: 65–70.

130 Li, J., Qu, X., Su, L. et al. (2016). Microplastics in mussels along the coastal waters of China. *Environmental Pollution* 214: 177–184.

131 Li, J., Yang, D., Li, L., Jabeen, K., and Shi, H. (2015). Microplastics in commercial bivalves from China. *Environmental Pollution* 207: 190–195.

132 Foley, C.J., Feiner, Z.S., Malinich, T.D., and Höök, T.O. (2018). A meta-analysis of the effects of exposure to microplastics on fish and aquatic invertebrates. *Science of the Total Environment* 631-632: 550–559.

133 Sussarellu, R., Suquet, M., Thomas, Y. et al. (2016). Oyster reproduction is affected by exposure to polystyrene microplastics. *Proceedings of the National Academy USA* 113 (9): 2430–2435.

134 Trestrail, C., Nugegoda, D., and Shimeta, J. (2020). Invertebrate responses to microplastic ingestion: reviewing the role of the antioxidant system. *Science of the Total Environment* 734: 138559.

135 Santana, M.F.M., Moreira, F.T., Pereira, C.D.S., Abessa, D.M.S., and Turra, A. (2018). Continuous Exposure to Microplastics Does Not Cause Physiological Effects in the Cultivated Mussel Perna perna. *Archives of Environmental Contamination and Toxicology* 74 (4): 594–604.

136 Weber, A., Scherer, C., Brennholt, N., Reifferscheid, G., and Wagner, M. (2018). PET microplastics do not negatively affect the survival, development, metabolism and feeding activity of the freshwater invertebrate Gammarus pulex. *Environmental Pollution* 234: 181–189.

137 Graham, P., Palazzo, L., Andrea de Lucia, G., Telfer, T.C., Baroli, M., and Carboni, S. (2019). Microplastics uptake and egestion dynamics in Pacific oysters, Magallana gigas (Thunberg, 1793), under controlled conditions. *Environmental Pollution* 252 (Pt A): 742–748.

138 Basto, M.N., Nicastro, K.R., Tavares, A.I. et al. (2019). Plastic ingestion in aquatic birds in Portugal. *Marine Pollution Bulletin* 138: 19–24.

139 Provencher, J.F., Vermaire, J.C., Avery-Gomm, S., Braune, B.M., and Mallory, M.L. (2018). Garbage in guano? Microplastic debris found in faecal precursors of seabirds known to ingest plastics. *Science of the Total Environment* 644: 1477–1484.

140 Cartraud, A.E., Le Corre, M., Turquet, J., and Tourmetz, J. (2019). Plastic ingestion in seabirds of the western Indian Ocean. *Marine Pollution Bulletin* 140: 308–314.

141 Ryan, P.G. (2008). Seabirds indicate changes in the composition of plastic litter in the Atlantic and south-western Indian Oceans. *Marine Pollution Bulletin* 56 (8): 1406–1409.

142 Nicastro, K.R., Lo Savio, R., McQuaid, C.D. et al. (2018). Plastic ingestion in aquatic-associated bird species in southern Portugal. *Marine Pollution Bulletin* 126: 413–418.

143 D'Souza, J.M., Windsor, F.M., Santillo, D., and Ormerod, S.J. (2020). Food web transfer of plastics to an apex riverine predator. *Global Change Biology* 26 (7): 3846–3857.

144 Furtado, R., Menezes, D., Santos, C.J., and Catry, P. (2016). White-faced storm-petrels Pelagodroma marina predated by gulls as biological monitors of plastic pollution in the pelagic subtropical Northeast Atlantic. *Marine Pollution Bulletin* 112 (1–2): 117–122.

145 Hammer, S., Nager, R.G., Johnson, P.C.D., Furness, R.W., and Provencher, J.F. (2016). Plastic debris in great skua (Stercorarius skua) pellets corresponds to seabird prey species. *Marine Pollution Bulletin* 103 (1–2): 206–210.

146 Shu-Li, L., Min-Fei, et al. (2019). Pollution Characteristics of Microplastics in Migratory Bird Habitats Located Within Poyang Lake Wetlands. *Huan jing ke xue= Huanjing kexue* 40 (6): 2639–2646.

147 Holland, E.R., Mallory, M.L., and Shutler, D. (2016). Plastics and other anthropogenic debris in freshwater birds from Canada. *Science of the Total Environment* 571: 251–258.

148 Reynolds, C. and Ryan, P.G. (2018). Micro-plastic ingestion by waterbirds from contaminated wetlands in South Africa. *Marine Pollution Bulletin* 126: 330–333.

149 Masiá, P., Ardura, A., and Garcia-Vazquez, E. (2019). Microplastics in special protected areas for migratory birds in the Bay of Biscay. *Marine Pollution Bulletin* 146: 993–1001.

150 Terepocki, A.K., Brush, A.T., Kleine, L.U., Shugart, G.W., and Hodum, P. (2017). Size and dynamics of microplastic in gastrointestinal tracts of Northern Fulmars (Fulmarus glacialis) and Sooty Shearwaters (Ardenna grisea). *Marine Pollution Bulletin* 116 (1–2): 143–150.

151 van Franeker, J.A., Blaize, C., Danielsen, J. et al. (2011). Monitoring plastic ingestion by the northern fulmar Fulmarus glacialis in the North Sea. *Environmental Pollution* 159 (10): 2609–2615.

152 Bessa, F., Barría, P., Neto, J.M. et al. (2018). Occurrence of microplastics in commercial fish from a natural estuarine environment. *Marine Pollution Bulletin* 128: 575–584.

153 Lavers, J.L., Hutton, I., and Bond, A.L. (2019). Clinical Pathology of Plastic Ingestion in Marine Birds and Relationships with Blood Chemistry. *Environmental Science & Technology* 53 (15): 9224–9231.

154 Roman, L., Lowenstine, L., Parsley, L.M. et al. (2019). Is plastic ingestion in birds as toxic as we think? Insights from a plastic feeding experiment. *Science of the Total Environment* 665: 660–667.

155 Herzke, D., Anker-Nilssen, T., Nost, T.H. et al. (2016). Negligible Impact of Ingested Microplastics on Tissue Concentrations of Persistent Organic Pollutants in Northern Fulmars off Coastal Norway. *Environmental Science & Technology* 50 (4): 1924–1933.

156 Coffin, S., Huang, G.Y., Lee, I., and Schlenk, D. (2019). Fish and Seabird Gut Conditions Enhance Desorption of Estrogenic Chemicals from Commonly-Ingested Plastic Items. *Environmental Science & Technology* 53 (8): 4588–4599.

157 Guo, H.Y., Zheng, X.B., Luo, X.J., and Mai, B.X. (2020). Leaching of brominated flame retardants (BFRs) from BFRs-incorporated plastics in digestive fluids and the influence of bird diets. *Journal of Hazardous Materials* 393.

158 Nelms, S.E., Barnett, J., Brownlow, A. et al. (2019). Microplastics in marine mammals stranded around the British coast: ubiquitous but transitory? *Scientific Reports* 9 (1): 1075.

159 Hudak, C.A. and Sette, L. (2019). Opportunistic detection of anthropogenic micro debris in harbor seal (Phoca vitulina vitulina) and gray seal (Halichoerus grypus atlantica) fecal samples from haul-outs in southeastern Massachusetts, USA. *Marine Pollution Bulletin* 145: 390–395.

160 Donohue, M.J., Masura, J., Gelatt, T. et al. (2019). Evaluating exposure of northern fur seals, Callorhinus ursinus, to microplastic pollution through fecal analysis. *Marine Pollution Bulletin* 138: 213–221.

161 Bravo Rebolledo, E.L., Van Franeker, J.A., Jansen, O.E., and Brasseur, S.M. (2013). Plastic ingestion by harbour seals (Phoca vitulina) in The Netherlands. *Marine Pollution Bulletin* 67 (1–2): 200–202.

162 Nelms, S.E., Galloway, T.S., Godley, B.J., Jarvis, D.S., and Lindeque, P.K. (2018). Investigating microplastic trophic transfer in marine top predators. *Environmental Pollution* 238: 999–1007.

163 van Franeker, J.A., Bravo Rebolledo, E.L., Hesse, E. et al. (2018). Plastic ingestion by harbour porpoises Phocoena phocoena in the Netherlands: establishing a standardised method. *Ambio* 47 (4): 387–397.

164 Xiong, X., Chen, X., Zhang, K. et al. (2018). Microplastics in the intestinal tracts of East Asian finless porpoises (Neophocaena asiaeorientalis sunameri) from Yellow Sea and Bohai Sea of China. *Marine Pollution Bulletin* 136: 55–60.

165 Hernandez-Gonzalez, A., Saavedra, C., Gago, J., Covelo, P., Santos, M.B., and Pierce, G.J. (2018). Microplastics in the stomach contents of common dolphin (Delphinus delphis) stranded on the Galician coasts (NW Spain, 2005-2010). *Marine Pollution Bulletin* 137: 526–532.

166 Lusher, A.L., Hernandez-Milian, G., O'Brien, J., Berrow, S., O'Connor, I., and Officer, R. (2015). Microplastic and macroplastic ingestion by a deep diving, oceanic cetacean: the True's beaked whale Mesoplodon mirus. *Environmental Pollution* 199: 185–191.

167 Fossi, M.C., Coppola, D., Baini, M. et al. (2014). Large filter feeding marine organisms as indicators of microplastic in the pelagic environment: the case studies of the Mediterranean basking shark (Cetorhinus maximus) and fin whale (Balaenoptera physalus). *Marine Environmental Research* 100: 17–24.

168 Alomar, C. and Deudero, S. (2017). Evidence of microplastic ingestion in the shark Galeus melastomus Rafinesque, 1810 in the continental shelf off the western Mediterranean Sea. *Environmental Pollution* 223: 223–229.

169 Germanov, E.S., Marshall, A.D., Hendrawan, I.G., Admiraal, R., and Lonergan, N.R. (2019). Microplastics on the menu: plastics pollute Indonesian Manta Ray and Whale Shark feeding grounds. *Frontiers in Marine Science* 6: 679.

170 Fossi, M.C., Baini, M., Panti, C. et al. (2017). Are whale sharks exposed to persistent organic pollutants and plastic pollution in the Gulf of California (Mexico)? First ecotoxicological investigation using skin biopsies. *Comparative Biochemistry and Physiology Part C: Toxicology & Pharmacology* 199: 48–58.

171 Fossi, M.C., Marsili, L., Baini, M. et al. (2016). Fin whales and microplastics: the Mediterranean Sea and the Sea of Cortez scenarios. *Environmental Pollution* 209: 68–78.

172 Fossi, M.C., Panti, C., Guerranti, C. et al. (2012). Are baleen whales exposed to the threat of microplastics? A case study of the Mediterranean fin whale (Balaenoptera physalus). *Marine Pollution Bulletin* 64 (11): 2374–2379.

173 Fossi, M.C., Romeo, T., Baini, M., Panti, C., and Lapucci, C. (2017). Plastic debris occurrence, convergence areas and fin whales feeding ground in the mediterranean marine protected area pelagos sanctuary: a modeling approach. *Frontiers in Marine Science* 4: 1.

174 Oberbeckmann, S., L?der, M.G.J., and Labrenz, M. (2015). Marine microplastic-associated biofilms – A review. *Environmental Chemistry* 12 (5): 551.

175 Shen, M., Zhu, Y., Zhang, Y. et al. (2019). Micro(nano)plastics: unignorable vectors for organisms. *Marine Pollution Bulletin* 139: 328–331.

176 Harrison, J.P., Schratzberger, M., Sapp, M., and Osborn, A.M. (2014). Rapid bacterial colonization of low-density polyethylene microplastics in coastal sediment microcosms. *BMC Microbiol* 14: 232.

177 Tuson, H.H. and Weibel, D.B. (2013). Bacteria-surface interactions. *Soft Matter* 9 (18): 4368–4380.

178 Hara, T., Takeda, T.A., Takagishi, T., Fukue, K., Kambe, T., and Fukada, T. (2017). Physiological roles of zinc transporters: molecular and genetic importance in zinc homeostasis. *The Journal of Physiological Sciences* 67 (2): 283–301.

179 Maret, W. (2016). The metals in the biological periodic system of the elements: concepts and conjectures. *International Journal of Molecular Sciences* 17: 1.

180 Bryant, J.A., Clemente, T.M., Viviani, D.A. et al. (2016). Diversity and activity of communities inhabiting plastic debris in the North Pacific Gyre. *mSystems* 1: 3.

181 Laganà, P., Caruso, G., Corsi, I. et al. (2019). Do plastics serve as a possible vector for the spread of antibiotic resistance? First insights from bacteria associated to a polystyrene piece from King George Island (Antarctica). *Int J Hyg Environ Health* 222 (1): 89–100.

182 McCormick, A., Hoellein, T.J., Mason, S.A., Schluep, J., and Kelly, J.J. (2014). Microplastic is an abundant and distinct microbial habitat in an urban river. *Environmental Science & Technology* 48 (20): 11863–11871.

183 Zettler, E.R., Mincer, T.J., and Amaral-Zettler, L.A. (2013). Life in the "plastisphere": microbial communities on plastic marine debris. *Environmental Science & Technology* 47 (13): 7137–7146.

184 Dussud, C., Meistertzheim, A.L., Conan, P. et al. (2018). Evidence of niche partitioning among bacteria living on plastics, organic particles and surrounding seawaters. *Environmental Pollution* 236: 807–816.

185 Viršek, M.K., Lovšin, M.N., Koren, Š., Kržan, A., and Peterlin, M. (2017). Microplastics as a vector for the transport of the bacterial fish pathogen species Aeromonas salmonicida. *Marine Pollution Bulletin* 125 (1–2): 301–309.

186 Miao, L., Guo, S., Liu, Z. et al. (2019). Effects of Nanoplastics on Freshwater Biofilm Microbial Metabolic Functions as Determined by BIOLOG ECO Microplates. *International Journal of Environmental Research and Public Health* 16: 23.

187 Oberbeckmann, S., Osborn, A.M., and Duhaime, M.B. (2016). Microbes on a bottle: substrate, season and geography influence community composition of microbes colonizing marine plastic debris. *PLoS ONE* 11 (8): e0159289.

188 Crespo, B.G., Pommier, T., Fernández-Gómez, B., and Pedrós-Alió, C. (2013). Taxonomic composition of the particle-attached and free-living bacterial assemblages in the Northwest Mediterranean Sea analyzed by pyrosequencing of the 16S rRNA. *Microbiologyopen* 2 (4): 541–552.

189 Mohit, V., Archambault, P., Toupoint, N., and Lovejoy, C. (2014). Phylogenetic differences in attached and free-living bacterial communities in a temperate coastal lagoon during summer, revealed via high-throughput 16S rRNA gene sequencing. *Applied and Environmental Microbiology* 80 (7): 2071–2083.

190 Lee, O.O., Wang, Y., Tian, R. et al. (2014). In situ environment rather than substrate type dictates microbial community structure of biofilms in a cold seep system. *Scientific Reports* 4: 3587.

191 Debroas, D., Mone, A., and Ter Halle, A. (2017). Plastics in the North Atlantic garbage patch: a boat-microbe for hitchhikers and plastic degraders. *Science of the Total Environment* 599-600: 1222–1232.

192 De Tender, C.A., Devriese, L.I., Haegeman, A., Maes, S., Ruttink, T., and Dawyndt, P. (2015). Bacterial community profiling of plastic litter in the belgian part of the North Sea. *Environmental Science & Technology* 49 (16): 9629–9638.

193 Fei, Y., Huang, S., Zhang, H. et al. (2020). Response of soil enzyme activities and bacterial communities to the accumulation of microplastics in an acid cropped soil. *Science of the Total Environment* 707: 135634.

194 Carson, H.S., Nerheim, M.S., Carroll, K.A., and Eriksen, M. (2013). The plastic-associated microorganisms of the North Pacific Gyre. *Marine Pollution Bulletin* 75 (1–2): 126–132.

195 George, S.C. and Thomas, S. (2001). Transport phenomena through polymeric systems. *Progress in Polymer Science* 26 (6): 985–1017.

196 Ogonowski, M., Motiei, A., Ininbergs, K. et al. (2018). Evidence for selective bacterial community structuring on microplastics. *Environmental Microbiology* 20 (8): 2796–2808.

197 Lobelle, D. and Cunliffe, M. (2011). Early microbial biofilm formation on marine plastic debris. *Marine Pollution Bulletin* 62 (1): 197–200.

198 Shen, J., Du, M., Wu, Z., Song, Y., and Zheng, Q. (2019). Strategy to construct polyzwitterionic hydrogel coating with antifouling, drag-reducing and weak swelling performance. *RSC Advances* 9.

199 Nauendorf, A., Krause, S., Bigalke, N.K. et al. (2016). Microbial colonization and degradation of polyethylene and biodegradable plastic bags in temperate fine-grained organic-rich marine sediments. *Marine Pollution Bulletin* 103 (1–2): 168–178.

200 Renner, L.D. and Weibel, D.B. (2011). Physicochemical regulation of biofilm formation. *MRS Bulletin* 36 (5): 347–355.

201 Xu, X., Wang, S., Gao, F. et al. (2019). Marine microplastic-associated bacterial community succession in response to geography, exposure time, and plastic type in China's coastal seawaters. *Marine Pollution Bulletin* 145: 278–286.

202 Oberbeckmann, S., Loeder, M.G., Gerdts, G., and Osborn, A.M. (2014). Spatial and seasonal variation in diversity and structure of microbial biofilms on marine plastics in Northern European waters. *FEMS Microbiology Ecology* 90 (2): 478–492.

203 Keswani, A., Oliver, D.M., Gutierrez, T., and Quilliam, R.S. (2016). Microbial hitchhikers on marine plastic debris: human exposure risks at bathing waters and beach environments. *Marine Environmental Research* 118: 10–19.

204 Arias-Andres, M., Klümper, U., Rojas-Jimenez, K., and Grossart, H.P. (2018). Microplastic pollution increases gene exchange in aquatic ecosystems. *Environmental Pollution* 237: 253–261.

205 Aminov, R.I. (2011). Horizontal gene exchange in environmental microbiota. *Frontiers in Microbiology* 2: 158.
206 Yang, Y., Liu, G., Song, W. et al. (2019). Plastics in the marine environment are reservoirs for antibiotic and metal resistance genes. *Environment International* 123: 79–86.
207 Laverty, A.L., Darr, K., and Dobbs, F.C. (eds.). Abundance and antibiotic susceptibility of Vibrio spp. isolated from microplastics. American Geophysical Union, Ocean Sciences Meeting; 2016.
208 Poole, K. (2017). At the Nexus of antibiotics and metals: the impact of Cu and Zn on antibiotic activity and resistance. *Trends Microbiol* 25 (10): 820–832.
209 Li, L.G., Xia, Y., and Zhang, T. (2017). Co-occurrence of antibiotic and metal resistance genes revealed in complete genome collection. *ISME Journal* 11 (3): 651–662.
210 Wang, X., Bolan, N., Tsang, D.C.W., Sarkar, B., Bradney, L., and Li, Y. (2021). A review of microplastics aggregation in aquatic environment: influence factors, analytical methods, and environmental implications. *Journal of Hazardous Materials* 402: 123496.
211 Silva, M.M., Maldonado, G.C., Castro, R.O. et al. (2019). Dispersal of potentially pathogenic bacteria by plastic debris in Guanabara Bay, RJ, Brazil. *Marine Pollution Bulletin* 141: 561–568.
212 Wu, X., Pan, J., Li, M., Li, Y., Bartlam, M., and Wang, Y. (2019). Selective enrichment of bacterial pathogens by microplastic biofilm. *Water Research* 165: 114979.
213 Kirstein, I.V., Wichels, A., Krohne, G., and Gerdts, G. (2018). Mature biofilm communities on synthetic polymers in seawater - Specific or general? *Marine Environmental Research* 142: 147–154.
214 Eckert, E.M., Di Cesare, A., Kettner, M.T. et al. (2018). Microplastics increase impact of treated wastewater on freshwater microbial community. *Environmental Pollution* 234: 495–502.
215 Imran, M., Das, K.R., and Naik, M.M. (2019). Co-selection of multi-antibiotic resistance in bacterial pathogens in metal and microplastic contaminated environments: an emerging health threat. *Chemosphere* 215: 846–857.
216 Corno, G., Coci, M., Giardina, M., Plechuk, S., Campanile, F., and Stefani, S. (2014). Antibiotics promote aggregation within aquatic bacterial communities. *Frontiers in Microbiology* 5: 297.
217 Barboza, L.G.A., Lopes, C., Oliveira, P. et al. (2020). Microplastics in wild fish from North East Atlantic Ocean and its potential for causing neurotoxic effects, lipid oxidative damage, and human health risks associated with ingestion exposure. *Science of the Total Environment* 717: 134625.
218 Catarino, A.I., Macchia, V., Sanderson, W.G., Thompson, R.C., and Henry, T.B. (2018). Low levels of microplastics (MP) in wild mussels indicate that MP

ingestion by humans is minimal compared to exposure via household fibres fallout during a meal. *Environmental Pollution* 237: 675–684.

219 Li, J., Lusher, A.L., Rotchell, J.M. et al. (2019). Using mussel as a global bioindicator of coastal microplastic pollution. *Environmental Pollution* 244: 522–533.

220 Oliveri Conti, G., Ferrante, M., Banni, M. et al. (2020). Micro- and nano-plastics in edible fruit and vegetables. The first diet risks assessment for the general population. *Environmental Research* 187: 109677.

221 Kosuth, M., Mason, S.A., and Wattenberg, E.V. (2018). Anthropogenic contamination of tap water, beer, and sea salt. *PLoS ONE* 13 (4): e0194970.

222 Li, L., Luo, Y., Li, R., Zhou, Q., and Zhang, Y. (2020). Effective uptake of submicrometre plastics by crop plants via a crack-entry mode. *Nature Sustainability* 3: 929.

223 Tong, H., Jiang, Q., Hu, X., and Zhong, X. (2020). Occurrence and identification of microplastics in tap water from China. *Chemosphere* 252: 126493.

224 Paredes, M., Castillo, T., Viteri, R., Fuentes, G., and Poveda, E.M.B. (2020). Microplastics in the drinking water of the Riobamba city, Ecuador. *Scientific Review Engineering and Environmental Sciences* 28 (4 (86)): 653–663.

225 Wang, Z., Lin, T., and Chen, W. (2020). Occurrence and removal of microplastics in an advanced drinking water treatment plant (ADWTP). *Science of the Total Environment* 700: 134520.

226 Xu, Y., He, Q., Liu, C., and Huangfu, X. (2019). Are Micro- or Nanoplastics Leached from Drinking Water Distribution Systems? *Environmental Science & Technology* 53 (16): 9339–9340.

227 Zhou, W., Han, Y., Tang, Y. et al. (2020). Microplastics aggravate the bioaccumulation of two waterborne veterinary antibiotics in an edible bivalve species: potential mechanisms and implications for human health. *Environmental Science & Technology* 54 (13): 8115–8122.

228 Gamarro, G., Ryder,, and Elvevoll, O. (2020). Microplastics in Fish and Shellfish – A Threat to Seafood Safety? *Journal of Aquatic Food Product Technology* 29: 417.

229 Renzi, M., Blašković, A., Bernardi, G., and Russo, G.F. (2018). Plastic litter transfer from sediments towards marine trophic webs: a case study on holothurians. *Marine Pollution Bulletin* 135: 376–385.

230 Renzi, M., Guerranti, C., and Blašković, A. (2018). Microplastic contents from maricultured and natural mussels. *Marine Pollution Bulletin* 131 (Pt A): 248–251.

231 Liebezeit, G. and Liebezeit, E. (2013). Non-pollen particulates in honey and sugar. *Food Additives & Contaminants: Part A: Chemistry, Analysis, Control, Exposure & Risk Assessment Foreword* 30 (12): 2136–2140.

232 Akhbarizadeh, R., Moore, F., and Keshavarzi, B. (2019). Investigating microplastics bioaccumulation and biomagnification in seafood from the

Persian Gulf: a threat to human health? *Food Addit Contam Part A Chem Anal Control Expo Risk Assess* 36 (11): 1696–1708.
233 Walkinshaw, C., Lindeque, P.K., Thompson, R., Tolhurst, T., and Cole, M. (2020). Microplastics and seafood: lower trophic organisms at highest risk of contamination. *Ecotoxicology and Environmental Safety* 190: 110066.
234 Mühlschlegel, P., Hauk, A., Walter, U., and Sieber, R. (2017). Lack of evidence for microplastic contamination in honey. *Food Addit Contam Part A Chem Anal Control Expo Risk Assess* 34 (11): 1982–1989.
235 Lehner, R., Weder, C., Petri-Fink, A., and Rothen-Rutishauser, B. (2019). Emergence of Nanoplastic in the Environment and Possible Impact on Human Health. *Environmental Science & Technology* 53 (4): 1748–1765.
236 Abbasi, S., Keshavarzi, B., Moore, F. et al. (2019). Distribution and potential health impacts of microplastics and microrubbers in air and street dusts from Asaluyeh County, Iran. *Environmental Pollution* 244: 153–164.
237 Allen, S., Allen, D., Phoenix, V.R. et al. (2019). Atmospheric transport and deposition of microplastics in a remote mountain catchment. *Nature Geoscience* 12: 339.
238 Zhang, Q., Zhao, Y., Du, F., Cai, H., Wang, G., and Shi, H. (2020). Microplastic Fallout in Different Indoor Environments. *Environmental Science & Technology* 54 (11): 6530–6539.
239 Wright, S.L. and Kelly, F.J. (2017). Plastic and human health: a micro issue? *Environmental Science & Technology* 51 (12): 6634–6647.
240 Vianello, A., Jensen, R.L., Liu, L., and Vollertsen, J. (2019). Simulating human exposure to indoor airborne microplastics using a Breathing Thermal Manikin. *Scientific Reports* 9 (1): 8670.
241 Enyoh, C.E., Verla, A.W., Verla, E.N., Ibe, F.C., and Amaobi, C.E. (2019). Airborne microplastics: a review study on method for analysis, occurrence, movement and risks. *Environmental Monitoring and Assessment* 191 (11): 668.
242 Revel, M., Chatel, A., and Mouneyrac, C. (2018). Micro(nano)plastics: a threat to human health? *Current Opinion in Environmental Ence & Health* 1: 17–23.
243 Rubio, L., Marcos, R., and Hernández, A. (2020). Potential adverse health effects of ingested micro- and nanoplastics on humans. Lessons learned from in vivo and in vitro mammalian models. *The Journal of Toxicology and Environmental Health: Part B, Critical Review* 23 (2): 51–68.
244 Stock, V., Böhmert, L., Lisicki, E. et al. (2019). Uptake and effects of orally ingested polystyrene microplastic particles in vitro and in vivo. *Archives of Toxicology* 93 (7): 1817–1833.
245 Schirinzi, G.F., Pérez-Pomeda, I., Sanchís, J., Rossini, C., Farré, M., and Barceló, D. (2017). Cytotoxic effects of commonly used nanomaterials and microplastics on cerebral and epithelial human cells. *Environmental Research* 159: 579–587.

246 Wu, S., Wu, M., Tian, D., Qiu, L., and Li, T. (2020). Effects of polystyrene microbeads on cytotoxicity and transcriptomic profiles in human Caco-2 cells. *Environmental Toxicology* 35 (4): 495–506.

247 Lehner, R., Wohlleben, W., Septiadi, D., Landsiedel, R., Petri-Fink, A., and Rothen-Rutishauser, B. (2020). A novel 3D intestine barrier model to study the immune response upon exposure to microplastics. *Archives of Toxicology* 94 (7): 2463–2479.

248 Browne, M.A., Dissanayake, A., Galloway, T.S., Lowe, D.M., and Thompson, R.C. (2008). Ingested microscopic plastic translocates to the circulatory system of the mussel, Mytilus edulis (L). *Environmental Science & Technology* 42 (13): 5026–5031.

249 Huang, D., Tao, J., Cheng, M. et al. (2020). Microplastics and nanoplastics in the environment: macroscopic transport and effects on creatures. *Journal of Hazardous Materials* 407: 124399.

250 Jeong, C.B., Kang, H.M., Lee, M.C. et al. (2017). Adverse effects of microplastics and oxidative stress-induced MAPK/Nrf2 pathway-mediated defense mechanisms in the marine copepod Paracyclopina nana. *Scientific Reports* 7: 41323.

251 Jeong, C.B., Won, E.J., Kang, H.M. et al. (2016). Microplastic Size-Dependent Toxicity, Oxidative Stress Induction, and p-JNK and p-p38 Activation in the Monogonont Rotifer (Brachionus koreanus). *Environmental Science & Technology* 50 (16): 8849–8857.

252 Tan, H., Yue, T., Xu, Y., Zhao, J., and Xing, B. (2020). Microplastics Reduce Lipid Digestion in Simulated Human Gastrointestinal System. *Environmental Science & Technology* 54 (19): 12285–12294.

253 Ju, P., Zhang, Y., Zheng, Y. et al. (2020). Probing the toxic interactions between polyvinyl chloride microplastics and Human Serum Albumin by multispectroscopic techniques. *Science of the Total Environment* 734: 139219.

254 Cortés, C., Domenech, J., Salazar, M., Pastor, S., Marcos, R., and Hernández, A. (2020). Nanoplastics as a potential environmental health factor: effects of polystyrene nanoparticles on human intestinal epithelial Caco-2 cells. *Environmental Science: Nano* 7 (1): 272–285.

255 Amereh, F., Babaei, M., Eslami, A., Fazelipour, S., and Rafiee, M. (2020). The emerging risk of exposure to nano(micro)plastics on endocrine disturbance and reproductive toxicity: from a hypothetical scenario to a global public health challenge. *Environmental Pollution* 261: 114158.

9

Effects of Microplastics on Algae in Receiving Waters

Dongbo Wang*, Qizi Fu, Xuemei Li, and Xuran Liu

College of Environmental Science and Engineering, Hunan University, Changsha, PR China
* Corresponding author

9.1 Introduction

With the development of society, plastics are widely used in industrial production and daily life because of efficient and convenient characteristics [1–3]. In recent years, the production of plastic has continued to increase and reached a large quantity. For example, global plastic production increased from 260 million tons in 2007 to 370 million tons in 2019, an increase of 42.3% [4]. The extensive use and improper disposal of plastic products lead to the widespread existence of plastics in the ecological environment. In particular, plastics continually accumulate in the ecosystem from long-lasting characteristics [5–7]. The released and accumulated plastics slowly degrade and break into smaller particles under the influence of environmental physical and chemical factors (e.g., ultraviolet radiation) and biotic factors (e.g., biological degradation). The breakdown of larger plastic waste generally produces many plastic debris of different sizes such as micro-size and nano-size fragments [8, 9]. Among them, plastic particles or fragments with particle size less than 5 mm are categorized as microplastics (MPs) [10]. In addition to production by degradation of larger plastic, MPs can also be directly manufactured, and are generally used as plastic pre-produced pellets or scrubbers in personal care products [11].

Recently, many studies have revealed that MPs widely exist in all aquatic ecosystems, including terrestrial aquatic ecosystems [12–14] and the global marine ecosystem (e.g., coastal, oceanic, deep-sea) [15–17]. Although MPs are prevalent in nature, wastewater treatment plants (WWTPs) are widely regarded as the primary source of MPs entering the environment [18]. Wastewater contains a high concentration of MPs, mainly derived from the microbeads or microfibers

---

Microplastics in Urban Water Management, First Edition. Edited by Bing-Jie Ni, Qiuxiang Xu, and Wei Wei.
© 2023 John Wiley & Sons, Inc. Published 2023 by John Wiley & Sons, Inc.

contained in personal care products and cosmetics, chemical fiber clothes washing, and the breakdown of larger plastic items [19, 20]. It was reported that the content of MPs in sewage was as high as 15.7 items/L [21]. Despite the fact that activated sludge or biofilms in WWTPs can effectively adsorb and intercept MPs [22–24], considerable amounts of MPs are still discharged from large sewage discharge [25,26]. For instance, it was documented that 10.4×10^{10} pieces of MPs contained in effluent were released daily from WWTPs in China [27]. Long et al. reported that the abundance of MPs in effluent from a WWTP of a coastal city in China was as high as 1.73 items /L [28]. Accordingly, MPs also present a high concentration in the receiving water (e.g., rivers, lakes, or other water bodies that receive discharged wastewater) [29,30]. It was found that there are obvious differences in the abundance of MPs in different receiving water environments. Generally, the abundance of MPs in Lake and reservoir water (240–15 600 items/m^3) is higher than that in river water (0–8550 items/m^3) [31].

The presence of MPs inevitably touches and affects organisms contained in receiving water, which might cause potential risks to the aquatic ecosystem. For example, it was reported that MPs have negative influences on aquatic organisms, such as nervous system necrosis, metabolic disorders and oxidative stress, and gastrointestinal obstruction and intestinal damage [32–34]. As the essential primary producers, algae play key roles in substance-recycling and energy-flowing in the aquatic ecosystems through providing oxygen, organic matters, or even as carriers for growth of other aquatic organisms [35, 36]. Therefore, the effects of MPs on algae, even if small, could have serious threats for the entire food web and even aquatic ecosystems. In particular, algae have a relatively short generation cycle and are highly sensitive to chemicals [37]. Many studies have reported that MPs have negative impacts on algae growth and populations that were closely related to MPs dosages and characteristics (e.g., size, materials, and surface charge) [38, 39]. In addition, the chemical additives (e.g., bisphenol A) released from MPs degradation or the organic pollutants/heavy metals gathered at MPs might pose combined effect to algae [40, 41].

In view of the increasing presence of MPs in the receiving water and the importance of algae, it is essential to better understand the ecological toxicity of MPs to algae. Based on analyzing more than 100 papers, we comprehensively summarize the impact of MPs on algae growth and populations, discuss key factors affecting toxicity, unveil the underlying mechanisms, and analyze the combined effects of MPs with gathered or released contaminants towards algae. Based on these, the knowledge gaps and future efforts are put forward. The findings discussed in this chapter could deepen the understanding of interaction between MPs and algae, arouse people's attention to the toxicity of MPs, and call for people to strengthen MPs treatment [42].

9.2 MPs Induced Effect on the Algae: Growth and Populations

The widespread presence of MPs in the receiving water inevitably contacts and even affect algae growth and populations. Algae can synthesize organics through photosynthesis in the process of growth and reproduction, to provide the basis for water productivity [43]. At the same time, changes in algae communities can also be used as important sensitive indicators for pollutant monitoring in an aquatic ecosystem owing to its large number, diversity and wide distribution of species and community structures [37, 44, 45]. Based on the importance of algae growth and populations, there have been many reports assessing the effects of MPs on algae through various laboratory or field studies [46–48]. To date, however, there is still a lack of systematic summary and critical thought about the effects of MPs on algae. Therefore, this part systematically summarizes the effects of MPs on algae from two aspects of algae growth and populations.

9.2.1 Effects of MPs on Algae Growth

Until now, the effect of MPs on algae growth has been widely reported, but the conclusions are inconsistent (Table 9.1). Some studies found that MPs did not affect the growth of algae [53, 54]. For example, Lagarde et al. reported that polypropylene and high-density polyethylene particles with high or low concentration had no effect on the growth of freshwater microalgae during 63 days of exposure [52]. Meanwhile, some papers have reported that the presence of MPs can promote the growth of algae [55]. Canniff and Hoang observed that the algal concentration was significantly higher in the exposure media with the present of plastic microbeads than without plastic microbeads, possibly related to the colonization of algae on MPs [56]. Algae was found to adhere and develop biofilms on plastic substrates through mucilaginous secretion of extracellular polymeric substances [57] that enhanced the stability of algal community [58, 59]. In addition, Chae et al. proposed that the additive chemicals (e.g., endocrine disruptors, phthalates, and stabilizers) leached from MPs may also be the reason for stimulating the growth of algae cells [49]. However, most studies found that MPs inhibited algae photosynthesis and growth [38, 60, 61]. Wang et al. found that the growth inhibition rate (GIR) of Chlamydomonas reinhardtii (a typical algae) respectively reached 51.8% and 49.1% after exposed to 24 h and 96 h of 200 mg/L virgin polyvinyl chloride microparticles (PVC-MPs) treatment [38]. Ansari et al. found that high-density polyethylene MPs (PE-MPs, 250 mg/L) significantly inhibited the growth of microalgae Acutodesmus obliquus, with the GIR was up to 42.7% after 21 d treatment [62]. The different effects of MPs on the growth of

Table 9.1 Effects of microplastics on algae growth.

Polymer type	Polymer features	Size of particle	Concentrations tested	Test duration	Species of algae	Parameters measured and results			References
						Cell growth	Photosynthetic activity	Other effects	
PE	/	200 μm	0, 50, 100, 150, 200, 250, 300, and 350 mg/L	6 d	Marine microalga Dunaliella salina	Increased	Increased	No effect on cell morphology	[49]
PVC	/	1 μm	0, 5, 25, 50 and 100 mg/L	96 h	Marine dinoflflagellate Karenia mikimotoi	Significant inhibition	Decreased	Aggregation occurred	[46]
PS	/	5 μm and 0.1 μm	0, 1, 10, 20, 30, 40 and 50 mg/L	96 h	Freshwater microalgae Euglena gracilis	Negative effect	Significant decrease	Cell membrane was damaged, hetero-aggregation and occurred, gene expression was inhibited, and antioxidant enzyme activity increased.	[50]
PS-NH$_2$	amino-modified	0.5 μm and 2 μm	2.5 μg/mL	72 h	Marine diatom Chaetoceros neogracile	Slight decrease	No significant alterations	Esterase activity and neutral lipid content diminished, aggregation occurred, and cell morphology unaltered	[51]
HDPE and PP	/	400 μm–1000 μm	400 mg/L	>78 d	Freshwater microalgae Chlamydomonas reinhardtii	No effect	/	Heteroaggregation occurred	[52]

PS, polystyrene; PP, polypropylene; PVC, polyvinyl chloride; PE, polyethylene; HDPE, high-density polyethylene algae may be attributed to the characteristics of MPs such as dosage and size. For example, it was found that low exposure concentrations (50 mg/L) of MPs would promote the growth of Microcystis aeruginosa while high exposure concentrations (200 mg/L) cause significant inhibition [63]. The key factors affecting the toxicity of MPs will be further analyzed in Section 3.

Although reports on the effects of MPs on the growth of algae have been inconsistent, most studies have confirmed that the presence of MPs inhibits the growth of algae [62, 63]. MPs presence induced oxidative stress, damaged the cell structure, blocked chlorophyll synthesis [64], destroyed protein complex and electron donor in optical system, and reduced photoelectron transport efficiency [37]. In addition, MPs have a large surface area and strong adsorption capacity, which promoting heterogeneous aggregation of algae and MPs, made algae in a passivation state [65, 66], and impeded the absorption and utilization of light energy and carbon dioxide by algae [67]. Together, these reduced the photosynthetic efficiency of algae and thereby inhibited algae growth. MPs also affected the enzyme activity contained in algae cells [38, 63]. It was found that MPs presence decreased the activity of esterase, related to the normal catalytic hydrolysis reaction in cells, thus disturbing the normal metabolism of algal cells [68]. These factors help explain why MPs presence would inhibit algae growth.

Recently, some researchers have taken a closer look at the effects of MPs on algae at the cellular and molecular levels. These studies found that MPs presence affected the composition and accumulation of lipids that play important roles in photosynthesis [69]. Seoane et al. found that the neutral lipid content of algal cells exposed to MPs after 72 h was 50% lower than that of the control group [51]. During the stationary phase of growth, algae usually accumulate lipids as energy reserves [70]. The reduction of lipid reserve occurs because algae cells need to adjust energy metabolism to adapt to toxic stress caused by MPs exposure, so as to maintain normal growth, photosynthesis, and even membrane integrity [71]. In addition, MPs presence also affects the expression of key genes, inhibits the synthesis of related enzymes, and thereby induces the adverse effects on algae [52]. Xiao et al. found that the expression of the CTR1 gene that encodes protein kinase was significantly down-regulated after exposure to 5 μm PS-MPs [50].

It is worth noting that in the face of the harmful effects of MPs, algae also self-regulate, enabling them to effectively overcome the harm caused by MPs. After exposure to MPs, algae thicken cell walls, promote homogenous aggregation, and form hetero-aggregates, thereby reducing the adverse effects of MPs on normal growth and photosynthetic activity of algae [60]. These countermeasures make the inhibitory effect of MPs on algae growth seem temporary and unsustainable. Li et al. observed that after exposure to MPs-containing toxic environment, the content of EPS (plays an important role in thickening the cell wall [72]) in Chlamydomonas reinhardtii was much higher than that in the control group at the later stage [61].

The increased EPS secretion also contributes to heterogeneous aggregation of algae and MPs, thus facilitating recovery and reproduction of algal cells in toxic environments. MPs presence also enhanced the activity of antioxidant enzymes (such as superoxide dismutase, catalase, and malondialdehyde synthesis) [73, 74], so as to protect algae cells from damage caused by excessive oxidative stress [75].

9.2.2 Effects of MPs on Algae Populations

MPs presence significantly affects algae populations including species composition, abundance, and biomass of an algae community. MPs can produce different toxicity to different algae from differing sensitivity. The growth of dominant algae under MPs stress may lead to population succession [76]. Li et al. found that MPs presence increased the abundance of cyanobacteria but reduced the abundance of brown algae compared with the control group. Moreover, the reduction of algal biomass by MPs was also observed [77]. Cunha et al. found that long-term exposure to MPs contamination influenced algal cell size and reduced biomass yield by 82% without affecting cell abundance [53]. The gradual change of community structure brings certain ecological risks to the receiving water [44]. For example, when cyanobacteria become the dominant population, they produce a large number of toxins and cause water blooms [78, 79].

In addition, MPs also affect the settlement and distribution of algae populations. For instance, polystyrene MPs had a significant negative effect on the colonization of ice algae Fragillariopsis cylindrus on sea ice. When cocultured with MPs, the number of algae cells colonized on sea ice (average of 6%) was less than that of the control group (average of 31%) [80]. Furthermore, the phenomenon that MPs can disturb the anthozoan-algae symbiotic relationship was observed, which may mean the loss of habitat and nutrient sources of certain specific algal species [81]. On the other hand, previous studies have shown that the surface of MPs can form biofilm communities [82], and algae are an important part of colonizing biological communities [58]. Miao et al. reported that MPs can be used as a unique habitat for the colonization and growth of cyanobacteria [83]. Meanwhile, other algae species such as green algae, diatoms and dinoflagellates have also been continuously found to be able to colonize the surface of MPs [84–86]. Many studies have reported that MPs can trigger algae to secrete protein-rich EPS [87, 88] that promote the formation of heterogeneous aggregates between algae and MPs because of the viscosity of EPS [66]. Long et al. proved that the sinking rate of aggregates increased when MPs were added to algae, although some large algae still sank as free cells [89]. The aggregates contained in a water environment are important food sources for high trophic levels, so the rapid settlement of hetero-aggregation between MPs and algae may have a significant impact on aquatic biota [90].

MPs in receiving water may also regulate the algae population by inhibiting the feeding capacity of aquatic organisms and reducing the nutrient availability rate of algae, both of which might cause the increase of algae population. The prevalence of MPs and their similar size and appearance to plankton significantly increased the possibility of MPs ingestion by aquatic fauna [56]. Cole et al. found that the ingested MPs may accumulate or even block the digestive tract of zooplankton, causing false satiety and reduced feeding rate [91]. Coppock et al. also reported that zooplankton exposed to MPs reduce their intake of algae similar to the shape or size of MPs [92]. The long-term exposure of MPs inevitably has an adverse impact on the body of herbivores, resulting in mechanical damage, inflammatory reaction [93], and even death [94]. A decrease in the population density of herbivores might increase algal populations by reducing the likelihood that plankton would be ingested [95].

9.3 Factors Affecting Toxicity

MPs are widely present in receiving water, but their effects on algae (i.e., growth and populations) may be inconsistent, depending on their different properties. According to their different origin, MPs can be divided into two types: (1) primary MPs ((diameter in the micron range), mainly derived from microparticles or microbeads in cosmetics, paints, and textiles [96]; and (2) MPs, largely from the breakdown of larger plastic waste (diameter greater than 5 mm) under the action of light and mechanical wear [97]. In addition, receiving water contains a variety of MPs of different materials such as polyethylene (PE), polypropylene (PP), polyvinyl chloride (PVC), polystyrene (PS), and polyester (PET) [98]. Several types and materials of MPs lead to different properties such as size and surface chemistry, which induced different influence on the algae growth and populations (Table 9.2).

9.3.1 Dosage

Multiple studies have revealed the effects of MPs on algae. Although the effects of MPs on algae are inconsistent, the toxic effects of MPs on algae are usually dose dependent. For example, Li et al. founded that with the increase of PS-MPs concentration from 5 to 100 mg/L, the growth inhibition rate of freshwater microalgae Chlamydomonas reinhardtii increased from 26.6 to 49.2% [61]. Zhang et al. also found that compared with 5 mg/L PVC treatment, 50 mg/L PVC treatment reduced chlorophyll content from 93 to 80% and reduced the electron transport efficiency of the chloroplast photosynthesis system from 95 to 68% [100]. Exposure of algae to high concentration of MPs can lead to intracellular heterogeneous

Table 9.2 Factors of microplastics affecting toxicity.

Polymer type	Size of particle	Concentrations tested	Test duration	Species of algae	Result	Factors of microplastics affecting toxicity			References
						size-dependent	dose-dependent	Type-correlation	
PS	1–5 μm	10 mg/L	96 h exposure	Green marine microalgae platymonas helgolandica var. tsingtaoensis and green freshwater microalgae Scenedesmus quadricauda	Cellular internalization	√			[65]
HDPE, PP, PVC	100 μm	0, 5, 10, 15, 25, 100, 125, 200, and 250 mg/L	21 days	Microalgae Acutodesmus obliquus	Inhibition of growth photosynthetic activity		√	√	[62]
PVC, PS, PE	3 μm	10, 25, 50, 100 and 200 mg/L	96 h	Freshwater microalgae Microcystis aeruginosa	Growth inhibition, antioxidant enzyme activity, cell membrane integrity		√	√	[63]
PS	1 μm and 100 nm	5 mg/L	96 h	Freshwater microalgae Microcystis aeruginosa	Oxidative stress and growth promotion	√			[55]
PS	300–600 nm	5, 25, 50 and 100 mg/L	10 d	freshwater microalgae Chlamydomonas reinhardtii	Inhibition of growth photosynthetic activity		√		[61]
PE, PP, PS, PVC, and PET	750 μm and 1500 μm	10, 250, 500, 750, and 1000 mg/L	3 d	freshwater microalgae Chlorella sp	Growth inhibition		√	√	[99]

PS, polystyrene; PP, polypropylene; PVC, polyvinyl chloride; PE, polyethylene; HDPE, high-density polyethylene; PET: polyethylene terephthalate aggregation and intracellular internalization, resulting in cell membrane damage and hindering the exchange of gases and nutrients, which may be the cause of the dose-dependent effect [39]. It is worth noting that current laboratory test concentrations of MPs are often higher than environmental concentrations [101]. Therefore, it remains to be further studied whether MPs have a substantial effect on algae at environmental concentration.

9.3.2 Size

Particle size is a key factor affecting the toxicity of MPs to algae, and with the decrease of particle size, the toxicity of MPs on algae is greater. Zhang et al. reported that m-PVC MPs (m-PVC, average diameter 1 μm) inhibited the growth of Marine microalgae Skeletonema costatum, with the maximum growth inhibition rate reached 39.7%, while b-PVC MPs (b-PVC, average diameter 1 mm) had no effect on the growth of algae. Similarly, Sjollema, Redondo-Hasselerharm (104) reported that smaller MPs (average diameter 0.5 μm) reduced the average algae cell density by 45% compared to the unexposed control, while larger MPs (average diameter 6 μm) reduced the average algae cell density only by 13%. The possible reason is that smaller MPs are more likely to enter the algae cell through the cell wall to cause cell internalization, or adhere to the cell surface to cause heterogeneous aggregation, thus affecting the normal nutrient absorption and material exchange between algal cells and the outside world [65]. However, it is worth noting that the current measurement of MPs particle size in water is not accurate, which may affect the prediction of cytotoxic effects of algae in water. More efforts should be made to improve the measurement of MPs in water.

9.3.3 Materials

Moreover, the composition and properties of materials also affect the toxicity of MPs. There are a variety of MPs of different materials such as polyethylene, polypropylene, polyvinyl chloride, polystyrene, and polyester contained in receiving water. These different material compositions result in different toxic effects of MPs on algae. For example, compared with high-density polyethylene (HDPE), polypropylene (PP) has higher cohesion and viscosity to extracellular polymer substances, so the heterogeneous aggregation ability of PP to algae is much stronger than HDPE [52]. In addition to the material composition of MPs during production, MPs also undergo various changes such as aging after release into the environment. Some environmental factors, such as physical effects (ultraviolet radiation, temperature change, weathering) and microbial decomposition, can accelerate the aging of MPs when released into the water environment [102,103], leading to a change of their surface morphology and characteristics such as color, structure, specific surface area, oxygen functional group [104–106]. For instance, Cai et al. reported that new polar oxidative functional groups (i.e., the hydroxyl and carbonyl groups) developed in MPs after three months of UV irradiation [103].

Among various environmental factors, photooxidation degradation is considered to be one of the main factors leading to plastics aging in the environment [107], which changes the surface properties of MPs and causes different charge polarities. The positive MPs have higher interaction and toxicity to algae cells than negative MPs [108–110], which might arise from the carboxyl and sulfate groups of anionic cellulose in cell walls repelling the negative charge MPs [111, 112] but adsorbs the positive charge MPs through electrostatic interaction, hydrogen bond and hydrophobic interaction [113, 114]. The adsorption of MPs prevents light and dioxide from entering the cells and inhibits algae photosynthesis [115]. MPs also promote ROS production, causing oxidative stress, apoptosis, and DNA damage [116]. These observations proved that the aging MPs might have stronger toxic effect on algae [117].

In addition, the potential influence of produced filtrate during the aging of MPs on algae cannot be ignored. Many additives such as plasticizer, heat stabilizer, lubricant, filler, colorant, and impact modifier released by MPs during aging have a great impact on the toxicity to algae cells [118]. Luo et al. reported that the leached fluorescent additives from polyurethane sponge MPs decreased photosynthetic activity of freshwater algae by 5.1%. The maximum quantum efficiency of photosystem II decreased with increasing leached concentrations of fluorescent additives [40]. All these results showed that different MPs have different chemical and physical properties, and several types or forms of MPs exhibits different toxic effects on algae. However, to date, there are few studies exploring the effects of distinct types of MPs on the same algae genus. More efforts are needed to fill this knowledge gap.

9.4 Combined Effects of MPs with Contaminants towards Algae

In addition to MPs, there are also many other contaminants such as heavy metals, antibiotics and other emerging contaminations contained in receiving water [119]. MPs are usually used as carriers for the migration of these pollutants in the water environment because of their small particle size, large specific surface area, and strong hydrophobicity. Moreover, MPs are vulnerable to mechanical erosion in water, resulting in pores and cracks on their surfaces that tend to adhere to and interact with certain pollutants in the environment [41, 120]. It is worth noting that the concentration of contaminants accumulated on some MPs surfaces can be much higher than the environmental concentration of contaminants as a result of the strong adsorption capacity of MPs. Hence, in receiving water, MPs including their exudates may form compound contamination with the adsorbed contaminants, thus increasing their toxicity to algae. The combined toxicity of MPs and their adsorbents is usually related to the properties of MPs, the adsorption and desorption capacity of MPs with contaminants, and the characteristics of contaminants (Table 9.3) [121].

9.4.1 Antibiotics

In recent years, antibiotics are not only widely used in the prevention and treatment of human and animal diseases as the main anti-infective drugs, but also used in livestock, poultry, and aquaculture as important growth promoters [127]. After use, antibiotics are typically incompletely absorbed and metabolized by organisms, and traditional wastewater treatment has low removal efficiency of antibiotics. These facts together result in the continuous accumulation of antibiotics in receiving water [128]. It was reported that the concentration of antibiotics in receiving water varies from few ng/L to hundreds of μg/L [129]. Sulfonamides and quinolones are the most common antibiotics found in the aquatic environment [130]. It was documented that the median concentration of these two most frequently detected antibiotics in China's surface water is higher than 5 ng/L [131]. The presence of antibiotics inevitably poses a risk to aquatic ecology, particularly vulnerable algae [132]. For example, sulfonamides and fluoroquinolones can inhibit protein biosynthesis and lead to growth retardation in algae, while tetracyclines and macrolides can affect chloroplast gene translation and thylakoid membrane protein synthesis, and reduce the content of chlorophyll in cells [133, 134].

The existence of large quantities of MPs and antibiotics in receiving water inevitably leads to their mutual interference and influence. Generally, MPs, as a carrier, adsorbs and aggregates various antibiotics. For example, PE-MPs can adsorb ciprofloxacin (CIP, a very commonly detected antibiotic) through electrostatic and hydrophobic interaction [135], while PET can adsorb sulfamethazine through van der Waals interaction [136]. Many studies have shown that the combined toxicity of MPs and antibiotics can be antagonistic or synergistic. For example, compared with single MPs systems containing 50 mg/L PP-MPs, the binary exposure groups for MPs and SMX (sulfamethoxazole) reduced the GIR values from 21.5 to 7.9% [137]. Similarly, Feng et al. found that positively charged polystyrene MPs (PS-NH$_2$) and tetracycline (TET) exhibited antagonistic growth inhibition on marine microalgae. The combination of TET and PS-NH$_2$ resulted in an 18.4% reduction of cell membrane damage, compared with that exposed to single PS-NPs (39.6%) [124].

Table 9.3 Studies of combined effects of microplastics and other contaminants on algae.

Species of algae	Microplastics			Contaminants			Exposure time	Combined effects	Comparison of combined toxicity and single toxicity of microplastics	References
	Type	Size	Concentration	Type	Concentration					
Marine microalgae Tetraselmis chuii	PE	1–5 μm	0.046–1.472 mg/L	Cu	0.02–0.64 mg/L		96 h	Reduction in microalgae population growth	No significant difference	[122]
Freshwater microalgae Chlorella vulgaris	PS	0.5 μm	1, 5, 50, 100, and 1000 mg/L	Metal mixture (Cu, Zn, Mn)	Mn: 0.35 mg/L; Cu: 0.285 mg/L; Zn: 0.130 mg/L		20 d	Reduced growth by 47.83–49.57% and chlorophyll a concentration by 44.75–50.25%	Synergism, combined toxicity is higher than single toxicity	[123]
Marine microalgae Skeletonema costatum	PS	1 μm	200 mg/L	Tetracycline (Tet)	100 mg/L		24 h	Cell membrane damaged	Antagonism, combined toxicity is lower than single toxicity	[124]

Marine microalgae Skeletonema costatum	PE, PS, PVC, PVC800	PE: 74 μm; PS: 74 μm; PVC, 74 μm; PVC800: 1 μm	0.05 g/L	Triclosan (TCS)	0.3 mg/L	96 h	Growth inhibition andoxidative stress	Antagonism	[75]
Microalgae Chlorella pyrenoidosa	PS	0.1, 0.55 and 5 μm	0.5, 1, 2, 4, 8, 16, 32, 64 mg/L	Dibutyl phthalate (DBP)	0.25, 0.5, 1, 2, 4, 8, 16 mg/L	96 h	Reduced growth and affect cell morphology	Antagonism at low concentrations of DBP; synergism at high concentrations of DBP	[125]
Microalgae Chlorella pyrenoidosa	PE1000, PE, PA1000, PA, PS,	PE1000: 13 μm; PE: 150 μm; PA1000: 13 μm; PA: 150 μm; PS: 150 μm	50 mg/L	Nonylphenol (NP)	2 mg/L	96 h	Growth inhibition; photosynthetic activity reduction; oxidative stress	Antagonism	[126]

PE: polyethylene; PS: polystyrene; PVC: polyvinylchloride; PA: polyamide

Interestingly, although MPs adsorb and aggregate various antibiotics, the combined toxicity of MPs and antibiotics is more antagonistic than synergistic. The adsorption behavior of MPs on antibiotics and the changes of surface characteristics between particles and cells (such as hydrophobicity and charge) are the two main antagonistic mechanisms. The adsorption of pollutants by MPs reduces the concentration of pollutants in water, resulting in the decline of bioavailability of pollutants by algae [138]. On the other hand, the adsorbed pollutants may enhance the hydrophobicity of MPs, leading to the aggregation and deposition of MPs in water and then reducing the possibility of contact with algae [137]. The adsorption of pollutants may also produce charge repulsion between MPs particles and algal cells and reduce the interaction between cells and particles, which alleviated the toxicity of pollutants [124].

9.4.2 Heavy Metals

In addition to antibiotics, heavy metals such as copper (Cu), lead (Pb), and zinc (Zn) are also widely found in receiving water [139], mainly produced from industrial and mining production [140]. In many parts of the world, the average concentrations of Cr, Ni, Hg. and Cd in surface water are much higher than the maximum allowable value of drinking water [141]. The presence of heavy metals would bring great harm to the survival and reproduction of aquatic organisms from the characteristics of nonbiodegradability, easy enrichment, high activity, and strong toxicity [142]. La Rocca et al. found that 5 ppm Cd almost completely inhibited algal growth and the activity of PSII after a week exposure [143]. Mechanism analysis showed that heavy metals can accumulate on the cell wall of green algae, inhibit the absorption of light and carbon dioxide, and lead to the growth retardation and even death of algae [144].

In recent years, the adsorption of heavy metals by MPs has attracted people's attention. Heavy metals such as Ni, Cd, Pb, Cu, Zn, and Cr were detected on the MPs collected from water environment. The enrichment of metals by MPs can reach several hundred mg/kg, and with the extension of time, the enrichment concentration is also greater [145]. For example, Liu et al. found that the MPs samples collected from waters around Hong Kong contained extremely high concentrations of Cd (676 mg/kg), Pb (1742 mg/kg) and Zn (1712 mg/kg), and the concentrations of Cr, Cu, and Ni were positively correlated with the specific surface area of the MPs [144].

Numerous studies have identified the adverse impacts of MPs-heavy metal combinations on algae, including growth inhibition, chlorophyll reduction, cell wall disruption, and cell membrane damage [123]. For instance, Liao et al. found that co-exposure of PS-MPs and cadmium (Cd) could induce morphological changes of algae through physical damage and oxidative stress, resulting in an increase in the number of intracellular vacuoles [146]. Like other pollutants, the combined toxicity of MPs and heavy metals to algae is also complex. MPs can adsorb different heavy metals, and the combined toxicity of various kinds of MPs and heavy metals is significantly different. In contrast to single toxicity tests (GIR_{Cd} = 85.14%), MPs significantly reduced the toxicity of heavy metals to algae in the combined groups, with the value of GIR $_{MPs\text{-}PS\ +Cd}$ = 27.55% [147]. Wang et al. reported that MPs + Pb treatments had a synergistic effect on promoting algae aggregation and growth compared with the corresponding concentration of Pb-only group [148]. In comparison, there were no differences on marine microalgae growth inhibition between exposure to copper only, and to copper combined with PS–MPs. This finding might indicate that MPs do not influence the copper induced toxicity on algae in the range of tested concentrations [122].

Similar to antibiotics, the effect of MPs and heavy metals on algae may be related to adsorption and aggregation. The combination of MPs and heavy metals might reduce the bioavailability of heavy metals to algae, especially if the content of MPs and heavy metals are poorly absorbed by algae, the toxicity of heavy metals in the presence of MPs would be reduced. However, if the binding particles produced are absorbed by algal cells, the incorporation and accumulation of heavy metals by MPs may also lead to increased exposure and toxicity to algae [122]. To date, the information about the comprehensive toxicity of MPs and heavy metals to algae is still limited. It is necessary to understand the adsorption/desorption mechanism of heavy metals on MPs, including several heavy metals and MPs. Overall, more endeavors are needed to explore the co-toxicity of MPs and heavy metals to algae.

9.4.3 Other Emerging Contaminations

In recent years, there have been a wide variety of emerging contaminations (ECs) such as pharmaceuticals and personal care products (PPCPs), endocrine disruptors (EDCs), disinfection by-products (DBPs), environmental hormones, and pesticides discharged into the receiving water [149]. It was reported that the concentrations of most ECs in the aquatic environment typically ranged from ng/L to μg/L, even up to mg/L [150]. These ECs are usually easily bioaccumulated in aquatic organisms, affecting organism growth and health even at low concentrations [151]. In particular, the discharged ECs in receiving water can be bioconcentrated in algae, which could cause oxidative damage, cell cycle arrest, even apoptosis [152, 153]. It was reported that the concentration of organochlorine pesticides in macroalgae can reach up to 67.9 ± 51.8 ng/g [154]. Algal cell chloroplasts are the main targets of pharmaceuticals attack. Machado et al. found that after 72 h exposure with 27 μg/L triclosan, the contents of chlorophyll a and chlorophyll b of freshwater algal cells decreased by 25% and 29% respectively [155].

The wide presence of MPs and ECs in receiving water also leads to their interaction. The small size and hydrophobic surface of MPs enhance their ability to adsorb ECs. There were many studies reported that the presence of a variety of ECs on MPs, such as nonylphenol, Phthalate, bisphenol, and 17β-estradiol [156–158]. For example, Hirai et al. found that many high concentrations of ECs were observed in plastic fragments from urban beaches, with bisphenol A up to 730 ng/g and nonylphenol up to 3936 ng/g [159]. However, compared with other pollutants such as antibiotics and heavy metals, there are few studies to unveil the joint toxicity of ECs and MPs to algae. Yang et al. studied the comprehensive toxicity of MPs and nonylphenol (NP) on Chlorella and observed the phenomenon of antagonism. Both single MPs and NP can inhibit the growth of algae and cause oxidative stress, but the comprehensive toxic effect shows that the existence of MPs has a positive effect on reducing the toxicity of NP, possibly from adsorption of NP by MPs [126]. On the contrary, Qu et al. found that the EC_{50} of methamphetamine to freshwater algae reduced from 0.77 to 0.32 mg/L when MPs were added as a co-pollutant of methamphetamine. This result indicated that the combined toxicity against algae in methamphetamine-MPs treatment was higher than that in methamphetamine treatment alone [42]. The enhancement of combined toxicity may be attributed to the fact that MPs damage algal cell structure and promote the absorption of pollutants by algal cells [160].

9.5 Research Gap and Perspective

Despite many studies that have been performed, the understanding of the effects of MPs on algae is still inadequate. Many studies have demonstrated that MPs in receiving water can interact with algae and have some beneficial/harmful effects. Given the importance of algae in aquatic ecosystems and the refractory nature of MPs, more research is needed to fill in existing knowledge gaps. In view of the current situation of MPs and algae research, further research on the following aspects is urgently needed:

1) At present most experiments are carried out in laboratories, using model plastics and model algae to assess toxicity. Most of the MPs used in the experiments are commercial plastics, which may differ from environmental plastic fragments in some properties such as weathering, composition, and size range. Many studies have confirmed that the properties of MPs have a significant impact on their toxicity, explaining why many studies have inconsistent results. Therefore, many current research results may not accurately reveal the toxicity of MPs to algae in the environment. In order to more accurately evaluate the toxicity effect of MPs on algae in the environment, on the one hand,

we need to further analyze the characteristics of MPs in the environment; on the other hand, it may be better to evaluate the toxicity of MPs directly in the real environment rather than in the laboratory.
2) In addition, the concentrations of MPs assessed in many studies are much higher than the current environmental concentrations, so it is impossible to accurately assess the toxic effects of MPs on algae in existing systems. Although MPs emission is expected to rise further as plastic use increases, many of the concentrations evaluated are still too high. More realistic environmental concentrations should be used when assessing the toxicity of MPs.
3) To date, many studies have explored the effects of MPs on algae through traditional methods. There are few studies to explore the toxicity mechanism of MPs to algal cells at the molecular level. Hence, more advanced techniques such as metagenomics and proteomics should be adopted for better analysis.
4) Many studies only revealed the impact of a single MP on single algae but lack research on the compound impact of multiple MPs and contaminants on algae and the cascade effect on the whole aquatic ecosystem. In fact, there are a variety of contaminants coexisting in receiving water, so it is necessary to explore the combined toxic effect of various contaminants to better evaluate the micro-toxic effect of MPs on algae in the actual water environment.

References

1 Jiang, C., Yin, L., Li, Z. et al. (2019). Microplastic pollution in the rivers of the Tibet Plateau. *Environmental Pollution* 249: 91–98.
2 Li, J., Lusher, A.L., Rotchell, J.M. et al. (2019). Using mussel as a global bioindicator of coastal microplastic pollution. *Environmental Pollution* 244: 522–533.
3 Zhang, D., Liu, X., Huang, W. et al. (2020). Microplastic pollution in deep-sea sediments and organisms of the Western Pacific Ocean. *Environmental Pollution* 259: 113948.
4 Peng, G., Zhu, B., Yang, D. et al. (2017). Microplastics in sediments of the Changjiang Estuary, China. *Environmental Pollution* 225: 283–290.
5 Eriksen, M., Lebreton, L.C., Carson, H.S. et al. (2014). Plastic pollution in the World's Oceans: more than 5 Trillion plastic pieces weighing over 250,000 Tons Afloat at Sea. *PLoS One* 9 (12): 111913.
6 Law, K.L. (2017). Plastics in the Marine environment. *Annual Review of Marine Science* 9 (1): 205–229.
7 Moharir, R.V. and Kumar, S. (2019). Challenges associated with plastic waste disposal and allied microbial routes for its effective degradation: a comprehensive review. *Journal of Cleaner Production* 208: 65–76.

8 Cole, M., Lindeque, P., Halsband, C. et al. (2011). Microplastics as contaminants in the marine environment: a review. *Marine Pollution Bulletin* 62 (12): 2588–2597.
9 Andrady, A.L. (2011). Microplastics in the marine environment. *Marine Pollution Bulletin* 62 (8): 1596–1605.
10 Lavender, L.K. and Thompson, R.C. (2014). Microplastics in the seas: concern is rising about widespread contamination of the marine environment by microplastics. *Science* 345: 61–93.
11 Wang, W.F., Gao, H., Jin, S.C. et al. (2019). The ecotoxicological effects of microplastics on aquatic food web, from primary producer to human: a review. *Ecotoxicology and Environmental Safety* 173: 110–117.
12 Ambrosini, R., Azzoni, R.S., Pittino, F. et al. (2019). First evidence of microplastic contamination in the supraglacial debris of an alpine glacier. *Environmental Pollution* 253: 297–301.
13 Matthias, T. and Kerstin, F.E. (2020). Microplastics in a deep, dimictic lake of the North German plain with special regard to vertical distribution patterns. *Environmental Pollution* 267: 115507.
14 Zhefan, R., Xiangyang, G., Xiaoyun, X. et al. (2021). Microplastics in the soil-groundwater environment: aging, migration, and co-transport of contaminants – a critical review. *Journal of Hazardous Materials* 419: 126455.
15 Desforges, J.P., Galbraith, M., Dangerfield, N. et al. (2014). Widespread distribution of microplastics in subsurface seawater in the NE Pacific Ocean. *Marine Pollution Bulletin* 79 (1–2): 94–99.
16 Lusher, A.L., Tirelli, V., O'Connor, I. et al. (2015). Microplastics in Arctic polar waters: the first reported values of particles in surface and sub-surface samples. *Scientific Reports* 5: 14947.
17 Woodall, L.C., Sanchez-Vidal, A., Canals, M. et al. (2014). The deep sea is a major sink for microplastic debris. *Royal Society Open Science* 1 (4): 140317.
18 Wei, W., Huang, Q.S., Sun, J. et al. (2019). Revealing the mechanisms of polyethylene microplastics affecting anaerobic digestion of waste activated sludge. *Environmental Science & Technology* 53 (16): 9604–9613.
19 Napper, I.E. and Thompson, R.C. (2016). Release of synthetic microplastic plastic fibres from domestic washing machines: effects of fabric type and washing conditions. *Marine Pollution Bulletin* 112 (1–2): 39–45.
20 So, W.K., Chan, K., and Not, C. (2018). Abundance of plastic microbeads in Hong Kong coastal water. *Marine Pollution Bulletin* 133: 500–505.
21 Zhang, Z. and Chen, Y. (2020). Effects of microplastics on wastewater and sewage sludge treatment and their removal: a review. *Chemical Engineering Journal* 382: 122955.
22 Murphy, F., Ewins, C., Carbonnier, F. et al. (2016). Wastewater treatment works (WwTW) as a source of microplastics in the aquatic environment. *Environmental Science & Technology* 50 (11): 5800–5808.

23 Franco, A.A., Arellano, J.M., Albendín, G. et al. (2021). Microplastic pollution in wastewater treatment plants in the city of Cádiz: abundance, removal efficiency and presence in receiving water body. *Science of the Total Environment* 776: 145795.

24 Carr, S.A., Liu, J., and Tesoro, A.G. (2016). Transport and fate of microplastic particles in wastewater treatment plants. *Water Research* 91: 174–182.

25 Menendez, D., Alvarez, A., Peon, P. et al. (2021). From the ocean to jellies forth and back? Microplastics along the commercial life cycle of red algae. *Marine Pollution Bulletin* 168: 112402.

26 Vardar, S., Onay, T.T., Demirel, B. et al. (2021). Evaluation of microplastics removal efficiency at a wastewater treatment plant discharging to the sea of Marmara. *Environmental Pollution* 289: 117862.

27 Tang, N., Liu, X., and Xing, W. (2020). Microplastics in wastewater treatment plants of Wuhan, central China: abundance, removal, and potential source in household wastewater. *Science of the Total Environment* 745: 141026.

28 Long, Z., Pan, Z., Wang, W. et al. (2019). Microplastic abundance, characteristics, and removal in wastewater treatment plants in a coastal city of China. *Water Research* 155: 255–265.

29 Yuyao, X., Shun, C.F.K., Matthew, J. et al. (2021). Microplastic pollution in Chinese urban rivers: the influence of urban factors. *Resources, Conservation & Recycling* 173: 105686.

30 Julia, D., David, G., Yves, P. et al. (2021). Microplastic pollution of worldwide lakes. *Environmental Pollution* 284: 117075.

31 Zeqian, Z., Chenning, D., Li, D. et al. (2021). Microplastic pollution in the Yangtze River Basin: heterogeneity of abundances and characteristics in different environments. *Environmental Pollution* 287: 117580.

32 Jia, T., Xingzhen, N., Zhi, Z. et al. (2018). Acute microplastic exposure raises stress response and suppresses detoxification and immune capacities in the scleractinian coral Pocillopora damicornis. *Environmental Pollution* 243: 66–74.

33 Rochman, C.M., Hoh, E., Kurobe, T. et al. (2013). Ingested plastic transfers hazardous chemicals to fish and induces hepatic stress. *Scientific Reports* 3 (1): 1–7.

34 Ziajahromi, S., Kumar, A., Neale, P.A. et al. (2018). Environmentally relevant concentrations of polyethylene microplastics negatively impact the survival, growth and emergence of sediment-dwelling invertebrates. *Environmental Pollution* 236: 425–431.

35 Stevenson, J. (2014). Ecological assessments with algae: a review and synthesis. *Journal of Phycology* 50 (3): 437–461.

36 Bai, X. and Acharya, K. (2017). Algae-mediated removal of selected pharmaceutical and personal care products (PPCPs) from Lake Mead water. *Science of the Total Environment* 581–582: 734–740.

37 Wu, Y., Guo, P., Zhang, X. et al. (2019). Effect of microplastics exposure on the photosynthesis system of freshwater algae. *Journal of Hazardous Materials* 374: 219–227.

38 Wang, Q., Wangjin, X., Zhang, Y. et al. (2020). The toxicity of virgin and UV-aged PVC microplastics on the growth of freshwater algae Chlamydomonas reinhardtii. *Science of the Total Environment* 749: 141603.

39 Sun, L., Sun, S., Bai, M. et al. (2021). Internalization of polystyrene microplastics in Euglena gracilis and its effects on the protozoan photosynthesis and motility. *Aquatic Toxicology* 236: 105840.

40 Luo, H., Xiang, Y., He, D. et al. (2019). Leaching behavior of fluorescent additives from microplastics and the toxicity of leachate to Chlorella vulgaris. *Science of the Total Environment* 678: 1–9.

41 Fotopoulou, K.N. and Karapanagioti, H.K. (2012). Surface properties of beached plastic pellets. *Marine Environmental Research* 81: 70–77.

42 Qu, H., Ma, R., Barrett, H. et al. (2020). How microplastics affect chiral illicit drug methamphetamine in aquatic food chain? From green alga (Chlorella pyrenoidosa) to freshwater snail (Cipangopaludian cathayensis). *Environment International* 136: 105480.

43 Xin, X., Huang, G., and Zhang, B. (2021). Review of aquatic toxicity of pharmaceuticals and personal care products to algae. *Journal of Hazardous Materials* 410: 124619.

44 Hallegraeff, G.M. (2010). Ocean climate change, Phytoplankton community responses, and harmful algal blooms: a formidable predictive challenge. *Journal of Phycology* 46 (2): 220–235.

45 Theroux, S., Mazor, R.D., Beck, M.W. et al. (2020). Predictive biological indices for algae populations in diverse stream environments. *Ecological Indicators* 119: 106421.

46 Zhao, T., Tan, L., Huang, W. et al. (2019). The interactions between micro polyvinyl chloride (mPVC) and marine dinoflagellate Karenia mikimotoi: the inhibition of growth, chlorophyll and photosynthetic efficiency. *Environmental Pollution* 247: 883–889.

47 Yu, H., Zhang, X., Hu, J. et al. (2020). Ecotoxicity of polystyrene microplastics to submerged carnivorous Utricularia vulgaris plants in freshwater ecosystems. *Environmental Pollution* 265: 114830.

48 Yang, W., Gao, P., Li, H. et al. (2021). Mechanism of the inhibition and detoxification effects of the interaction between nanoplastics and microalgae Chlorella pyrenoidosa. *Science of the Total Environment* 783: 146919.

49 Chae, Y., Kim, D., and An, Y.J. (2019). Effects of micro-sized polyethylene spheres on the marine microalga Dunaliella salina: focusing on the algal cell to plastic particle size ratio. *Aquatic Toxicology* 216: 105296.

50 Xiao, Y., Jiang, X., Liao, Y. et al. (2020). Adverse physiological and molecular level effects of polystyrene microplastics on freshwater microalgae. *Chemosphere* 255: 126914.

51 Seoane, M., Gonzalez-Fernandez, C., Soudant, P. et al. (2019). Polystyrene microbeads modulate the energy metabolism of the marine diatom Chaetoceros neogracile. *Environmental Pollution* 251: 363–371.

52 Lagarde, F., Olivier, O., Zanella, M. et al. (2016). Microplastic interactions with freshwater microalgae: Hetero-aggregation and changes in plastic density appear strongly dependent on polymer type. *Environmental Pollution* 215: 331–339.

53 Cunha, C., Lopes, J., Paulo, J. et al. (2020). The effect of microplastics pollution in microalgal biomass production: a biochemical study. *Water Research* 186: 116370.

54 Sjollema, S.B., Redondo-Hasselerharm, P., Leslie, H.A. et al. (2016). Do plastic particles affect microalgal photosynthesis and growth? *Aquatic Toxicology* 170: 259–261.

55 Di, W., Ting, W., Jing, W. et al. (2021). Size-dependent toxic effects of polystyrene microplastic exposure on Microcystis aeruginosa growth and microcystin production. *Science of the Total Environment* 761: 143265.

56 Canniff, P.M. and Hoang, T.C. (2018). Microplastic ingestion by Daphnia magna and its enhancement on algal growth. *Science of the Total Environment* 633: 500–507.

57 Sarmah, P. and Rout, J. (2018). Efficient biodegradation of low-density polyethylene by cyanobacteria isolated from submerged polyethylene surface in domestic sewage water. *Environmental Science And Pollution Research* 25 (33): 33508–33520.

58 Gross, M., Zhao, X., Mascarenhas, V. et al. (2016). Effects of the surface physico-chemical properties and the surface textures on the initial colonization and the attached growth in algal biofilm. *Biotechnology for Biofuels* 9 (1): 1–14.

59 Yokota, K., Waterfield, H., Hastings, C. et al. (2017). Finding the missing piece of the aquatic plastic pollution puzzle: Interaction between primary producers and microplastics. *Limnology and Oceanography Letters* 2 (4): 91–104.

60 Mao, Y., Ai, H., Chen, Y. et al. (2018). Phytoplankton response to polystyrene microplastics: perspective from an entire growth period. *Chemosphere* 208: 59–68.

61 Li, S., Wang, P., Zhang, C. et al. (2020). Influence of polystyrene microplastics on the growth, photosynthetic efficiency and aggregation of freshwater microalgae Chlamydomonas reinhardtii. *Science of the Total Environment* 714: 136767.

62 Ansari, F.A., Ratha, S.K., Renuka, N. et al. (2021). Effect of microplastics on growth and biochemical composition of microalga Acutodesmus obliquus. *Algal Research* 56: 102296.

63 Zheng, X., Zhang, W., Yuan, Y. et al. (2021). Growth inhibition, toxin production and oxidative stress caused by three microplastics in Microcystis aeruginosa. *Ecotoxicology and Environmental Safety* 208: 111575.

64 Geoffroy, L., Dewez, D., Vernet, G. et al. (2003). Oxyfluorfen toxic effect on S. obliquus evaluated by different photosynthetic and enzymatic biomarkers. *Archives of Environmental Contamination and Toxicology* 45 (4): 445–452.

65 Chen, Y., Ling, Y., Li, X. et al. (2020). Size-dependent cellular internalization and effects of polystyrene microplastics in microalgae P. helgolandica var. tsingtaoensis and S. quadricauda. *Journal of Hazardous Materials.* 399: 123092.

66 Wang, S., Wang, Y., Liang, Y. et al. (2020). The interactions between microplastic polyvinyl chloride and marine diatoms: physiological, morphological, and growth effects. *Ecotoxicology and Environmental Safety* 203: 111000.

67 Long, M., Paul-Pont, I., Hégaret, H. et al. (2017). Interactions between polystyrene microplastics and marine phytoplankton lead to species-specific hetero-aggregation. *Environmental Pollution* 228: 454–463.

68 Zhao, W., Zheng, Z., Zhang, J. et al. (2019). Allelopathically inhibitory effects of eucalyptus extracts on the growth of Microcystis aeruginosa. *Chemosphere* 225: 424–433.

69 GI, A., HA, J., and OS, J. (2020). Polystyrene microplastics decrease accumulation of essential fatty acids in common freshwater algae. *Environmental Pollution* 263: 114425.

70 Zienkiewicz, K., Du, Z.Y., Ma, W. et al. (2016). Stress-induced neutral lipid biosynthesis in microalgae – molecular, cellular and physiological insights. *Biochimica et Biophysica Acta (BBA)-Molecular and Cell Biology of Lipids* 1861 (9): 1269–1281.

71 Xu, Z., Yan, X., Pei, L. et al. (2007). Changes in fatty acids and sterols during batch growth of Pavlova viridis in photobioreactor. *Journal of Applied Phycology* 20 (3): 237–243.

72 Mathimani, T. and Mallick, N. (2018). A comprehensive review on harvesting of microalgae for biodiesel – key challenges and future directions. *Renewable and Sustainable Energy Reviews* 91: 1103–1120.

73 Lu, T., Zhu, Y., Xu, J. et al. (2018). Evaluation of the toxic response induced by azoxystrobin in the non-target green alga Chlorella pyrenoidosa. *Environmental Pollution* 234: 379–388.

74 Fan, G., Zhou, J., Zheng, X. et al. (2018). Growth inhibition of microcystis aeruginosa by Copper-based MOFs: performance and physiological effect on algal cells. *Applied Organometallic Chemistry* 32 (12): 4600.

75 Zhu, Z.-L., Wang, S.-C., Zhao, F.-F. et al. (2019). Joint toxicity of microplastics with triclosan to marine microalgae Skeletonema costatum. *Environmental Pollution* 246: 509–517.

76 Lim, Y.K., Baek, S.H., Seo, M.H. et al. (2020). Succession of a phytoplankton and mesozooplankton community in a coastal area with frequently occurring algal blooms. *Journal of Sea Research* 166: 101961.

77 Troost, T.A., Desclaux, T., Leslie, H.A. et al. (2018). Do microplastics affect marine ecosystem productivity? *Marine Pollution Bulletin* 135: 17–29.

78 Rastogi, R.P., Madamwar, D., and Incharoensakdi, A. (2015). Bloom dynamics of cyanobacteria and their toxins: environmental health impacts and mitigation strategies. *Frontiers In Microbiology* 6: 1254.

79 McAllister, T.G., Wood, S.A., and Hawes, I. (2016). The rise of toxic benthic Phormidium proliferations: a review of their taxonomy, distribution, toxin content and factors regulating prevalence and increased severity. *Harmful Algae* 55: 282–294.

80 Hoffmann, L., Eggers, S.L., Allhusen, E. et al. (2020). Interactions between the ice algae Fragillariopsis cylindrus and microplastics in sea ice. *Environment International* 139: 105697.

81 Okubo, N., Takahashi, S., and Nakano, Y. (2018). Microplastics disturb the anthozoan-algae symbiotic relationship. *Marine Pollution Bulletin* 135: 83–89.

82 Zettler, E.R., Mincer, T.J., and Amaral-Zettler, L.A. (2013). Life in the "plastisphere": microbial communities on plastic marine debris. *Environmental Science & Technology* 47 (13): 7137–7146.

83 Miao, L., Wang, P., Hou, J. et al. (2019). Distinct community structure and microbial functions of biofilms colonizing microplastics. *Science of the Total Environment* 650: 2395–2402.

84 Demestre, M., Masó, M., Fortuño, J.M. et al. (2016). Microfouling communities from pelagic and benthic marine plastic debris sampled across Mediterranean coastal waters. *Scientia Marina* 80 (1): 117–127.

85 Reisser, J., Shaw, J., Hallegraeff, G. et al. (2014). Millimeter-sized marine plastics: a new pelagic habitat for microorganisms and invertebrates. *PLoS One* 9 (6): 100289.

86 Davarpanah, E. and Guilhermino, L. (2019). Are gold nanoparticles and microplastics mixtures more toxic to the marine microalgae Tetraselmis chuii than the substances individually? *Ecotoxicology and Environmental Safety* 181: 60–68.

87 Shiu, R.F., Vazquez, C.I., Chiang, C.Y. et al. (2020). Nano- and microplastics trigger secretion of protein-rich extracellular polymeric substances from phytoplankton. *Science of the Total Environment* 748: 141469.

88 Song, C., Liu, Z., Wang, C. et al. (2020). Different interaction performance between microplastics and microalgae: the bio-elimination potential of Chlorella sp. L38 and Phaeodactylum tricornutum MASCC-0025. *Science of the Total Environment* 723: 138146.

89 Long, M., Moriceau, B., Gallinari, M. et al. (2015). Interactions between microplastics and phytoplankton aggregates: impact on their respective fates. *Marine Chemistry* 175: 39–46.

90 Boerger, C.M., Lattin, G.L., Moore, S.L. et al. (2010). Plastic ingestion by planktivorous fishes in the North Pacific Central Gyre. *Marine Pollution Bulletin* 60 (12): 2275–2278.

91 Cole, M., Lindeque, P., Fileman, E. et al. (2013). Microplastic ingestion by zooplankton. *Environmental Science & Technology* 47 (12): 6646–6655.

92 Coppock, R.L., Galloway, T.S., Cole, M. et al. (2019). Microplastics alter feeding selectivity and faecal density in the copepod, Calanus helgolandicus. *Science of the Total Environment* 687: 780–789.

93 Cole, M., Lindeque, P., Fileman, E. et al. (2015). The impact of polystyrene microplastics on feeding, function and fecundity in the marine copepod Calanus helgolandicus. *Environmental Science & Technology* 49 (2): 1130–1137.

94 Jemec, A., Horvat, P., Kunej, U. et al. (2016). Uptake and effects of microplastic textile fibers on freshwater crustacean Daphnia magna. *Environmental Pollution* 219: 201–209.

95 Jiang, J., Pang, S.Y., Ma, J. et al. (2012). Oxidation of phenolic endocrine disrupting chemicals by potassium permanganate in synthetic and real waters. *Environmental Science & Technology* 46 (3): 1774–1781.

96 Napper, I.E., Bakir, A., Rowland, S.J. et al. (2015). Characterisation, quantity and sorptive properties of microplastics extracted from cosmetics. *Marine Pollution Bulletin* 99 (1–2): 178–185.

97 Laskar, N. and Kumar, U. (2019). Plastics and microplastics: a threat to environment. *Environmental Technology & Innovation* 14: 100352.

98 Zhang, Z., Gao, S.H., Luo, G. et al. (2022). The contamination of microplastics in China's aquatic environment: occurrence, detection and implications for ecological risk. *Environmental Pollution* 296: 118737.

99 Miloloža, M., Bule, K., Ukić, Š. et al. (2021). Ecotoxicological determination of microplastic toxicity on algae chlorella sp.: response surface modeling approach. *Water, Air, & Soil Pollution* 232 (8): 1–16.

100 Zhang, C., Chen, X., Wang, J. et al. (2017). Toxic effects of microplastic on marine microalgae Skeletonema costatum: interactions between microplastic and algae. *Environmental Pollution* 220: 1282–1288.

101 Prata, J.C., da Costa, J.P., Lopes, I. et al. (2019). Effects of microplastics on microalgae populations: a critical review. *Science of the Total Environment* 665: 400–405.

102 de Souza Machado, A.A., Lau, C.W., Kloas, W. et al. (2019). Microplastics can change soil properties and affect plant performance. *Environmental Science & Technology* 53 (10): 6044–6052.

103 Cai, L., Wang, J., Peng, J. et al. (2018). Observation of the degradation of three types of plastic pellets exposed to UV irradiation in three different environments. *Science of the Total Environment* 628: 740–747.

104 Ge, J., Li, H., Liu, P. et al. (2021). Review of the toxic effect of microplastics on terrestrial and aquatic plants. *Science of the Total Environment* 791: 148333.

105 Veerasingam, S., Mugilarasan, M., Venkatachalapathy, R. et al. (2016). Influence of 2015 flood on the distribution and occurrence of microplastic pellets along the Chennai coast, India. *Marine Pollution Bulletin* 109 (1): 196–204.

106 Liu, Y., Zhang, J., Cai, C. et al. (2020). Occurrence and characteristics of microplastics in the Haihe River: an investigation of a seagoing river flowing through a megacity in northern China. *Environmental Pollution* 262: 114261.

107 Song, Y.K., Hong, S.H., Jang, M. et al. (2017). Combined effects of UV exposure duration and mechanical abrasion on microplastic fragmentation by polymer type. *Environmental Science & Technology* 51 (8): 4368–4376.

108 Bergami, E., Pugnalini, S., Vannuccini, M.L. et al. (2017). Long-term toxicity of surface-charged polystyrene nanoplastics to marine planktonic species Dunaliella tertiolecta and Artemia franciscana. *Aquatic Toxicology* 189: 159–169.

109 Nel, A.E., Madler, L., Velegol, D. et al. (2009). Understanding biophysicochemical interactions at the nano-bio interface. *Nature Materials* 8 (7): 543–557.

110 Thiagarajan, V., Iswarya, V.P.A.J. et al. (2019). Influence of differently functionalized polystyrene microplastics on the toxic effects of P25 TiO2 NPs towards marine algae chlorella sp. *Aquatic Toxicology* 207: 208–216.

111 Liu, Y., Li, W., Lao, F. et al. (2011). Intracellular dynamics of cationic and anionic polystyrene nanoparticles without direct interaction with mitotic spindle and chromosomes. *Biomaterials* 32 (32): 8291–8303.

112 Dausend, J., Musyanovych, A., Dass, M. et al. (2008). Uptake mechanism of oppositely charged fluorescent nanoparticles in HeLa cells. *Macromolecular Bioscience* 8 (12): 1135–1143.

113 Casado, M.P., Macken, A., and Byrne, H.J. (2013). Ecotoxicological assessment of silica and polystyrene nanoparticles assessed by a multitrophic test battery. *Environment International* 51: 97–105.

114 Atul, A., Santimukul, S., Charalambos, K. et al. (2010). Surface-charge-dependent cell localization and cytotoxicity of cerium oxide nanoparticles. *ACS nano* 4 (9): 5321–5331.

115 Bhattacharya, P., Lin, S., Turner, J.P. et al. (2010). Physical adsorption of charged plastic nanoparticles affects algal photosynthesis. *The Journal of Physical Chemistry C* 114 (39): 16556–16561.

116 Mukherjee, S.P., Lyng, F.M., Garcia, A. et al. (2010). Mechanistic studies of in vitro cytotoxicity of poly(amidoamine) dendrimers in mammalian cells. *Toxicology and Applied Pharmacology* 248 (3): 259–268.

117 Fu, D., Zhang, Q., Fan, Z. et al. (2019). Aged microplastics polyvinyl chloride interact with copper and cause oxidative stress towards microalgae Chlorella vulgaris. *Aquatic Toxicology* 216: 105319.

118 Gewert, B., Plassmann, M.M., and MacLeod, M. (2015). Pathways for degradation of plastic polymers floating in the marine environment. *Environmental Science-Processes & Impacts* 17 (9): 1513–1521.

119 Rathi, B.S., Kumar, P.S., and Vo, D.N. (2021). Critical review on hazardous pollutants in water environment: occurrence, monitoring, fate, removal technologies and risk assessment. *Science of the Total Environment* 797: 149134.

120 Beckingham, B. and Ghosh, U. (2017). Differential bioavailability of polychlorinated biphenyls associated with environmental particles: microplastic in comparison to wood, coal and biochar. *Environmental Pollution* 220: 150–158.

121 Dokyung, K., Yooeun, C., and Youn-Joo, A. (2017). Mixture toxicity of Nickel and microplastics with different functional groups on Daphnia magna. *Environmental Science & Technology* 51 (21): 12852–12858.

122 Davarpanah, E. and Guilhermino, L. (2015). Single and combined effects of microplastics and copper on the population growth of the marine microalgae Tetraselmis chuii. *Estuarine, Coastal and Shelf Science* 167: 269–275.

123 Tunali, M., Uzoefuna, E.N., Tunali, M.M. et al. (2020). Effect of microplastics and microplastic-metal combinations on growth and chlorophyll a concentration of Chlorella vulgaris. *Science of the Total Environment* 743: 140479.

124 Feng, L.J., Shi, Y., Li, X.Y. et al. (2020). Behavior of tetracycline and polystyrene nanoparticles in estuaries and their joint toxicity on marine microalgae Skeletonema costatum. *Environmental Pollution* 263: 114453.

125 Li, Z., Yi, X., Zhou, H. et al. (2020). Combined effect of polystyrene microplastics and dibutyl phthalate on the microalgae Chlorella pyrenoidosa. *Environmental Pollution* 257: 113604.

126 Yang, W., Gao, X., Wu, Y. et al. (2020). The combined toxicity influence of microplastics and nonylphenol on microalgae Chlorella pyrenoidosa. *Ecotoxicology and Environmental Safety* 195: 110484.

127 Bilal, M., Mehmood, S., Rasheed, T. et al. (2020). Antibiotics traces in the aquatic environment: Persistence and adverse environmental impact. *Current Opinion in Environmental Science & Health* 13: 68–74.

128 Grenni, P., Ancona, V., and Barra Caracciolo, A. (2018). Ecological effects of antibiotics on natural ecosystems: a review. *Microchemical Journal* 136: 25–39.

129 Huang, F., An, Z., Moran, M.J. et al. (2020). Recognition of typical antibiotic residues in environmental media related to groundwater in China (2009–2019). *Journal of Hazardous Materials* 399: 122813.

130 Li, S., Shi, W., Liu, W. et al. (2018). A duodecennial national synthesis of antibiotics in China's major rivers and seas (2005–2016). *Science of the Total Environment* 615: 906–917.

131 Li, Z., Li, M., Zhang, Z. et al. (2020). Antibiotics in aquatic environments of China: a review and meta-analysis. *Ecotoxicology and Environmental Safety* 199: 110668.

132 Zhang, T., Jiang, B., Xing, Y. et al. (2022). Current status of microplastics pollution in the aquatic environment, interaction with other pollutants, and effects on aquatic organisms. *Environmental Science And Pollution Research* 29 (12): 16830–16859.

133 Chen, J.Q. and Guo, R.X. (2012). Access the toxic effect of the antibiotic cefradine and its UV light degradation products on two freshwater algae. *Journal of Hazardous Materials* 209: 520–523.

134 Gomaa, M., Zien-Elabdeen, A., Hifney, A.F. et al. (2021). Phycotoxicity of antibiotics and non-steroidal anti-inflammatory drugs to green algae Chlorella sp. and Desmodesmus spinosus: assessment of combined toxicity by Box–Behnken experimental design. *Environmental Technology & Innovation* 23: 101586.

135 Atugoda, T., Wijesekara, H., Werellagama, D.R.I.B. et al. (2020). Adsorptive interaction of antibiotic ciprofloxacin on polyethylene microplastics: implications for vector transport in water. *Environmental Technology & Innovation* 19: 100971.

136 Guo, X., Liu, Y., and Wang, J. (2019). Sorption of sulfamethazine onto different types of microplastics: a combined experimental and molecular dynamics simulation study. *Marine Pollution Bulletin* 145: 547–554.

137 Li, X., Luo, J., Zeng, H. et al. (2022). Microplastics decrease the toxicity of sulfamethoxazole to marine algae (Skeletonema costatum) at the cellular and molecular levels. *Science of the Total Environment* 824: 153855.

138 You, X., Cao, X., Zhang, X. et al. (2021). Unraveling individual and combined toxicity of nano/microplastics and ciprofloxacin to Synechocystis sp. at the cellular and molecular levels. *Environment International* 157: 106842.

139 Yan, C., Qu, Z., Wang, J. et al. (2022). Microalgal bioremediation of heavy metal pollution in water: recent advances, challenges, and prospects. *Chemosphere* 286: 131870.

140 Luo, M., Zhang, Y., Li, H. et al. (2022). Pollution assessment and sources of dissolved heavy metals in coastal water of a highly urbanized coastal area: the role of groundwater discharge. *Science of the Total Environment* 807: 151070.

141 Zamora-Ledezma, C., Negrete-Bolagay, D., Figueroa, F. et al. (2021). Heavy metal water pollution: a fresh look about hazards, novel and conventional remediation methods. *Environmental Technology & Innovation* 22: 101504.

142 Niu, L., Li, J., Luo, X. et al. (2021). Identification of heavy metal pollution in estuarine sediments under long-term reclamation: ecological toxicity, sources and implications for estuary management. *Environmental Pollution* 290: 118126.

143 Rocca, N., Andreoli, C., Giacometti, G.M. et al. (2009). Responses of the Antarctic microalga Koliella antarctica (Trebouxiophyceae, Chlorophyta) to cadmium contamination. *Photosynthetica* 47 (3): 471–479.

144 Liu, Y., Hu, H., Wang, Y. et al. (2020). Effects of heavy metals released from sediment accelerated by artificial sweeteners and humic acid on a green algae Scenedesmus obliquus. *Science of the Total Environment* 729: 138960.

145 Liu, S., Huang, J., Zhang, W. et al. (2022). Microplastics as a vehicle of heavy metals in aquatic environments: a review of adsorption factors, mechanisms, and biological effects. *Journal of Environmental Management* 302: 113995.

146 Liao, Y., Jiang, X., Xiao, Y. et al. (2020). Exposure of microalgae Euglena gracilis to polystyrene microbeads and cadmium: perspective from the physiological and transcriptional responses. *Aquatic Toxicology* 228: 105650.

147 Wang, Z., Fu, D., Gao, L. et al. (2021). Aged microplastics decrease the bioavailability of coexisting heavy metals to microalga Chlorella vulgaris. *Ecotoxicology and Environmental Safety* 217: 112199.

148 Wang, S., Li, Q., Huang, S. et al. (2021). Single and combined effects of microplastics and lead on the freshwater algae Microcystis aeruginosa. *Ecotoxicology and Environmental Safety* 208: 111664.

149 Alvarez-Ruiz, R., Pico, Y., and Campo, J. (2021). Bioaccumulation of emerging contaminants in mussel (Mytilus galloprovincialis): influence of microplastics. *Science of the Total Environment* 796: 149006.

150 Klosterhaus, S.L., Grace, R., Hamilton, M.C. et al. (2013). Method validation and reconnaissance of pharmaceuticals, personal care products, and alkylphenols in surface waters, sediments, and mussels in an urban estuary. *Environment International* 54: 92–99.

151 Chen, Y., Lin, M., and Zhuang, D. (2022). Wastewater treatment and emerging contaminants: Bibliometric analysis. *Chemosphere* 297: 133932.

152 Lin, W., Zhang, Z., Chen, Y. et al. (2023). The mechanism of different cyanobacterial responses to glyphosate. *Journal of Environmental Sciences* 125: 258–265.

153 Anton, F.A., Ariz, M., and Alia, M. (1993). Ecotoxic effects of four herbicides (glyphosate, alachlor, chlortoluron and isoproturon) on the algae Chlorella pyrenoidosa Chick. *Science of the Total Environment* 134: 845–851.

154 Qiu, Y.W., Zeng, E.Y., Qiu, H. et al. (2017). Bioconcentration of polybrominated diphenyl ethers and organochlorine pesticides in algae is an important contaminant route to higher trophic levels. *Science of the Total Environment* 579: 1885–1893.

155 Machado, M.D. and Soares, E.V. (2021). Toxicological effects induced by the biocide triclosan on Pseudokirchneriella subcapitata. *Aquatic Toxicology* 230: 105706.

156 Liu, X., Xu, J., Zhao, Y. et al. (2019). Hydrophobic sorption behaviors of 17beta-Estradiol on environmental microplastics. *Chemosphere* 226: 726–735.

157 Mato, Y., Isobe, T., Takada, H. et al. (2001). Plastic resin pellets as a transport medium for toxic chemicals in the marine environment. *Environmental Science & Technology* 35 (2): 318–324.

158 Santana-Viera, S., Montesdeoca-Esponda, S., Torres-Padron, M.E. et al. (2021). An assessment of the concentration of pharmaceuticals adsorbed on microplastics. *Chemosphere* 266: 129007.

159 Hirai, H., Takada, H., Ogata, Y. et al. (2011). Organic micropollutants in marine plastics debris from the open ocean and remote and urban beaches. *Marine Pollution Bulletin* 62 (8): 1683–1692.

160 Yi, X., Chi, T., Li, Z. et al. (2019). Combined effect of polystyrene plastics and triphenyltin chloride on the green algae Chlorella pyrenoidosa. *Environmental Science And Pollution Research* 26 (15): 15011–15018.

10

Effects of Microplastics on Aquatic Organisms in Receiving Waters

Gabriela Kalčíková and Ula Rozman

University of Ljubljana, Faculty of Chemistry and Chemical Technology, Ljubljana, Slovenia

10.1 Introduction

Research on microplastics began half a century ago when E. J. Carpenter and K. L. Smith accidentally found plastic pellets in a neuston net in the western Sargasso Sea [1]. Since then, much of the attention has focused on the presence, fate, and effects of microplastics in the marine environment. Recently, researchers have begun to focus their attention on the sources and pathways of microplastics into the environment and how various ecosystems may be impacted before microplastics reach their final sink: the oceans. As a result, current research efforts have shifted to terrestrial and freshwater ecosystems [2].

Many of the microplastics found in aquatic ecosystems originate from the land and are transported into freshwaters by runoff [3] or via atmospheric deposition [4]. Land-based sources contribute to 80% of marine plastic litter [5], with improperly disposed plastics [6], abrasion of plastic coatings [7], and car tires [8] among the most important sources. Wastewater discharges also contribute significantly to the input of microplastics into freshwaters [9]. Fragments, microbeads, and fibers (produced during laundry) are often found downstream of wastewater treatment plants [10].

Once microplastics occur in freshwaters, most of them are transported to the oceans via rivers, while some remain in freshwater ecosystems [11]. In freshwaters, they are distributed according to their density, but the majority of microplastics are made of lightweight polymers (e.g., polyethylene (PE), polypropylene (PP)) and float on the water surface [12]. Over time, their density increases due to aging, so they sink and eventually reach the sediment [13]. On their way from the water surface to the sediment, they can interact with and affect a variety of aquatic organisms.

Microplastics in Urban Water Management, First Edition. Edited by Bing-Jie Ni, Qiuxiang Xu, and Wei Wei.
© 2023 John Wiley & Sons, Inc. Published 2023 by John Wiley & Sons, Inc.

While most organisms studied in microplastics research are (in)vertebrates, much less research has been done on interactions with microorganisms and plants [14]. However, these interactions may be critical to the fate and effects of microplastics in the aquatic environment. For example, microorganisms colonize microplastics and form a biofilm on their surface [15]. As a result, microplastics become more bioavailable and may be preferentially ingested by invertebrates or fish [16]. Aquatic plants also interact with microplastics, and although recent research indicates that they are not strongly affected [17, 18], plants can accumulate high numbers of microplastics [19]. This means that microplastics trapped by plant biomass can then be readily eaten by herbivores, making plants a viable pathway for microplastics to enter aquatic food webs [20, 21].

In addition to the negative effects caused by mechanical stress (e.g., gut blockage [22]), microplastics can also affect organisms through chemical stress. Chemicals added during their manufacture (e.g., additives [23]) or adsorbed to their surface from the surrounding environment can be transported together with microplastics and released when environmental conditions change (e.g., lowering of pH support leaching of metals from microplastics [24]). The presence of such chemicals can significantly alter their ecotoxicological profile [24, 25].

The aim of this chapter is to summarize the current knowledge on the occurrence of microplastics in aquatic ecosystems, their abundance in water bodies and sediments, their interactions, and effects on aquatic organisms, to highlight the sources of toxicity, and to identify research gaps and future priorities. Microplastics occur in a variety of forms, are composed of a range of polymers, differ in the degree of degradation, and change their properties over time and in the presence of contaminants. Therefore, it is important to study interactions and impacts of diverse types of microplastics on organisms along the food web to understand the complexity of the overall problem.

10.1.1 Occurrences in Water and Sediment

While marine microplastics have received the most public and scientific attention in recent decades, the study of microplastic pollution of freshwater ecosystems is important as well because inland waters are closer to the sites of plastic production, use, and disposal [2] and therefore they can contain and carry a significant number of microplastics [26].

Microplastics in aquatic ecosystems have been found virtually everywhere; floating on the water surface, in the water body or in sediments of lakes/ponds, rivers/streams, and estuaries [27, 28]. They are made of various polymers, mainly polyethylene (PE), polypropylene (PP), polystyrene (PS), polyamide (PA), polyvinyl chloride (PVC) and they occur in a variety of shapes and sizes. Common shapes include pellets, spheres, fibers, fragments, foams, or films/sheets [29, 30]. Their shapes and sizes depend on the original form of the primary microplastic particle or the plastic item from which microplastics were generated, the processes that caused the degradation/fragmentation, and time spent in the environment [31].

Their concentration in receiving waters is highly heterogeneous, temporally fluctuating [32], and highly variable even within one site [29]. It ranges from almost none (e.g., 0.028/m^3 in the Tamar estuary, England [33]) to several thousand particles/m^3 (e.g., 100 000/m^3 in the canal waters of Amsterdam, the Netherlands [34]). Similarly, concentrations of microplastics in sediments reach several thousand particles/kg [35]. Microplastics with higher density than water settle, while microplastics with lower density preferentially float on the water surface or in the water column [36]. However, biofouling (i.e., growth of microorganisms and development of biofilm) and adsorption of organic matter can increase particle density, so low-density microplastics may also be present in sediments [37]. Therefore, sediment is considered a long-term sink for microplastics [35].

The presence of microplastics in water and sediment depends on many factors, but high population densities and industrial activities have been assessed as the critical factors affecting the abundance of microplastics in aquatic ecosystems [26, 38] (Table 10.1). Consequently, the plastics industry, urban dust, and municipal wastewaters contribute significantly to the pollution of freshwaters by microplastics.

Table 10.1 Recent studies on monitoring of microplastics in urban waters.

Location	Sampling mesh size	Microplastics concentrations	References
Tuojiang River basin, China	50 μm	From 912 to 3395 items/m^3	[52]
Urban lagoon – Bizerte lagoon, Southern Mediterranean Sea	300 μm	Mean concentration 453 items/m^3	[53]
20 urban lakes and urban reaches of the Hanjiang River and Yangtze River of Wuhan, China	50 μm	From 1660 to 8925 items/m^3	[54]
Outlet of the Sucy-en-Brie catchment, Greater Paris region, France	80 μm	*29 000 items/m^3	[55]
Tsurumi River, Japan	10 μm	From 300 to 1240 items/m^3	[56]
Lake Ontario, Canada	10 μm	*800 items/m^3	[57]
Chesapeake Bay, USA	330 μm	From 0.007 to 1.245 items/m^3	[58]
Qing river, China	10 μm	*170 –260 items/m^3	[59]
Tecolutla estuary, Gulf of Mexico, Mexico	1.2 μm	*151 000 items/ m^3	[60]

*Concentrations originally expressed as items/L.

Locations around facilities of plastic industry are often highly polluted by plastic pellets. These preproduction pellets are the building blocks for further plastic production [39]. They are often released into the environment from production, during transportation, and before the final product manufacturing by careless handling [40].

One of the most important sources of microplastics in water bodies is urban dust. This dust may contain a high number of microplastics originating from abrasion of road paints, abrasion of car tires, construction and building materials, and road littering [7, 41]. Microplastics from car tires contribute up to 50% of the total microplastics [42], making them the dominant microplastic species in road dust. Stormwater runoff can carry these microplastics into water bodies during rain events [43], or they can enter water bodies through wet deposition [44].

In addition, municipal wastewaters can contain high numbers of microplastics, and they represent one of the most important pathway for microplastics to enter aquatic ecosystems [10]. Microplastics in municipal wastewaters are mainly composed of synthetic textile fibers, cosmetic microbeads, and consumer products that are accidentally or intentionally flushed down the toilet [45, 46]. Wastewater treatment plants can efficiently retain microplastics, but the efficiency depends on the treatment technology [47]. Effluents from tertiary treatment may contain almost no microplastics [48], while effluents from secondary treatment can still carry a significant number of them [49]. Consequently, higher concentrations of microplastics have been found in rivers and streams downstream of wastewater treatment plants compared to upstream [50, 51].

The key factors affecting the assessment of the occurrence and distribution of microplastics in water bodies include sampling locations and sampling methodology, as well as sample processing and microplastic identification. Different sampling and detection methods have been used to date, making it difficult to compare the data obtained. Overview of sampling methods has shown that microplastics of 1 μm in size can be detected, but sampling below 50 μm is rarely performed [26, 28]. The selection of a sampling method is crucial for the assessment of microplastic abundance in the environment, as, for example, sampling 80 μm instead of 330 μm resulted in a 250-fold higher concentration of microplastics [29].

Although research on microplastics in water bodies is progressing intensively, there are still considerable gaps in our knowledge. Harmonization of sampling and identification methods, large-scale sampling, long-term monitoring, and a global perspective of microplastic pollution have not yet been implemented.

10.1.2 The Concerns about Potential Ecological Risks

The growing concern about potential ecological risks of microplastic pollution calls for extensive ecological risk assessment, however the risk assessment process for microplastics remains considerably difficult because of the high heterogeneity of microplastics in the environment [61].

To conduct risk assessment, it is necessary to properly quantify exposures and effects. However, the concentration and effects of various microplastics contain high uncertainty. Sampling of specific fractions of microplastics (e.g., preferably large microplastics) as well as expressing results as number of particles per surface area does not provide sufficient information about exposure of organisms [61]. Similarly, the effects of microplastics on various organisms are based on results of laboratory studies where microplastics with limited environmental relevance are often used (e.g., small spherical microplastics) in very high concentration that also do not meet environmentally relevant threshold [62]. Consequently, the current ecological risk assessment is calculated from exposure data where large particles were considered while the hazard assessment used data for smaller microplastics [6].

In addition, the effects of microplastics on organisms can significantly vary and it depends on microplastics size, shape, aging, presence of additives, and adsorbed pollutants. Therefore, microplastics can cause physical or/and chemical stress in organisms as a result of their chemical and physical characteristics that also vary with time [23, 24]. For example, 100 mg/L PE microplastics in the form of fragments (100 μm) slightly affected roots of duckweed *Lemna minor* by the mechanical abrasion. When the same particles were aged in natural waters their surface become smooth and they did not affect roots to any extent. However, when the aged particles were exposed to pollution (in this study silver ion) their toxicity significantly increased and the effect was noticed also under environmentally relevant concentrations [24]. In this case, it suggests that PE microplastics do not represent hazard while the same particles after aging and with adsorbed pollutants do.

Because of the high variability and uncertainty, it is difficult to draw a direct conclusion about the overall microplastics ecological risks. However, it is very likely, that there are already hotspots where risk can occur. If microplastic emissions to the environment will remain the same, the ecological risks of microplastics may be widespread within a century [6].

10.2 Into the Food Chain of Aquatic Animals

Because of the widespread occurrence of microplastics across aquatic ecosystems, microplastics can interact with and affect many organisms of different trophic levels of a food web. Microplastics have been detected in a number of aquatic organisms, including zooplankton, mollusks, crustaceans, and fish [63]. They are highly bioavailable as they occupy the same size fraction as sediments and some planktonic organisms, making them accessible to a wide range of organisms despite the feeding strategy. Organisms can therefore ingest microplastics if they nonselectively filter substantial amounts of water and sediment, or inadvertently if they mistake it for food [64]. In addition, some recent studies suggest that microplastics can be transferred between trophic levels from plants to herbivores and from herbivores to carnivores [65] (Figure 10.1).

10.2.1 Accumulation

In contrast to the considerable body of research on the occurrence of microplastics in aquatic organisms, information on the mechanisms of microplastic uptake, translocation to organs, cellular transport pathways, and elimination kinetics is scarce. Consequently, conclusions on their bioaccumulation in organisms and biomagnification between trophic levels are limited [63].

Microplastics can be taken up from the water column or sediment via several physiological pathways. They can enter an organism through dermal surfaces, respiratory surfaces (e.g., gills), or directly by ingestion [66]. Their ingestion is proportional to their concentration [67–69] and there is no significant difference in the amount of ingested particles composed of natural materials (e.g., cellulose) or synthetic ones [70].

The size of microplastics is crucial for ingestion, as larger microplastics cannot enter the buccal cavity and thus cannot be ingested [71]. Particle shape can also

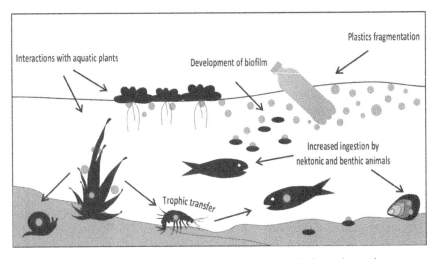

Figure 10.1 Schematic view of interactions between microplastics and aquatic organisms.

affect the fate of microplastics in organisms. Irregularly shaped particles such as fragments and fibers are more easily attached and retained in the digestive tract compared to smooth particles [72, 73]. For example, shape-dependent accumulation was observed in the gut of zebrafish in the order of fibers (8.0 mg/mg) > fragments (1.7 mg/mg) > beads (0.5 mg/mg) [74]. The life stage of the organism may influence uptake, as larger specimens may ingest more and larger microplastics [69]. In addition, the presence of food appears to play a key role in microplastic ingestion and excretion. Microplastic ingestion is significantly reduced when food is present, and on the other hand, excretion is significantly increased in presence of food [21, 69, 75].

The distribution of microplastics depends on the physiology of an organism and how they are taken up. Subsequently, they are translocated to different tissues or organs, but most commonly to the digestive tract [63]. For example, in bivalves, microplastics are captured by the gill surface. On the gills and palps, they may be rejected as pseudofeces or assimilated directly by the gill epithelium or transported to the mouth and digestive system. In the stomach, they may be transported to the digestive gland or intestine and further excreted as feces. From the digestive gland, microplastics can be transported to the lysosomes and circulatory system and further to other tissues [76]. Similarly, in fish, microplastics are found in the digestive tract, but can also be found in the liver [77]. Very small microplastics (below 20 μm) can be found in a tissue area of the gills and head of a fish [78], but information on the translocation of microplastics to other tissues and organs is limited.

Ingestion of microplastics by aquatic organisms is usually followed by rapid excretion. For example, more than 90% of microplastics were removed from the gut of *Daphnia magna* after 9–12 min of depuration [72]. Mussel *Mytilus edulis* depurated majority of particles after 1 h of exposure to particle-free seawater, suggesting a high capacity for self-cleaning [79]. Similarly, fish (*Oryzias latipes*) can rapidly eliminate microplastics from the digestive system, however, small particles (2 μm) could still be detected in the digestive system of fish after 24 days of depuration [78]. Thus, the longer residence time of small microplastics may further increase their potential for long-term persistence in aquatic organisms.

10.2.2 Transfer within the Organizations

Accumulated microplastics can be translocated from one organism to another by trophic transfer. The first level of the food web are producers – algae and plants – and they provide food for primary consumers (herbivores) [65, 80]. Consequently, the first possible transfer of microplastics within organizations is when microplastics are adsorbed on the surface of a plant or algae [19, 21, 81–83]. However, it is still unclear whether primary consumers can distinguish between contaminated food with microplastics and food without microplastics. For example, *Gammarus* species have previously been shown to ingest a lower number of microplastics in the presence of food than in the absence of food, but it is unclear whether this is incidental ingestion or *Gammarus* species avoid microplastics [69]. Furthermore, the periwinkle *Littorina littorea* fed with the seaweed *Fucus vesiculosus* did not discriminate between uncontaminated and contaminated seaweed [20], while Yardy and Callaghan performed a choice experiment with *Gammarus pulex* and concluded that algal wafers with microplastics were visited less frequently than uncontaminated ones [75].

Furthermore, trophic transfer of microplastics from primary (herbivores) to secondary consumers (carnivores) can also occur. The number of microplastics transferred depending on their sizes, shapes, and concentrations, and corresponds with the number of microplastics accumulated in the prey [84, 85]. Mussels are important prey for crabs in the aquatic environment, but less than 1% of the microplastics to which mussels were exposed were found in crab tissues [86]. Similarly, less than 1% of microplastics were found in fish *Pimephales promelas* transferred by ingestion of contaminated crustaceans *Daphnia magna* [87]. The number of microplastics in a predator is usually determined at the end of the experiment, so the low percentages of bioaccumulated microplastics after predator-prey exposure may also be a result of the continuous excretion of microplastic during the exposure period [84]. Comparing the ingestion of microplastics by crabs through water filtration (direct exposure) or food consumption (predator-prey exposure), the concentration of microplastics in the soft tissues of crabs is practically the same [88]. On the other hand, the fish *Myoxocephalus brandti* can ingest 3–11 times more microplastics by ingesting contaminated prey (mysids *Neomysis* sp.) than by direct ingestion from the water column [85]. In addition, there may be trophic transfer of microplastics between top aquatic predators (e.g., from Atlantic mackerel *Scombar scombrus* to grey seal *Halichoerus grypus* [89]), as well as from aquatic to terrestrial organisms (from *tambatinga* fish to mice [90]).

10.3 Toxicity to Aquatic Organisms

When microplastics interact with aquatic organisms, either through ingestion or adsorption on their surfaces, they can cause a range of negative effects. These effects are not limited to animal species (consumers); microplastics can also affect aquatic plants (producers) and interact with microorganisms (decomposers).

10.3.1 Decomposers

Decomposers such as bacteria and fungi play a critical role in the flow of energy through an ecosystem, and therefore can be a critical link in the life cycle of microplastics in aquatic ecosystems. The effects of microplastics on aquatic decomposers have not yet been systematically studied, but some studies suggested that the acute effects may be rather low or none [91, 92].

On the other hand, the interactions between microplastics and microorganisms may have important consequences for the fate of microplastics in aquatic ecosystems. The microplastic surface attracts diverse microorganisms (bacteria, fungi, and protozoa, but also microscopic algae), which in turn form biofilms [15]. The developed biofilm increases the density and size of the particles [24] and they sink easily and become more bioavailable to nektonic and benthic organisms [13]. In addition to their increased bioavailability by their presence in the water column, the developed biofilm may specifically attract some organisms and increase their uptake [93]. Biofilms on microplastic surfaces often contain plastic-degrading microorganisms in high abundance [51], potentially contributing to their degradation/fragmentation in the environment [94].

10.3.2 Producers

Primary producers such as algae and vascular plants form the basis of the aquatic food web. Although aquatic plants are an important food source for many fish and birds, serve as rearing habitat for many species, and provide both protection from predators and enhanced feeding opportunities [95], they are rarely studied within microplastics research. The majority of research focusing on the effects of microplastics on producers concerns algae (Chapter 9).

However, in recent studies, microplastics have been shown to interact with aquatic plants. These interactions can occur when microplastics enter an aquatic ecosystem and float on the water surface. There, they can adhere to roots and leaves of floating plants, as shown in laboratory experiments with duckweed *Lemna minor* [18]. After sinking, they can further interact with submerged plants. Monitoring of microplastics on seagrass *Thalassia testudinum* revealed high concentrations of microplastics trapped by their leaves [19]. The possible mechanisms of interactions between microplastics and plants were recently summarized by Kalčíková: the first interactions arise from electrostatic forces between negatively charged cellulose components of plant cells and positively charged microplastics. Furthermore, retention of microplastics is enhanced by complex leaf morphology and strong binding of microplastics occurs when periphyton develops on a plant surface or overgrows retained microplastics [14].

The effects of microplastics on aquatic plants depend on the positioning of plant roots and leaves in an aquatic ecosystem. Floating plants such as duckweed may be affected by floating microplastics with sharp edges. They affect root length and root cell viability rather than growth rate and chlorophyll content in leaves [18]. However, when sharp microplastics are overgrown by a biofilm, their abrasive effect is reduced and they do not cause damage to duckweed [24, 96]. Similarly, the relative growth rates and functional properties of leaves of the free-floating carnivorous aquatic plant *Utricularia vulgaris* were not affected by spherical microplastics up to 70 mg/L. The highest concentrations (140 mg/L) affected the growth of the plant, but the effect can most likely be linked to the ingestion of microplastics by bladders used for feeding by this species [83]. In the case of sediment-rooted plants such as *Myriophyllum spicatum* and *Elodea* sp. the effect on their roots and shoots is reported to be minimal when microplastics are present in the sediment up to 10% of dry weight [17]. However, in all these studies, the effects occurred at much higher than environmentally realistic concentrations, suggesting low ecotoxicological consequences of microplastic-plant interactions.

10.3.3 Consumers

Interactions between microplastics and consumers occurs through ingestion. However, some studies also reported adhesion of microplastics on the surface of exposed organisms, but the number of ingested microplastic was usually higher than the number of adhered microplastics, suggesting that ingestion is the main source of exposure for consumers [97–99].

In general, microplastics do not cause acute harm to consumers. The survival of the planktonic crustacean *Daphnia magna* appears to be more dependent on the food concentration than on the presence of microplastics [100–102], therefore, microplastics did not cause an acute negative effect on juvenile and adult *Daphnia magna* in most studies [103–107]. However, increased mortality was found when microplastic concentrations were very high [108, 109], or when microplastics were adhered to algae (the food of *Daphnia magna*) and a large number of microplastics were ingested [110]. Higher concentrations of microplastics also caused oxidative stress in *Daphnia magna*, as induced formation of reactive oxygen species (ROS) and consequently increased activity of the antioxidant system in the organisms, as evidenced by increased lipid peroxidation activity, catalase activity, or increased total antioxidant capacity [111, 112]. Reproduction of *Daphnia magna* was not affected in the majority of studies [100, 102, 103, 113], but the number of offspring was reduced at higher microplastic concentrations [108]. Furthermore, the negative effects of microplastics can be transmitted up to the fourth generation of offspring with significantly reduced reproductive and population growth rates [114]. In addition, multigenerational exposure can lead to the extinction of an organism [108, 115].

Similar to *Daphnia magna*, microplastics do not have a major toxic effect on *Gammarus* species, despite the high ingestion rate [116, 117]. Chronic exposure may result in a slightly higher mortality rate compared to the control, but this occurred rather randomly [116–119]. No negative effect of microplastics on energy reserves, development, or feeding rate was found [116–118]. On the other hand, assimilation efficiency (ratio of ingested to assimilated food), length, and weight of organisms were significantly reduced when exposed to microplastics. However, the impact is most probably shape-dependent [116, 117]. Irregularly shaped particles (fragments, fibers) generally have a longer passage through the gastrointestinal tract [72], consequently lower amounts of food are absorbed, which in turn negatively affected *Gammarus* species [117, 120]. On the other hand, spherical microplastics have a smooth surface and can be excreted more easily, allowing the organism to consume more food [72]. Thus, assimilation efficiency and weight were reduced when *Gammarus* were exposed to fibers and fragments, but not when exposed to spherical microplastics [117, 120], and similarly, body length of *Gammarus* was significantly reduced by ingested microplastic fragments [120].

Zebrafish *Danio rerio* is an important secondary consumer often used in ecotoxicological research and the uptake of microplastics can occur directly or through the food web [97, 121], causing intestinal damage [74, 98, 122, 123], neurotoxicity [97, 122], oxidative stress [124], and reproductive toxicity [125, 126]. Microplastics generally do not penetrate the reproductive organs of zebrafish but may affect them indirectly. For example, Qiang and Cheng reported that the cell apoptosis rate in zebrafish testes was significantly higher after microplastic exposure, also confirmed by the expression of apoptosis-related genes [126]. However, the female gonads were not affected [126]. Furthermore, Qiang et al. determined the effects of microplastics on zebrafish sex hormones and also concluded that male reproductive organs were more sensitive than females, but there was no overall effect on fertilization rates [125]. Zebrafish behavior and locomotor activity play an important role in feeding, social and defensive activities [127]. When zebrafish are exposed to microplastics, they may become hyperactive, most likely caused by oxidative stress or particle stimulation [122, 124]. In addition, microplastics may induce antipredatory behavior in fish after exposure, which may affect the population, consequent interspecies interactions, and overall community structure [80, 97].

Mussels are by far the most studied marine consumers within microplastic research. They are filter-feeding organisms and can inadvertently ingest high numbers of microplastics. In some cases, the accumulation of microplastics in mussels can lead to physical damage, such as abnormalities of the gills and digestive tract [128, 129], affect the structure of the gut microbiota [130], or reduce the filtration/depuration rate, which can lead to starvation, growth impairment and even death [128, 131, 132]. Similarly to other consumers, microplastics cause oxidative stress to mussels, as evidenced by increased activity of antioxidant enzymes, e.g., lipid peroxidation, catalase, superoxide dismutase, and glutathione peroxidase [133–135], but to a greater extent (longer exposure time, higher microplastic concentration, etc.), microplastics can cause disruptions in the antioxidant defense system or ultimately irreversible oxidative damage [131, 136, 137].

The effects of microplastics on other marine consumers such as seahorses, oysters, and crabs have also been studied. Many of the organisms were not acutely affected by microplastics [138–140]. Gardon et al. reported an immune response and oxidative stress in the pearl oyster *Pictada margaritifera*, but only at higher concentrations of microplastics (10^4 particles/L) [139]. On the other hand, crabs seem to be more susceptible to the presence of microplastics, as studies reported increased adult mortality [141], growth impairment [142], and deformities in reproductivity [141]. However, additional research needs to be done to better understand interactions between marine consumers and microplastics.

10.4 The Sources of Toxicity

As described in detail in the previous section of this chapter, microplastics can interact with and affect different organisms in aquatic ecosystems. In general, microplastics can cause mechanical and/or chemical stress. Therefore, when microplastics attach to organisms or are ingested, they can cause physical (mechanical) damage such as abrasion or blockage. In addition, toxicity may also result from additives that leach from the microplastic or contaminants that have been adsorbed onto the surface of the microplastics during their lifetime. Leaching of chemicals from microplastics can represent a potential exposure pathway for aquatic organisms [22].

10.4.1 The Release of Plasticizers and Other Additives

Plastic materials contain, beside the main polymer, multiple additives. They are added to the material to improve desirable properties and reduce undesirable properties. Additives are very common to improve the processability, performance, and appearance of plastics during manufacture and use, such as pigments, plasticizers, stabilizers, and optical brighteners [143]. In addition, plastics may contain residual chemicals such as reactants and solvents, as well as non-intentionally added substances such as impurities and byproducts of polymerization [144].

According to recently established *Chemicals associated with Plastic Packaging* database, 906 chemicals are likely associated with plastics used for packaging. Of those, 63 rank highest for human health hazards and 68 for environmental hazards according to the harmonized hazard classifications assigned by the European Chemicals Agency within the Classification, Labeling and Packaging (CLP) regulation implementing the United Nations Globally Harmonized System (GHS). In addition, seven substances are classified as persistent, bioaccumulative, and toxic, or very persistent and very bioaccumulative, and 15 as endocrine disrupting chemicals. These chemicals are widely used in plastic industry as monomers, intermediates, solvents, surfactants, plasticizers, stabilizers, biocides, flame retardants, accelerators, and colorants [145].

The most produced polymers are polypropylene (PP), polyethylene (PE, high and low density), polyvinyl chloride (PVC), and polyurethane (PUR). PP and PE contain additives such as antioxidants (most commonly bisphenol A, octylphenol, or nonylphenol) and flame retardants (for cable insulation applications, brominated flame retardant, boric acid; tris(2-chloroethyl) phosphate). PVC may contain stabilizers (such as bisphenol A, nonylphenol) and high concentrations of plasticizers [146], typically 10–60 wt% phthalates [147]. PUR may contain various flame retardants (most commonly brominated flame retardant, boric acid; tris(2-chloroethyl) phosphate). Phthalates, nonylphenol, bisphenol A are known endocrine disruptors and brominated flame retardants are potential endocrine disruptors [146].

These additives are often not chemically bound in the polymer matrix and can leach from the plastics into surrounding environment [148]. Leaching of additives is enhanced by fragmentation of plastics when plastic parts are broken down into microplastics [149]. Turbulence can further increase the leaching of additives, while salinity or the presence of UV radiation seems to have a minor impact on the release of additives from plastics [150]. Plastic additives have been detected worldwide, e.g., concentrations of brominated flame retardants exceeding 10 ng/L have been detected in harbors [151], while various phthalates, bisphenol A and nonylphenol have been detected in several µg/L worldwide [146]. Consequently, aquatic organisms are frequently contaminated by these additives. Phthalates have been detected in 18 marine species from four trophic levels [152] and the concentrations in fish can be as high as 1 µg/kg. The high concentrations of plastic additives in organisms are a consequence of exposure through water/food or ingestion of (micro)plastics [146].

In addition, laboratory studies showed that plastic additives can leach from some microplastics to a considerable extent. Recent results suggested that the toxicity of PVC microplastics may come from their leachates rather than from the particles themselves. More specifically, PVC without additives had no effect on *Daphnia magna* compared to PVC leachates containing phthalates [106]. Similarly, Bakelite microplastics leach toxic chemicals that affected *Lemna minor* more than particles [23]. On the other hand, many microplastics from plastics routinely used for food storage, clothing, and cosmetics, such as PE and PET, do not leach additives to such an extent that they could harm tested organisms [18, 23] or be detected by analytical approaches [153]. Moreover, in many laboratory leaching tests, the liquid-to-solid ratio (i.e., medium-to-microplastics ratio) is chosen to be 10 (e.g., [154]). This means that the concentration of microplastics in the leaching test is 100 g/L, far from any environmentally relevant concentrations. Therefore, the contribution of additives released from microplastics may be of less importance compared to the total plastic debris.

10.4.2 The Adsorbed Pollutants

Microplastics in aquatic ecosystems interact with organic and inorganic matter along with anthropogenic pollutants that can be adsorbed on the surface of microplastics. A number of studies have reported increased adsorption of various hydrophobic and hydrophilic pollutants including polychlorinated biphenyls (PCBs), polycyclic aromatic hydrocarbons (PAHs), perfluorinated alkyl substances (PFASs), polybrominated diethers (PBDs), pesticides, pharmaceuticals and personal care products (PPCPs), and metals [155–157]. These pollutants can accumulate on the surface of microplastics at concentrations several orders of magnitude higher than in the surrounding water or sediment [158, 159].

Previous studies on the adsorption of pollutants on microplastics have focused on hydrophobic pollutants such as PAHs, PCBs, and chlorinated pesticides. They are usually linked to old pollution and the interactions with microplastics persist in marine environment. Interactions of microplastics with hydrophilic pesticides, PPCPs, and metals can be of an immense importance in urban waters [66].

The mechanism of pollutant adsorption on microplastic surface is diverse and complex. During the adsorption process, multiple adsorption mechanisms occur and interact. Fu et al. summarized mechanisms of adsorption of organic pollutants on microplastic surface and they primarily include hydrophobic interactions, partitioning, electrostatic interactions, and other non-covalent interactions such as hydrogen bonding, halogen bonding, π-π interactions, and van der Waals forces [156]. Cao et al. focused on the interaction of microplastics and heavy metals and suggested that the mechanisms mainly include single electrostatic interactions or electrostatic interactions along with surface complexation, formation of new complexes with biofilms and natural organics previously adsorbed on microplastics, and other interactions involving precipitation and coprecipitation with hydrous oxides of Fe and Mn [160].

The extent to which pollutants are adsorbed onto microplastics depends on the physico-chemical properties of the microplastic particles (material, size, age, crystallinity, functional groups, and polarity), the properties of the adsorbed pollutant (hydrophobicity, ionic properties) and the physico-chemical properties of the surrounding environment (pH, ionic strength, temperature). In this context, different polymers showed significant differences in the adsorption capacity of pollutants under different conditions [156]. For example, PS had a higher absorption capacity for dichlorophenol from deionized water compared to PET, PP, or HDPE [161]. On the other hand, HDPE, LDPE, and PP tended to have much greater capacity for adsorption of PCBs from seawater compared to PET and PVC [162] and in the case of metals, the adsorption capacity appeared to be driven mainly by the associated biofilm on their surface, regardless of the polymer [157]. Microplastics with biofilm can adsorb significantly higher amounts of metals compared to virgin microplastics [24, 163]. The size of the microplastic may also play a key role and smaller particles showed the highest adsorption capacity [161]. The natural environment can also influence adsorption and, for example, increased pH and salinity can lead to a decrease in adsorption capacity for organic pollutants [156].

Adsorption of pollutants on microplastics may have great ecological consequences. First, it may increase their ecotoxicological profile [24]. Several studies showed no effect of microplastics on various organisms [92, 164], while study using microplastics with adsorbed pollutants showed the opposite [24, 165]. Adsorbed pollutants can ultimately leach from microplastics under certain conditions, such as low pH in the gut of organisms [166] or in acidic waters [24]. On the other hand, several studies suggested, that when organisms are at the same time exposed to microplastics in combination with pollutants, the overall ecotoxicity of the pollutant may be reduced [167, 168]. This is because some microplastics can function as contaminant sinks, reducing the bioavailability and consequently the ecotoxicity of the contaminant [169].

10.4.3 Physical Damage

Physical damage from plastic debris (such as plastic ropes and fragments) has been widely reported in marine vertebrates, causing severe effects such as internal and external abrasions and blockages of the digestive tract. Similarly, microplastics can cause physical damage to other aquatic organisms [22].

The most studied physical effects of microplastics occur after consumer ingestion. Many studies have shown that microplastics can cause severe intestinal damage, thereby affecting food intake, leading to starvation [117]. These effects are usually associated with irregular and sharp particles, while spherical particles are often excreted from the digestive tract without significant effects [117, 120]. In the absence of additives, the effects of microplastics do not seem to depend on the material of the microplastics. For example, Lei et al. studied the effects of unplasticized microplastic fragments from PA, PE, PP, PVC, and PS on the fish *Danio rerio* and the nematode *Caenorhabditis elegans*. All microplastics tested caused intestinal damage to a similar extent [123]. In fish, microplastics caused cracking of villi and splitting of enterocytes, while in nematodes, they caused oxidative stress and changes in intestinal calcium levels. In addition, damage to the intestine may result in increased intestinal permeability and cause inflammation. The latter is associated with induced microbiota dysbiosis and a change in specific gut bacteria, which can also lead to metabolic dysfunction [98, 122, 123, 170] and depletion of energy reserves [122].

Physical damage can also result from external mechanical abrasion by adhered microplastics on tissues of various aquatic organisms. A mixture of microplastic fragments can directly damage coral tissue by causing tears and creating incisions down to the skeleton and sclerites. In addition, the tissue abrasions allow opportunistic bacteria to proliferate [171]. Microplastics can also adhere to plant tissues and cause mechanical damage. The sharp fragments reduced root length and root cell viability, while rounded microplastics of the same size and material did not [18]. Furthermore, wood fragments had no effect on the root length of *Lemna minor*, although they had sharp surface and a comparable size and shape to PE microplastics causing a 25% reduction in root length. It seems that the softness of natural materials such as wood is responsible for the lack of abrasive/mechanical effects on roots compared to the hard PE particles [24]. In conclusion, the physical damage caused by external abrasion of tissues of organisms depends on the material and shape of the microplastic.

10.5 Summary and Outlook

In this chapter, we have provided a brief overview of the occurrence of microplastics in urban waters and summarized their interaction with different aquatic organisms, from simple uptake or adsorption of microplastics, to their accumulation in organisms and possible effects. We also discussed trophic transfer from primary producers to primary and secondary consumers. In addition, we focused on the source of toxicity of microplastics, which has different modes of action depending on their size, shape, material, ageing, and presence of additives and adsorbed chemicals. Therefore, the topic of microplastics is complex and offers a number of research questions that need to be answered in the future.

To understand the ecological consequences of microplastics in the environment, it is first necessary to determine their actual concentrations in different aquatic ecosystems. This requires improving sampling and measurement methods for their monitoring. In addition, there is a need to study the interactions and effects of microplastics with different species from different trophic levels (e.g., plants and bacteria), which can help to understand the possible pathways into the food chain and assess the fate of microplastics in the environment. Research with environmentally relevant microplastics and concentrations seems to be one of the essential future tasks for ecotoxicologists. The greatest challenge is undoubtedly the development of a microplastic risk assessment method that would cover the great diversity of microplastics and associated pollutants.

References

1. Carpenter, E.J. and Smith, K.L. (1972). Plastics on the Sargasso sea surface. *Science* 175 (4027): 1240–1241.
2. Rochman, C.M. (2018). Microplastics research-from sink to source. *Science* 360 (6384): 28–29.
3. Liu, F., Olesen, K.B., Borregaard, A.R. et al. (2019). Microplastics in urban and highway stormwater retention ponds. *Science of the Total Environment* 671: 992–1000.
4. Allen, S., Allen, D., Phoenix, V.R. et al. (2019). Atmospheric transport and deposition of microplastics in a remote mountain catchment. *Nature Geoscience* 12 (5): 339–344.
5. Jambeck, J.R., Geyer, R., Wilcox, C. et al. (2015). Plastic waste inputs from land into the ocean. *Science* 347 (6223): 768.
6. SAPEA (2019). A scientific perspective on microplastics in nature and society.
7. Horton, A.A., Svendsen, C., Williams, R.J. et al. (2017). Large microplastic particles in sediments of tributaries of the River Thames, UK – abundance, sources and methods for effective quantification. *Marine Pollution Bulletin* 114 (1): 218–226.
8. Kole, P.J., Löhr, A.J., Van Belleghem, F.G.A.J. et al. (2017). Wear and tear of tires: a stealthy source of microplastics in the environment. *International Journal of Environmental Research and Public Health* 14 (10): 1265.
9. Kalčíková, G., Alič, B., Skalar, T. et al. (2017). Wastewater treatment plant effluents as source of cosmetic polyethylene microbeads to freshwater. *Chemosphere* 188: 25–31.
10. Kay, P., Hiscoe, R., Moberley, I. et al. (2018). Wastewater treatment plants as a source of microplastics in river catchments. *Environmental Science and Pollution Research* 25 (20): 20264–20267.
11. Browne, M.A. (2015). Sources and pathways of microplastics to habitats. In: *Marine Anthropogenic Litter* ((M. Bergmann, L. Gutow, M. Klages eds)), 229–244. Berlin: Springer
12. Li, Y., Zhang, H., and Tang, C. (2020). A review of possible pathways of marine microplastics transport in the ocean. *Anthropocene Coasts* 3 (1): 6–13.
13. Kooi, M., Nes, E.Hv., Scheffer, M. et al. (2017). Ups and downs in the ocean: Effects of biofouling on vertical transport of microplastics. *Environmental Science & Technology* 51 (14): 7963–7971.
14. Kalčíková, G. (2020). Aquatic vascular plants – a forgotten piece of nature in microplastic research. *Environmental Pollution* 262: 114354.
15. Rummel, C.D., Jahnke, A., Gorokhova, E. et al. (2017). Impacts of biofilm formation on the fate and potential effects of microplastic in the aquatic environment. *Environmental Science & Technology Letters* 4 (7): 258–267.

16 Vroom, R.J.E., Koelmans, A.A., Besseling, E. et al. (2017). Aging of microplastics promotes their ingestion by marine zooplankton. *Environmental Pollution* 231: 987–996.

17 van Weert, S., Redondo-Hasselerharm, P.E., Diepens, N.J. et al. (2019). Effects of nanoplastics and microplastics on the growth of sediment-rooted macrophytes. *Science of the Total Environment* 654: 1040–1047.

18 Kalčíková, G., Žgajnar Gotvajn, A., Kladnik, A. et al. (2017). Impact of polyethylene microbeads on the floating freshwater plant duckweed *Lemna minor*. *Environmental Pollution* 230: 1108–1115.

19 Goss, H., Jaskiel, J., and Rotjan, R. (2018). *Thalassia testudinum* as a potential vector for incorporating microplastics into benthic marine food webs. *Marine Pollution Bulletin* 135: 1085–1089.

20 Gutow, L., Eckerlebe, A., Giménez, L. et al. (2016). Experimental evaluation of seaweeds as a vector for microplastics into marine food webs. *Environmental Science & Technology* 50 (2): 915–923.

21 Mateos-Cárdenas, A., Scott, D.T., Seitmaganbetova, G. et al. (2019). Polyethylene microplastics adhere to *Lemna minor* (L.), yet have no effects on plant growth or feeding *by Gammarus duebeni* (Lillj.). *Science of the Total Environment* 689: 413–421.

22 Wright, S.L., Thompson, R.C., and Galloway, T.S. (2013). The physical impacts of microplastics on marine organisms: a review. *Environmental Pollution* 178: 483–492.

23 Rozman, U., Turk, T., Skalar, T. et al. (2021). An extensive characterization of various environmentally relevant microplastics – material properties, leaching and ecotoxicity testing. *Science of the Total Environment* 773: 145576.

24 Kalčíková, G., Skalar, T., Marolt, G. et al. (2020). An environmental concentration of aged microplastics with adsorbed silver significantly affects aquatic organisms. *Water Research* 175: 115644.

25 Rainieri, S., Conlledo, N., Larsen, B.K. et al. (2018). Combined effects of microplastics and chemical contaminants on the organ toxicity of zebrafish (*Danio rerio*). *Environmental Research* 162: 135–143.

26 Elizalde-Velázquez, G.A. and Gómez-Oliván, L.M. (2021). Microplastics in aquatic environments: a review on occurrence, distribution, toxic effects, and implications for human health. *Science of the Total Environment* 780: 146551.

27 Eerkes-Medrano, D., Thompson, R.C., and Aldridge, D.C. (2015). Microplastics in freshwater systems: A review of the emerging threats, identification of knowledge gaps and prioritisation of research needs. *Water Research* 75: 63–82.

28 Li, J., Liu, H., and Paul Chen, J. (2018). Microplastics in freshwater systems: a review on occurrence, environmental effects, and methods for microplastics detection. *Water Research* 137: 362–374.

29 Dris, R., Gasperi, J., Rocher, V. et al. (2015). Microplastic contamination in an urban area: a case study in greater Paris. *Environmental Chemistry* 12 (5): 592–599.

30 Sighicelli, M., Pietrelli, L., Lecce, F. et al. (2018). Microplastic pollution in the surface waters of Italian Subalpine Lakes. *Environmental Pollution* 236: 645–651.

31 Zhang, K., Hamidian, A.H., Tubić, A. et al. (2021). Understanding plastic degradation and microplastic formation in the environment: a review. *Environmental Pollution* 274: 116554.

32 Rodrigues, M.O., Abrantes, N., Gonçalves, F.J.M. et al. (2018). Spatial and temporal distribution of microplastics in water and sediments of a freshwater system (Antuã River, Portugal). *Science of the Total Environment* 633: 1549–1559.

33 Sadri, S.S. and Thompson, R.C. (2014). On the quantity and composition of floating plastic debris entering and leaving the Tamar Estuary, Southwest England. *Marine Pollution Bulletin* 81 (1): 55–60.

34 Leslie, H.A., Brandsma, S.H., van Velzen, M.J.M. et al. (2017). Microplastics en route: Field measurements in the Dutch river delta and Amsterdam canals, wastewater treatment plants, North Sea sediments and biota. *Environment International* 101: 133–142.

35 Yang, L., Zhang, Y., Kang, S. et al. (2021). Microplastics in freshwater sediment: a review on methods, occurrence, and sources. *Science of the Total Environment* 754: 141948.

36 Alam, F.C., Sembiring, E., Muntalif, B.S. et al. (2019). Microplastic distribution in surface water and sediment river around slum and industrial area (case study: Ciwalengke River, Majalaya district, Indonesia). *Chemosphere* 224: 637–645.

37 Wu, N., Zhang, Y., Li, W. et al. (2020). Co-effects of biofouling and inorganic matters increased the density of environmental microplastics in the sediments of Bohai Bay coast. *Science of the Total Environment* 717: 134431.

38 Rezania, S., Park, J., Md Din, M.F. et al. (2018). Microplastics pollution in different aquatic environments and biota: a review of recent studies. *Marine Pollution Bulletin* 133: 191–208.

39 Acosta-Coley, I., Mendez-Cuadro, D., Rodriguez-Cavallo, E. et al. (2019). Trace elements in microplastics in Cartagena: a hotspot for plastic pollution at the Caribbean. *Marine Pollution Bulletin* 139: 402–411.

40 Antunes, J., Frias, J., and Sobral, P. (2018). Microplastics on the Portuguese coast. *Marine Pollution Bulletin* 131: 294–302.

41 Patchaiyappan, A., Dowarah, K., Zaki Ahmed, S. et al. (2021). Prevalence and characteristics of microplastics present in the street dust collected from Chennai metropolitan city, India. *Chemosphere* 269: 128757.

42 Baensch-Baltruschat, B., Kocher, B., Stock, F. et al. (2020). Tyre and road wear particles (TRWP) – a review of generation, properties, emissions, human health risk, ecotoxicity, and fate in the environment. *Science of the Total Environment* 733: 137823.

43 Piñon-Colin, Td.J., Rodriguez-Jimenez, R., Rogel-Hernandez, E. et al. (2020). Microplastics in stormwater runoff in a semiarid region, Tijuana, Mexico. *Science of the Total Environment* 704: 135411.

44 Abbasi, S. and Turner, A. (2021). Dry and wet deposition of microplastics in a semi-arid region (Shiraz, Iran). *Science of the Total Environment* 786: 147358.

45 Murphy, F., Ewins, C., Carbonnier, F. et al. (2016). Wastewater treatment works (WwTW) as a source of microplastics in the aquatic environment. *Environmental Science & Technology* 50 (11): 5800–5808.

46 Prata, J.C. (2018). Microplastics in wastewater: state of the knowledge on sources, fate and solutions. *Marine Pollution Bulletin* 129 (1): 262–265.

47 Talvitie, J., Mikola, A., Setälä, O. et al. (2017). How well is microlitter purified from wastewater? – a detailed study on the stepwise removal of microlitter in a tertiary level wastewater treatment plant. *Water Research* 109: 164–172.

48 Carr, S.A., Liu, J., and Tesoro, A.G. (2016). Transport and fate of microplastic particles in wastewater treatment plants. *Water Research* 91: 174–182.

49 Schmidt, C., Kumar, R., Yang, S. et al. (2020). Microplastic particle emission from wastewater treatment plant effluents into river networks in Germany: loads, spatial patterns of concentrations and potential toxicity. *Science of the Total Environment* 737: 139544.

50 Estahbanati, S. and Fahrenfeld, N.L. (2016). Influence of wastewater treatment plant discharges on microplastic concentrations in surface water. *Chemosphere* 162: 277–284.

51 McCormick, A., Hoellein, T.J., Mason, S.A. et al. (2014). Microplastic is an abundant and distinct microbial habitat in an Urban River. *Environmental Science & Technology* 48 (20): 11863–11871.

52 Zhou, G., Wang, Q., Zhang, J. et al. (2020). Distribution and characteristics of microplastics in urban waters of seven cities in the Tuojiang River basin, China. *Environmental Research* 189: 109893.

53 Wakkaf, T., El Zrelli, R., Kedzierski, M. et al. (2020). Characterization of microplastics in the surface waters of an urban lagoon (Bizerte lagoon, Southern Mediterranean Sea): composition, density, distribution, and influence of environmental factors. *Marine Pollution Bulletin* 160: 111625.

54 Wang, W., Ndungu, A.W., Li, Z. et al. (2017). Microplastics pollution in inland freshwaters of China: a case study in urban surface waters of Wuhan, China. *Science of the Total Environment* 575: 1369–1374.

55 Treilles, R., Gasperi, J., Gallard, A. et al. (2021). Microplastics and microfibers in urban runoff from a suburban catchment of Greater Paris. *Environmental Pollution* 287: 117352.

56 Kameda, Y., Yamada, N., and Fujita, E. (2021). Source- and polymer-specific size distributions of fine microplastics in surface water in an urban river. *Environmental Pollution* 284: 117516.
57 Grbić, J., Helm, P., Athey, S. et al. (2020). Microplastics entering northwestern Lake Ontario are diverse and linked to urban sources. *Water Research* 174: 115623.
58 Bikker, J., Lawson, J., Wilson, S. et al. (2020). Microplastics and other anthropogenic particles in the surface waters of the Chesapeake Bay. *Marine Pollution Bulletin* 156: 111257.
59 Wang, C., Xing, R., Sun, M. et al. (2020). Microplastics profile in a typical urban river in Beijing. *Science of the Total Environment* 743: 140708.
60 Sánchez-Hernández, L.J., Ramírez-Romero, P., Rodríguez-González, F. et al. (2021). Seasonal evidences of microplastics in environmental matrices of a tourist dominated urban estuary in Gulf of Mexico, Mexico. *Chemosphere* 277: 130261.
61 Syberg, K., Khan, F.R., Selck, H. et al. (2015). Microplastics: addressing ecological risk through lessons learned. *Environ Toxicol Chem* 34 (5): 945–953.
62 Xu, P., Peng, G., Su, L. et al. (2018). Microplastic risk assessment in surface waters: a case study in the Changjiang Estuary, China. *Marine Pollution Bulletin* 133: 647–654.
63 Xu, S., Ma, J., Ji, R. et al. (2020). Microplastics in aquatic environments: occurrence, accumulation, and biological effects. *Science of the Total Environment* 703: 134699.
64 Carbery, M., O'Connor, W., and Palanisami, T. (2018). Trophic transfer of microplastics and mixed contaminants in the marine food web and implications for human health. *Environment International* 115: 400–409.
65 Tang, Y., Liu, Y., Chen, Y. et al. (2021). A review: Research progress on microplastic pollutants in aquatic environments. *Science of the Total Environment* 766: 142572.
66 Lambert, S. and Wagner, M. (2018). Microplastics are contaminants of emerging concern in freshwater environments: An overview. In: *Freshwater Microplastics: Emerging Environmental Contaminants?* (ed. T. Page), 1–23.
67 Mateos-Cárdenas, A., O'Halloran, J., van Pelt, F.N.A.M. et al. (2020). Rapid fragmentation of microplastics by the freshwater amphipod *Gammarus duebeni* (Lillj.). *Scientific Reports* 10 (1): 12799.
68 Redondo-Hasselerharm, P.E., de Ruijter, V.N., Mintenig, S.M. et al. (2018). Ingestion and chronic effects of car tire tread particles on freshwater Benthic macroinvertebrates. *Environmental Science & Technology* 52 (23): 13986–13994.
69 Scherer, C., Brennholt, N., Reifferscheid, G. et al. (2017). Feeding type and development drive the ingestion of microplastics by freshwater invertebrates. *Scientific Reports* 7 (1): 17006.
70 Mateos-Cárdenas, A., O'Halloran, J., van Pelt, F.N.A.M. et al. (2021). Beyond plastic microbeads – short-term feeding of cellulose and polyester microfibers to the freshwater amphipod *Gammarus duebeni*. *Science of the Total Environment* 753: 141859.

71 Fueser, H., Mueller, M.-T., Weiss, L. et al. (2019). Ingestion of microplastics by nematodes depends on feeding strategy and buccal cavity size. *Environmental Pollution* 255: 113227.

72 Ogonowski, M., Schür, C., Jarsén, Å. et al. (2016). The effects of natural and anthropogenic microparticles on individual fitness in *Daphnia magna*. *PLOS ONE* 11 (5): 0155063.

73 Gray, A.D., Weinstein, J.E. et al. (2017). Size- and shape-dependent effects of microplastic particles on adult daggerblade grass shrimp (*Palaemonetes pugio*). *Environmental Toxicology and Chemistry* 36 (11): 3074–3080.

74 Qiao, R., Deng, Y., Zhang, S. et al. (2019). Accumulation of different shapes of microplastics initiates intestinal injury and gut microbiota dysbiosis in the gut of zebrafish. *Chemosphere* 236: 124334.

75 Yardy, L. and Callaghan, A. (2020). What the fluff is this? – *Gammarus pulex* prefer food sources without plastic microfibers. *Science of the Total Environment* 715: 136815.

76 Li, J., Wang, Z., Rotchell, J.M. et al. (2021). Where are we? Towards an understanding of the selective accumulation of microplastics in mussels. *Environmental Pollution* 286: 117543.

77 Avio, C.G., Gorbi, S., and Regoli, F. (2015). Experimental development of a new protocol for extraction and characterization of microplastics in fish tissues: first observations in commercial species from Adriatic Sea. *Marine Environmental Research* 111: 18–26.

78 Liu, Y., Qiu, X., Xu, X. et al. (2021). Uptake and depuration kinetics of microplastics with different polymer types and particle sizes in Japanese medaka (*Oryzias latipes*). *Ecotoxicology and Environmental Safety* 212: 112007.

79 Woods, M.N., Stack, M.E., Fields, D.M. et al. (2018). Microplastic fiber uptake, ingestion, and egestion rates in the blue mussel (*Mytilus edulis*). *Marine Pollution Bulletin* 137: 638–645.

80 Ockenden, A., Tremblay, L.A., Dikareva, N. et al. (2021). Towards more ecologically relevant investigations of the impacts of microplastic pollution in freshwater ecosystems. *Science of the Total Environment* 792: 148507.

81 Huang, Y., Xiao, X., Xu, C. et al. (2020). Seagrass beds acting as a trap of microplastics – emerging hotspot in the coastal region? *Environmental Pollution* 257: 113450.

82 Jones, K.L., Hartl, M.G.J., Bell, M.C. et al. (2020). Microplastic accumulation in a *Zostera marina* L. bed at Deerness Sound, Orkney, Scotland. *Marine Pollution Bulletin* 152: 110883.

83 Yu, H., Zhang, X., Hu, J. et al. (2020). Ecotoxicity of polystyrene microplastics to submerged carnivorous *Utricularia vulgaris* plants in freshwater ecosystems. *Environmental Pollution* 265: 114830.

84 Crooks, N., Parker, H., and Pernetta, A.P. (2019). Brain food? Trophic transfer and tissue retention of microplastics by the velvet swimming crab (*Necora puber*). *Journal of Experimental Marine Biology and Ecology* 519: 151187.

85 Hasegawa, T. and Nakaoka, M. (2021). Trophic transfer of microplastics from mysids to fish greatly exceeds direct ingestion from the water column. *Environmental Pollution* 273: 116468.

86 Farrell, P. and Nelson, K. (2013). Trophic level transfer of microplastic: *Mytilus edulis* (L.) to *Carcinus maenas* (L.). *Environmental Pollution* 177: 1–3.

87 Elizalde-Velázquez, A., Carcano, A.M., Crago, J. et al. (2020). Translocation, trophic transfer, accumulation and depuration of polystyrene microplastics in *Daphnia magna* and *Pimephales Promelas*. *Environmental Pollution* 259: 113937.

88 Wang, T., Hu, M., Xu, G. et al. (2021). Microplastic accumulation via trophic transfer: can a predatory crab counter the adverse effects of microplastics by body defence? *Science of the Total Environment* 754: 142099.

89 Nelms, S.E., Galloway, T.S., Godley, B.J. et al. (2018). Investigating microplastic trophic transfer in marine top predators. *Environmental Pollution* 238: 999–1007.

90 da Costa Araújo, A.P. and Malafaia, G. (2021). Microplastic ingestion induces behavioral disorders in mice: a preliminary study on the trophic transfer effects via tadpoles and fish. *Journal of Hazardous Materials* 401: 123263.

91 Gambardella, C., Morgana, S., Bramini, M. et al. (2018). Ecotoxicological effects of polystyrene microbeads in a battery of marine organisms belonging to different trophic levels. *Marine Environmental Research* 141: 313–321.

92 Gambardella, C., Piazza, V., Albentosa, M. et al. (2019). Microplastics do not affect standard ecotoxicological endpoints in marine unicellular organisms. *Marine Pollution Bulletin* 143: 140–143.

93 Procter, J., Hopkins, F.E., Fileman, E.S. et al. (2019). Smells good enough to eat: Dimethyl sulfide (DMS) enhances copepod ingestion of microplastics. *Marine Pollution Bulletin* 138: 1–6.

94 Han, Y.N., Wei, M., Han, F. et al. (2020). Greater biofilm formation and increased biodegradation of polyethylene film by a microbial consortium of *Arthrobacter* sp. and *Streptomyces* sp. *Microorganisms* 8 (12): 1979.

95 Van Hoeck, A., Horemans, N., Monsieurs, P. et al. (2015). The first draft genome of the aquatic model plant *Lemna minor* opens the route for future stress physiology research and biotechnological applications. *Biotechnology for Biofuels* 8 (1): 188.

96 Jemec Kokalj, A., Kuehnel, D., Puntar, B. et al. (2019). An exploratory ecotoxicity study of primary microplastics versus aged in natural waters and wastewaters. *Environmental Pollution* 254: 112980.

97 da Costa Araújo, A.P., de Andrade Vieira, J.E., and Malafaia, G. (2020). Toxicity and trophic transfer of polyethylene microplastics from *Poecilia reticulata* to *Danio rerio*. *Science of the Total Environment* 742: 140217.

98 Limonta, G., Mancia, A., Benkhalqui, A. et al. (2019). Microplastics induce transcriptional changes, immune response and behavioral alterations in adult zebrafish. *Scientific Reports* 9 (1): 15775.

99 Kolandhasamy, P., Su, L., Li, J. et al. (2018). Adherence of microplastics to soft tissue of mussels: a novel way to uptake microplastics beyond ingestion. *Science of the Total Environment* 610–611: 635–640.

100 Aljaibachi, R. and Callaghan, A. (2018). Impact of polystyrene microplastics on *Daphnia magna* mortality and reproduction in relation to food availability. *PeerJ* 6: e4601–e.

101 Aljaibachi, R., Laird, W.B., Stevens, F. et al. (2020). Impacts of polystyrene microplastics on *Daphnia magna*: a laboratory and a mesocosm study. *Science of the Total Environment* 705: 135800.

102 Hiltunen, M., Vehniäinen, E.-R., and Kukkonen, J.V.K. (2021). Interacting effects of simulated eutrophication, temperature increase, and microplastic exposure on *Daphnia*. *Environmental Research* 192: 110304.

103 Canniff, P.M. and Hoang, T.C. (2018). Microplastic ingestion by *Daphnia magna* and its enhancement on algal growth. *Science of the Total Environment* 633: 500–507.

104 De Felice, B., Sabatini, V., Antenucci, S. et al. (2019). Polystyrene microplastics ingestion induced behavioral effects to the cladoceran *Daphnia magna*. *Chemosphere* 231: 423–431.

105 Jemec Kokalj, A., Kunej, U., and Skalar, T. (2018). Screening study of four environmentally relevant microplastic pollutants: Uptake and effects on *Daphnia magna* and *Artemia franciscana*. *Chemosphere* 208: 522–529.

106 Schrank, I., Trotter, B., Dummert, J. et al. (2019). Effects of microplastic particles and leaching additive on the life history and morphology of *Daphnia magna*. *Environmental Pollution* 255: 113233.

107 Horton, A.A., Vijver, M.G., Lahive, E. et al. (2018). Acute toxicity of organic pesticides to *Daphnia magna* is unchanged by co-exposure to polystyrene microplastics. *Ecotoxicology and Environmental Safety* 166: 26–34.

108 Schür, C., Zipp, S., Thalau, T. et al. (2020). Microplastics but not natural particles induce multigenerational effects in *Daphnia magna*. *Environmental Pollution* 260: 113904.

109 Zimmermann, L., Göttlich, S., Oehlmann, J. et al. (2020). What are the drivers of microplastic toxicity? Comparing the toxicity of plastic chemicals and particles to *Daphnia magna*. *Environmental Pollution* 267: 115392.

110 Besseling, E., Wang, B., Lürling, M. et al. (2014). Nanoplastic affects growth of *S. obliquus* and reproduction of *D. magna*. *Environmental Science & Technology* 48 (20): 12336–12343.

111 Na, J., Song, J., Achar, J.C. et al. (2021). Synergistic effect of microplastic fragments and benzophenone-3 additives on lethal and sublethal Daphnia magna toxicity. *Journal of Hazardous Materials* 402: 123845.

112 Zhang, P., Yan, Z., Lu, G., and Ji, Y. (2019). Single and combined effects of microplastics and roxithromycin on Daphnia magna. *Environmental Science and Pollution Research* 26 (17): 17010–17020.

113 Imhof, H.K., Rusek, J., Thiel, M. et al. (2017). Do microplastic particles affect *Daphnia magna* at the morphological, life history and molecular level? *PLOS ONE* 12 (11): e0187590.

114 Martins, A. and Guilhermino, L. (2018). Transgenerational effects and recovery of microplastics exposure in model populations of the freshwater cladoceran Daphnia magna Straus. *Science of the Total Environment* 631–632: 421–428.

115 Schür, C., Weil, C., Baum, M. et al. (2021). Incubation in wastewater reduces the multigenerational effects of microplastics in *Daphnia magna*. *Environmental Science & Technology* 55 (4): 2491–2499.

116 Straub, S., Hirsch, P.E., and Burkhardt-Holm, P. (2017). Biodegradable and Petroleum-based microplastics do not differ in their ingestion and excretion but in their biological effects in a freshwater invertebrate *Gammarus fossarum*. *International Journal of Environmental Research and Public Health* 14 (7): 774.

117 Blarer, P. and Burkhardt-Holm, P. (2016). Microplastics affect assimilation efficiency in the freshwater amphipod *Gammarus fossarum*. *Environmental Science and Pollution Research* 23 (23): 23522–23532.

118 Weber, A., Scherer, C., Brennholt, N. et al. (2018). PET microplastics do not negatively affect the survival, development, metabolism and feeding activity of the freshwater invertebrate *Gammarus pulex*. *Environmental Pollution* 234: 181–189.

119 Gerhardt, A., Schaefer, M., Blum, T. et al. (2020). Toxicity of microplastic particles with and without adsorbed tributyltin (TBT) in Gammarus fossarum (Koch, 1835). *Foundamental and Applied Limnology* 194 (1): 57–65.

120 Redondo-Hasselerharm, P.E., Falahudin, D., Peeters, E.T.H.M. et al. (2018). Microplastic effect thresholds for freshwater benthic macroinvertebrates. *Environmental Science & Technology* 52 (4): 2278–2286.

121 De Sales-Ribeiro, C., Brito-Casillas, Y., Fernandez, A. et al. (2020). An end to the controversy over the microscopic detection and effects of pristine microplastics in fish organs. *Scientific Reports* 10 (1): 12434.

122 Chen, Q., Lackmann, C., Wang, W. et al. (2020). Microplastics lead to hyperactive swimming behaviour in adult Zebrafish. *Aquatic Toxicology* 224: 105521.

123 Lei, L., Wu, S., Lu, S. et al. (2018). Microplastic particles cause intestinal damage and other adverse effects in zebrafish Danio rerio and nematode *Caenorhabditis elegans*. *Science of the Total Environment* 619–620: 1–8.

124 Lu, Y., Zhang, Y., Deng, Y. et al. (2016). Uptake and accumulation of polystyrene microplastics in Zebrafish (*Danio rerio*) and toxic effects in liver. *Environmental Science & Technology* 50 (7): 4054–4060.

125 Qiang, L., Lo, L.S.H., Gao, Y. et al. (2020). Parental exposure to polystyrene microplastics at environmentally relevant concentrations has negligible transgenerational effects on zebrafish (*Danio rerio*). *Ecotoxicology and Environmental Safety* 206: 111382.

126 Qiang, L. and Cheng, J. (2021). Exposure to polystyrene microplastics impairs gonads of zebrafish (*Danio rerio*). *Chemosphere* 263: 128161.

127 Colwill, R.M. and Creton, R. (2011). Locomotor behaviors in zebrafish (*Danio rerio*) larvae. *Behavioural Processes* 86 (2): 222–229.

128 Alnajar, N., Jha, A.N., and Turner, A. (2021). Impacts of microplastic fibers on the marine mussel, *Mytilus galloprovinciallis*. *Chemosphere* 262: 128290.

129 Hariharan, G., Purvaja, R., Anandavelu, I. et al. (2021). Accumulation and ecotoxicological risk of weathered polyethylene (wPE) microplastics on green mussel (*Perna viridis*). *Ecotoxicology and Environmental Safety* 208: 111765.

130 Li, -L.-L., Amara, R., Souissi, S. et al. (2020). Impacts of microplastics exposure on mussel (*Mytilus edulis*) gut microbiota. *Science of the Total Environment* 745: 141018.

131 Abidli, S., Pinheiro, M., Lahbib, Y. et al. (2021). Effects of environmentally relevant levels of polyethylene microplastic on *Mytilus galloprovincialis* (Mollusca: Bivalvia): filtration rate and oxidative stress. *Environmental Science and Pollution Research* 28 (21): 26643–26652.

132 Gu, H., Wei, S., Hu, M. et al. (2020). Microplastics aggravate the adverse effects of BDE-47 on physiological and defense performance in mussels. *Journal of Hazardous Materials* 398: 122909.

133 Capolupo, M., Valbonesi, P., and Fabbri, E. (2021). A comparative assessment of the chronic effects of Micro- and Nano-plastics on the physiology of the mediterranean mussel *Mytilus galloprovincialis*. *Nanomaterials* 11 (3): 649.

134 Cole, M., Liddle, C., Consolandi, G. et al. (2020). Microplastics, microfibers and nanoplastics cause variable sub-lethal responses in mussels (*Mytilus* spp.). *Marine Pollution Bulletin* 160: 111552.

135 Provenza, F., Piccardo, M., Terlizzi, A. et al. (2020). Exposure to pet-made microplastics: particle size and pH effects on biomolecular responses in mussels. *Marine Pollution Bulletin* 156: 111228.

136 Hamm, T. and Lenz, M. (2021). Negative impacts of realistic doses of spherical and irregular microplastics emerged late during a 42 weeks-long exposure experiment with blue mussels. *Science of the Total Environment* 778: 146088.

137 Cappello, T., De Marco, G., Oliveri Conti, G. et al. (2021). Time-dependent metabolic disorders induced by short-term exposure to polystyrene microplastics in the Mediterranean mussel *Mytilus galloprovincialis*. *Ecotoxicology and Environmental Safety* 209: 111780.

138 Jinhui, S., Sudong, X., Yan, N. et al. (2019). Effects of microplastics and attached heavy metals on growth, immunity, and heavy metal accumulation in the yellow seahorse, *Hippocampus kuda* Bleeker. *Marine Pollution Bulletin* 149: 110510.

139 Gardon, T., Morvan, L., Huvet, A. et al. (2020). Microplastics induce dose-specific transcriptomic disruptions in energy metabolism and immunity of the pearl oyster *Pinctada margaritifera*. *Environmental Pollution* 266: 115180.

140 Revel, M., Châtel, A., Perrein-Ettajani, H. et al. (2020). Realistic environmental exposure to microplastics does not induce biological effects in the Pacific oyster *Crassostrea gigas*. *Marine Pollution Bulletin* 150: 110627.

141 Horn, D.A., Granek, E.F., and Steele, C.L. (2020). Effects of environmentally relevant concentrations of microplastic fibers on Pacific mole crab (*Emerita analoga*) mortality and reproduction. *Limnology and Oceanography Letters* 5 (1): 74–83.

142 Yu, P., Liu, Z., Wu, D. et al. (2018). Accumulation of polystyrene microplastics in juvenile *Eriocheir sinensis* and oxidative stress effects in the liver. *Aquatic Toxicology* 200: 28–36.

143 Murphy, J. (2001). An overview of additives. In: *Additives for Plastics Handbook*, 2e (ed. T. Page), 1–3. Amsterdam: Elsevier Science.

144 Nerin, C., Alfaro, P., Aznar, M. et al. (2013). The challenge of identifying non-intentionally added substances from food packaging materials: a review. *Analytica Chimica Acta* 775: 14–24.

145 Groh, K.J., Backhaus, T., Carney-Almroth, B. et al. (2019). Overview of known plastic packaging-associated chemicals and their hazards. *Science of the Total Environment* 651: 3253–3268.

146 Hermabessiere, L., Dehaut, A., Paul-Pont, I. et al. (2017). Occurrence and effects of plastic additives on marine environments and organisms: a review. *Chemosphere* 182: 781–793.

147 Net, S., Sempéré, R., Delmont, A. et al. (2015). Occurrence, fate, behavior and ecotoxicological state of phthalates in different environmental matrices. *Environmental Science & Technology* 49 (7): 4019–4035.

148 Bridson, J.H., Gaugler, E.C., Smith, D.A. et al. (2021). Leaching and extraction of additives from plastic pollution to inform environmental risk: a multidisciplinary review of analytical approaches. *Journal of Hazardous Materials* 414: 125571.

149 Boyle, D., Catarino, A.I., Clark, N.J. et al. (2020). Polyvinyl chloride (PVC) plastic fragments release Pb additives that are bioavailable in zebrafish. *Environmental Pollution* 263: 114422.

150 Suhrhoff, T.J. and Scholz-Böttcher, B.M. (2016). Qualitative impact of salinity, UV radiation and turbulence on leaching of organic plastic additives from four common plastics — a lab experiment. *Marine Pollution Bulletin* 102 (1): 84–94.

151 Sánchez-Avila, J., Tauler, R., and Lacorte, S. (2012). Organic micropollutants in coastal waters from NW Mediterranean Sea: sources distribution and potential risk. *Environment International* 46: 50–62.

152 Mackintosh, C.E., Maldonado, J., Hongwu, J. et al. (2004). Distribution of phthalate esters in a marine aquatic food web: comparison to polychlorinated biphenyls. *Environmental Science & Technology* 38 (7): 2011–2020.

153 Jemec, A., Horvat, P., Kunej, U. et al. (2016). Uptake and effects of microplastic textile fibers on freshwater crustacean *Daphnia magna*. *Environmental Pollution* 219: 201–209.

154 Bejgarn, S., MacLeod, M., Bogdal, C. et al. (2015). Toxicity of leachate from weathering plastics: an exploratory screening study with *Nitocra spinipes*. *Chemosphere* 132: 114–119.

155 Atugoda, T., Vithanage, M., Wijesekara, H. et al. (2021). Interactions between microplastics, pharmaceuticals and personal care products: implications for vector transport. *Environment International* 149: 106367.

156 Fu, L., Li, J., Wang, G. et al. (2021). Adsorption behavior of organic pollutants on microplastics. *Ecotoxicology and Environmental Safety* 217: 112207.

157 Yu, F., Yang, C., Zhu, Z. et al. (2019). Adsorption behavior of organic pollutants and metals on micro/nanoplastics in the aquatic environment. *Science of the Total Environment* 694: 133643.

158 Mato, Y., Isobe, T., Takada, H. et al. (2001). Plastic resin pellets as a transport medium for toxic chemicals in the marine environment. *Environmental Science & Technology* 35 (2): 318–324.

159 Teuten, E.L., Rowland, S.J., Galloway, T.S. et al. (2007). Potential for plastics to transport hydrophobic contaminants. *Environmental Science & Technology* 41 (22): 7759–7764.

160 Cao, Y., Zhao, M., Ma, X. et al. (2021). A critical review on the interactions of microplastics with heavy metals: mechanism and their combined effect on organisms and humans. *Science of the Total Environment* 788: 147620.

161 Munoz, M., Ortiz, D., Nieto-Sandoval, J. et al. (2021). Adsorption of micropollutants onto realistic microplastics: role of microplastic nature, size, age, and NOM fouling. *Chemosphere* 283: 131085.

162 Rochman, C.M., Hoh, E., Hentschel, B.T. et al. (2013). Long-term field measurement of sorption of organic contaminants to five types of plastic pellets: implications for plastic marine debris. *Environmental Science & Technology* 47 (3): 1646–1654.

163 Holmes, L.A., Turner, A., and Thompson, R.C. (2014). Interactions between trace metals and plastic production pellets under estuarine conditions. *Marine Chemistry* 167: 25–32.

164 Jemec Kokalj, A., Horvat, P., Skalar, T. et al. (2018). Plastic bag and facial cleanser derived microplastic do not affect feeding behaviour and energy reserves of terrestrial isopods. *Science of the Total Environment* 615: 761–766.

165 Qi, K., Lu, N., Zhang, S. et al. (2021). Uptake of Pb(II) onto microplastic-associated biofilms in freshwater: adsorption and combined toxicity in comparison to natural solid substrates. *Journal of Hazardous Materials* 411: 125115.

166 Lee, H., Lee, H.-J., and Kwon, J.-H. (2019). Estimating microplastic-bound intake of hydrophobic organic chemicals by fish using measured desorption rates to artificial gut fluid. *Science of the Total Environment* 651: 162–170.

167 Wakkaf, T., Allouche, M., Harrath, A.H. et al. (2020). The individual and combined effects of cadmium, polyvinyl chloride (PVC) microplastics and their polyalkylamines modified forms on meiobenthic features in a microcosm. *Environmental Pollution* 266: 115263.

168 Zhang, R., Wang, M., Chen, X. et al. (2020). Combined toxicity of microplastics and cadmium on the zebrafish embryos (*Danio rerio*). *Science of the Total Environment* 743: 140638.

169 Liu, X., Shi, H., Xie, B. et al. (2019). Microplastics as both a sink and a source of bisphenol A in the marine environment. *Environmental Science & Technology* 53 (17): 10188–10196.

170 Qiao, R., Sheng, C., Lu, Y. et al. (2019). Microplastics induce intestinal inflammation, oxidative stress, and disorders of metabolome and microbiome in zebrafish. *Science of the Total Environment* 662: 246–253.

171 Corinaldesi, C., Canensi, S., Dell'Anno, A. et al. (2021). Multiple impacts of microplastics can threaten marine habitat-forming species. *Communications Biology* 4 (1): 431.

11

Chemicals Associated with Microplastics in Urban Waters

Yali Wang[1,2,3,*]

[1] *Hebei Key Laboratory of Close-to-Nature Restoration Technology of Wetlands, School of Eco-Environment, Hebei University, Baoding, Hebei, China*
[2] *Xiong'an Institute of Eco-Environment, Hebei University, Baoding, China*
[3] *Institute of Life Science and Green Development, Hebei University, China*
[*] *Corresponding author*

11.1 Introduction

Plastic is considered one of the greatest inventions of the last century but has since become a serious environmental pollution issue. Until the 1920s, low-density polyethylene (LDPEs) and polybrominated diphenyl ethers (PBDEs) were found in the ecosystem since plastic was invented [1]. Since the 1950s, plastics have surged consumption in commercially and were widely applied in food packaging, medical equipment, agriculture, construction industry, etc. with the rapid development of synthetic plastics [2]. Approximately 8.3 billion tons of plastic have been produced worldwide since it was invented. Plastics are used in all walks of life, such as transportation, catering services, clothing, and construction, because of their characteristics of light, portable, good insulation, wear resistance etc. Therefore, their production has increased exponentially, and load of plastic waste discharge will continue to grow, especially developing countries. It was reported that plastics production was 5 Mega ton (Mt) in the 1950s, while they increased to over 300 Mt in 2015 [3, 4]. What's more, it is speculated that the calculation of plastics in the environment would be an additional 155–265 Mt by 2060 [5].

Statistically, approximately 5.8 million metric tons plastic were discharged into the oceans around the world in the mid-1970s. At that time, the United Nations Environment Assembly recognized plastic waste as causing potential toxicity to marine life [6, 7], therefore, marine plastic waste first attracted scientists attention. However, it was not until the 1990s that studies began to widely report the environmental behavior and toxicity of plastics [8]. The ecotoxicity and environmental influence of raw materials for plastic production was first summarized by the European Union's Existing Chemicals Regulation in 1993 [9]. Soon afterwards,

Microplastics in Urban Water Management, First Edition. Edited by Bing-Jie Ni, Qiuxiang Xu, and Wei Wei.
© 2023 John Wiley & Sons, Inc. Published 2023 by John Wiley & Sons, Inc.

Berkeley's Ecology Center published a plastics task force report, in which it stated different types of plastics would have biological effects on human health [10]. Although many scientists and environmental groups in the world have declared the potential ecological risks of plastics, it was not until 2014 that marine plastic waste pollution was a consensus, and was then recognized as an international agenda issue at the 45th G7 summit [11]. Recently, a definition of microplastics (MPs) was proposed based on size <5 mm by the European Commission.

There are two types of MPs discussed based on origin (primary microplastics and secondary microplastics). The first category refers to a tiny size (micrometer or even nanoscale) of plastics from original manufacture, typically textiles, personal care products, and drug components, including woolen yarn, facial cleansers, and drug vectors [12]. Another type of MPs is secondary microplastics, derived from the breakdown of larger plastic debris and is considered the dominant type of MPs detected in the water eco-environment [13]. In 1994, the World Health Organization (WHO) identified polybrominated diphenyl oxides (PBDPOs, a main raw material for plastic synthesis) as one of the potential contaminants. The next year, it was registered in the Organization for Economic Cooperation [14, 15]. According to the previous documents, abrasion by mechanical forces, such as waves and irradiation by ultraviolet light, are the two dominating physicochemical processes for breaking large plastics into MPs [16]. Relative to primary plastics, secondary MPs are usually discovered in chemical raw material, agricultural equipment, daily necessities, and plastic garbage, i.e., fishing nets, agricultural films, and shopping bags [17].

With the increasing use and application of MPs, they have been detected in various environments. The US National Research Council reported that approximately 580 Mt garbage has been discharged into global oceans, and 70–80% was estimated to have resulted from effluents. Moore et al. first reported MPs in a freshwater environment in 2005. Since then, there has been increasing interest in MPs in the aquatic environment, as witnessed by the number of publications [1]. Moreover, it was stated that high content of MPs and phthalic acid esters (PAEs) were generally found in areas surrounding anthropogenic activities. Microplastics have been found in all samples taken from an aquatic system located in the Indian subcontinent with 2–64 particles/L present in water samples and 15–632 particles/kg/dw in sediment samples [18]. Interestingly, scientists recently detected certain chemicals on MPs, such as polyhydroxyalkanoates, bisphenol A, nonylphenol, and phenanthrene, classified in two types. One type is additives for plastics production, and the other comes from the surrounding environment sorbed by plastics [19]. However, because of the limitations of detection and extraction methods, it is difficult to distinguish whether these chemicals on MPs are additives or sorption from the surroundings. Many countries around the world have found this phenomenon to a different extent, such as the USA, China, and Korea. Moreover, researchers have begun to pay more and more attention to chemicals associated with MPs [20]. It was summarized that the concentrations of PCBs, PAHs, and HCHs on MPs received from large cities were much higher than that from suburbs or small cities. Specially, over 2000 ng/g polychlorinated biphenyls (PCBs) were detected on MPs in Ookushi beaches and the São Paulo State coast [21].

What's more, environmental scientists focus on the sorption correlation and the influence factors for MPs and chemicals [22]. It was found that the type of MPs and the properties of contaminants are the two main influential factors that can directly determine the sorption level and efficiency. Hydrophobic organic pollutants, such as biphenyls and polycyclic aromatic hydrocarbons (PAHs), are more easily adsorbed on MPs, shown by their high octanol partition coefficient. Polystyrene (PS), polypropylene (PP), polyethylene (PE), and polyvinyl chloride (PVC) are the main types of MPs in the environment [8]. Generally, PE has higher sorption affinities for organic chemicals than PP and PS because of their different polarity and functional groups [23]. Studies reported pH, plastic aging, salinity, and other external factors also impacted the sorption behavior between MPs and organic pollutants [24–26]. Salinity is beneficial to the adsorption of chemicals by MPs, and its influence was concerned with chemicals' properties, while pH, as another important environmental factor, could also impact adsorption by changing the structure of MPs and their electrostatic interactions [27]. The adsorption kinetics of MPs for chemicals are mostly associated with the surrounding environment and influencing factors, but there is little research on interactions between chemicals in MPs and its fragments, the release of chemical additives for MPs, and their environmental behaviors. Thus, this chapter aims to review the chemical species, concentration, release in MPs and its fragments, their relationship (adsorption or resolution), and coupled environmental toxicity [28].

11.1.1 Chemicals in Microplastics and Its Fragments

Recently, increasing attention has been paid to chemicals in plastic particles. There are three primary types of chemical contaminants on MPs, namely, organic pollutants, heavy metals, and the additives from plastics [4, 9]. Generally, the first two types were more commonly reported.

Microplastics have two characteristics (large surface area and hydrophobicity) that contribute to adhesion of environmental contaminants, especially hydrophobic organic pollutants, likely antibiotics, paregoric, personal care products, polychlorinated biphenyls, polybrominated diethers, polycyclic aromatic hydrocarbons, etc. [7]. The sum of concentrations of PCBs, PAHs, and OCs respectively were 9.0 – 29.8, 806 – 7528, 199 – 1202 mg/L in microplastics fragments sampled from Fabeiro beach, while PAHs concentration ranged 104–3595 ng/g dw in southwestern of Taiwan [3, 4]. Other reports have shown similar findings. In

Guangdong eastern beaches, 16 PAHs and 8 OCPs on microplastics were detected at 11.2–7710 and 2.2–1970 ng/g, respectively [29]. Relative to other microplastic particles (pellets or fragments), the mean PHA concentration on foam was the highest, ranging 81–7714 ng/g [11]. Moreover, it was declared that there is ubiquitous pollution of microplastics associated with PAEs. In Renuka Lake, it can be observed that the concentrations of MPs and PAEs in surface sediment vary from 15 to 632 n/kg and 6 to 357 ng/g, respectively, and the greatest abundance of microplastics and highest dose of PAEs were detected in the western part of the investigated lake [9]. China, as a major economic development country, has detected many plastics and chemical pollutants in various environments. In the Qiantang River (QR) and Hangzhou Bay (HZ), Maria et al. inspected the interrelation between PCBs and microplastics [6]. Over 800 particles were identified as microplastics and the dominant polymer type was PE at 60% of the total microplastics. Meanwhile, all sampling sites detected different concentrations of PCBs ranging 1.13–1.65 ng/g sediment [1]. Content analysis concluded that microplastic concentration showed a contrary tendency in the vertical distribution and microplastic shapes played an important role in the adsorbing PCBs. However, the environmental concentration relationship between microplastics and chemicals was not uniform, and showed different behavior in various sites [10]. PAH and its homologs were detected in all microplastics in Muskegon Lake, but smaller size microplastics contributed a negative effect, illustrating PAH concentrations in sediment was several orders of magnitude lower than polyethylene chlorinated. The concentration of PCBs associated with PEC was 676 ng/g, suggesting PCBs are unlikely to present a significant toxicity threat to aquatic organisms [15].

It was demonstrated that most of the adsorption processes of chemicals on microplastics were spontaneous and exothermic [30]. Adsorption kinetics models can systematically simulate adsorption characteristics of chemicals on microplastics, and are generally described by Lagrange and Freundlich adsorption isotherms. Dibutyl phthalate (DBP) adsorption isotherms on microplastics of polystyrene and polypropylene were well suited with the Freundlich model, with R^2 over 0.95 at temperatures 15–30°C. However, the sorption capacity of DBP were higher on polystyrene than on polypropylene [31]. For amoxicillin and phenol, polyethylene terephthalate, polyethylene, polyvinyl chloride, and polystyrene played an important role on their adsorption to plastics. The Langmuir model better described the adsorption equilibrium of phenol and amoxicillin with maximum capacities from 1.25 to 2.80 mg/g and 4.03 to 8.80 mg/g [32]. Li et al. explored pesticides (difenoconazole, buprofezin, imidacloprid) adsorption on microplastics through a series of adsorption tests and concluded that they were all well fitted by pseudo-first-order kinetics and the Freundlich isotherm model [24]. As we know, hydroxyl group content of pollutants has great effects on in their environmental behavior. Anthracene (ANT)-attached microplastics experiments revealed that hydroxy derivatives (OHAs) of anthracene adsorption on polyvinyl chloride microplastics were well fitted by the linear model, Langmuir model, and Freundlich model with R^2 ranging 0.905–0.993, 0.876–0.998, and 0.835–0.996, suggesting their interactions were well described as mono- or multi-layer adsorption [30]. In addition, the monohydroxy derivatives of phenanthrene adsorption on microplastics showed a good fit for Freundlich isotherm and pseudo-second-order models with R^2 both over 0.9 [32]. Antibiotics, as widespread pollutants in the surroundings, have been reported to adsorb on microplastics in large quantities. Bao et al. investigated the adsorption of tetracyclines by polyethylene with adsorption capacities ranging from 53.5 ± 3.4 to 64.4 ± 2.3 mg/g, and adsorption kinetics experiments showed that the three typical tetracyclines adsorption processes were well matched by the Freundlich isotherm and pseudo-second-order models [33]. While the adsorption isotherm of norfloxacin on microplastics showed a good fit by linear model in simulated natural water but was well explained by the Freundlich model in surface water [34]. Other researchers had obtained and demonstrated similar phenomena, like the adsorption of oxytetracycline, chlortetracycline, and amoxicillin on foams and polyethylene microplastics [35–37]. Except the traditional mathematical models, new modified mathematical models were also proposed to deeply elucidate the adsorption mechanism. Lončarski et al. applied a modified mathematical model to investigate the transfer kinetics of four typical polycyclic aromatic hydrocarbons and their adsorption processes on microplastics [38].

It was well known that the interadsorption relation between chemicals and microplastics depends on two kinds of factors: one is the characteristic of chemicals and microplastics (such as octanol-water partition coefficients, hydrophobicity for chemicals, type, surface area, colors, aging, composition, morphology, and degree of crystallinity of microplastics); the other is the adsorption environmental conditions (such as pH, salinity, and temperature) [25]. Microplastic particles possessed analogous properties to sediment particles with strong adsorption to hydrophobic pollutants. Therefore, microplastics contain a great adsorption ability for hydrophobic pollutants, especially hydrophobic persistent organic pollutants, e.g., polycyclic aromatic hydrocarbons (PAHs), polychlorinated biphenyls (PCBs), and antibiotics. The order of adsorption capacity of PCBs by different microplastics was polyethylene>polyvinyl chloride>polystyrene>polyethylene terephthalate [38]. Along with the surface area and particle size of microplastics, categories of polymers also have a great influence on the adsorption relationship between organic pollutant and microplastics [39]. Moreover, the adsorption partition coefficient (K_d) and functional groups were the crucial factors for the adsorption process of hydrophilic pollutants [40]. Adsorption isotherms for perfluorooctanesulfonate (PFOS) and perfluorooctanesulfonamide (FOSA), the most typical hydrophilic pollutants, on polyethylene, polyvinyl chloride, and polystyrene microplastics fit the linear model well [41]. On the other hand, microplastics contain a variety of shapes and colors according to their source and ingredients, primarily including pellets, foam, fibers, and fragments. Particularly, it has been demonstrated that microplastic shapes would seriously impact the adsorption process for polychlorinated biphenyls [25]. For example, polyhydroxyalkanoates were in higher concentrations on foam microplastics than fragments or films microplastics, but they were detected at lower concentrations on white microplastics than on yellow microplastics [41, 42]. Plastic exposed to the environment, damaged by water wave, light, heat, or other factors, is called aged plastic and leads to the adsorption binding sites increasing and the functional groups (such as hydrogen bonding among oxygen-containing functional groups) on the surface of microplastics during in the aging process [43]. Therefore, aged microplastics have a large surface area and their adsorption mechanisms mainly are the partition, electrostatic interaction, and hydrogen bonding [44]. It was reported aged polyethylene microplastics possessed more active adsorption sites that have greater ability to adsorb methylene blue with pseudo-second-order adsorption kinetics. Experimental results revealed a rate-limiting chemical adsorption process in methylene blue and microplastics [38]. Pristine and aged microplastics have different adsorption capacities and mechanisms. The adsorption capacities of ciprofloxacin by aged polystyrene and polyvinyl chloride were enhanced ca. 23.3% and 20.4% compared with the virgin microplastics [45].

pH is another important environmental factor. The effect of pH on the adsorption of chemicals by microplastics contributes to electrostatic interactions. Alkaline pH was beneficial to the adsorption process between chemicals and microplastics because high alkalinity can break the surface of the microplastics and expand their surface area. However, desorption often appeared under acidic conditions [34, 46]. As reported by You et al., 29.7% of methylene blue was adsorbed at pH 3.0, but reached 90.3% at pH 11 [39]. Other researchers have confirmed similar results. However, there was a certain degree of difference for the optimum adsorption pH with different pollutants and microplastics. Previous studies have reported the maximum adsorption of oxytetracycline on microplastic occurred at pH 5, but the maximal adsorption capacity of tetracycline on polyethylene occurred at pH 6 [33, 34]. Salinity in the environment plays an important role in the entire ecosystem and also greatly impacts the adsorption process between chemicals and microplastics [25]. Some studies reported high solution salinity caused an increase in the adsorption capacity of hydrophobic organic pollutants on microplastics. On the contrary, high NaCl showed a severe inhibiting effect on the adsorption of pesticides by polyethylene film microplastics [24]. It was stated that the presence of salinity and dissolved organic matter (DOM) greatly inhibited norfloxacin adsorption by microplastics by competition with adsorption sites. In the Jiang'an River, the adsorption capacity of norfloxacin decreased ca. 19.7–41.2% in the presence of coexisting NaCl and higher pH condition [34].

The above-mentioned influencing factors were the most commonly investigated indicators, but beyond that, temperature, ionic substances (likely heavy metals, surfactants, and humic acid), and biological films also impact the adsorption relationship between chemical pollutant and microplastics [47, 48]. The adsorption experiments for anthracene and polyvinyl chloride were investigated at 25, 35, and 45°C, and results showed that lower temperature was beneficial to anthracene adsorption by microplastics [30]. Similar experimental phenomena were observed during the adsorption process for bisphenol and polyvinyl chloride microplastics [50]. Adsorption capacity was enhanced when humic acid was abundant in the adsorption environment. π-π conjugation between humic acid and the microplastic surface increased electrostatic attraction between oxytetracycline and microplastics [35]. Furthermore, biofilm and extracellular polymeric substances have a higher ability to adsorb tetracycline and polycyclic aromatic hydrocarbons on microplastics, from increased competition interactions and surface area [48].

Chemisorption and physisorption function simultaneously occur in the adsorption process of chemicals and microplastics [25, 36]. In general, chemical sorption is irreversible as a result of covalent bonds forming between chemicals and solid phase. On the contrary, the desorption phenomenon for chemical pollutants and microplastics contributes to physical adsorption. Organic pollutants were adsorbed on the microplastic with different interactions, mainly including electrostatic interactions, partition, hydrophobic interactions, hydrogen bonding, van der Waals forces, and π-π interactions [51, 52]. Electrostatic interactions between pollution and microplastics are closely related to pH. Microplastics electrostatically adsorb positive organic matter when their zero point of charge is lower than the pH in the adsorption environment, while microplastics show an electrostatic repulsion to the organic pollutants with acid dissociation constants (pK_a) lower than the pH of the adsorption environment [33]. This adsorption mechanism primarily exists in microplastic and ionizable compounds, such as tetracycline, chloramphenicol, clarithromycin, and other antibiotics. It is well known that the pH_{pzc} of polyvinyl chloride is 3.41 and polystyrene and polyvinyl chloride are negative at the solution for adsorption pH<7.0. The influence of pH on tylosin (TYL) adsorption by polystyrene and polyvinyl chloride was investigated by Guo et al. [53]. The proportion of TYL^+ decreased when adsorption environmental pH increased from 3.0 to 7.0, indicating that pH reduces the adsorption capacity of TYL on polystyrene and polyvinyl chloride microplastics [53]. Wu et al. reported bisphenols incur partial ionization and even had a higher ionization for bisphenol F because of its 7.55 pK_a. The adsorption of bisphenol F by microplastics was restrained by electrostatic repulsion between microplastics and a partial anion dissociated from bisphenols [54]. The partition effect often appears on an adjacent water layer at the surface between pollutant and microplastic owing to van der Waals forces. This adsorption mechanism usually demostrates a linear adsorption isotherm with the n value near one. Strongly linear adsorption was observed in aliphatic and aromatic organic compounds on polyethylene microplastics [54]. In addition, the partition effect also played a significant role on the adsorption of antibiotics by microplastics. Cuo et al. verified polystyrene and polyethylene linearly adsorbed antibiotics by the Freundlich model in freshwater systems. Additionally, diethyl phthalate (DEP) and dibutyl phthalate (DBP) adsorption on microplastics was also described by the linear adsorption model [55].

As we know, most organic pollutants have low water solubility and high fat solubility shown by the high octanol/water partition coefficient (K_{ow}/$Log_{K_{ow}}$). However, plastic is characterized by high hydrophobicity and the non-wetting surface has similar characteristics to sediment [3, 4]. Thus, microplastic surfaces are extremely susceptible to adsorption of organic pollutants. Gui et al. investigated 13 organic chemicals adsorption on polyethylene and chlorinated polyethylene microplastics in a freshwater environment, and certified the octanol/water partition coefficient values of these chemicals played an important role in their adsorption model [43, 56]. Hüffer and Hofmann also affirmed the hydrophobicity of compounds in the microplastic adsorption process is a crucial factor [57]. As well, some studies reported the adsorption of bisphenol, polyaromatic hydrocarbons, organochlorinated pesticides, and diphenyl dichloroethane by different types of microplastics is impacted by hydrophobic interactions [23, 57–59]. As shown in Figure 11.1, van der Waals forces, covalent bonds (hydrogen bonding, halogen bonding), and π-π interactions are the interaction mechanisms involved in the adsorption process for contaminants and microplastics [48]. Van der Waals forces are recognized as the weak interactions between molecules, typically forming at the surface of aliphatic polymers, e.g., polyethylene or polyvinyl chloride. Guo et al. found microplastics adsorb phenol, naphthalene, lindane, and 1-naphthol through van der Waals forces [60]. It was reported that weak van der Waals forces occupied a significant position in the adsorption process between polyethylene microplastics and alkylphenol polyoxyethylene ether sulfate contaminants (such as tributyl phosphate and trichloroethyl phosphate) [60]. π-π interaction is another weak adsorption function that specially occurs on aromatic compounds, thus it is a critically important force in the adsorption of benzene ring-containing organic contaminants by microplastics when both possess similar chemical structure [61]. Zhang et al. observed that π-π interaction is one of the representative adsorption mechanisms for norfloxacin adsorption by microplastics in real surface water [36]. Ghaffar et al. demonstrated that π-π interactions promoted the adsorption of dialkyl phthalates on microplastics, but π-π interactions weaken with microplastics aging, resulting from the benzene ring-containing molecules of microplastic falling off [62]. Like π-π interactions, hydrogen bonds are a kind of noncovalent bond interaction, and also greatly affect the adsorption behavior between organic contaminants and microplastics. Hydrogen bonds and multivalent cationic bridging were the two ionic effects in the

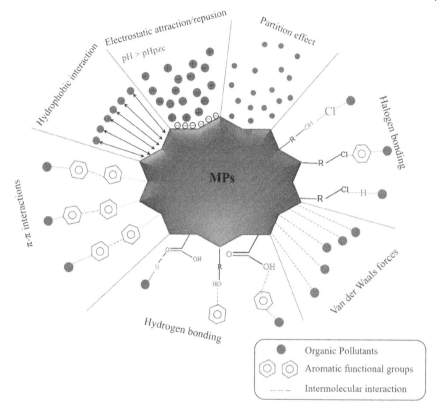

Figure 11.1 Mechanisms of organic pollutant adsorption by microplastics [63] / with permission of Elsevier.

adsorption process of oxytetracycline to beached foams [35]. A large number of studies investigated the adsorption characteristics and mechanism of organic compounds (e.g., polychlorinated biphenyls, dichlorodiphenyltrichloroethane, and phenanthrene and its monohydroxy derivative) on various of microplastics (such as polystyrene, polyethylene, and polyvinyl chloride) and found hydrogen bonds mainly acted on the polar polyamide and hydrophilic organics. This observation was also speculated by Liu et al. and this experiment analyzed the adsorption behavior of ciprofloxacin on microplastics through Fourier-transform infrared (FTIR) spectrometry [13]. Moreover, it was reported the binding energy orders for adsorption of phenanthrene by polyvinyl chloride microplastic were 2-monohydroxy derivatives > 1-monohydroxy derivatives > 9- and 4-monohydroxy derivatives > phenanthrene itself, indicating the hydrogen bonding of CH/π interaction impacted this adsorption behavior to a certain extent.

Aside from organic pollutants, plastic polymers are also regarded as an adsorber and transfer device for metal ions. Up to now, there are abundant reports demonstrating that microplastics adsorb heavy metals from surroundings, such as rivers, estuaries, sediments, soils, etc. The commonly detected metals on microplastics include Cu, Pb, Mn, Zn, Cd, and Hg, and their concentrations were closely associated with the corresponding environmental concentration. It was reviewed that the content of heavy metals in microplastics were usually higher than that in surface water. In the Musi River, the two highest concentration metals were found to be Pb and Cu. Their concentrations in surface water were detected from 0.152 to 2218 and 0.012 to 0.365 mg/kg, and these average concentrations respectively were 0.470 and 0.091 mg/kg for microplastics [17]. Similar phenomena (metals attached to microplastics) were found in the coastal areas of the Bohai Sea and Yellow Sea of China. In the Malaysian Coast, seven metals (Cd, As, Ni, Zn, Cu, Pb, and Hg) were detected adhered in microplastics [64]. Based on literature investigation, it was found that most researchers mainly discussed the adsorption behavior of metals on plastics and their interaction mechanisms. The adsorption isotherms of metals on microplastic were usually nonlinear, and under normal conditions the processes were well fitted by Langmuir and Freundlich isotherms. Shen et al. found Pb^{2+} adsorption on aged nylon microplastics were perfectly simulated by the Langmuir and Freundlich isotherms [49]. Wang et al. demonstrated biofilms played a positive role in the migration of Cu^{2+} in microplastics and impacted its adsorption characteristics on microplastics [51]. In this situation, the Freundlich model perfectly fitted the adsorption and desorption isotherms of Cu^{2+} on microplastics, further verifying the benefit of biofilm in the interrelation between metal and microplastics. In addition, other mathematical models have also been applied to analyze the adsorption process, promoting researchers further understanding the interrelation between metal and microplastics. It was reported that the Pb^{2+} chemisorption process on microplastics could be well simulated by the Elovich and pseudo-second-order equations with a maximum adsorption concentration of 1.05 mg/g on aged nylon [65]. Shen et al. had concluded that the second-order kinetic model also could appropriately explain Pb^{2+} adsorption kinetics on microplastics [49].

Just as organic contaminants on microplastics, the properties of microplastics/metals and their adsorption environmental conditions, such as plastic aging, types, pH, concentration of organic matter, and others, the adsorption capacities of metals were high on aged microplastics because of their large surface area and more adsorption sites. Generally, the adsorption capacity is aged polyvinyl chloride > aged polystyrene > unaged microplastics. The presence of functional groups (e.g., C–H and -OH) in microplastics is the decisive reason for the adsorption of exogenous chemicals. When many ions coexist, microplastics selectively adsorb chemicals with competition at active adsorption sites [65]. The adsorption capacity of Pb^{2+} on polymethyl methacrylate (PMMA), polyethylene (PE), and polypropylene (PP) respectively were 4.21, 2.01, and 1.57 mg/g in a condition without interference by pH or organic matter, while the maximum adsorption capacity was increased to 7.3 mg/g on microplastics with surfactant addition [65]. On the contrary, the fluvic acid concentration brought a negative impact on the adsorption capacity of Pb^{2+} on microplastics. Adsorption environmental conditions are another important influence factor like solution pH and salinity. It was reported that increasing the pH is conductive to the adsorption of metals on microplastics. Some adsorption experiments showed that the adsorption efficiency of Pb^{2+} reached 91% at pH 6, but Shen et al. concluded the neutral or weak alkaline environment was the optimum condition for Pb^{2+} adsorption by microplastics [49]. Additionally, the adsorption capability of Cr^{6+} was 0.03 mg/g on 2 g/L polyethylene at pH 5, while its adsorption amount increased 10.7 and 13 times when PE dosage increased to 14 g/L with sodium dodecyl benzene sulfonate (SDBS) addition. Furthermore, the concentration of sodium chloride had a remarkable influence on the metal adsorption on microplastics. The adsorption of Pb^{2+} on microplastics observably decreased with increasing solution salinity from competition of metal ions and external ions for adsorption sites of microplastics [49]. Furthermore, the adsorption mechanism of heavy metals on microplastics significantly contained surface complexation (with electrostatic interactions of functional groups) and precipitation/coprecipitation (See Figure 11.2). Microplastic aging increased the content of carboxyl functional groups, such as C-OH, C=O, C-O-C, O-C=O, impacting the adsorption capacity [66]. Simultaneously, the adsorption capacity and adsorption mechanism of Cd^{2+} by polystyrene were analyzed via adsorption kinetics experiments and Fourier

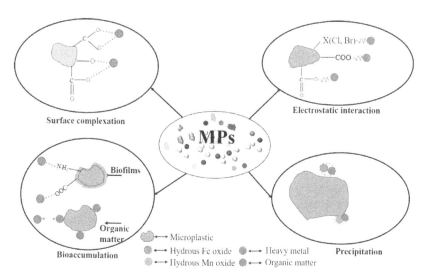

Figure 11.2 The main adsorption mechanism of heavy metals on microplastics [48] / with permission of Elsevier.

transform infrared spectroscopy analysis, indicating Cd^{2+} adsorption on polystyrene was a complex process with physisorption and chemisorption coexisting [67]. Similar adsorption behavior was observed in other studies. It was reported the adsorption peaks (C=O, C-O-C) of Pb^{2+} on polymethylmethacrylate were changed with surfactant addition [48].

11.1.2 Chemical Additives in Plastic Consumer Products

Plastic is a polymer compound made from monomeric raw materials via addition polymerization or polycondensation, and its deformation resistance is moderate, between fiber and rubber. Since plastics are light, convenient, and economic, they have become essential commodities in our daily life and their production increases year by year [7, 9]. It is well known that plastics are malleable at elevated temperature and curing or cooling under low temperature, sometimes called thermoplastics. Meanwhile, it suffers irreversible plasticity and forms a three-dimensional polymer network structure in the process of plastic products making. Therefore, plastic products not only contain some residual monomers, catalysts, and solvents from the synthesis process, but also certain intentionally added chemical substances. Most chemical additives in plastic consumer products are designed to enhance or confer specific properties on final plastic products [68–70].

Plastic additives determine the main function of plastic by adjusting and modifying plastic physical and chemical characteristics. According to plastic production requirements, plasticizers, curing agents, flame retardants, stabilizers, and antioxidants are the five most common types of plastic additive ingredients [71]. Figure 11.3 describes the most commonly applied plastic additives. Plasticizers are one of crucial chemical materials that improve the durability and flexibility of polymers. They weaken the secondary valence bond between resin molecules and reduce crystallinity, but increase plasticity and mobility of resin molecules, and finally enhance the flexibility of plastic production [68]. Among these compounds, phthalates and adipates are the most typical plasticizers. Curing agents are essential additives for plastic adhesives, coatings, and castables, and seriously affect the mechanical properties, heat resistance, water resistance, and corrosion resistance [72]. 4,4-diamino-diphenylmethane and 2,2-dichloro-4,4-methylenedianiline are representative curing agents. Flame retardants are mainly designed for polymer materials via a mechanical hybrid method for addition and their fundamental aim is to retard or inhibit plastic ignition [73]. According to the application of flame-retardant classification, there are largely reaction and additive type. To date, additive flame retardants consist of organic flame retardants (e.g., hexabromocyclododecanes, polybrominated diphenyl ethers, and triphenyl phosphate) and inorganic flame retardants (e.g., antimony trioxide, aluminum hydroxide, and magnesium hydroxide) [74]. Furthermore, stabilizers and antioxidants are both functional reagents to keep and improve persistent

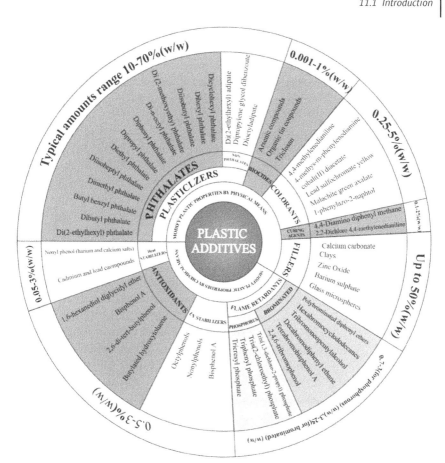

Figure 11.3 The dominating types and components of plastic additives [69] / with permission of Elsevier.

stability of polymer plastic, rubber, and synthetic fiber to prevent decomposition and aging by ultraviolet, temperature, or moderate oxidation [70]. With the final plastic product demand, stabilizers with different functions have been developed, such as heat and ultraviolet stabilizers [75]. Antioxidants, like ultraviolet stabilizers, are used to delay the oxidation process of plastic polymers and extend the service life of plastic products. Octylphenols and nonylphenols, cadmium and lead compounds, and 2,6- di-tert-butylphenol and butylated hydroxytoluene respectively are the plastic additives used as ultraviolet stabilizers, heat stabilizers, and antioxidants. It is worth noting that bisphenol A has multiple functions and identities as a plastic additive, and is commonly applied as antioxidant, UV stabilizer, and even plasticizer [76, 77]. Except for the above-mentioned additive categories, other functions and classes of additives have also been invented and used, including colorants, slip agents, reinforcements, filler materials, biocides, and blowing agents. The specific colors of plastic products arise from colorants, a group of insoluble organic or inorganic particles.

Recent studies have explored and confirmed microplastic chemical additives can leach from plastics consumer products or plastic itself into real environment since most of them are noncovalently bonded to plastics and easily released and transported into the surroundings [15]. Plastic additives release throughout the life of consumer products, bringing potential ecological risk and toxicity to organisms, even humans [68]. Plastic additive release behavior has three key processes: firstly, they diffuse from inside polymer substances, then partition at the interface of solid and liquid, and finally disperse into the surrounding medium. Hexabromocyclododecane (HBCDs) is the most important component for expanded polystyrene buoys. Analysis and detection found its concentration was lower on the outside than in inside, suggesting the release of HBCDs started at the surface of polystyrene buoys and continued throughout its life [69]. Plastic additives were detected in ubiquitous environmental regions, e.g., municipal wastewater, atmospheric dust, river, runoff, sediment, sludge, and even human exposure [7, 9, 17]. León et al. investigated the environmental distribution of frequent pollutants from Calblanque, Cape Cope, and La Llana (located on the southeastern Spanish Mediterranean coast) and found plastic additives and personal care products were the most detected pollutants, and urban and tourism activities were one of the important pollutant sources [72]. The migration and transformation of plastic additives might represent another important environmental pollutant. Besides, these chemicals can be transferred to microorganisms, aquatic plants and animals, soil microorganisms, crops, and even humans in the process of plastic use, disposal, and recycling, indicating they are a significant source of hazardous chemicals to organisms in the surrounding environment. Jang et al. found extracellular polymeric substance generated in marine debris was enriched in abundant hexabromocyclododecane. Furthermore, the relevant mussels also exhibited these chemical additives, suggesting chemical additives in plastic debris were migrated into inhabitants [78]. Similar results and phenomena have been found by other researchers [79, 80]. As a result, it is necessary to attract attention to probing and investigating plastic additive hazards.

11.2 The Release of Chemicals from Microplastics and Environmental Levels

Microplastics contain many chemical additives (e.g., plasticizers, antioxidants, heat stabilizers), and these additives act in great roles for the final plastic products. Polyvinyl chloride products are one of the largest general plastic productions in the world, widely used in building materials, industrial products, plastic flooring, artificial leather, pipe, wire and cable, packaging film, sealing materials, fibers, and they constitute approximately 80% plasticizer. Plasticizer producers spend billions of dollars worldwide with most applied in Asia, especially in China. Approximately 85% of plasticizers are phthalic acid esters, accounting for ca. 70% of the global market [80]. Phthalic acid esters (PAEs), also referred to phthalates, are typical plasticizers made up of aromatic diesters with a variety of branched or linear alcohols. Certain chemical additives, likely heat stabilizers, mixed with some metallic compounds (such as $CaCO_3$, $Mg(OH)_2$, $PbSO_4$) are incorporated into plastics to improve or control the specific functionality of plastic products. Moreover, potentially harmful additives to ecosystems generally are not chemically but physically bound in polymer molecules and are easily released from plastics by abrasive and oxide surfaces [81]. Compared to whole bulk plastics, microplastics can potentially release more toxic additives because of the shorter diffusion distance from free field to the surface [82]. On the other hand, studies have also verified that organotin stabilizers, phenolic antioxidants, and phthalate plasticizers were released, transferred, transformed, and leached from commercial polyvinyl chloride, polypropylene, and polyethylene microplastics during the plastic ageing process. Based on investigation and research about plastic additives, the following content mainly describes the typical additives released from plastics and their relevant environmental concentration.

11.2.1 Phthalic Acid Esters (PAEs)

Plasticizers are known as one of the important additives for flexible or semi-rigid plastic products, such as food packaging and personal care products, raincoats, medical devices. Their dosages are higher than other additive substances, with 10–70% abundance in plastic materials. The US, China, and Western Europe were the three biggest consumers of plasticizers accounting for 11, 42, and 14%, respectively [83]. The main function of plasticizers is to decrease the hardness, modulus, and the embrittlement temperature of the polymer, while enhancing the elongation, flexure, and flexibility of plastic products [81]. There are two classes of plasticizers, namely phthalates and nonphthalates. Their chemical structures are basically diesters, triesters, and heterojunctions. Nonphthalates are less commonly used than phthalates and usually serve as a replacement for phthalate, mainly including adipates, benzoates, phosphoric esters, trimellitic esters, and sebacates. Among these, adipates are the most widely used plasticizers for polyvinyl chloride production or niche materials because they are uneconomical for commercial products [82]. It was reported that the application of nonphthalates would make plastic products flexible. Phthalates account for the vast majority of the plasticizer market and are the most popular and dominant plasticizers. PAEs are used in every aspect of our daily life, and widely used in a variety of household products, such as shampoos, perfumes, face wash, creams, wrapping paper, toothpaste, cosmetics, etc. [84]. The total global PAEs production exceeds 8 000 000 tons per year with annual demand increasing, and over 235 000 tons are produced or imported in the USA [85]. For China, according to the statistics, ca. 1.36×10^6 tons was used and consumed in 2010, followed by a 7.7% increase in consumption rate from 2010 to 2015 [86, 87]. Di-2-ethylhexyl phthalate, diethyl phthalate, di-2-methoxyethyl phthalate, diisoheptyl phthalate, di-2-ethylhexyl terephthalate, dibutylbenzyl phthalate, and dipentyl phthalate are the most typically utilized chemical constituents for PAEs. Among these chemical substances, di-2-ethylhexyl phthalate was the largest proportion for PAEs (accounting for ca. 37%), which is a 390.5 g/mol high molecular weight phthalate. Di-2-ethylhexyl phthalate is reported as the majority of the plasticizer market with above 3 000 000 tons annual production, and is usually applied for PVC production via hydrogen bonding or Van der Waals forces [83]. Because of their special combination to polymeric matrix, PAEs are easily released into the ecological environment during the manufacture, application, and discharge through direct discharge or leaching, resulting in potential ecological risk [88]. PAEs have been listed as priority and restrictive organic pollutants by the United States Environmental Protection Agency (USEPA) according to their toxicity, persistence, and abundance. In addition, China's Environmental Monitoring Center and USEPA both considered dimethyl phthalate and dibutyl phthalate as priority pollutants [89]. The European Food Safety Authority suggested that the daily intake for di-n-octyl phthalate, dibutyl phthalate, di-2-ethylhexyl phthalate, and butyl benzyl phthalate cannot respectively be more than 0.15, 0.01, 0.01, and 0.5 μg/kg/day [88]. Besides, most studies have confirmed PAEs are also endocrine disruptors. Their release from plastics is derived from pharmaceutical, medical, waste management, and construction industries. Therefore, PAEs release and environmental content have attracted attention and become a hot spot in the ecological environment field.

As we know, PAEs are widely applied abundant products, resulting in frequent detection in water environment, sediment, sludge, soil, atmosphere, biota, and even the human body [90–93]. PAEs are used as an additive in polyolefin, polyvinyl chloride, vinyl acetate, or metallocene polyethylene plastic with concentrations ranging 2.59–282 000 mg/kg. The release rate of PAEs from plastics largely depended on the film thickness and most of these processes follow a first-order kinetic model [91]. An investigation of most delivery companies of China found approximately 181.44–5320.64 ng/g of PAEs was released from plastic express packaging bags after 48 h leaching [92]. Di-2-ethylhexyl phthalate was reported as the most widely used plasticizer among these PAEs. In addition, it was reported that di-2-ethylhexyl phthalate was one of the most refractory phthalates because of its long hydrocarbon chain. Di-2-ethylhexyl phthalate and di-n-butyl phthalate are the primary PAE congeners and the most abundant pollutants in the global ecosystem.

In the past decade, PAEs have been constantly detected in different aquatic environment. In drinking water obtained from polyethylene terephthalate bottles, twenty-one PAEs were detected, particularly dibutyl phthalate, di-n-butyl phthalate, and dimethyl phthalate [82]. Additionally, abundant studies have reported and documented PAEs existence in surface water, lakes, or rivers, such as Yangtze River Delta, Jiulong River, Taihu Lake, and Chaohu Lake. It was found that environmental concentrations have a higher magnitude and a much greater mean in the lakes located in China than that in Asan Lake and Kaveri River. However, relative to these two sites, there are much higher PAEs concentration in the river on the border between France and Belgium [94]. The concentration of PAEs ranged 226.70–7086.62 ng/g dry weight (dw) in the surface sediment of Songhua river, reported by Yang et al. Di-2-ethylhexyl phthalate and di-n-butyl phthalate were the main detected PAEs, accounting for 65.02–99.07% and 1.50–55.43%, respectively [95]. A breadth of reports on concentration of PAEs, and it has been reported that the sludge environments were the most common and frequent detection sites. According to literature research on PAEs, the concentration ranges in sludge respectively were 69 000–320 000, 7400–138 600, and 632–68 700 μg/kg TSS in Swedish, TiaWang, and the mainland of China [95]. PAEs distribution and content are diverse. Moreover, some studies have explored and reported their spatial distribution in sediment. As the most frequently contaminated PAEs, the concentration for di-2-ethylhexyl phthalate and dibutyl phthalate were as high as 1000 μg/L in groundwater [96]. The distribution levels were up to mg/L in freshwater systems. Abundant studies have found concentrations of di-2-ethylhexyl phthalate were 55, 28, and 5.4 mg/L in the Wuhan section of the Yangtze River, the Yangtze River Delta region, and the Three Gorges Reservoir, which suggests the Yangtze River is widely contaminated by PAEs [97].

Since it is difficult to degrade PAEs under natural conditions, as expected, soil, sludge or sediment has become permanently contaminated by PAEs [97]. It was found the PAEs detected concentration was up to 33.39 mg/kg in agricultural land long covered with agricultural film. In China, the concentration of PAEs was higher in the North than in the East. For example, studies reported the concentration of PAEs were mean 3738 ng/g in Hebei province, while ranging 654–2603 ng/g in Hangzhou Bay, Taizhou Bay, and Wenzhou Bay [96, 98, 99]. Recent studies revealed that di-2-ethylhexyl phthalate accounted for the highest proportion in PAEs pollution, followed by di-isobutyl phthalate, then di-n-butyl phthalate, and dimethyl phthalate with 52, 22, 15, and 5.0%, respectively. Lü et al. comprehensively summarized and reviewed the soil contamination by PAEs all over China, and concluded their total concentration were 0–35.4 mg/kg in agricultural soil environments [100]. It is worth noting that dibutyl phthalate and dimethyl phthalate had a higher detection content with almost 100% detection frequencies. Similar phenomena were also reported by Luo et al. and results showed more than 50% sampling sites detected dimethyl phthalate, dibutyl phthalate and di 2-ethyl hexyl phthalate [101]. The concentrations of PAEs in sediment sampled from coastal estuaries of southeast China were slightly higher than that in the Yellow Sea or Asan Lake of Korea, but were greatly lower than that in the Pearl River or the Yellow River [101, 102]. Di-2-ethylhexyl phthalate and dibutyl phthalate pollutants not only were serious in soil or sediment, but their detected levels were particularly high in activated sludge. For example, approximately 25.4 µg/L and 67.8 µg/g dw of di-2-ethylhexyl phthalate were found in wastewater treatment plant effluent and sludge, and the reported dibutyl phthalate concentrations were over 97.4 mg/kg in activated sludge [103-105]. According to the European Union's identification for organic pollutants, most PAEs are volatile or semi-volatile, resulting in dispersion into the air by means of vapors or particulates. Studies reported detected di-2-ethylhexyl phthalate concentration up to 1439 ng/m^3 in urban air in Europe. Besides, in the household environment, di-2-ethylhexyl phthalate is most commonly detected in bedrooms and living rooms, where the concentrations respectively were 455 and 462 mg/g [96].

11.2.2 Bisphenol A

During the plastic processing process, stability and oxidation resistance of relative products are maintained by adding certain chemical reagents. Bisphenol A is one of the most important plastic additives, and is applied to make polycarbonate and epoxy resins. Bisphenol A is widely applied in food packaging, paper coatings, beverage bottles, water pipes, toys, electronic equipment, thermal paper receipts, and inner lining of cans [106]. Bisphenol S, bisphenol AF, and bisphenol F are the most commonly applied structural analogs for bisphenol A. They are reportedly endocrine-disrupting compounds, and the structure and properties lead to a ubiquitous presence in the environment [107]. Bisphenol S is typically used for food packaging, thermal paper, cans, and nursing bottles as a heat resistant structure. Bisphenol AF can be found in polymer synthesis for polyesters, fluoroelastomers, polyamides, and polycarbonate copolymers, while bisphenol F is used for food packaging, water pipes, dental adhesives, and restoration [90].

Because of its low hydrophobicity ($\log_{Kow} = 3.40$), bisphenol A is not easily sorbed to plastic. On the other hand, with the application increasing, the manufacture of bisphenol A was more than 8 million tons per year worldwide, and is expected to reach 10.2 million metric tons by 2022 [108]. According to previous literature, bisphenol A can rapidly release and leach into the surroundings. Bisphenol A was released from polycarbonate bottles at 2–10, 17.6–324, and 200–300 ng/L, and its release was related to temperature [95]. Casajuana and Lacorte reported bisphenol A release concentration ca. 3–8 ng/L at 30°C, but the concentration rose to 119 ± 9.7 ng/L at 70°C [109]. Thus, bisphenol A and its homologs were frequently detected in environment. Researchers have recently given increased attention to environmental levels [110, 111]. Selvaraj et al. has reported the concentration of bisphenol A ranged 9.82–36.0, 6.63–136, and 2.83–6.00 ng/L, respectively, in Tamiraparani River, Kaveri River, and Vellar River [112]. Data showed that approximately 41.9, 29.7, and 67.6% of sampling sites found bisphenol F, bisphenol A, and bisphenol S, with the highest concentration of 333, 438, and 14 800 ng/L, respectively, in the Yamuna River [112]. Similar concentration distribution of these bisphenol analogues was found on 11 beaches in Hong Kong (82.4–989 ng/g). Aside from rivers and beaches, bisphenol A is also ubiquitously and frequently detected in surface water and drinking water. The concentration distributions of bisphenol A were reported ranging 4.4–8000 ng/L in Asia, Europe, and North America [113]. These pollutants in the above-mentioned locations were mainly derived from wastewater discharge. Studies reported the detection level and detection frequency of bisphenol A (16.7–8920 ng/L) were both higher in wastewater than its homologs (16.7–438 ng/L for bisphenol S, 16.7–333 ng/L for bisphenol F) [112]. Although bisphenol A contamination is frequently reported and described in aquatic ecosystem, it was more serious in the land ecosystem [114]. Bisphenol A and its homologs were detected in the soil environment via the ultra-performance liquid chromatography-tandem mass spectrometry, with detected concentrations of 78.2, 166.0, and 212.9 ng/g dry weight for bisphenol P, bisphenol A, and bisphenol F, respectively [115]. Bisphenols also demonstrated widespread presence in sludge where organic matter content and properties affect the adsorption and accumulation of bisphenol A [116, 117]. Besides water and soil environments, air is also one of the major pollution sources for bisphenol A, resulting from the thermal destruction process of the related materials. Current reports on bisphenols distribution in air are associated with particles because of the low volatility. Vasiljevic and Harner summarized the properties, sources, and global content of bisphenols in air, and found bisphenol A levels was the highest with 1.1×10^6 pg/m^3 in a low-tech e-waste recycling site located in China [118].

References

1 Moore, C.J., Lattin, G.L., and Zellers, A.F. (2005). Working our way upstream-A snapshot of land-based contributions of plastic and other trash to coastal waters and beaches of Southern California. Algalita Marine Research Foundation, Long Beach.

2 Fred, O.H., Bhagwat, G., Oluyoye, I. et al. (2020). Interaction of chemical contaminants with microplastics: Principles and perspectives. *Science of the Total Environment* 706: 135978.

3 Geyer, R., Jambeck, J., and Law, K. (2017). Production, use, and fate of all plastics ever made. *Science of Advanced Materials* 3: 1–5.
4 Wang, F., Wong, C.S., Chen, D. et al. (2018). Interaction of toxic chemicals with microplastics: A critical review. *Water Research* 139: 208–219.
5 Khalid, N., Aqeel, M., Noman, A. et al. (2021). Linking effects of microplastics to ecological impacts in marine environments. *Chemosphere* 264 (Pt 2): 128541.
6 Fraser, M., Chen, L., Ashar, M. et al. (2020). Occurrence and distribution of microplastics and polychlorinated biphenyls in sediments from the Qiantang River and Hangzhou Bay, China. *Ecotoxicology and Environmental Safety* 196: 110536.
7 Rodríguez, C., Fossatti, M., Carrizo, D. et al. (2020). Mesoplastics and large microplastics along a use gradient on the Uruguay Atlantic coast: Types, sources, fates, and chemical loads. *Science of the Total Environment* 721: 137734.
8 Chen, C., Ju, Y., Lin, Y. et al. (2020). Microplastics and their affiliated PAHs in the sea surface connected to the southwest coast of Taiwan. *Chemosphere* 254: 126818.
9 Kumar, A., Behera, D., Bhattacharya, S. et al. (2021). Distribution and characteristics of microplastics and phthalate esters from a fresh water lake system in Lesser Himalayas. *Chemosphere* 15: 131132.
10 Klasios, N., Frond, H., Miller, E. et al. (2020). Microplastics and other anthropogenic particles are prevalent in mussels from San Francisco Bay, and show no correlation with PAHs. *Environmental Pollution* 271: 116260.
11 Shi, J., Sanganyado, E., Wang, L. et al. (2020). Organic pollutants in sedimentary microplastics from eastern Guangdong: Spatial distribution and source identification. *Ecotoxicology and Environmental Safety* 193: 110356.
12 Purwiyanto, A., Suteja, Y., Ningrum, P. et al. (2020). Concentration and adsorption of Pb and Cu in microplastics: Case study in aquatic environment. *Marine Pollution Bulletin* 158: 111380.
13 Liu, G., Zhu, Z., Yang, Y. et al. (2019). Sorption behavior and mechanism of hydrophilic organic chemicals to virgin and aged microplastics in freshwater and seawater. *Environmental Pollution* 246: 26–33.
14 Lo, H., Po, B., Li, L. et al. (2021). Bisphenol A and its analogues in sedimentary microplastics of Hong Kong. *Marine Pollution Bulletin* 164: 112090.
15 Steinman, A., Scott, J., Green, L. et al. (2020). Persistent organic pollutants, metals, and the bacterial community composition associated with microplastics in Muskegon Lake (MI). *Journal of Great Lakes Research* 46.
16 Niu, L., Li, Y., Li, Y. et al. (2021). New insights into the vertical distribution and microbial degradation of microplastics in urban river sediments. *Water Research* 188: 116449.

17 Zhou, Y., Liu, X., and Wang, J. (2019). Characterization of microplastics and the association of heavy metals with microplastics in suburban soil of central China. *The Science of the Total Environment* 694: 133798–133798.

18 Patchaiyappan, A., Dowarah, K., Ahmed, S. et al. (2021). Prevalence and characteristics of microplastics present in the street dust collected from Chennai metropolitan city, India. *Chemosphere* 269: 1287571.

19 Yu, F., Yang, C., Zhu, Z. et al. (2019). Adsorption behavior of organic pollutants and metals on micro/ nanoplastics in the aquatic environment. *Science of the Total Environment* 694: 133643.

20 Ravit, B., Cooper, K., Buckley, B. et al. (2019). Organic compounds associated with microplastic pollutants in New Jersey, U.S.A. surface waters. *AIMS Environmental Science* 6: 445–459.

21 Wang, T., Wang, L., Chen, Q. et al. (2020). Interactions between microplastics and organic pollutants: Effects on toxicity, bioaccumulation, degradation, and transport. *Science of the Total Environment* 748: 1424271.

22 Xia, Y., Zhou, J., Gong, Y. et al. (2020). Strong influence of surfactants on virgin hydrophobic microplastics adsorbing ionic organic pollutants. *Environmental Pollution* 265: 1150611.

23 Godoy, V., Martín-Lara, M., and Calero, M. (2020). The relevance of interaction of chemicals/pollutants and microplastic samples as route for transporting contaminants. *Process Safety and Environmental Protection* 138: 312–323.

24 Li, H., Wang, F., Li, J. et al. (2021). Adsorption of three pesticides on polyethylene microplastics in aqueous solutions: Kinetics, isotherms, thermodynamics, and molecular dynamics simulation. *Chemosphere* 264: 128556.

25 Gao, X., Hassan, I., Peng, Y. et al. (2021). Behaviors and influencing factors of the heavy metals adsorption onto microplastics: A review. *Journal of Cleaner Production* 319: 128777.

26 Mammol, F., Amoah, D., Gani, K. et al. (2020). Microplastics in the environment: Interactions with microbes and chemical contaminants. *Science of the Total Environment* 743: 140518.

27 Roman, M., Gutierrez, L., Dijk, L. et al. (2020). Effect of pH on the transport and adsorption of organic micropollutants in ion-exchange membranes in electrodialysis-based desalination. *Separation and Purification Technology* 252: 117487.

28 Luo, H., Liu, C., He, D. et al. (2022). Environmental behaviors of microplastics in aquatic systems: A systematic review on degradation, adsorption, toxicity and biofilm under aging conditions. *Journal of Hazardous Materials* 423: 126915.

29 Chen, C., Ju, Y., Lim, Y. et al. (2020). Microplastics and their affifiliated PAHs in the sea surface connected to the southwest coast of Taiwan. *Chemosphere* 254: 126818.

30 Bao, Z., Chen, Z., Lu, S. et al. (2021). Effects of hydroxyl group content on adsorption and desorption of anthracene and anthrol by polyvinyl chloride microplastics. *Science of the Total Environment* 790: 148077.

31 Chi, J., Zhang, H., and Zhao, D. (2021). Impact of microplastic addition on degradation of dibutyl phthalate in offshore sediments. *Marine Pollution Bulletin* 162: 111881.

32 Hu, D. and Wang, L. (2016). Adsorption of amoxicillin onto quaternized cellulose from flax noil: Kinetic, equilibrium and thermodynamic study. *Journal of the Taiwan Institute of Chemical Engineers* 64: 227–234.

33 Bao, Z., Chen, Z., Zhong, Y. et al. (2021). Adsorption of phenanthrene and its monohydroxy derivatives on polyvinyl chloride microplastics in aqueous solution: Model fitting and mechanism analysis. *Science of the Total Environment* 764: 142889.

34 Chen, Y., Li, J., Wang, F. et al. (2020). Adsorption of tetracyclines onto polyethylene microplastics: A combined study of experiment and molecular dynamics simulation. *Chemosphere* 265: 129133.

35 Zhang, Y., Ni, F., He, J. et al. (2021). Mechanistic insight into different adsorption of norfloxacin on microplastics in simulated natural water and real surface water. *Environmental Pollution* 284: 117537.

36 Zhang, H., Wang, J., Zhou, B. et al. (2018). Enhanced adsorption of oxytetracycline to weathered microplastic polystyrene: Kinetics, isotherms and influencing factors. *Environmental Pollution* 243: 1550–1557.

37 Fan, X., Gan, R., Liu, J. et al. (2021). Adsorption and desorption behaviors of antibiotics by tire wear particles and polyethylene microplastics with or without aging processes. *Science of the Total Environment* 771: 145451.

38 Lončarski, M., Gvoić, V., and Prica, M. (2021). Sorption behavior of polycyclic aromatic hydrocarbons on biodegradable polylactic acid and various nondegradable microplastics: Model fitting and mechanism analysis. *Science of the Total Environment* 771: 145451.

39 You, H., Huang, B., Cao, C. et al. (2021). Adsorption–desorption behavior of methylene blue onto aged polyethylene microplastics in aqueous environments. *Marine Pollution Bulletin* 167: 112287.

40 Yu, Y., Mo, W., and Luukkonen, T. (2021). Adsorption behaviour and interaction of organic micropollutants with nano and microplastics – A review. *Science of the Total Environment* 797: 149140.

41 Vieira, Y., Lima, E., Foletto, E. et al. (2021). Microplastics physicochemical properties, specific adsorption modeling and their interaction with pharmaceuticals and other emerging contaminants. *Science of the Total Environment* 753: 141981.

42 Wang, F., Shih, K., and Li, X. (2015). The partition behavior of perfluorooctanesulfonate (PFOS) and perfluorooctanesulfonamide (FOSA) on microplastics. *Chemosphere* 119: 841–847.

43 Gui, B., Xu, X., Zhang, S. et al. (2021). Prediction of organic compounds adsorbed by polyethylene and chlorinated polyethylene microplastics in freshwater using QSAR. *Environmental Research* 197: 111001.

44 Hanun, J., Hassan, F., and Jiang, J. (2021). Occurrence, fate, and sorption behavior of contaminants of emerging concern to microplastics: Influence of the weathering/aging process. *Journal of Environmental Chemical Engineering* 9: 106290.

45 Lin, L., Tang, S., Wang, X. et al. (2020). Adsorption of malachite green from aqueous solution by nylon microplastics: Reaction mechanism and the optimum conditions by response surface methodology. *Process Safety and Environmental Protection* 140: 2.

46 Liu, P., Lu, K., Li, J. et al. (2020). Effect of aging on adsorption behavior of polystyrene microplastics for pharmaceuticals: Adsorption mechanism and role of aging intermediates. *Journal of Hazardous Materials* 384: 121193.

47 Beiras, R., Verdejo, E., Campoy-Lopez, P. et al. (2021). Aquatic toxicity of chemically defined microplastics can be explained by functional additives. *Journal of Hazardous Materials* 406: 124338.

48 Cao, Y., Zhao, M., Ma, X. et al. (2021). A critical review on the interactions of microplastics with heavy metals: Mechanism and their combined effect on organisms and humans. *Science of the Total Environment* 788: 147620.

49 Shen, M., Song, B., Zeng, G. et al. (2021). Surfactant changes lead adsorption behaviors and mechanisms on microplastics. *Chemical Engineering Journal* 405: 126989.

50 Wu, P., Tang, Y., Jin, H. et al. (2020). Consequential fate of bisphenol-attached PVC microplastics in water and simulated intestinal fluids. *Environmental Science and Ecotechnology* 2: 100027.

51 Wang, Y., Wang, X., Li, Y. et al. (2020). Biofilm alters tetracycline and copper adsorption behaviors onto polyethylene microplastics. *Chemical Engineering Journal* 392: 123808.

52 Cortés-Arriagada, D. (2021). Elucidating the co-transport of bisphenol A with polyethylene terephthalate (PET) nanoplastics: A theoretical study of the adsorption mechanism. *Environmental Pollution* 270: 116192.

53 Guo, X., Pang, J., Chen, S. et al. (2018). Sorption properties of tylosin on four different microplastics. *Chemosphere* 209: 240–245.

54 Wu, P., Cai, Z., Jin, H. et al. (2019). Adsorption mechanisms of five bisphenol analogues on PVC microplastics. *Science of the Total Environment* 650: 671–678.

55 Guo, X., Liu, Y., and Wang, J. (2019). Sorption of sulfamethazine onto different types of microplastics: A combined experimental and molecular dynamics simulation study. *Marine Pollution Bulletin* 145: 547–554.

56 Liu, F., Liu, G., Zhu, Z. et al. (2019). Interactions between microplastics and phthalate esters as affected by microplastics characteristics and solution chemistry. *Chemosphere* 214: 688–694.

57 Hüffer, T. and Hofmann, T. (2016). Adsorption of non-polar organic compounds by microsized plastic particles in aqueous solution. *Environmental Pollution* 214: 194–201.

58 Lo, H., Wong, C., Tam, N. et al. (2019). Spatial distribution and source identifification of hydrophobic organic compounds (HOCs) on sedimentary microplastic in Hong Kong. *Chemosphere* 219: 418–426.

59 Beiras, R. and Tato, T. (2019). Microplastics do not increase toxicity of a hydrophobic organic chemical to marine plankton. *Marine Pollution Bulletin* 138: 58–62.

60 Guo, X., Wang, X., Zhou, X. et al. (2012). Sorption of four hydrophobic organic compounds by three chemically distinct polymers: Role of chemical and physical composition. *Environmental Science Technology* 46: 7252–7259.

61 Chen, S., Tan, Z., Qi, Y. et al. (2019). Sorption of tri-n-butyl phosphate and tris(2-chloroethyl) phosphate on polyethylene and polyvinyl chloride microplastics in seawater. *Marine Pollution Bulletin* 149: 110490.

62 Ghaffar, A., Ghosh, S., Li, F. et al. (2015). Effect of biochar aging on surface characteristics and adsorption behavior of dialkyl phthalates. *Environmental Pollution* 206: 502–509.

63 Torres, F., Salinas, D., Pizarro, I. et al. (2020). Sorption of chemical contaminants on degradable and non-degradable microplastics: Recent progress and research trends. *Science of the Total Environment* 757: 143875.

64 Fu, L., Li, J., Wang, G. et al. (2021). Adsorption behavior of organic pollutants on microplastics. *Ecotoxicology and Environmental Safety* 217: 112207.

65 Tang, S., Lin, L., Wang, X. et al. (2020). Pb (II) uptake onto nylon microplastics: Interaction mechanism and adsorption performance. *Journal of Hazardous Materials* 386: 121960.

66 Gao, L., Fu, D., Zhao, J. et al. (2021). Microplastics aged in various environmental media exhibited strong sorption to heavy metals in seawater. *Marine Pollution Bulletin* 169: 112480.

67 Huang, D., Xu, Y., Yu, X. et al. (2021). Effect of cadmium on the sorption of tylosin by polystyrene microplastics. *Ecotoxicology and Environmental Safety* 207: 111255.

68 Lang, M., Yu, X., Liu, J. et al. (2020). Fenton aging significantly affects the heavy metal adsorption capacity of polystyrene microplastics. *Science of the Total Environment* 722: 137762.

69 Carmen, S. (2021). Microbial capability for the degradation of chemical additives present in petroleum-based plastic products: A review on current status and perspectives. *Journal of Hazardous Materials* 402: 123534.

70 Al-Odaini, N., Shim, W., Han, G. et al. (2015). Enrichment of hexabromocyclododecanes in coastal sediments near aquaculture areas and a wastewater treatment plant in a semi-enclosed bay in South Korea. *Science of the Total Environment* 505: 290–298.

71 Groh, K., Backhaus, T., Carney-Almroth, B. et al. (2019). Overview of known plastic packaging-associated chemicals and their hazards. *Science of the Total Environment* 651: 3253–3268.

72 León, V., García-Agüera, I., Moltó, V. et al. (2019). PAHs, pesticides, personal care products and plastic additives in plastic debris from Spanish Mediterranean beaches. *Science of the Total Environment* 670: 672–684.

73 Yang, J., Wang, F., He, X. et al. (2021). Potential usage of porous autoclaved aerated concrete waste as eco-friendly internal curing agent for shrinkage compensation. *Journal of Cleaner Production* 320: 128894.

74 Zhang, S., Chu, F., Xu, Z. et al. (2021). Interfacial flame retardant unsaturated polyester composites with simultaneously improved fire safety and mechanical properties. *Chemical Engineering Journal* 426: 131313.

75 Xie, W., Guo, S., Liu, Y. et al. (2020). Organic-inorganic hybrid strategy based on ternary copolymerization to prepare flame retardant poly(methyl methacrylate) with high performance. *Composites Part B: Engineering* 203: 108437.

76 Ouyang, X., Liang, R., Hu, Y. et al. (2021). Hollow tube covalent organic framework for syringe filter-based extraction of ultraviolet stabilizer in food contact materials. *Journal of Chromatography A* 1656: 462538.

77 Barboza, L., Cunha, S., Fernandes, C. et al. (2020). Bisphenol A and its analogs in muscle and liver of fish from the North East Atlantic Ocean in relation to microplastic contamination. Exposure and risk to human consumers. *Journal of Hazardous Materials* 393: 122419.

78 Jang, M., Shim, W., Han, G. et al. (2016). Styrofoam debris as a source of hazardous additives for marine organisms. *Environmental Science Technology* 50: 4951–4960.

79 Jang, M., Shim, W., Han, G. et al. (2017). Widespread detection of a brominated flame retardant, hexabromocyclododecane, in expanded polystyrene marine debris and microplastics from South Korea and the Asia-Pacifific coastal region. *Environmental Pollution* 231: 785–794.

80 Jang, M., Shim, W., Han, G. et al. (2021). Relative importance of aqueous leachate versus particle ingestion as uptake routes for microplastic additives (hexabromocyclododecane) to mussels. *Environmental Pollution* 270: 116272.

81 Yao, S., Cao, H., Arp, H. et al. (2021). The role of crystallinity and particle morphology on the sorption of dibutyl phthalate on polyethylene microplastics: Implications for the behavior of phthalate plastic additives. *Environmental Pollution* 283: 117393.

82 Mo, L., Wang, Q., and Bi, E. (2021). Effects of endogenous and exogenous dissolved organic matter on sorption behaviors of bisphenol A onto soils. *Journal of Environmental Management* 287: 112312.

83 Pang, X., Skillen, N., Gunaratne, N. et al. (2021). Removal of phthalates from aqueous solution by semiconductor photocatalysis: A review. *Journal of Hazardous Materials* 402: 123461.

84 Ren, L., Wang, G., Huang, Y. et al. (2021). Phthalic acid esters degradation by a novel marine bacterial strain Mycolicibacterium phocaicum RL-HY01: Characterization, metabolic pathway and bioaugmentation. *Science of the Total Environment* 791: 148303.

85 Xiong, Y. and Pei, D. (2021). A review on efficient removal of phthalic acid esters via biochars and transition metals-activated persulfate systems. *Chemosphere* 277: 130256.

86 Agency, U.E.P. (2012). Phthalates action plan. Environmental Protection Agency, EPA US.

87 Emanuel, C. (2011). Plasticizer market update. In: 22nd Annual Vinyl Compounding Conference.

88 Wang, P., Wang, S., and Fan, C. (2008). Atmospheric distribution of particulate- and gas-phase phthalic esters (PAEs) in a Metropolitan City, Nanjing, East China. *Chemosphere* 72: 1567–1572.

89 Hu, H., Mao, L., Fang, S. et al. (2020). Occurrence of phthalic acid esters in marine organisms from Hangzhou Bay, China: Implications for human exposure. *Science of the Total Environment* 721: 137605.

90 You, Y., Wang, Z., Xu, W. et al. (2019). Phthalic acid esters disturbed the genetic information processing and improved the carbon metabolism in black soils. *Science of the Total Environment* 653: 212–222.

91 Gireeshkumar, A., Rahman, K., Balachandran, M. et al. (2018). Distribution and contamination status of phthalic acid esters in the sediments of a tropical monsoonal estuary, Cochin – India. *Chemosphere* 210: 232–238.

92 Wang, Y., Wang, Y., Xiang, F. et al. (2021). Risk assessment of agricultural plastic films based on release kinetics of phthalate acid esters. *Environmental Science Technology* 55: 3676–3685.

93 Zhang, L., Liu, J., He, J. et al. (2009). The occurrence and ecological risk assessment of phthalate esters (PAEs) in urban aquatic environments of China. *Asian Journal of Ecotoxicology* 24: 1–18.

94 Lee, Y., Lee, J., Choe, W. et al. (2019). Distribution of phthalate esters in air, water, sediments, and fish in the Asan lake of Korea. *Environment International* 126: 635–643.

95 Net, S., Rabodonirina, S., Sghaier, R. et al. (2015). Distribution of phthalates, pesticides and drug residues in the dissolved, particulate and sedimentary phases from transboundary rivers (France-Belgium). *Science of the Total Environment* 521: 152–159.

96 Yang, Y., Wang, H., Chang, Y. et al. (2020). Distributions, compositions, and ecological risk assessment of polycyclic aromatic hydrocarbons and phthalic acid esters in surface sediment of Songhua river, China. *Marine Pollution Bulletin* 152: 110923.

97 Chen, F., Chen, Y., Chen, C. et al. (2021). High-effifificiency degradation of phthalic acid esters (PAEs) by Pseudarthrobacter defluvii E5: Performance, degradative pathway, and key genes. *Science of the Total Environment* 794: 148719.

98 Li, B., Liu, R., Gao, H. et al. (2016). Spatial distribution and ecological risk assessment of phthalic acid esters and phenols in surface sediment from urban rivers in Northeast China. *Environmental Pollution* 409.

99 Hu, H., Fang, S., Zhao, M. et al. (2020). Occurrence of phthalic acid esters in sediment samples from East China Sea. *Science of the Total Environment* 722: 137997.

100 Lü, H., Mo, C., Zhao, H. et al. (2018). Soil contamination and sources of phthalates and its health risk in China: A review. *Environmental Research* 164: 417–429.

101 Luo, X., Shu, S., Feng, H. et al. (2021). Seasonal distribution and ecological risks of phthalic acid esters in surface water of Taihu Lake, China. *Science of the Total Environment* 768: 144517.

102 Zhang, Z., Zhang, H., Zhang, J. et al. (2018). Occurrence, distribution, and ecological risks of phthalate esters in the seawater and sediment of Changjiang River Estuary and its adjacent area. *Science of the Total Environment* 620: 93–102.

103 Liu, H., Cui, K., Zeng, F. et al. (2014). Occurrence and distribution of phthalate esters in riverine sediments from the Pearl River Delta region, South China. *Marine Pollution Bulletin* 83: 358–365.

104 Zhang, Z., Yang, G., Zhang, H. et al. (2020). Phthalic acid esters in the sea-surface microlayer, seawater and sediments of the East China Sea: Spatiotemporal variation and ecological risk assessment. *Environmental Pollution* 259: 113802.

105 Al-Saleh, I., Elkhatib, R., Al-Rajoudi, T. et al. (2017). Assessing the concentration of phthalate esters (PAEs) and bisphenol a (BPA) and the genotoxic potential of treated wastewater (final effluent) in Saudi Arabia. *Science of the Total Environment* 578: 440.

106 Gao, D., Li, Z., Wen, Z. et al. (2014). Occurrence and fate of phthalate esters in full-scale domestic wastewater treatment plants and their impact on receiving waters along the Songhua River in China. *Chemosphere* 95: 24–32.

107 Zhang, H., Quan, Q., Zhang, M. et al. (2020). Occurrence of bisphenol A and its alternatives in paired urine and indoor dust from Chinese university students: Implications for human exposure. *Chemosphere* 247: 125987.

108 Frankowski, R., Płatkiewicz, J., Stanisz, E. et al. (2021). Biodegradation and photo-Fenton degradation of bisphenol A, bisphenol S and fluconazole in water. *Environmental Pollution* 289: 117947.

109 Lin, L., Dong, L., Meng, X. et al. (2018). Distribution and sources of polycyclic aromatic hydrocarbons and phthalic acid esters in water and surface sediment from the Three Gorges Reservoir. *Journal of Environmental Sciences* 69.

110 Casajuana, N. and Lacorte, S. (2003). Presence and release of phthalic esters and other endocrine disrupting compounds in drinking water. *Chromatographia* 57: 649–655.

111 Czarny, K., Krawczyk, B., and Szczukocki, D. (2021). Toxic effects of bisphenol A and its analogues on cyanobacteria Anabaena variabilis and Microcystis aeruginosa. *Chemosphere* 263: 128299.

112 Selvaraj, K., Shanmugam, G., Sampath, S. et al. (2014). GC – MS determination of bisphenol A and alkylphenol ethoxylates in river water from India and their ecotoxicological risk assessment. *Ecotoxicology and Environmental Safety* 99: 13–20.

113 Lalwani, D., Ruan, Y., Taniyasu, S. et al. (2020). Nationwide distribution and potential risk of bisphenol analogues in Indian waters. *Ecotoxicology and Environmental Safety* 200.

114 Selvaraj, K., Shanmugam, G., Sampath, S. et al. (2018). Distributions of concentrations of bisphenol A in North American and European surface waters and sediments determined from 19 years of monitoring data. *Chemosphere* 201: 448–458.

115 Bavinck, M. (2018). Enhancing the wellbeing of Tamil fishing communities (and government bureaucrats too): The role of ur panchayats along the coromandel coast, India. In: *Social Wellbeing and the Values of Small-Scale Fisheries*, (D.S. Johnson, T.G. Acott, N. Stacey, J. Urquhart eds.), 175–194. Cham: Springer.

116 Xu, Y., Hu, A., Li, Y. et al. (2021). Determination and occurrence of bisphenol A and thirteen structural analogs in soil. *Chemosphere* 277: 130232.

117 Peng, G., Lu, Y., You, W. et al. (2020). Analysis of five bisphenol compounds in sewage sludge by dispersive solid-phase extraction with magnetic montmorillonite. *Microchemical Journal* 2020 157: 105040.

118 Vasiljevic, T. and Harner, T. (2021). Bisphenol A and its analogues in outdoor and indoor air: Properties, sources and global levels. *Science of the Total Environment* 789: 148013.

12

Interactions between Microplastics and Contaminants in Urban Waters

Tianyi Luo[1], Xiaohu Dai[1], and Bing-Jie Ni[2,*]

[1] *State Key Laboratory of Pollution Control and Resources Reuse, College of Environmental Science and Engineering, Tongji University, Shanghai, PR China*
[2] *Centre for Technology in Water and Wastewater, School of Civil and Environmental Engineering, University of Technology Sydney, Sydney, NSW, Australia*
* *Corresponding author*

Abbreviations

AChE	Acetylcholinesterase
AMX	Amoxicillin
ARB	Antibiotic-resistant bacteria
ARG	Antibiotic resistance gene
BaP	Benzo(a)pyrene
BeP	Benzo(e)pyrene
BPA	Bisphenol A
Chr	Chrysene
CIP	Ciprofloxacin
DT	Digestive tubules
Fla	Fluoranthene
FQL	fluoroquinolones
HDPE	High density polyethylene
HGT	Horizontal gene transfer
HOCs	Hydrophobic organic chemicals
MCL	Macrolides
MPs	Microplastics
NOR	Norfloxacin
PAHs	Polycyclic aromatic hydrocarbons
PBS	Polybutylene succinate

Microplastics in Urban Water Management, First Edition. Edited by Bing-Jie Ni, Qiuxiang Xu, and Wei Wei.
© 2023 John Wiley & Sons, Inc. Published 2023 by John Wiley & Sons, Inc.

PCBs	Polychlorinated biphenyls
PE	Polyethylene
PFOS	Perfluorooctane sulfonate
Phe	Phenanthrene
POPs	Persistent organic pollutants
PP	Polypropylene
PS	Polystyrene
PVC	Polyvinyl chloride
Pyr	Pyrene
ROS	Reactive oxygen species
SDZ	Sulfadiazine
SMX	Sulfamethoxazole
SSA	Specific surface area
TC	Tetracycline
TCS	Triclosan
TMP	Trimethoprim
UV	Ultraviolet

12.1 Introduction

Plastic has been in our daily lives for more than 120 years, and it was calculated that the plastic production was ca. 227 million tons in 2009, and increased to 368 million tons in 2019 [1]. As more than eight million tons of plastic litter are discharged into the worldwide aquatics environment each year [2], plastic pollution requires global attention. Following collision, fragmentation, and natural chemical aging processes, most plastic debris in freshwater and marine environments consequently gets converted to microplastics (MPs) [3]. MPs with a particle size < 5 mm have been extensively detected in the aquatic environment, which has led to rising concern as emerging environmental issue [1, 4].

Given the unique surface including porosity, large specific surface areas, and high hydrophobicity, MPs attract, adsorb, and accumulate diverse contaminants from aquatic environments [5]. Thus, MPs are widely regarded as a potential vector of various contaminants and can transport contaminants over long distances into the environment. Anthropological contaminants, including antibiotics, heavy metals, and organic compounds, are massively produced by industry and increasingly released into the aquatic system with human advancement and have an adverse effect on ecosystems and human health [6, 7]. Increasing studies concentrated on the coexistence of MPs and environmental contaminants have pointed out their interaction creates potential risk for the environment and human health [8-10]. Therefore, it is essential to summarize published studies about the combination of MPs and contaminants and their combined potential risk, thereby evaluating the environmental implication of both types of contaminants.

Pollution containing antibiotic resistance genes (ARGs) induced by antibiotic abuse has raised global concerns [11]. Specifically, gene exchange between antibiotic-resistant bacteria (ARB) and environmental microorganisms further exacerbates this problem [12]. Similar to MPs, ARGs and ARB are extensively distributed in various aquatic systems, such as rivers, marine ecosystems, and wastewater treatment plants (WWTPs) [13-16]. Therefore, the coexistence of MPs and ARGs are necessarily observed in multiple environments. Multiple studies found that MPs had ability to select ARB and ARGs to adsorb on their surface [17, 18], thereby influencing the distribution and dissemination of ARGs. Other studies pointed out that some environmental contaminants could induce co-selection on the enrichment of ARB and ARGs, further increasing ARGs pollution in the aquatic environment. Accordingly, the single selection and co-selection of MPs on ARB and ARGs will also be discussed in this chapter.

The purpose of this chapter is to summarize recent studies regarding the adsorption behavior of MPs as environmental contaminants and their potential threats to the ecosystem. For this objective, related studies are summarized, and the interaction of MPs and environmental contaminants (including antibiotics, heavy metals, and organic compounds) are discussed. The role of virgin and contaminant-adsorbed MPs on selective enrichment of ARB and ARGs is elaborated. Additionally, the associated influencing factors of adsorptive interaction, such as pH, temperature, salinity, and aging/weathering processes are summarized. Furthermore, joint potential risks of MPs and contaminants on the environment, organisms, and human beings are explained. Lastly, recommendations for further research are proposed to assess the potential risk of MPs and contaminants in the ecosystem.

12.2 Sorption of Contaminants on Microplastics

The adsorption process is vital for the interaction of MPs and contaminants in an aqueous system. Therefore, it is essential to summarize research regarding the adsorption behavior of MPs towards environmental contaminants. In this section, antibiotics, heavy metals, and organic compounds were selected as representative contaminants to illustrate the adsorption behavior of MPs towards environmental contaminants.

12.2.1 Antibiotics

Since they are used as pharmaceutical and personal care products, antibiotics existing in various environments have received extensive attention because of their environmental pollution and ecotoxicity [19, 20]. It was reported that in China alone, the amount of discharged antibiotics reached 91 100 tons in 2016 [21]. Among all antibiotic types, tetracyclines (TC), macrolides (MCL), fluoroquinolones (FQL), and sulfonamides (SDZ) were extensively detected in aquatic environments around the world [22]. Both antibiotics and MPs are ubiquitous; hence the combination of MPs and antibiotics is inevitable and has negative effects on aquatic organisms. Recently, additional researchers have focused on the interaction of MPs and antibiotics.

Many studies have concentrated on the role of MPs as antibiotic adsorbents. The mechanisms of adsorption of MPs for antibiotics primarily include hydrophobic interactions, hydrogen-bonding, and electrostatic interactions. Additionally, several new mechanisms are also proposed to reveal the antibiotics adsorption, such as halogen bonding, CH-π interactions, cation-π interactions, and negative charge-assisted hydrogen bonds [1]. Sorption of antibiotics on MPs varies with the types of MPs and antibiotics (Table 12.1). Multiple antibiotics can be adsorbed by MPs, including amoxicillin (AMX), ciprofloxacin (CIP), trimethoprim (TMP), SDZ, TC, etc. For instance, Li et al. studied the adsorption capacity of five types of MPs for five antibiotics in freshwater and seawater and observed that the adsorption amounts of MPs decreased with the hydrophobicity of the antibiotics (CIP > AMX > TMP > SDZ > TC) [22]. Their result also showed that PS showed stronger adsorption qualities for CIP, TMP, and SDZ than PE (Table 12.1), probably because π-π interactions at the aromatic surface existed with PS but not with PE [22]. Likewise, Guo et al. investigated the sulfamethoxazole (SMX) adsorption of six MPs, in which the strongest adsorption capacity for SMX was observed on PA (Table 12.1) [23]. Notably, other studies further pointed out that the adsorption of certain contaminants (including dissolved organic matter and heavy metals) can lead to the greater affinity of antibiotics to MPs [24, 25], thereby enhancing the interaction of MPs and antibiotics.

12.2.2 Heavy Metals

Heavy metals are also adsorbed on MPs. Heavy metal pollution in the environment has been regarded as a major worldwide threat. It was reported that ca. 40% of the lakes and rivers on earth were threatened by heavy metal pollution [28]. Massive quantities of heavy metals can induce numerous adverse effects, such as damaging ecosystems, influencing the growth and metabolism of organisms, and even posing threats on downstream animals and human health via the food chain [29]. Multiple studies demonstrated that heavy metals could be adsorbed by MPs

Table 12.1 Adsorption behavior of MPs for antibiotics.

MPs Type	Target antibiotic	Matrix	Q_{max} (ug/g)	References
PP	TMP	seawater	0.0597 ± 0.009	[22]
PS	TMP	seawater	0.166 ± 0.122	
PVC	TMP	seawater	0.034 ± 0.014	
PE	TMP	seawater	0.087 ± 0.005	
PA	TMP	seawater	0130 ± 0.0320	
PA	TC	seawater	0.088 ± 0.019	
PP	CIP	freshwater	0.615 ± 0.030	
PS	CIP	freshwater	0.416 ± 0.043	
PVC	CIP	freshwater	0.453 ± 0.009	
PE	CIP	freshwater	0.200 ± 0.014	
PA	CIP	freshwater	2.200 ± 0.657	
PP	TMP	freshwater	0.102 ± 0.014	
PS	TMP	freshwater	0.174 ± 0.039	
PVC	TMP	freshwater	0.481 ± 0.496	
PE	TMP	freshwater	0.154 ± 0.041	
PA	TMP	freshwater	0.468 ± 0.128	
PP	AMX	freshwater	0.294 ± 0.070	
PVC	AMX	freshwater	0.523 ± 0.368	
PE	AMX	freshwater	0.131 ± 0.028	
PA	AMX	freshwater	22.700 ± 22.600	
PA	TC	freshwater	3.840 ± 0.839	
PE	TC	2 mg/L TC solution	0.231	[26]
PS	TC	2 mg/L TC solution	0.055	
PVC	TC	2 mg/L TC solution	0.016	

(Continued)

Table 12.1 (Continued)

MPs Type	Target antibiotic	Matrix	Q_{max} (ug/g)	References
PE	TC	10 mg/L TC solution	0.053 ± 0.003	[27]
PE	Chlortetracycline hydrochloride (CTC)	10 mg/L CTC solution	0.063 ± 0.005	
PE	Oxytetracycline hydrochloride (OTC)	10 mg/L OTC solution	0.064 ± 0.002	
PA	SMX	2.4 mg/L SMX solution	96.400	[23]
PE	SMX	2.4 mg/L SMX solution	0.660	
PS	SMX	2.4 mg/L SMX solution	0.712	
PET	SMX	2.4 mg/L SMX solution	114.000	
PVC	SMX	2.4 mg/L SMX solution	2.800	
PP	SMX	2.4 mg/L SMX solution	6.900	

from the surrounding environment then transported to aquatic biota [30–32]. Therefore, it is essential to understand the behavior and mechanism of the adsorption of MPs for heavy metals.

Environmental monitoring programs indicated the occurrence and accumulation of heavy metals on MPs (Table 12.2). Compared to the surrounding aquatic environment, MPs surface showed a greater concentration of heavy metals because of larger surface area and greater polarity [33]. Zn, Fe, Pb, and Cu were the most commonly detected heavy metals on MPs, exceeding by 2–130 400 times the concentration of the surrounding aquatic environment [34–36]. It was found that concentrations of heavy metals on MPs varied with region, possibly depending on the heavy metal source. It is worth noting that of all the metals in Beijiang River (China) and Lake Garda (Italy), Ti demonstrated the highest adsorption concentration of MPs. In addition, Cr was the one of the most popular heavy metals on MPs in India but not China nor Thailand. Collectively, these studies prove that environmental heavy metals can be absorbed by MPs, and the adsorption behavior varies with the location of MPs.

Table 12.2 Interactions between MPs and heavy metals.

Polymer type	Metals (maximum amount)		Location	References
	Water/Sediments	MPs surface		
PE, PP, PS, PET, PVC		Pb, 13.5; Sn, 0.081	Ookushi Beach, Goto Islands, Japan	[37]
PE, PA, PET, PS, PVC		Ni, 1.1; Cd, 23.6; Pb, 219.7; Cu, 19.6; Ti, 1046.0	Lake Garda, Italy	[38]
PE, PP, PS, PVC		Cd, 3390; Pb, 5330	Sandy beaches in southwest England	[39]
PP, PE	Ni, 0.07; Cd, 2.458; Pb, 57.93; Cu, 79.5; Zn, 349.2; Ti, 28 575.2	Ni, 2.39; Cd, 17.56; Pb, 131.1; Cu, 500.6; Zn, 14 815.3; Ti, 38 823.7	Beijiang River, China	[33]
PP, PE, HDPE		Al, 45; Cu, 1.0; Fe, 228.0; Mn, 9.0; Zn, 8.0; Ti, 3.0	Beaches along the coast of São Paulo State in southeastern Brazil	[40]
PP, PVC, PE		Pb, 0.095	Sandy beaches of Plymouth, England	[41]
PE, PP, PS, PET, PA, PVC		As, 0.00; Cd, 0.00; Cu, 0.89; Fe, 302.0; Mn, 18.6; Ni, 0.15; Zn, 19.6	Sandy beaches in Hong Kong	[42]
PP, PE, PES, PVC, nylon	Pb, 0.068; Cu, 0.018	Pb, 2.21; Cu, 0.36	Musi River, South Sumatera Province, Indonesia	[30]
PP, PE, PS, Polyester	Cr, 1.1; Cu, 0.05; Ni, 0.05; Pb, 0.01	Cr, 2.95; Cu, 13.02; Ni, 0.78; Pb, 17.61	Chao Phraya River at the Tha Pra Chan area	[43]
PP, PE, PS	Cr, 0.01; Cu, 0.005; Pb, 0.005; Cd, 0.004; Zn, 0.003	Cr, 17.3; Cu, 14.5; Ni, 3.52; Pb, 38.67; Cd, 2.81; Zn, 391.2	Chao Phraya River Estuary, Thailand	[34]

(Continued)

Table 12.2 (Continued)

Polymer type	Metals (maximum amount)		Location	References
	Water/Sediments	MPs surface		
PE, PP, PET	Cr, 42.17; Ni, 27.1; Cu, 41.55; Zn, 199.2; Pb, 107.5; As, 8.68; Cd, 0.78; Hg, 1.13	Cr, 15.70; Ni, 6.0; Cu, 23.25; Zn, 185.3; Pb, 51.5; As, 6.53; Cd, 0.85; Hg, 0.08	Jinjiang Estuarine Mangrove Reserve, China	[35]
PE, PET	Sediments: As, 24.74; Cd, 6.87; Cr, 273.8; Cu, 268.5; Ni, 87.6; Pb, 11.13; Zn, 383.2; Surface water: As, 0.24; Cd, 1.23;0 Cr, 23.65; Cu, 23.14; Ni, 42.36; Pb, 12.15; Zn, 286.3	As, 4.51; Cd, 5.78; Cr, 342.2; Cu, 119.5; Ni, 75.5; Pb, 104.6; Zn, 1191.5	Wetlands in Eastern India	[36]

Laboratory experiments were performed to explore the affinity of MPs towards heavy metals. For example, Guan et al. studied the adsorptive properties of six trace metals on MPs and indicated that Cu showed the largest adsorption capacity (643.1 μg/g) on MPs, whereas Ni and Co were the lowest (148.6 and 197.7 μg/g), with intraparticle diffusion dominating the adsorption process [31]. Purwiyanto et al. explored the adsorptive capacities of two heavy metals (Pb and Cu) on MPs collected in the Musi River. Their results showed that Pb had a higher adsorptive concentrations on MPs (0.470 mg/kg) than Cu (0.091 mg/kg), and pointed out that the combination of MPs and heavy metals was physical sorption with weak bonds and the metals were easily desorbed in aqueous system [30]. The determining mechanisms for adsorption process were depicted by Gao et al. (shown as Figure 12.1) [44]. The above results are clearly of great concern as MPs can absorb heavy metals from their surroundings onto their surface because of their tiny size and ubiquity in aquatic environments. This suggests that MPs act as a carrier of heavy metals and play a key role in transportation of heavy metals in the environment.

12.2.3 Organic Pollutants

Numerous studies have demonstrated that MPs showed strong adsorption capability to different organic pollutant. Environmental organic pollutants include pharmaceuticals, personal care products, endocrine disruptors, pesticides, and common industrial organic wastes like phenolics, halogens, and aromatics [45]. Organic pollutants in water bodies are varied, complicated in composition, and

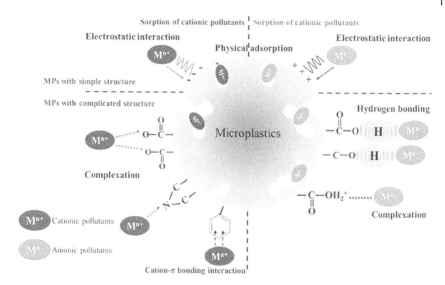

Figure 12.1 Potential interaction mechanisms of MPs and heavy metals [44] / with permission of Elsevier.

mostly hazardous [46]. Therefore, the adsorption behavior of MPs for organic pollutants should be discussed to better understand the ecological risk of their combination.

Numerous studies have indicated that considerably higher concentrations of organic pollutants than is environmentally relevant were observed on MPs surface because of adsorption processes [52–54]. Table 12.3 shows the maximum adsorption capacity of different MPs for some organic pollutants based on the Langmuir adsorption model. Like antibiotics and heavy metals, the adsorption capability of MPs varied with both type of MPs and organic pollutants. LonĈarski et al. investigated the interaction of four types of polycyclic aromatic hydrocarbons (PAHs) and six MPs in water. They pointed out that the strongest adsorption capacity of MPs was observed with pyrene and confirmed that chemical reactions occurred during the adsorption process [51]. The MPs type is also a key determinant of its absorption capability for organic pollutants. Sorensen et al. presented that PE had a higher adsorption capability for PAHs than PS at the same particle size [55]. Hüffer et al. investigated the adsorption quantity of MPs for seven organic pollutants (n-hexane, cyclohexane, benzene, toluene, chlorobenzene, ethyl benzoate, and naphthalene), and showed that adsorption concentration increased with the order of PA < PE < PVC < PS. They further indicated that particle size was not a key factor, but the hydrophobicity of organic pollutants had a strong relationship with the adsorptive capability of MPs [56].

Table 12.3 Interactions between MPs and organic pollutants.

MPs type	Organic Pollutants	Q_{max} in experiment	References
PS	Oxytetracycline	1.520	[47]
PE	Trichlorobenzenes	0.227	[48]
PE	Trifluralin	0.333	
PE	Imidacloprid	2.630	[49]
PE	Buprofezin	1.892	
PE	Difenoconazole	2.365	
PE	Bisphenol A (BPA)	0.208	[50]
PVC	Bisphenol AF (BPAF)	0.244	
PVC	Bisphenol B (BPB)	0.241	
Powdered PE	Naphthalene	0.5051	[51]
PE	Naphthalene	0.095	
PET	Naphthalene	0.460	
PP	Naphthalene	0.122	
Polylactic acid	Naphthalene	0.030	
Powdered PE	Fluorene	1.069	
PE	Fluorene	0.100	
PET	Fluorene	0.307	
PP	Fluorene	0.161	
Polylactic acid	Fluorene	0.045	
Powdered PE	Fluoranthene	0.625	
PE	Fluoranthene	0.169	
PET	Fluoranthene	0.759	
PP	Fluoranthene	0.156	
Polylactic acid	Fluoranthene	0.034	
Powdered PE	Pyrene	1.676	
PE	Pyrene	0.253	
PET	Pyrene	2.597	
PP	Pyrene	0.188	
Polylactic acid	Pyrene	0.052	

Collectively, MPs not only act as a "gathering point" to increase the concentration of contaminants on their surface, but also function as a carrier of OPs, thus enhancing biological exposure to and the bioavailability of contaminants. In addition, the co-occurrence of microplastics and contaminants might exhibit a synergistic toxic effect, causing greater damage to organisms and the environment.

12.3 Enrichment of Antibiotic-Resistant Bacteria and Antibiotic Resistance Genes

Antibiotic resistance has become a worldwide health threat, as it can increase human morbidity and mortality [57]. It was reported that more than 3.1 million patients from 94 regions worldwide were infected by antibiotic-resistant pathogens in 2020, and this number is expected to increase in the future [58]. Especially, gene exchange between antibiotic-resistant bacteria (ARB) and environmental microorganisms further exacerbates this problem [12]. As the genetic vector of antibiotic resistance, antibiotic resistance genes (ARGs) are considered emerging contaminants and have been widely detected in aquatic systems, including marine ecosystems, freshwater, and wastewater [13–15]. It was found that MPs could enrich ARB and ARGs on their surface, thereby enhancing ARB and ARGs propagation and posing a great threat to the aquatic system. Thus, the adsorption/enrichment performance and mechanism of ARGs and ARB on MPs should be explained.

12.3.1 Single Selection

Given non-degradability and large specific surface area, MPs can provide a novel niche for microorganism in the environment. Microorganisms will colonize the MPs surface and generate a biofilm, the so called plastisphere [59]. Recently, the plastisphere in aquatic ecosystems is of widespread concern, revealing that the composition and diversity of plastisphere microbiota are distinctive from the surrounding environment and even selective for ARB and ARGs [18, 60]. Numerous studies have demonstrated that MPs have strong adsorption of ARGs. Su et al. found that the amount of ARGs detected on MPs surface was $6.7–10^3$ times in surrounding leachate, and aged MPs even exhibited an increased enrichment of ARGs [61]. Wu et al. also found that the abundance of ARGs detected on MPs was approximately four times that of river water [18]. They also indicated that the number of ARG subtypes in the plastisphere was lower than that in river water, showing a specific antibiotic resistance on MPs biofilm compared with water [18].

MPs also influence the composition and distribution of microbial community in a system, especially for ARB. Wang et al. compared ARGs abundance on MPs with the surrounding water and pointed out their relationship with the bacterial community. Their results indicated that ARGs abundance on MPs was higher than the surrounding environment. A strong relationship between ARG profiles and microbial diversity was also observed [60], demonstrating that MPs enriched ARGs and ARB on their surface. Wu et al. further observed that more distinctive ARB, such as Pseudomonas monteilii, Pseudomonas mendocina, and Pseudomonas syringae, were only observed in plastisphere microbiota, but not in biofilms covering other natural materials (i.e., rock and leaf) [18], indicating that MPs acted as a special niche for selectively enriching ARB, rather than other materials. What's more, it was indicated that MPs enhanced the growth and metabolism of plastisphere microbiota [62], meaning the enriched ARB on plastisphere perform a faster proliferation than in the surrounding environment. By enrichment of ARB and the enhanced microbial turnover in plastisphere, the abundances of ARB and ARGs in an aquatic system are increased, thereby creating aquatic environmental risk, and posing adverse effects to human health.

In addition, ARGs propagation can be enhanced by gene exchange (i.e., horizontal gene transfer (HGT)). MPs possibly affect aquatic ecosystem gene exchange, because their surface biofilm provides an ideal environment for cell-to-cell communication and ARGs exchange between ARB and environmental bacteria [63]. For ARGs exchange, mobile genetic elements (MGEs, such as plasmids and integrons) are primarily responsible for the HGT, considered a primary contributor of ARGs spread [64]. Arias-Andres et al. demonstrated that microplastic-associated bacteria exhibited a higher transfer frequency of plasmids than free bacteria or natural-aggregated bacteria [65]. Other studies also found various MGEs were also detected on MPs surface, including intI1, IS613 and tnpA-04. Their abundances on the MPs surface were higher than in surrounding water [66, 67].

Collectively, MPs could appeal to ARB in their plastisphere and form a distinctive microbiota that have a higher abundance of ARB than the surrounding environment. Moreover, MPs provide a suitable environment for absorbed ARB growth, proliferation, and ARGs exchange, thereby promoting ARB and ARGs propagation in the aquatic ecosystem.

12.3.2 Co-Selection

Anthropogenic activities contribute to environmental co-pollution with antibiotics, heavy metals, organic pollutants, and MPs. The co-exposure to environmental contaminants may induce selection and co-selection towards antibiotic resistance in bacteria [1, 25]. According to the above discussion, MPs can adsorb and gather environmental contaminants on their surface, possibly leading to selection of contaminants that cause antibiotic resistance in microbiota on MPs.

The primary contaminant concern is antibiotics. The presence of antibiotics can induce a selective pressure, contribute to the emergence and dissemination of ARGs and ARB through stimulating related functions, and shift the microbial community to a greater extent [1, 68, 69]. As stated above, MPs can absorb antibiotics from the surrounding environment. Therefore, bacteria in the plastisphere are exposed to antibiotics at a higher concentration than in a surrounding system. Therefore, we can speculate that antibiotics attached on MPs surface can further cause the enrichment of ARB and ARGs in the plastisphere.

Heavy metals are also recognized as a driver of the emergence and spread of ARB and ARGs. Bacterial antibiotic resistance induced by co-selection of heavy metals is contributing by co-resistance or cross resistance [70]. Presumably this occurs by similar resistance strategies to both antibiotics and metals/metalloids such as reduction in membrane permeability and antibiotic/metal inactivation [71]. After adsorbing heavy metals, MPs act as a novel hot spot for co-selection of ARB/ARGs induced by heavy metals. Yang et al. observed that the abundances of ARGs and metal resistance genes (MRGs) in the plastisphere were observably higher than those in seawater microbiota. They further point out that the frequency of nonrandom co-presence of ARGs and MRGs detected in the plastisphere was higher than in seawater, indicating that the joint effects of selection for ARGs and MRGs are nonnegligible factors influencing ARB distribution on MPs [72]. Liu et al. observed that heavy metals facilitated the adsorption of MPs for ARGs based on correlation analysis [73], indicating that heavy metals exerted co-selection with MPs to enrich ARB and ARGs in the plastisphere. In addition, heavy metals were proven to promote the adsorption behavior of antibiotics on MPs [25], further stimulating MPs selection and enrichment of ARB and ARGs on their surface.

With regard to organic pollutants, multiple studies proved that they also facilitated enrichment of ARGs and ARB and accelerated ARGs propagation in various systems [74, 75]. After being adsorbed on MPs from the surroundings, organic pollutants also seem to exert co-selection with MPs on ARGs and ARB. So far, there have been only limited studies focused on the role of organic pollutants on ARGs and ARB enrichment on MPs. Hence, more attention should be given to the impact of the combination of MPs and organic pollutants on the prevalence of ARGs and ARB in the aquatic environment.

Collectively, contaminants attached on MPs were demonstrated to facilitate the enrichment of ARB and ARGs in the plastisphere. Adsorbed antibiotics directly present antibiotic resistance pressure on bacteria in the plastisphere. Meanwhile, adsorbed heavy metals and organic pollutants exert co-selective or cross-selective effects towards ARGs in bacteria. The co-selective effect of organic pollutants and MPs on ARGs/ARB remains to be explored.

12.4 The Effects of Environmental Conditions

12.4.1 pH

In aqueous systems, pH plays a key role for adsorption as it can affect the biochemical reactions and equilibria in the system. Freshwater aquatic life requires a pH between 6.5 and 9.0 for existence, and the pH value of seawater ranges 8.08–8.33 [10]. Numerous studies have highlighted the impacts of pH in solution on the contaminant adsorption behavior of MPs.

For antibiotics, Sun et al. pointed that the norfloxacin (NOR) adsorption qualities of MPs at pH 5 and 7 were observably higher than at other pH values [76]. Atugoda et al. observed that the adsorption of CIP on PE elevated with increasing pH and reached the maximum adsorption at pH 6.5–7.5, then subsequently decreased [77]. Similar trends were observed with heavy metals. The interaction process of MPs and heavy metals is highly pH-dependent, since pH determines not only the ionization/speciation/formation of the heavy metals in an aqueous system, but also the available sorption positions of the adsorbent. For instance, Zhou et al. indicated that the adsorptive capability of MPs for Cd(II) first increased at pH 2–6 then subsequently declined with pH from 6 to 9 [78] in an aqueous system.

The underlying mechanism is explained as follows: as the pH rises, the quantity of attached cationic contaminants also increases because of precipitation, enhanced electrostatic forces, and less competitive H^+ in the aqueous system [44]. Meanwhile, the adsorption capacity of anionic pollutants decreases gradually when the solution pH elevates. Another explanation is that the zeta potential of MPs is negative when pH > pH_{pzc} (point of zero charge): the higher the pH value, the more negatively charged the surface will be, and this may create more electrostatic attraction to the metal cation [28]. From a field test, Zou et al. indicated that the adsorption quality of MPs for Cu^{2+}, Pb^{2+}, and Cd^{2+} increased with increasing solution pH [79], confirming the above demonstration. However, as the solution pH increases further, a charge repulsion may be formed subsequently lowering the adsorption quality of MPs.

However, some exceptions were observed. For instance, Wang et al. indicated that the adsorption capacity of MPs for several organic pollutants (such as PFOS, a kind of POPs) increased with a decline in pH from 7 to 3 [80]. Tizaoui et al. also pointed out that elevating pH can decrease the adsorption amount of endocrine disrupting chemicals (EDCs) on MPs observed when pHs > EDCs pK_a (ca. 10.5) [81]. The above observed results might be caused by the fact that the microplastic surface can be protonated when the pH value declines. Therefore, a more anionic molecule is easily attached on the positive surface of MPs at lower pH [80].

12.4.2 Temperature

Temperature is also a vital factor for the adsorption behaviors of MPs. Previous studies have proved that the adsorption of heavy metals was endothermic [55, 82].

Thus, the adsorption behaviors of MPs generally improve with rising temperature. An increase in temperature also results in a decline in critical coagulation concentration at high ionic strength, i.e., decreases the stability by promoting aggregation by an increase in Brownian motion at elevated temperature. In turn, this enhances the number of effective collusions required for aggregation [5]. Accordingly, the adsorption capability of MPs for organic pollutants increases with elevating temperature. Wang et al. point out that the adsorptive capability of MPs for Cu^{2+} and Zn^{2+} increased with increasing temperature from 288 to 308 K [82]. Table 12.4 represents the equilibrium partition coefficient values (log K_{PE-w}) of some organic contaminants at various temperatures. When temperature rises from 18 to 24°C, the log K_{PE-w} of phenanthrene (Phe) rises from 4.2 to 4.3, but is slightly lower at 30°C (4.16) than at 24°C. For most hydrophobic organic chemicals (HOCs), such as Phe, fluoranthene (Fla), pyrene (Pyr), chrysene (Chr), and

Table 12.4 The log K_{PE-w} of HOCs at different temperature[a].

HOCs	Log K_{PE-w} T = 16.4–18.7°C	Log K_{PE-w} T~18°C	Log K_{PE-w} T = 23 ± 1°C	Log K_{PE-w} T = 24 ± 1°C	Log K_{PE-w} T = 30°C
Phe		4.2	4.23 ± 0.02	4.3 ± 0.1	4.16 ± 0.02
2-methyl phenanthrene				4.8 ± 0.2	
Fla	4.52			4.9 ± 0.1	4.75 ± 0.02
Pyr	4.62		5.02 ± 0.03	5.0 ± 0.1	4.90 ± 0.01
BaA				5.7 ± 0.1	
D12-benz(a)-anthracene				5.7 ± 0.1	
Chr				5.7 ± 0.1	5.53 ± 0.02
BeP				6.2 ± 0.1	5.94 ± 0.04
Per				6.5 ± 0.2	
PCB-29				5.1 ± 0.1	
PCB-52		4.6	5.4 ± 0.1		5.55 ± 0.01
PCB-69				5.6 ± 0.2	
PCB-97				6.3 ± 0.1	
PCB-118					6.18 ± 0.05
PCB-143				6.8 ± 0.2	

[a] Data extracted from Adams et al. [83].

Benzo(e)pyrene (BeP), the log K_{PE-w} values at 30°C were lower than those at 24°C. However, PCB-52 is an exception, as it's log K_{PE-w} at 30°C was slightly higher than that at 23°C.

However, for the adsorption of antibiotics, temperature was shown to have little effect on the adsorption of tetracycline (TC) on MPs [84]. A previous study demonstrated that the partition coefficient (K_w) for chlorobenzenes, PCBs, PAHs, and DDE on MPs was presumably temperature independent [85].

12.4.3 Salinity

Salinity is another influencing factor. The impact of salinity on the combination of MPs and contaminants is primarily contributed by variations in pH and ion type. Generally, higher salinity causes adsorption behavior at a lower level. Guo et al. studied the impact of salinity on adsorption behavior of five MPs (PA, PE, PET, PVC, and PP) for SMX, and found that the adsorption amounts of SMX dramatically decreased at the presence of salts [23]. Similar results were also found with heavy metals and organic pollutants. Lin et al. tested the Pb(II) adsorption behaviors of MPs with 0.01 and 0.1 M sodium chloride and showed that high salinity inhibited the adsorption of Pb(II) on MPs. Correspondingly, Llorca et al. also pointed that the introduction of salt decreased the absorption amount of PFASs on MPs [86].

The potential mechanisms of the altered adsorption capability of MPs for contaminants are twofold. Firstly, the ionic strength of Na^+ from NaCl may compete with cations (taking Cd^{2+} for example) for the adsorption position. Correspondingly, the Cl^- can limit Cd^{2+} by generating complexes such as $[CdCl]^+$, $CdCl_2$, $[CdCl_3]^-$, and Cd(OH)Cl, when they coexist [28]. Additionally, salinity also can induce electrostatic shielding to a metal and influence its electrostatic adsorption [87], thereby reducing its adsorption capability.

12.4.4 Weathering/Aging Effect

MPs in the environment inevitably undergo weathering/aging processes, including oxidation, UV irradiation, and thermal irradiation [88]. After an aging process, MPs featuring a rough surface and larger surface area have a stronger adsorption ability for contaminants compared to virgin MPs. Liu et al. explored the impacts of the UV aging of two types of MPs (i.e., PS and PVC) on their adsorption abilities for CIP. They observed that the adsorptive amounts of aged PS and PVC were 2.23- and 1.20-times virgin MPs, respectively [89]. Similar behavior was also shown with heavy metals. Wang et al. observed that the MPs aged by UV radiation showed increased adsorption capacities of Cu^{2+} and Zn^{2+} by 30.2 and 57.2%, compared to the original MPs, and a higher temperature and a higher pH value could induce more effective adsorption of metal ions on MPs [82]. Similarly, the adsorption affinity of MPs towards organic pollutants is also enhanced by the aging process. Bhagat et al. evaluated the influence of original and aged MPs on adsorption amounts of two model organic pollutants (phenanthrene and methylene blue). They found that the adsorptive capability on aged MPs for organic pollutants was higher than on virgin MPs, and methylene blue obtained the highest adsorption amount of aged MPs, at 5.7-times of virgin MPs [88].

Many studies have pointed out that the adsorption capability of MPs is dominated by the properties of MPs [1, 44, 90]. However, the aging processes can change the properties of MPs, such as the surface structure, hydrophobicity, crystallinity, and specific surface area (SSA), and increase the amount of some oxygen-containing groups, such as C–O, C=O, or C–OH, on MPs surface (shown as Figure 12.2) [1]. After aging, the hydrophilicity of MPs and the generation of hydrogen-bonding interactions are enhanced by the freshly formed oxygen-containing functional groups, thereby promoting contaminant adsorption. During the aging process, cracks might occur on the surface as a particle collides with another substance, and even smaller particles (i.e., nanoscale plastics) might be separated from MPs, thereby increasing the SSA of MPs. In addition, the time, condition, route of aging process, and the biofilm formed on MPs also effect the adsorption of contaminants on MPs. The adsorption capacities of MPs for contaminants were proved to be enhanced with prolonged aging time [82, 91]. For aging condition, compared to water-aged MPs, MPs aged in air have greater changes on

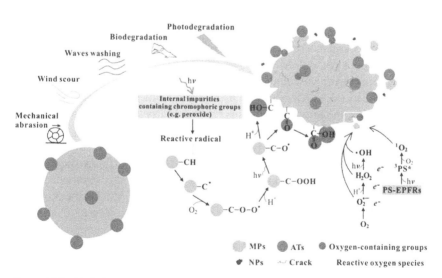

Figure 12.2 Variations in aged MPs and potential generation routes of oxygen-containing groups on MPs under UV radiation [1] / with permission of American Chemical Society.

the surface structure, increasing adsorption capacities of MPs [92]. Turning to the route of the aging process, the adsorption capacity of MPs aged by Fenton's reagent is much stronger than aging by H_2O_2 [93]. In addition, Bhagwat et al. observed that naturally aged, biofilm-covered MPs showed larger surface areas from the accumulation of biomass, compared to pristine MPs [94]. Also, the amounts of adsorbed Pb and perfluorooctane sulfonate (PFOS) in aged MPs increased by 4–25% and 20–85%, compared to pristine MPs, demonstrating that biofilms play a vital role in the adsorption performance of MPs [94]. Therefore, as aged MPs are universal in the natural environment, more attention should be paid to the ecological risks of aged MPs and their combination with contaminants.

12.5 Joint Potential Risks

The capability of MPs to adsorb environmental contaminants is presumably harmful for ecosystem homeostasis, as the adsorbed contaminants can then be transported and introduced into other places. Additionally, when exposed to organisms, the interaction of MPs with contaminants may alter the toxicity to aquatic organisms and even pose a threat to human health. Many studies have demonstrated the potential joint risks, as discussed in this section.

12.5.1 For Contaminants Distribution in Aquatic Environment

Interaction of MPs with contaminants may disrupt contaminant distribution in aquatic systems around the world. Because of the inertia and mobility of MPs, the absorbed contaminants could drift and be transferred along with MPs in the environment. In this research field, the role that MPs could play as a carrier for contaminants in marine environments was demonstrated by Brennecke et al., as they found that Cu and Zn released from antifouling paints were adsorbed on microplastic surfaces in sea water [95]. This result confirmed that heavy metals were adsorbed and stored on MPs locally and then transported by MPs with runoff and rivers to other places [96]. This transfer and introduction of contaminants caused by MPs have impacts on the ecosystem, including the microbial community, plant growth, and organism responses. The presence of MPs was reported to alter microbial community composition in water and sludge [4, 97]. Furthermore, the heavy metals affiliated with MPs, such as vanadium (V), copper (Cu), and lead (Pb), were reported to affect the compositions of a microbial community in the plastisphere [98, 99]. Therefore, the contaminant-adsorbed MPs pose a potential impact to microbial diversity in both freshwater and sludge systems.

Plants in an aquatic system were also influenced by contaminant-adsorbed MPs. It was found that the growth and photosynthesis of microalgae were limited by MPs [100]. Tunali et al. demonstrated that the attached contaminants exert a synergistic effect with MPs. The results showed that MPs attached to heavy metals greatly inhibited growth and amount of chlorophyll A in Chlorella vulgaris, compared to uncontaminated MPs, and the observed adsorption and accumulation of MPs on the algal cell provided probable evidence of the toxic effect [101]. On the other hand, a few studies showed that the joint toxicity towards algae had diverse results, as the combination of MPs and heavy metals can even have a positive impact on algae. Fu et al. pointed out that the growth of microalgae Chlorella vulgaris was inhibited in single treatment with either MPs or Cu, but was facilitated by joint treatment with Cu and MPs, possibly from Cu that was partially adsorbed on MPs and therefore the toxic effect was limited [102]. Similarly, Li et al. studied the individual and combined toxicity of SMX and five MPs on marine algae (Skeletonema costatum) and found that the five MPs and SMX inhibited the growth of algae, but their joint toxicities showed an antagonistic effect because of the "shelter" role of MPs adsorption [103].

12.5.2 For ARGs and ARB Distribution in Aquatic Environment

MPs impact the distribution of ARGs in the environment, as they can enrich ARGs and ARB from surrounding substrates and alter their distributions in the aquatic environment [17, 60]. ARGs-adsorbed MPs can be digested by aquatic plants and animals, facilitating the transportation of ARGs and ARB in the aquatic environment (shown as Figure 12.3). This disturbing behavior can be enhanced by the interaction of MPs and environmental contaminants, because certain contaminants (including antibiotics and heavy metals) were proved to be critical contributors to the emergence of ARGs and ARB in a different system [104, 105]. After adsorbing contaminants, MPs can concentrate more ARGs and ARB on their surface, thereby accelerating the spread of ARGs and ARB in the aquatic system. Given strong floatability and mobility [106], MPs can speed up the transportation and prevalence of ARGs and ARB in the aquatic environment, and the contaminant-adsorbed MPs even increase the risk of ARGs transportation. Therefore, compared to virgin MPs, the contaminant-adsorbed MPs are a more hazardous pollutant that functions both as a hotspot for the development of ARB and ARGs via co-selection and as a carrier for propagation of ARGs in aquatic system.

Furthermore, numerous types of environmental contaminants could facilitate horizontal transfer of ARGs, demonstrated by numerous studies. For instance, Guo et al. revealed that under metallic nanoparticles and ions pressure, the natural transformation frequency of ARGs was enhanced by 11.0-fold, almost as effective as antibiotics [107]. Wang et al. showed that an antiepileptic drug also promoted horizontal transfer of ARGs through inducing a series of acute responses of bacterial genera [108]. Collectively, MPs combined with contaminants not only

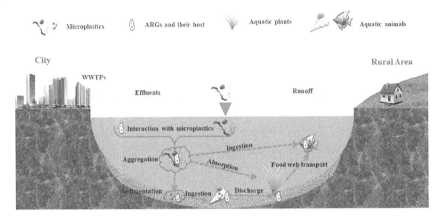

Figure 12.3 The transport of ARGs on MPs in the aquatic environment [109] / with permission of Elsevier.

concentrate ARGs and ARB on surface biofilm, but also provide an attractive environment for ARGs dissemination between ARB and other bacteria, thereby increasing the total amount of ARGs in the system.

12.5.3 For Aquatic Organisms

MPs can be ingested by different aquatic organisms including mussels, bivalve mollusks, and zooplankton, and further enter into macroaquatic organisms like fish through the food chain. If the combination of MPs and contaminants in the surrounding environment is digested by organisms, it could cause increased bioaccumulation of contaminants in organisms, as confirmed by multiple studies [110–112]. Besides, contaminant-adsorbed MPs could change the distribution of contaminants in an organism. For example, Nor et al. demonstrated that MPs could bring polychlorinated biphenyls (PCBs) into the relatively clean gut systems of lugworms and cod in many cases [113]. A similar result was observed by Gonzalez-Soto et al., who further found that small MPs (0.5 μm) exerted an enhanced toxicity regarding the transfer of benzo(a)pyrene (BaP) to tissues relative to larger MPs (4.5 μm) [114]. The released degree of heavy metals from MPs in the body is different depending on the type of metal and tissue environment, thus causing uneven distribution of heavy metals in the organism. Godoy et al. found that Cr attached to MPs was released early in the gastric phase but precipitated rapidly, while Pb was released more slowly and in lower amounts and did not precipitate until the beginning of the duodenal phase [115]. Consequently, the intake of contaminant-adsorbed MPs could not only increase concentrations, but also change the distribution of contaminants in the organisms.

Currently, some laboratory tests have shown the toxicity of MPs combined with contaminants on aquatic organisms. Most mixture toxicity studies explored whether the adverse effect follows the simple additive approach, or it deviates to form a synergistic, antagonistic, or potentiating effect.

Synergistic effect: Most studies showed that the interaction of MPs and contaminants could cause more serious toxicity for an organism than MPs or contaminants alone, as shown by fatality rate, cytotoxicity, immune system impairment, disturbed metabolism, neurotoxicity, and abnormal behavior [114, 116–118]. For example, Wang et al. found that Cd-adsorbed MPs caused an increased mortality of M. monogolica (a zooplankton), compared to non-adsorbed MPs treatment [116]. González-Soto et al. also indicated that MPs affiliated with benzo[a]pyrene (BaP) were more toxic than individual MPs based on hemocyte viability, catalase activity, and the quantitative structure of digestive tubules (DT) epithelium. Smaller MPs also showed higher toxicity than larger ones [114].

Antagonistic toxicity: Some studies have indicated an antagonistic action between MPs and contaminants on aquatic organisms. Liao et al. pointed that PFOS produced an antagonistic toxicity with nanoscale MPs to thermophilic bacteria, as the combination led to decreased reactive oxygen species (ROS) in organisms [119]. Miranda et al. compared the toxicities of Cd and Cd-MPs in Pomatoschistus microps and indicated that MPs exerted antagonism with Cd on the acetylcholinesterase (AchE) effect [120]. They proposed three possible hypotheses about the antagonism of MPs and Cd: (1) Cd is tightly attached on MPs leading to its lower bioavailability; (2) Cd may be sheltered by MPs and eliminated from the organism without being adsorbed, (3) the anticholinesterase effects of Cd may be reduced by combination of MPs and Cd that slowed Cd access to acetylcholinesterase (AChE) sites.

Potentiating effect: Potentiating action occurs when one contaminant that does not originally have a toxic impact is added to another contaminant, making the second chemical much more toxic [121]. Certain articles have pointed that potentiating action of MPs on contaminants. Syberg et al. pointed that the co-exposure of triclosan (TCS) and MPs was more toxic than either TCS or MPs alone on the mortality of marine copepod Acartia tonsa, showing that MPs presumedly potentiated TCS-mediated toxicity [122]. A similar result was found with heavy metals by Luís et al. They pointed out that AChE suppression was not observed in the individual treatment of Cr(VI) or MPs for fish (Pomatoschistus microps), whereas the combination of Cr(VI) and MPs showed a distinct decline of predatory behavior and an apparent impediment of AChE activity for fish [123].

12.5.4 For Human Health

The uptake of MPs by the human body could occur via various exposure routes, such as ingestion, inhalation, and dermal contact [124–126]. It was estimated that

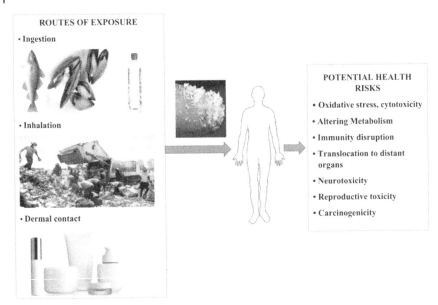

Figure 12.4 The exposure routes of MPs and potential risks for human [9] / with permission of Elsevier.

74 000–121 000 particles MPs are consumed by Americans each year through dietary intake and air inhalation [127]. The MPs or nanoscale plastics have been indicated to possibly cause cytotoxicity, alteration of metabolism, and neurotoxicity, as shown as Figure 12.4 [9]. Many studies point out that the hazard of MPs was size-dependent. Smaller particles could more adversely affect human health [128, 129].

MPs in the environment inevitably coexist with other contaminants, therefore, contaminants may also be transported by MPs into the human body through ingestion, air inhalation, and dermal contact. Previous studies also identified the potential for contaminants by transfer of MPs to higher trophic levels along the food chain to food products ingested by humans [130, 131], thereby increasing the intake of harmful substances by human beings. A few researchers have studied the adsorption of contaminants from MPs in the human body. Liao et al. evaluated the behavior of Cr-affiliated MPs using the whole digestive system in-vitro method and found that Cr from Cr-affiliated MPs was primarily released in the gastric digestive phase with the acidic condition of the stomach regarded as the major contributor in the releasing process. Additionally, they also calculated the maximum daily total Cr intake for different human groups through consuming different Cr-adsorbed MPs to be 0.50 to 1.18 µg/day [132]. Godoy et al. also observed that heavy metals adsorbed on MPs can be accepted by the human gastrointestinal tract. They pointed that 23.11% of the attached Cr and 23.17% of the attached Pb on MPs were able to cross the tubular membrane, facilitating intestinal absorption [115], thereby increasing the ingestion of contaminants by human beings. These contaminants are easily accumulated in the body and pose serious threats to human health [133, 134]. In addition, ARGs and ARB adsorbed on MPs also have obvious adverse effects on human health. MPs with heavy metals and antibiotics attached can enter the human body via the food chain. Then, potential ARB already presented in the human gut will be in contact with contaminants, which may further lead to contaminant-driven co-selection of antibiotic resistance and thus increase a global health threat [71]. Nevertheless, knowledge of the toxicity of the combination of MPs and contaminants (including antibiotics, heavy metals, and organic pollutants) to humans is still poorly understood. In depth research on whether and how contaminant-adsorbed MPs influence human health is urgently needed.

12.6 Conclusion and Recommendations

This chapter summarized the adsorption behavior of MPs for several contaminants (including antibiotics, heavy metals, and organic pollutants) and discussed influencing factors on the interaction of MPs and contaminants. Many studies regarding the joint effect of MPs and contaminants indicate that the contaminant-adsorbed MPs can trigger different effects on aquatic ecosystems, organisms, and human beings. As the actual threat from combination of MPs and contaminants towards humans is complex, the true impact is largely unknown. Based on previous works, we propose future research suggestions for researchers to examine.

1) Despite numerous studies showing that MPs can adsorb environmental contaminants, the mechanisms underlying adsorption behavior in aquatic systems are still lacking. Additionally, since existing studies were primarily based on temporary adsorption tests in the laboratory, long-term adsorption behavior of MPs for environmental contaminants should be explored.
2) Waste activated sludge (WAS) is an important reservoir of MPs and contaminants. However, little information on the interaction of MPs and contaminants in WAS has been obtained thus far. To better elucidate the ecological risk of MPs and contaminants scientifically, studies on the emergence and transfer of contaminant-adsorbed MPs in WAS treatment are badly needed. Additionally, joint impacts on WAS biochemical characteristics need to be examined to assess any loss caused by the co-exposure to energy recovery in sludge.
3) Since many studies have concentrated on the single impacts of MPs or contaminants on specific species, the joint toxicity of MPs and affiliated contaminants is possibly unique for contaminant or species. Which property (such as particle scale, polymer shapes, or materials) of MPs elevates or alleviates the toxicity of

contaminants to organisms is largely unknown. An understanding of the key factors and potential mechanisms of observed antagonistic or synergistic behavior is urgently needed.

4) Whether contaminant-adsorbed MPs are detrimental to the diversity of an ecosystem is still unclear. For one thing, MPs may be transported via the food chain, which facilitates accumulation and amplification of the hazard of contaminants in organisms. For another, MPs may also be transferred to descendants through proliferation. Thus, the presence of contaminant-adsorbed MPs can influence populations and communities in an ecosystem. Therefore, the potential joint risk of MPs and contaminants on organisms in the food chain and on offspring in the propagation process need to be explored.

5) At present, research on adsorption and toxicity behavior of MPs mainly focuses on several common types of MPs. Other types of MPs may require more attention. Because of intricate natural circumstances, pristine MPs and aged MPs treated by chemical methods in the laboratory do not represent the natural adsorption capabilities of MPs for contaminants and their combined effects on the ecosystem. Consequently, MPs obtained from natural circumstances can be a preferable choice to conduct adsorption and toxicological tests. Additionally, compared to particles, MPs in the shapes of fibers and fragments are more universal in the environment and should be strongly considered in future studies.

6) With regard to the exposure route of MPs and contaminants on humans, most studies concentrated on the effect of their pressure on marine organisms, whereby the combination is ingested by humans via the food chain. However, most exposure to contaminant-adsorbed MPs possibly originates from our plastic necessities. Therefore, more exploration of routes of exposure is needed evaluate the real risk of contaminant-adsorbed MPs on human health.

References

1 Wang, Y., Yang, Y., Liu, X. et al. (2021). Interaction of microplastics with antibiotics in aquatic environment: Distribution, adsorption, and toxicity. *Environmental Science & Technology* 55 (23): 15579–15595.
2 Erni-Cassola, G., Zadjelovic, V., Gibson, M.I. et al. (2019). Distribution of plastic polymer types in the marine environment; A meta-analysis. *Journal of Hazardous Materials* 369: 691–698.
3 Naik, R.K., Naik, M.M., D'Costa, P.M. et al. (2019). Microplastics in ballast water as an emerging source and vector for harmful chemicals, antibiotics, metals, bacterial pathogens and HAB species: A potential risk to the marine environment and human health. *Marine Pollution Bulletin* 149: 110525.

4 Wei, W., Huang, Q.S., Sun, J. et al. (2019). Revealing the mechanisms of polyethylene microplastics affecting anaerobic digestion of waste activated sludge. *Environment Science and Technology* 53 (16): 9604–9613.

5 Maity, S., Biswas, C., Banerjee, S. et al. (2021). Interaction of plastic particles with heavy metals and the resulting toxicological impacts: A review. *Environmental Science and Pollution Research* 28 (43): 60291–60307.

6 Khalid, N., Aqeel, M., Noman, A. et al. (2021). Interactions and effects of microplastics with heavy metals in aquatic and terrestrial environments. *Environmental Pollution* 290: 118104.

7 Rodrigues, J.P., Duarte, A.C., Santos-Echeandía, J. et al. (2019). Significance of interactions between microplastics and POPs in the marine environment: A critical overview. *TrAC Trends in Analytical Chemistry* 111: 252–260.

8 Wang, F., Wong, C.S., Chen, D. et al. (2018). Interaction of toxic chemicals with microplastics: A critical review. *Water Research* 139: 208–219.

9 Rahman, A., Sarkar, A., Yadav, O.P. et al. (2021). Potential human health risks due to environmental exposure to nano- and microplastics and knowledge gaps: A scoping review. *Science of the Total Environment* 757: 143872.

10 Fred-Ahmadu, O.H., Bhagwat, G., Oluyoye, I. et al. (2020). Interaction of chemical contaminants with microplastics: Principles and perspectives. *Science of the Total Environment* 706: 12.

11 Zhao, Q., Guo, W., Luo, H. et al. (2021). Deciphering the transfers of antibiotic resistance genes under antibiotic exposure conditions: Driven by functional modules and bacterial community. *Water Research* 205: 117672.

12 Zhou, S., Zhu, Y., Yan, Y. et al. (2019). Deciphering extracellular antibiotic resistance genes (eARGs) in activated sludge by metagenome. *Water Research* 161: 610–620.

13 Yang, J., Wang, C., Shu, C. et al. (2013). Marine sediment bacteria harbor antibiotic resistance genes highly similar to those found in human pathogens. *Microbial Ecology* 65 (4): 975–981.

14 Moon, K., Jeon, J.H., Kang, I. et al. (2020). Freshwater viral metagenome reveals novel and functional phage-borne antibiotic resistance genes. *Microbiome* 8 (1): 75.

15 Du, S., Shen, J.-P., Hu, H.-W. et al. (2020). Large-scale patterns of soil antibiotic resistome in Chinese croplands. *Science of the Total Environment* 712: 136418.

16 Chen, J., Wang, T., Zhang, K. et al. (2021). The fate of antibiotic resistance genes (ARGs) and mobile genetic elements (MGEs) from livestock wastewater (dominated by quinolone antibiotics) treated by microbial fuel cell (MFC). *Ecotoxicology and Environmental Safety* 218: 112267.

17 Wang, S., Xue, N., Li, W. et al. (2020). Selectively enrichment of antibiotics and ARGs by microplastics in river, estuary and marine waters. *Science of the Total Environment* 708: 134594.

18 Wu, X., Pan, J., Li, M. et al. (2019). Selective enrichment of bacterial pathogens by microplastic biofilm. *Water Research* 165: 114979.

19 Zhao, R.X., Feng, J., Liu, J. et al. (2019). Deciphering of microbial community and antibiotic resistance genes in activated sludge reactors under high selective pressure of different antibiotics. *Water Research* 151: 388–402.

20 Grenni, P., Ancona, V., and Caracciolo, A.B. (2018). Ecological effects of antibiotics on natural ecosystems: A review. *Microchemical Journal* 136: 25–39.

21 Wang, B., Yan, J., Li, G. et al. (2020). Risk of penicillin fermentation dreg: Increase of antibiotic resistance genes after soil discharge. *Environmental Pollution* 259: 113956.

22 Li, J., Zhang, K., and Zhang, H. (2018). Adsorption of antibiotics on microplastics. *Environmental Pollution* 237: 460–467.

23 Guo, X., Chen, C., and Wang, J. (2019). Sorption of sulfamethoxazole onto six types of microplastics. *Chemosphere* 228: 300–308.

24 Xu, B., Liu, F., Brookes, P.C. et al. (2018). The sorption kinetics and isotherms of sulfamethoxazole with polyethylene microplastics. *Marine Pollution Bulletin* 131: 191–196.

25 Yu, F., Li, Y., Huang, G. et al. (2020). Adsorption behavior of the antibiotic levofloxacin on microplastics in the presence of different heavy metals in an aqueous solution. *Chemosphere* 260: 127650.

26 Yu, F., Yang, C., Huang, G. et al. (2020). Interfacial interaction between diverse microplastics and tetracycline by adsorption in an aqueous solution. *Science of the Total Environment* 721: 137729.

27 Chen, Y., Li, J., Wang, F. et al. (2021). Adsorption of tetracyclines onto polyethylene microplastics: A combined study of experiment and molecular dynamics simulation. *Chemosphere* 265: 129133.

28 Cao, Y., Zhao, M., Ma, X. et al. (2021). A critical review on the interactions of microplastics with heavy metals: Mechanism and their combined effect on organisms and humans. *Science of the Total Environment* 147620.

29 Jarup, L. (2003). Hazards of heavy metal contamination. *British Medical Bulletin* 68: 167–182.

30 Purwiyanto, A.I.S., Suteja, Y., Trisno et al. (2020). Concentration and adsorption of Pb and Cu in microplastics: Case study in aquatic environment. *Marine Pollution Bulletin* 158.

31 Guan, J., Qi, K., Wang, J. et al. (2020). Microplastics as an emerging anthropogenic vector of trace metals in freshwater: Significance of biofilms and comparison with natural substrates. *Water Research* 184: 116205.

32 Li, X., Mei, Q., Chen, L. et al. (2019). Enhancement in adsorption potential of microplastics in sewage sludge for metal pollutants after the wastewater treatment process. *Water Research* 157: 228–237.

33 Wang, J., Peng, J., Tan, Z. et al. (2017). Microplastics in the surface sediments from the Beijiang River littoral zone: Composition, abundance, surface textures and interaction with heavy metals. *Chemosphere* 171: 248–258.

34 Ta, A.T. and Babel, S. (2020). Microplastic contamination on the lower Chao Phraya: Abundance, characteristic and interaction with heavy metals. *Chemosphere* 257.

35 Deng, J., Guo, P., Zhang, X. et al. (2020). Microplastics and accumulated heavy metals in restored mangrove wetland surface sediments at Jinjiang Estuary (Fujian, China). *Marine Pollution Bulletin* 159: 111482.

36 Sarkar, D.J., Das Sarkar, S., Das, B.K. et al. (2021). Occurrence, fate and removal of microplastics as heavy metal vector in natural wastewater treatment wetland system. *Water Research* 192: 116853.

37 Nakashima, E., Isobe, A., Kako, S.I. et al. (2012). Quantification of toxic metals derived from macroplastic litter on Ookushi Beach, Japan. *Environmental Science & Technology* 46 (18): 10099–10105.

38 Imhof, H.K., Laforsch, C., Wiesheu, A.C. et al. (2016). Pigments and plastic in limnetic ecosystems: A qualitative and quantitative study on microparticles of different size classes. *Water Research* 98: 64–74.

39 Massos, A. and Turner, A. (2017). Cadmium, lead and bromine in beached microplastics. *Environmental Pollution* 227: 139–145.

40 Vedolin, M.C., Teophilo, C.Y.S., Turra, A. et al. (2018). Spatial variability in the concentrations of metals in beached microplastics. *Marine Pollution Bulletin* 129 (2): 487–493.

41 Turner, A., Holmes, L., Thompson, R.C. et al. (2020). Metals and marine microplastics: Adsorption from the environment versus addition during manufacture, exemplified with lead. *Water Research* 173.

42 Li, W.J., Lo, H.S., Wong, H.M. et al. (2020). Heavy metals contamination of sedimentary microplastics in Hong Kong. *Marine Pollution Bulletin* 153: 7.

43 Ta, A.T. and Babel, S. (2020). Microplastics pollution with heavy metals in the aquaculture zone of the Chao Phraya River Estuary, Thailand. *Marine Pollution Bulletin* 161: 111747.

44 Gao, X., Hassan, I., Peng, Y. et al. (2021). Behaviors and influencing factors of the heavy metals adsorption onto microplastics: A review. *Journal of Cleaner Production* 319: 128777.

45 Lu, F. and Astruc, D. (2020). Nanocatalysts and other nanomaterials for water remediation from organic pollutants. *Coordination Chemistry Reviews* 408: 213180.

46 Luo, H., Liu, C., He, D. et al. (2022). Environmental behaviors of microplastics in aquatic systems: A systematic review on degradation, adsorption, toxicity and biofilm under aging conditions. *Journal of Hazardous Materials* 423: 126915.

47 Zhang, H., Wang, J., Zhou, B. et al. (2018). Enhanced adsorption of oxytetracycline to weathered microplastic polystyrene: Kinetics, isotherms and influencing factors. *Environmental Pollution* 243 (Pt B): 1550–1557.

48 Tubić, A., Lončarski, M., Apostolović, T. et al. (2021). Adsorption mechanisms of chlorobenzenes and trifluralin on primary polyethylene microplastics in the aquatic environment. *Environmental Science and Pollution Research* 28 (42): 59416–59429.

49 Li, H., Wang, F., Li, J. et al. (2021). Adsorption of three pesticides on polyethylene microplastics in aqueous solutions: Kinetics, isotherms, thermodynamics, and molecular dynamics simulation. *Chemosphere* 264: 128556.

50 Wu, P., Cai, Z., Jin, H. et al. (2019). Adsorption mechanisms of five bisphenol analogues on PVC microplastics. *Science of the Total Environment* 650 (Pt 1): 671–678.

51 Lončarski, M., Gvoić, V., Prica, M. et al. (2021). Sorption behavior of polycyclic aromatic hydrocarbons on biodegradable polylactic acid and various nondegradable microplastics: Model fitting and mechanism analysis. *Science of the Total Environment* 785: 147289.

52 Chen, C.-F., Ju, Y.-R., Lim, Y.C. et al. (2020). Microplastics and their affiliated PAHs in the sea surface connected to the southwest coast of Taiwan. *Chemosphere* 254.

53 Mai, L., Bao, L.-J., Shi, L. et al. (2018). Polycyclic aromatic hydrocarbons affiliated with microplastics in surface waters of Bohai and Huanghai Seas, China. *Environmental Pollution* 241: 834–840.

54 Abdurahman, A., Cui, K.Y., Wu, J. et al. (2020). Adsorption of dissolved organic matter (DOM) on polystyrene microplastics in aquatic environments: Kinetic, isotherm and site energy distribution analysis. *Ecotoxicology and Environmental Safety* 198: 6.

55 Sorensen, L., Rogers, E., Altin, D. et al. (2020). Sorption of PAHs to microplastic and their bioavailability and toxicity to marine copepods under co-exposure conditions. *Environmental Pollution* 258.

56 Hueffer, T. and Hofmann, T. (2016). Sorption of non-polar organic compounds by micro-sized plastic particles in aqueous solution. *Environmental Pollution* 214: 194–201.

57 Yang, Y., Li, B., Ju, F. et al. (2013). Exploring variation of antibiotic resistance genes in activated sludge over a four-year period through a metagenomic approach. *Environmental Science & Technology* 47 (18): 10197–10205.

58 Organization, W.H. (2021). Global antimicrobial resistance and use surveillance system (GLASS) report: 2021.

59 Zhu, D., Ma, J., Li, G. et al. (2021). Soil plastispheres as hotpots of antibiotic resistance genes and potential pathogens. *The ISME Journal* 16: 615-.

60 Wang, J., Qin, X., Guo, J. et al. (2020). Evidence of selective enrichment of bacterial assemblages and antibiotic resistant genes by microplastics in urban rivers. *Water Research* 183: 116113.

61 Su, Y., Zhang, Z., Zhu, J. et al. (2021). Microplastics act as vectors for antibiotic resistance genes in landfill leachate: The enhanced roles of the long-term aging process. *Environmental Pollution* 270.

62 Zhou, J., Gui, H., Banfield, C.C. et al. (2021). The microplastisphere: Biodegradable microplastics addition alters soil microbial community structure and function. *Soil Biology and Biochemistry* 156: 108211.

63 Kaur, K., Reddy, S., Barathe, P. et al. (2021). Microplastic-associated pathogens and antimicrobial resistance in environment. *Chemosphere* 291: 133005.

64 Syranidou, E. and Kalogerakis, N. (2022). Interactions of microplastics, antibiotics and antibiotic resistant genes within WWTPs. *Science of the Total Environment* 804: 150141.

65 Arias-Andres, M., Kluemper, U., Rojas-Jimenez, K. et al. (2018). Microplastic pollution increases gene exchange in aquatic ecosystems. *Environmental Pollution* 237: 253–261.

66 Lu, J., Zhang, Y., Wu, J. et al. (2019). Effects of microplastics on distribution of antibiotic resistance genes in recirculating aquaculture system. *Ecotoxicology and Environmental Safety* 184: 109631.

67 Wang, Z., Gao, J., Zhao, Y. et al. (2021). Plastisphere enrich antibiotic resistance genes and potential pathogenic bacteria in sewage with pharmaceuticals. *Science of the Total Environmental* 768: 144663.

68 Ma, J., Sheng, G.D., and O'Connor, P. (2020). Microplastics combined with tetracycline in soils facilitate the formation of antibiotic resistance in the Enchytraeus crypticus microbiome. *Environmental Pollution* 264: 114689.

69 Willmann, M., Vehreschild, M.J.G.T., Biehl, L.M. et al. (2019). Distinct impact of antibiotics on the gut microbiome and resistome: A longitudinal multicenter cohort study. *BMC Biology* 17: 1.

70 Henriques, I., Tacão, M., Leite, L. et al. (2016). Co-selection of antibiotic and metal(loid) resistance in gram-negative epiphytic bacteria from contaminated salt marshes. *Marine Pollution Bulletin* 109 (1): 427–434.

71 Imran, M., Das, K.R., and Naik, M.M. (2019). Co-selection of multi-antibiotic resistance in bacterial pathogens in metal and microplastic contaminated environments: An emerging health threat. *Chemosphere* 215: 846–857.

72 Yang, Y., Liu, G., Song, W. et al. (2019). Plastics in the marine environment are reservoirs for antibiotic and metal resistance genes. *Environment International* 123: 79–86.

73 Liu, X., Wang, H., Li, L. et al. (2022). Do microplastic biofilms promote the evolution and co-selection of antibiotic and metal resistance genes and their associations with bacterial communities under antibiotic and metal pressures? *Journal of Hazardous Materials* 424: 127285.

74 Chen, B., He, R., Yuan, K. et al. (2017). Polycyclic aromatic hydrocarbons (PAHs) enriching antibiotic resistance genes (ARGs) in the soils. *Environmental Pollution* 220: 1005–1013.

75 Wang, J., Wang, J., Zhao, Z. et al. (2017). PAHs accelerate the propagation of antibiotic resistance genes in coastal water microbial community. *Environmental Pollution* 231: 1145–1152.

76 Sun, M., Yang, Y., Huang, M. et al. (2022). Adsorption behaviors and mechanisms of antibiotic norfloxacin on degradable and nondegradable microplastics. *Science of the Total Environment* 807: 151042.

77 Atugoda, T., Wijesekara, H., Werellagama, D.R.I.B. et al. (2020). Adsorptive interaction of antibiotic ciprofloxacin on polyethylene microplastics: Implications for vector transport in water. *Environmental Technology and Innovation* 19: 100971.

78 Zhou, Y., Yang, Y., Liu, G. et al. (2020). Adsorption mechanism of cadmium on microplastics and their desorption behavior in sediment and gut environments: The roles of water pH, lead ions, natural organic matter and phenanthrene. *Water Research* 184: 116209.

79 Zou, J.Y., Liu, X.P., Zhang, D.M. et al. (2020). Adsorption of three bivalent metals by four chemical distinct microplastics. *Chemosphere* 248: 12.

80 Wang, F., Shih, K.M., and Li, X.Y. (2015). The partition behavior of perfluorooctanesulfonate (PFOS) and perfluorooctanesulfonamide (FOSA) on microplastics. *Chemosphere* 119: 841–847.

81 Tizaoui, C., Fredj, S.B., and Monser, L. (2017). Polyamide-6 for the removal and recovery of the estrogenic endocrine disruptors estrone, 17β-estradiol, 17α-ethinylestradiol and the oxidation product 2-hydroxyestradiol in water. *Chemical Engineering Journal* 328: 98–105.

82 Wang, Q., Zhang, Y., Wangjin, X. et al. (2020). The adsorption behavior of metals in aqueous solution by microplastics effected by UV radiation. *Journal of Environmental Sciences* 87: 272–280.

83 Adams, R.G., Lohmann, R., Fernandez, L.A. et al. (2007). Polyethylene devices: Passive samplers for measuring dissolved hydrophobic organic compounds in aquatic environments. *Environmental Science & Technology* 41 (4): 1317–1323.

84 Shen, X.-C., Li, D.-C., Sima, X.-F. et al. (2018). The effects of environmental conditions on the enrichment of antibiotics on microplastics in simulated natural water column. *Environmental Research* 166: 377–383.

85 Lohmann, R. (2012). Critical review of low-density polyethylene's partitioning and diffusion coefficients for trace organic contaminants and implications for its use as a passive sampler. *Environmental Science & Technology* 46 (2): 606–618.

86 Llorca, M., Schirinzi, G., Martinez, M. et al. (2018). Adsorption of perfluoroalkyl substances on microplastics under environmental conditions. *Environmental Pollution* 235: 680–691.

87 Lin, Z., Hu, Y., Yuan, Y. et al. (2021). Comparative analysis of kinetics and mechanisms for Pb(II) sorption onto three kinds of microplastics. *Ecotoxicology and Environmental Safety* 208: 111451.

88 Bhagat, K., Barrios, A.C., Rajwade, K. et al. (2022). Aging of microplastics increases their adsorption affinity towards organic contaminants. *Chemosphere* 298: 134238.

89 Liu, G., Zhu, Z., Yang, Y. et al. (2019). Sorption behavior and mechanism of hydrophilic organic chemicals to virgin and aged microplastics in freshwater and seawater. *Environmental Pollution* 246: 26–33.

90 Liu, S., Huang, J., Zhang, W. et al. (2022). Microplastics as a vehicle of heavy metals in aquatic environments: A review of adsorption factors, mechanisms, and biological effects. *Journal of Environmental Management* 302: 113995.

91 Guan, Y., Gong, J., Song, B. et al. (2022). The effect of UV exposure on conventional and degradable microplastics adsorption for Pb (II) in sediment. *Chemosphere* 286: 131777.

92 Ding, L., Mao, R., Ma, S. et al. (2020). High temperature depended on the ageing mechanism of microplastics under different environmental conditions and its effect on the distribution of organic pollutants. *Water Research* 174.

93 Lang, M., Yu, X., Liu, J. et al. (2020). Fenton aging significantly affects the heavy metal adsorption capacity of polystyrene microplastics. *The Science of the Total Environment* 722: 137762-.

94 Bhagwat, G., Tran, T.K.A., Lamb, D. et al. (2021). Biofilms enhance the adsorption of toxic contaminants on plastic microfibers under environmentally relevant conditions. *Environmental Science & Technology* 55 (13): 8877–8887.

95 Brennecke, D., Duarte, B., Paiva, F. et al. (2016). Microplastics as vector for heavy metal contamination from the marine environment. *Estuarine Coastal and Shelf Science* 178: 189–195.

96 Nizzetto, L., Bussi, G., Futter, M.N. et al. (2016). A theoretical assessment of microplastic transport in river catchments and their retention by soils and river sediments. *Environmental Science-Processes & Impacts* 18 (8): 1050–1059.

97 Zhang, Y.-T., Wei, W., Huang, Q.-S. et al. (2020). Insights into the microbial response of anaerobic granular sludge during long-term exposure to polyethylene terephthalate microplastics. *Water Research* 179: 115898.

98 Yin, W., Zhang, B., Shi, J. et al. (2022). Microbial adaptation to co-occurring vanadium and microplastics in marine and riverine environments. *Journal of Hazardous Materials* 424: 127646.

99 Qiongjie, W., Yong, Z., Yangyang, Z. et al. (2022). Effects of biofilm on metal adsorption behavior and microbial community of microplastics. *Journal of Hazardous Materials* 424: 127340.

100 Zhang, C., Chen, X., Wang, J. et al. (2017). Toxic effects of microplastic on marine microalgae Skeletonema costatum: Interactions between microplastic and algae. *Environmental Pollution* 220: 1282–1288.

101 Tunali, M., Uzoefuna, E.N., Tunali, M.M. et al. (2020). Effect of microplastics and microplastic-metal combinations on growth and chlorophyll a concentration of Chlorella vulgaris. *The Science of the Total Environment* 743: 140479.

102 Fu, D., Zhang, Q., Fan, Z. et al. (2019). Aged microplastics polyvinyl chloride interact with copper and cause oxidative stress towards microalgae Chlorella vulgaris. *Aquatic Toxicology* 216: 105319.

103 Li, X., Luo, J., Zeng, H. et al. (2022). Microplastics decrease the toxicity of sulfamethoxazole to marine algae (Skeletonema costatum) at the cellular and molecular levels. *Science of the Total Environment* 824: 153855.

104 Lu, J., Wang, Y., Zhang, S. et al. (2020). Triclosan at environmental concentrations can enhance the spread of extracellular antibiotic resistance genes through transformation. *Science of the Total Environment* 713: 136621.

105 Lu, J., Wang, Y., Jin, M. et al. (2020). Both silver ions and silver nanoparticles facilitate the horizontal transfer of plasmid-mediated antibiotic resistance genes. *Water Research* 169: 115229.

106 Shen, M., Zhu, Y., Zhang, Y. et al. (2019). Micro(nano)plastics: Unignorable vectors for organisms. *Marine Pollution Bulletin* 139: 328–331.

107 Guo, J., Zhang, S., Lu, J. et al. (2020). Metallic nanoparticles and ions accelerate the uptake of extracellular antibiotic resistance genes through transformation.

108 Wang, Y., Lu, J., Mao, L. et al. (2019). Antiepileptic drug carbamazepine promotes horizontal transfer of plasmid-borne multi-antibiotic resistance genes within and across bacterial genera. *The ISME Journal* 13 (2): 509–522.

109 Liu, Y., Liu, W., Yang, X. et al. (2021). Microplastics are a hotspot for antibiotic resistance genes: Progress and perspective. *Science of the Total Environment* 773: 145643.

110 Menendez-Pedriza, A. and Jaumot, J. (2020). Interaction of environmental pollutants with microplastics: A critical review of sorption factors, bioaccumulation and ecotoxicological effects. *Toxics* 8 (2).

111 Lin, W., Jiang, R., Xiao, X. et al. (2020). Joint effect of nanoplastics and humic acid on the uptake of PAHs for Daphnia magna: A model study. *Journal of Hazardous Materials* 391.

112 Sleight, V.A., Bakir, A., Thompson, R.C. et al. (2017). Assessment of microplastic-sorbed contaminant bioavailability through analysis of biomarker gene expression in larval zebrafish. *Marine Pollution Bulletin* 116 (1–2): 291–297.

113 Nor, N.H.M. and Koelmans, A.A. (2019). Transfer of PCBs from microplastics under simulated gut fluid conditions is biphasic and reversible. *Environmental Science & Technology* 53 (4): 1874–1883.

114 Gonzalez-Soto, N., Hatfield, J., Katsumiti, A. et al. (2019). Impacts of dietary exposure to different sized polystyrene microplastics alone and with sorbed benzo a pyrene on biomarkers and whole organism responses in mussels Mytilus galloprovincialis. *Science of the Total Environment* 684: 548–566.

115 Godoy, V., Martinez-Ferez, A., Angeles Martin-Lara, M. et al. (2020). Microplastics as vectors of chromium and lead during dynamic simulation of the human gastrointestinal tract. *Sustainability* 12 (11).

116 Wang, Z., Dong, H., Wang, Y. et al. (2020). Effects of microplastics and their adsorption of cadmium as vectors on the cladoceran Moina monogolica Daday: Implications for plastic-ingesting organisms. *Journal of Hazardous Materials* 400: 123239.

117 Banaee, M., Soltanian, S., Sureda, A. et al. (2019). Evaluation of single and combined effects of cadmium and micro-plastic particles on biochemical and immunological parameters of common carp (Cyprinus carpio). *Chemosphere* 236.

118 Verla, A.W., Enyoh, C.E., Verla, E.N. et al. (2019). Microplastic–toxic chemical interaction: A review study on quantified levels, mechanism and implication. *SN Applied Sciences* 1 (11).

119 Liao, Y., Jiang, X., Xiao, Y. et al. (2020). Exposure of microalgae Euglena gracilis to polystyrene microbeads and cadmium: Perspective from the physiological and transcriptional responses. *Aquatic Toxicology* 228: 105650.

120 Miranda, T., Vieira, L.R., and Guilhermino, L. (2019). Neurotoxicity, behavior, and lethal effects of cadmium, microplastics, and their mixtures on pomatoschistus microps juveniles from two wild populations exposed under laboratory conditions? Implications to environmental and human risk assessment. *International Journal of Environmental Research and Public Health* 16 (16): 24.

121 Bhagat, J., Nishimura, N., and Shimada, Y. (2020). Toxicological interactions of microplastics/nanoplastics and environmental contaminants: Current knowledge and future perspectives. *Journal of Hazardous Materials* 405: 123913.

122 Syberg, K., Nielsen, A., Khan, F.R. et al. (2017). Microplastic potentiates triclosan toxicity to the marine copepod Acartia tonsa (Dana). *Journal of Toxicology and Environmental Health, Part A* 80 (23–24): 1369–1371.

123 Luis, L.G., Ferreira, P., Fonte, E. et al. (2015). Does the presence of microplastics influence the acute toxicity of chromium(VI) to early juveniles of the common goby (Pomatoschistus microps)? A study with juveniles from two wild estuarine populations. *Aquatic Toxicology* 164: 163–174.

124 Farrell, P. and Nelson, K. (2013). Trophic level transfer of microplastic: Mytilus edulis (L.) to Carcinus maenas (L.). *Environmental Pollution* 177: 1–3.

125 Koelmans, A.A., Mohamed Nor, N.H., Hermsen, E. et al. (2019). Microplastics in freshwaters and drinking water: Critical review and assessment of data quality. *Water Research* 155: 410–422.

126 Yukioka, S., Tanaka, S., Nabetani, Y. et al. (2020). Occurrence and characteristics of microplastics in surface road dust in Kusatsu (Japan), Da Nang (Vietnam), and Kathmandu (Nepal). *Environmental Pollution* 256: 113447.

127 Cox, K.D., Covernton, G.A., Davies, H.L. et al. (2019). Human consumption of microplastics. *Environmental Science & Technology* 53 (12): 7068–7074.

128 Rist, S., Carney Almroth, B., Hartmann, N.B. et al. (2018). A critical perspective on early communications concerning human health aspects of microplastics. *Science of the Total Environment* 626: 720–726.

129 Jeong, C.-B., Won, E.-J., Kang, H.-M. et al. (2016). Microplastic size-dependent toxicity, oxidative stress induction, and p-JNK and p-p38 activation in the Monogonont Rotifer (Brachionus koreanus). *Environmental Science & Technology* 50 (16): 8849–8857.

130 Chen, Q., Reisser, J., Cunsolo, S. et al. (2018). Pollutants in plastics within the North Pacific subtropical gyre. *Environmental Science & Technology* 52 (2): 446–456.

131 Toussaint, B., Raffael, B., Angers-Loustau, A. et al. (2019). Review of micro- and nanoplastic contamination in the food chain. *Food Additives & Contaminants: Part A* 36 (5): 639–673.

132 Liao, Y.L. and Yang, J.Y. (2020). Microplastic serves as a potential vector for Cr in an in-vitro human digestive model. *Science of the Total Environment* 703: 134805.

133 Mitra, S., Chakraborty, A.J., Tareq, A.M. et al. (2022). Impact of heavy metals on the environment and human health: Novel therapeutic insights to counter the toxicity. *Journal of King Saud University – Science* 34 (3): 101865.

134 Li, Z., Zhang, W., and Shan, B. (2022). Effects of organic matter on polycyclic aromatic hydrocarbons in riverine sediments affected by human activities. *Science of the Total Environment* 815: 152570.

13

Nanoplastics in Urban Waters

Recent Advances in the Knowledge Base

Ilaria Corsi[1,*], Elisa Bergami[1,2], Ian J. Allan[3], and Julien Gigault[4]

[1] Department of Physical, Earth and Environmental Sciences, University of Siena, Siena, Italy
[2] British Antarctic Survey, Natural Environment Research Council, Cambridge, UK
[3] Norwegian Institute for Water Research (NIVA), Oslo, Norway
[4] TAKUVIK Laboratory, CNRS/Université Laval, Quebec City, QC, Canada
* Corresponding author

13.1 Introduction

The world is enthusiastic for nanotechnologies, but there is now a heavy load of nanoplastics (< 1 µm) in the environment, because of waste products from the production, usage, and disposal of nanoplastics (primary sources), as well as the fragmentation of bulk materials (secondary sources). Whether nanoplastics begin in the air, soil, sewage, or industrial effluent, many of them make their way into aquatic environments [1, 2]. Synthetic nanopolymers are used in industrial and commercial products, including biosensors, photonics [3], cosmetics [4, 5], food nanocomposites [6], and pharmacological drug nanocarriers [7].

Nanoplastics end up in aquatic systems following the same routes as many other chemical pollutants: 1) through domestic and industrial discharges into urban waters, 2) wastewater treatment plants (WWTPs) and 3) mismanaged waste disposal [8]. Nanoplastics usually refer to polymers with various colors and shapes (e.g., fiber, film, sphere, and fragment) and size below 1 µm, thus comprehensive of the sub-micron and nanometric fraction [9].

The following studies have traced the occurrences and sources of nanoplastics in the environment:

- Environmental breakdown of larger polymer-based products by mechanical processes [10, 11], weathering agents (e.g., UV radiation, high temperature), [12] and biota [13] provide a secondary source of nanoplastics.
- Breakdown of macro- and microplastics in the rivers, brackish, and marine coastal areas can produce concentrations of nanoplastics at levels that are highly toxic to aquatic organisms [14].

Microplastics in Urban Water Management, First Edition. Edited by Bing-Jie Ni, Qiuxiang Xu, and Wei Wei.
© 2023 John Wiley & Sons, Inc. Published 2023 by John Wiley & Sons, Inc.

- Bench-scale studies and field campaigns have documented the continuous degradation of plastic debris in water down to the nano-fraction [6, 12, 15–19], raising concerns because of the potential hazard to the aquatic biota associated to their nanoscale dimension, as reviewed by [20].
- Modelled predicted environmental concentrations (PECs) for nanoplastics in surface waters, based on production, usage, disposal, and fragmentation have been estimated at < 20 µg/L [21].
- However, PECs are likely to increase in the near future, particularly in hot spot areas, because of sewage release and fragmentation of larger plastics already present in the aquatic environment [2].

This chapter describes our best current definitions for nanoplastics in terms of specific origin, size and properties, behavior in water media, sources, and ecotoxic interactions with other aquatic pollutants and living beings. We describe the overall large-scale transformations that occur in aquatic systems as a function of the intrinsic properties of nanoplastics and the receiving aquatic media. We also explore the role of biomolecular coronas formation (surface coverage of nanoplastics by biogenic material and/or contaminants) in nanoplastic uptake into cellular pathways, leading to toxic effects. They may be known as "ecological corona," or "eco-corona" when they originate from biomolecules present outside the organism from aquatic media, or as "protein coronas" when they are proteins originating from biological fluids inside the body of an exposed organism. Finally, we summarize the documented hazards posed by nano-polymeric particles used in bench-scale studies as a proxy for nanoplastics on aquatic biota (including planktonic species, invertebrates, and fish).

13.2 Nanoplastics in the Aquatic Environment

13.2.1 Nanoplastics or Polymeric Nanoparticles

Before investigating the sources of nanoplastics, we need to clearly define them. In the past five years (2016–2021), several opinions and perspective papers have been published on the definition of nanoplastics. As illustrated in [9], there is no consensus on their definition based on the size. Typically, two size classes occur in the literature: nanoparticles (NPs) and colloidal nanoparticles (NPs). The size can also help researchers to understand the sources of the nanoplastics.

According to the ISO normalization policy, **nanoparticles (NPs)** are *within 1–100 nm* along one of its dimensions. This definition was set more than ten years ago to help regulation agencies better define a nanomaterial for environmental and health protection policies.

For nanoparticles, some polymeric NPs are directly synthesized within this range and applied in manufactured products such as cosmetics, biomedical applications, or other construction materials [22]. These NPs are not plastic-based materials as we classically refer to, and they are usually defined as polymeric engineered nanoparticles (PENPs).

The second definition is for **nanoplastics**, which are any *colloidal and polymeric materials* found in the environment [23] that fall within the colloidal size range, i.e., *1–1000 nm*. The varied size distribution may seem to imply diverse sources, but in fact corresponds to an uncontrolled synthesis process, resulting in considerable heterogeneity in size and shape.

Koelmans et al. first proposed defining nanoplastics as 1–100 nm as in 2015, no clear definition of nanoplastics was available. The authors typically refer to the possibility to have nanoplastics either as a raw material or resulting from the last step of degradation of the polymer in the environment [24]. However, other studies have compared nanoplastics to very tiny microplastics leading to a shift in the size definition and misunderstanding in their source. For this reason, in 2018, Gigault et al. proposed a definition for nanoplastics according to a specific origin and size, i.e., *as a particle coming from the degradation of plastic debris and which demonstrates colloidal behavior in an aqueous system* [23]. For the first time, the source and the size were placed in the middle of the debate. This opened the door to new considerations about nanoplastics' environmental fate and ecotoxicological investigations.

Interestingly, since this clear definition, the number of papers using the 1–100 nm definition started to decrease, compared to those studying particles of 1–1000 nm. To bypass this lack of consensus about the nanoplastic definition, the term "submicron" or microplastics was used to indicate the size of the particle. However, while the size will always be a part of the debate on nanoplastics and their sources in the environment, we proposed an addition to the definition based on properties rather than size [25]. The size cutoff is mainly imposed or required by regulatory agencies to define and develop recommendations or environmental surveys on plastic issues [26]. Recently, guidance has been published by several governments to regulate the release of plastic debris in the environment. While large plastic debris can be easily described in terms of size and composition [9], when plastics degrade to sizes that show colloidal properties, several related properties specific to the nanoscale are enhanced [23, 25]. These properties define nanoplastics according to their various origin, formation, and transformation pathways in natural media. The principal properties are related to their Brownian motion, typical in aqueous media; biological interaction characteristic to nanomaterials, discussed later on; their surface specificity; their systematic heteroaggregation with other colloids including the natural organic matter (NOM) that is controlling the analytical challenge related to their final detection and quantification in environmental media [27]. In urban waters, the formation pathway of the nanoplastics will control its source, fate, and impact [8].

13.2.2 Formation Pathways of Nanoplastics

Two different release types are possible according to nanoplastic particle size and the other factors listed above [9, 28]. The first type of particles released are polymeric engineered nanoparticles, PENPs. The second type, nanoplastics, are formed by the degradation of plastic debris. The following papers address the synthesis and formation of various PENPs:

- The synthesis of PENPs, as for instance polystyrene nanoparticles (PSNPs), generally starts with the monomer constituting the final polymer. The most common approach to synthesize them is to use the emulsion-based method [29, 30], such as the one employed to synthesize PS nanobeads, primarily used in the literature as proxy for nanoplastics over the last five years. The emulsion polymerization consists of using a synthetic surfactant to create a micellar environment and then polymerize the polymer in a defined volume (i.e., controlled by the micelle).
- PENPs with well-defined properties are manufactured for various applications such as cosmetics, drug delivery, paints, etc. The particles are synthesized through various methods such as micro-emulsion, surfactant-free emulsion, and interfacial polymerization [31–33].
- The synthetic surfactant may be removed by eco-friendly or natural methods. In a recent study by Pikuda et al., it was demonstrated that the toxicity attributed to the PSNPs arose from the surfactant present in the stock suspension, rather than the polymeric materials [34].
- More recently, nanoprecipitation was used to synthesize spherical PENPs. This approach consists of dissolving a plastic in a compatible organic solvent and then inducing precipitation of the monomer in the water system using the solvent exchange process. This method was used to produce and disperse fullerene NPs in water [16].
- With this method, Balakrishnan et al. produced polyethylene (PE) nanoplastics with sizes ranging in the colloidal size range [35].

The second type of nanoplastics arises from the uncontrolled degradation of plastic debris in nature [25]. The following studies examine this phenomenon:

- It is common sense that nanoplastics result from the degradation of microplastics that come from degradation of large plastic debris in the environment [24].
- In 2016, we demonstrated that microplastics collected from the North Atlantic Ocean could release a large quantity of nanoplastics [16]. We placed the collected microplastics into a UV reactor. After several days of exposure, significant

amounts of nanoplastics were detected and characterized by in situ dynamic light scattering. However, nanoplastics are released all along the degradation pathways of plastic debris.

- Lambert and Wagner demonstrated that pristine plastic cups could rapidly release PS nano-size fragments up to 10^8 items/mL after 56 d of UV irradiation [36]. The same authors investigated the possible formation of nanoscale PET, PS, and polylactic acid (PLA) using a weathering chamber by Nanoparticle Tracking Analyzer [12].
- Other mechanical processes, including abrasion and laser ablation, can degrade plastics into nanoplastics for the principal family of plastic materials [37].
- The recent development of 3D printers is likely to increase the unintentional release of nanoplastics. Stephen et al. demonstrated the emission of up to 2×10^{11} PS particles per minute during the formulation of the plastic piece [38].
- The synthesis process makes particles within the sub-100 nm size range. As these kinds of nanoplastics are generated from plastic materials, it is challenging to classify them as PENPs or nanoplastics. However, their accidental formation induces in these particles a significant heterogeneity in their physical and chemical properties, making them closer to nanoplastics rather than polymeric PENPs. When this system is operating, a considerable nanoplastics can be transferred to air and then further deposited onto a hydrophobic surface [39].

13.2.3 Source of Nanoplastics

Plastic pollution in aqueous media can come from several sources, including poor waste management, landfill leaching, and biosolid application in agriculture [40–42]. Recent evidence of atmospheric deposition of microplastics has brought up new questions concerning this potential source in an aqueous system [43, 44]. While large microplastics are globally well recovered from the WWTPs, tiny particles of plastics can pass through these treatments and reach the river and marine systems [45–47].

After reviewing several published studies, significant reductions in the concentrations of microplastics in the effluent from various stations compared to the tributary were found. Concentrations ranged from 1–10 044 particles/L to 0–447 particles/L in the effluent. Recent work has estimated that 1000 rivers across the Earth account for 80% of global annual incomes in the marine system, which can be translated from 0.8 to 2.7 million metric tons per year, with urban rivers as the largest contributors [48]. However, though nanoplastics have been detected in soil [49], coastal [50], and marine waters [17], there is still no data concerning the estimation of nanoplastic flux in urban waters.

It is now well known that there is no industrial large-scale process to recover nanoscale particles from urban waters. To get an initial idea of the source and estimate the flux, we need to consider both PENPs and nanoplastics [51].

- First, if PENPs are contained in commercial products, they can be directly released in urban waters by simple use (use-rinse-release). Of course, accidental release of these manufactured NPs in the environment is possible but negligible compared to the first case. Hernandez et al. demonstrated by electron microscopy that different facial scrub products contain both PE microplastics (with a diameter ca. 0.2 mm) and NPs (24–52 nm) [52].
- They diluted the scrub into deionized water and then applied a sequential filtration approach in the nanoscale range (1–1000 nm) to isolate the microplastics and NPs. Fourier Transform Infrared Spectroscopy (FTIR) and X-ray photoelectron spectroscopy (XPS) determined the PE composition of these microplastics and NPs. By analogy with the engineered nanomaterials (ENMs), it is possible to have a range of the possible concentration and localization of the PENPs in the environment [53].
- Keller et al. estimated that ENMs generally end up between water media and soils [54]. These old estimations had helped to establish different emission scenarios [55]. The models predict that NPs will be released either to wastewater or directly to environmental systems by ineffective waste management and irresponsible disposal of plastic wastes.
- For PENPs, by analogy to other NPs (silver or other metal oxides), their presence in urban waters is principally because of their transportation pathways along with their use and life cycle [51–53].
- However, for nanoplastics, the release is principally induced by the aging or weathering of the plastic materials and their use and waste management. Using the same analytical approach used for scrub leaching of PENPs, Hernandez et al. demonstrated the release of polyethylene terephthalate (PET) and nylon micro- and nanoplastics from plastic teabags [6]. They show that plastic teabags can be altered under high temperatures and release billion of nanoplastics in the beverage. This result is consistent with the possibility of nanoplastics release or formation from plastic cups under UV conditions.

As previously explained, weathered plastics can directly release nanoplastics in water bodies under UV oxidation [12, 36]. However, in urban water, plastic debris and microplastics are considerably less weathered than in the marine system. Such variation will induce the relative efficiency of nanoplastics released along the plastics debris degradation pathways.

13.2.4 The Behavior and Environmental Fate of Nanoplastics

Learning about where nanoplastics end up in the environment, and how they behave, is crucial to help develop strategies to deal with them. The following studies aimed to trace the distribution and behavior of plastics in the environment.

13.2 Nanoplastics in the Aquatic Environment

- Plastic debris transportation in urban waters often follows hydrological pathways [56], as rivers are the primary carrier [40].
- However, plastic debris transportation' mechanisms are poorly understood because of wide size distribution and related properties.
- Intuitively, these mechanisms are mainly governed by the particle shape, density, size, and surface reactivity (i.e., oxidation degree) [57].
- It is now well known that the behavior of large plastic debris (size > 5 mm) differs from that of microplastics, as it would take more energy to transport larger plastics through an ecosystem for a single transport mechanism [56].
- Similarly, as colloidal particles, nanoplastics are highly diffusive species, and their aggregation will govern their transportation pathways [51, 58]. The final aggregation state of the nanoplastics condition their final size and shape, and therefore their transportation vs. accumulation in urban waters.

In the colloid science, ionic strength is the main parameter controlling the electrical double layer on the NP surface. This double layer is rapidly screened when ionic strength reaches 1 mmol/L. It is even more pronounced when divalent cations and ions are involved, as nanoplastics are generally negatively charged (e.g., carboxylated (-COOH) surface from the oxidation process in the natural system). Various authors have demonstrated the change in the nanoplastics while the ionic strength increases [59–61]. However, to our knowledge, there are neither aggregation nor changes in particle properties when ionic strength is below 20–30 mmol/L. However, in urban and freshwaters, the ionic strength is generally below 10 mmol/L. Therefore, the aggregation of nanoplastics can be explained by their ability to associate with natural colloids such as clay or other NOM [62, 63].

The following studies contain reports of nanoplastic interactions with NOM in both static and closed conditions:

- The NOM stabilizes nanoplastics by increasing the ionic strength. A different phenomenon is expected to occur: stabilizing the medium with humic-based materials [62] or wrapping the nanoplastics using polysaccharide-based materials [60].
- In urban waters, there is a large distribution of natural and colloidal organic matter: viruses, bacteria, humic substances, proteins, lipids, and inorganic NPs. It is well known that these species could interact with NPs and nanoplastics in natural media [40].
- Even if this interaction is generally not characterized, there is a strong correlation of their fate and transportation through similar hydrological conditions. While the stability in static conditions offers precious knowledge of the nanoplastics' physical and chemical properties compared to other NPs, no indication of their transportation pathways can be obtained. As explained before, the

dynamic of the system, especially urban waters, is crucial to determine the environmental fate and behavior of nanoplastics.
- Recently, using microfluidic approaches, Venel et al. demonstrated that the change ionic strength and the flow conditions are the principal parameters that govern their aggregation pathways [64].
- In urban waters, from the source to the WWTPs, the flow conditions change seasonally and need to be considered for fate and impact assessment as well as for remediation plans [8]. This final hetero-aggregation form and transportation pathways of nanoplastics governs their reactivity with other pollutants.

13.2.5 Interaction of Nanoplastics with Contaminants

Much of the knowledge about the interactions of organic chemicals and inorganic species with microplastics has been developed over the last decade, as reviewed by [65–68]. However, less is known about the interactions of these chemicals with smaller nanoplastics. The problem arises because the information obtained from studies with microplastics may not necessarily be applicable to nanoplastics. The properties of plastic debris can change at the nano-level.

The ability of plastic debris to sorb organic and inorganic contaminants from the aquatic environment depends on certain factors, such as the characteristics of the plastic debris (polymer type, structure and crystallinity, and size, age, and weathering of the debris); and the environment the plastic debris is exposed to (pH, salinity, ionic strength, and presence of dissolved and/or particulate organic matter). When the contaminant can diffuse into the polymer (most often seen with amorphous or rubbery polymers), the process is known as absorption, while in the case of glassy polymers with a high degree of crystallinity, sorption is a surface process (known as adsorption).

In the case of contaminant absorption into plastic debris, the concentration in the plastic at equilibrium will be related to the freely dissolved concentration through a polymer-water partition coefficient (K_{pw}). The adsorption process is governed by the formation of ionic, van der Waals, or steric and covalent bonds, and by plastic surface charge, surface area, and presence of functional groups at the surface. Adsorption is characterized by nonlinear sorption isotherms with proportionally higher sorption at lower environmental concentrations.

A wide range of chemicals have been found to sorb to microplastics, and these include legacy persistent organic pollutants such as polychlorinated biphenyls (PCBs), flame retardants such as polybrominated diphenyl ether (PBDEs), polycyclic aromatic hydrocarbons (PAHs) [69, 70], or other relatively more contemporary chemicals such as pharmaceuticals or perfluorochemicals.

Chemicals are added to plastics during their production to enhance specific properties, e.g., flexibility, flame retardancy, or lifetime. The leaching capability of additives such as plasticizers, flame retardants, UV stabilizers, pigments, or residual curing agents depends on the type of plastic and the chemical agent itself [71].

The release of chemical additives from plastic debris, and their fate in the environment is linked to the fate of plastic debris (emission, particle size, or breakdown). Plastic debris may release chemical additives to the aquatic environment then into organisms. While Bridson et al. reviewed the literature for protocols for assessing the leaching of additives from plastic debris, none of these studies were applied to nanoplastics [72].

In general, studies involving the interaction of chemicals with nanoplastics remain relatively scarce. As detailed below, most studies have focused on a limited number of metals, organic contaminants (PAHs, bisphenol A (BPA), and selected pharmaceutical products). The main lines of work have been to evaluate sorption isotherms, usually with model nanoplastics (PSNPs) and exposure of aquatic organisms to nanoplastics and free/sorbed chemicals to assess the contribution of sorption to nanoplastics on contaminant bioavailability, uptake, and toxicity. The impact of various factors has been investigated, including nanoplastic aging, solution ionic strength or salinity, pH, and organic matter on the sorption of contaminants to nanoplastics.

Town and Van Leeuwen calculated the exchange of chemicals between micro- and nanoplastics and the water they are in, by showing the clear impact of particle size and radius on contaminant uptake and release rate constants. The fastest, highest contaminant exchange kinetics can be expected for nanoplastics. As expected from this modelling or the passive sampling theory developed over the last three decades, the time required for a contaminant concentration in a plastic particle to reach equilibrium with the water will be the shortest for the smallest particles. Based on this modelling, it can be expected that contaminants will exchange between nanoplastics and the water phase more readily than for larger particles of the same polymer [73].

Plastic particles at the nano-size can effectively sorb metals from water. Monikh et al. conducted isotherm sorption experiments with silver ions (Ag^+), PSNPs, and PENPs in the presence and absence of NOMs. Both particle size and polymer type affect Ag^+ sorption. NOM coatings on nanoplastic particles increased sorption and inhibited toxicity towards daphnids through a decrease of Ag^+ release [74]. This, in turn, would tend to reduce the trojan horse effect of nanoplastics in freshwater environments.

To improve the representativeness of the evaluation, Davranche et al. studied the sorption of lead (Pb) to nanoplastics produced from microplastics originating from the North Atlantic gyre. A high Freundlich isotherm sorption coefficient was measured for Pb sorption that occurred mostly as a surface process involving oxygenated binding sites (following environmental UV oxidation of the plastic debris) and to a lesser extent through intraparticle diffusion [75]. In another study, the sorption isotherms of Cd, Cu, Ni, Pb, and Zn to 100 nm-PS following aging with UV irradiation under different conditions demonstrated increased sorption of these metals to PS with nanoplastic aging [76]. These studies confirm the potential of nanoplastic debris to be vectors of ad/absorbed contaminants.

Wastewater, effluent, and freshwater now also include a wide range of pharmaceuticals at the ng to µg/L range. In urban waters, nanoplastics may react or interact with pharmaceuticals to alter the risk these pose to freshwater organisms. The following studies all observed this effect:

- Barreto et al. found that 60 nm PSNPs affected the toxicity of the anticholesterolemic drug simvastatin to zebrafish (Danio rerio) [77].
- Chen et al. observed an increased uptake of BPA in zebrafish upon co-exposure with PSNPs [78].
- Parenti et al. observed a modification of the toxicological effects of triclosan towards zebrafish larvae, depending on whether the compound was free in solution or sorbed onto PS nanobeads [79].
- Yilimulati et al. reported the adsorption of the antibiotic ciprofloxacin to PSNPs exhibiting carboxylic functional groups mainly through electrostatic and hydrophobic interactions and hydrogen bonding [80]. The nanoplastic-adsorbed antibiotic appeared more toxic to the nematode Caenorhabditis elegans than either component on its own.
- Xiong et al. also investigated the sorption of ciprofloxacin to PSNPs but synthesized in house for this study. Aging of the nanoplastics through UV treatment and measured through the carbonyl index increased the sorption of ciprofloxacin and BPA [81].
- Zhang et al. reported higher sorption of two fluoroquinolones, norfloxacin and levofloxacin, on PSNPs with carboxyl-functionalization than without. Owing to their multiple pK_as, and to the mechanisms of sorption, the solution pH was shown to affect binding to nanoplastics. Salinity and the presence of organic matter also altered their sorption [82].
- Cortés-Arriagada et al. investigated the sorption mechanisms of BPA onto nanosized PET using density functional theory calculations. The outer surface of nanoscale PET was found to be nucleophilic with dispersion and electrostatic interactions as major sorption mechanisms to the outer and inner surface adsorption, respectively. Adsorption was found to be less significant in waters with higher ionic strength [83].
- Liu et al. observed differences in enhanced contaminant transport with PSNPs for nonpolar chemicals and more polar ones. The authors attributed these differences to sorption/desorption hysteresis for nonpolar chemicals from irreversible adsorption to glassy parts of the nanoplastics [84].
- In a follow-up study, these authors demonstrated that the aging of PSNPs (with UV or ozone) increased the mobility of the particles through an increase in surface hydrophilicity [85]. Aging also increased the sorption of contaminants to

13.2 Nanoplastics in the Aquatic Environment

the nanoplastics, albeit through different processes for the non-polar and more polar chemicals.

PAHs, as products of incomplete combustion processes, are ubiquitous in the aquatic environment and are also the subject of sorption studies with nanoplastics. The following studies have increased our knowledge of PAH sorption:

- The measurement of adsorption isotherms of PAHs and PCBs onto PSNPs were recently reported [86, 87]. In both studies, passive sampling methods were put in place to estimate freely dissolved concentrations in water during the experiments. A stronger sorption of PCBs to PSNPs than to micro-PE was attributed to the higher aromaticity and surface-to-volume ratio of PSNPs. As opposed to linear sorption to the PE (through hydrophobic partitioning), the sorption isotherm for PS was nonlinear. $\pi-\pi$ interactions were expected to be prevalent in the adsorption of planar PAHs to the aromatic surface of PS.
- Liu et al. suggested that aggregation did not affect the ability of PAHs to reach adsorption sites on the surface of the nanoplastics [86].
- Passive sampling techniques were also used in an attempt to distinguish the mechanisms of PCB uptake into the freshwater crustacean Daphnia magna in the presence of PSNPs [88].
- Ma et al. showed the additive toxicological effect of PSNPs and phenanthrene on D. magna [89]. Nanoplastics increase the bioaccumulation of phenanthrene into daphnids.
- Lin et al. developed a biodynamic model to evaluate the impact of dissolved organic matter (DOM) and nanoplastic on the uptake of PAHs from water into daphnids [90]. Results showed that PSNP on its own delayed intestinal uptake while the DOM-nanoplastic combination facilitated the transfer of PAHs in the gut.
- The effects of PSNPs-associated PAHs on zebrafish were investigated by Trevisan et al. [91]. The exposure of PAHs adsorbed onto nanoplastics appeared to affect the internal distribution of PAHs in the fish larvae.
- Another study by these authors showed a decrease in bioavailability, uptake, and toxicity of free PAHs to zebrafish larvae upon concomitant exposure to PSNPs [92]. PAH uptake may also be affected by nanoplastic aggregation. While in a finite volume of water, sorption to nanoplastics is likely to affect the contaminant concentration in water, this remains to be demonstrated in real environmental scenarios.

Studies focusing on the interaction between nanoplastics and co-contaminants are limited to one or two polymers only in most cases (i.e., PS and PE). There is therefore a need to extend the research effort to other polymer types. Since the adsorption of model chemicals to nanoplastics has been identified, the resulting *Trojan horse* effect is possible and some of the results mentioned above tend to point in that direction. However, to confirm this, more robust experiments will be needed. There is a need for greater understanding of the distribution and redistribution of contaminants during combined exposure to the organisms. The relative contribution of this *Trojan horse* effect to contaminant accumulation and toxicity in organisms or transport of chemicals over long(er) distances will also depend on the nanoplastic concentrations in the environment. The characteristic travel distance of chemicals may indeed be affected by their sorption to nanoplastics. This interaction of chemicals with nanoplastics can legitimately be expected to affect nanoplastics' transformation in the environment or how they behave in receiving waters or WWTPs.

Until now, most of this work has been conducted with PENPs as a proxy for nanoplastics, or with nanoplastics produced in the laboratory. A longer-term objective to improve the assessment of the risk posed by nanoplastics in the aquatic environment is to measure the concentrations of contaminants sorbed or those present as additives from production in nanoplastics in the field. When considering chemical additives present in nanoplastics, it will be useful to understand the mechanisms of release of these chemicals during particle breakdown, erosion, or UV exposure. Ultimately, the presence of nanoplastics in urban and natural waters may lead to interactions with other types of pathogens; by this mechanism, nanoplastics could become vectors of environmental DNA or antibiotic resistance genes. This transport pathway will also need to be elucidated in the future.

13.3 Interactions between Nanoplastics and Aquatic Organisms

The most recent findings on the adverse effects of nanoplastics on aquatic organisms rely on a mechanistic approach employed in human nanotoxicology that aims to unravel cause-and-effect relationships based on their nanoscale properties. Major challenges remain to identify "nano" and biological bases behind the biological effects arising from the complex interplay occurring in the natural environment which link to toxicity [93].

Nanoplastics enter various parts of aquatic ecosystems before they reach organisms. Bio-nano interactions occur with biomolecules and chemicals either in the external environment (i.e., freshwater, brackish or sea waters) or within organisms. Nanoscale properties, such as the surface area-to-volume ratio, dictate how these interactions proceed. The formation of the well-known biomolecular "corona," including proteins, lipids, and saccharides, creates a coating of a continuously evolving adsorbed entities which derive from a biological environment [94–96]. Either long-lived and tightly bound (i.e., hard corona) or as a cloud of loosely bound attachments (i.e., soft corona), coronas are key determinants of nanoscale particles' physico-chemical features and bio-interactions [97]. The following studies have looked at types of coronas on nanoplastics:

- Biomolecular corona formation has been incorporated into the field of ecotoxicology, for example, in the internal body fluids of invertebrate species as a means of understanding their biological outcomes [98, 99].
- Protein-corona formation around PSNPs has been documented recently in several studies in which major proteins involved in opsonization and cell-adhesion functions have been identified (i.e., extrapallial protein precursor, toposome precursor, nectin, actins) [100–103].
- Eco-corona formation in natural waters from the interaction with NOM including algal exudates and their polysaccharides have also recently been documented and their role in affecting bio-nano interactions and ultimate toxicity recently elucidated [93].

13.3.1 Effects on Aquatic Organisms: From Microalgae to Fish

Microalgae sustain aquatic ecosystems. As primary producers, microalgae provide energy and matter to the ecosystem while they consume atmospheric CO_2. Considering the potential ubiquity of nanoplastics in the aquatic environment, the understanding of their interactions with microalgae is crucial to predict single and cumulative impacts not only on microalgae populations and communities but also at the ecosystem level [104].

As shown by a recent meta-analysis [105], several ecotoxicity studies conducted between 2010 and 2020 have reported no or slight negative effect of model NPs (mainly PSNPs) on microalgae growth during their exponential phase in standard 48–96 h tests. Exposure to plain and negatively charged NPs usually led to EC_{50} values > 25 mg/L, while positively charged NPs were often associated with higher toxicity (EC_{50} < 1 mg/L) [106].

Many studies have shown a strong interaction between PENPs and microalgae surface and associated with negative effects such shading, cell membrane destabilization, decrease in pigment content and photosynthetic capacity, enhancing oxidative stress and production of antioxidant enzymes [93, 106]. However, positive effects have also been reported, including reduction in infection disease and growth stimulation [105] (Figure 13.1).

Schampera et al. found that the formation of heteroaggregates of negatively charged PSNPs with the filamentous cyanobacterium Planktothrix agardhii, a common component of harmful algal blooms in temperate freshwater ecosystems, had a protective role towards the attack by its fungal parasite. In 10 d experiments, PSNPs (10 and 100 mg/L) inhibited phytoplankton growth, thus reducing the parasitic pressure in terms of number of infected hosts and disease transmission [107]. These observations suggest that nanoplastics in the environment can alter host-parasite interactions leading to possible cascading effects on ecosystem functioning.

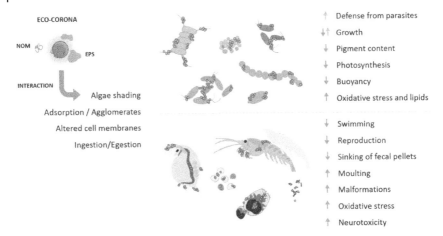

Figure 13.1 Interactions and effects of model nanoplastics to phytoplankton (upper part) and zooplankton (lower part) in natural waters: eco-corona formation (left) and impact (right), based on the cited literature. Positive and negative effects are reported as green dotted and red plain arrows, respectively. Not to scale.

A first study performed on Rhodomonas baltica, a red microalgae widespread in the North Atlantic, showed the effects of two functionalized polymethyl methacrylate (PMMA) NPs (50 nm), both plain and -COOH, on this cryptophyte [108]. PMMA-COOH did not form aggregates in the medium and were found to significantly affect cell viability of the cryptophytes at concentrations higher than 25 mg/L, with impairment in the metabolic and photosynthetic activities. In contrast, microscale aggregates of plain PMMA mainly caused alterations in microalgal cell size and subcellular components with increase in reactive oxygen species (ROS) and lipid peroxidation levels and a decreased in photosynthetic capacity. The use of multiple end points allowed efficiently discrimination between the mechanistic effects of two PENPs with the same core but different functionalization and behavior. In terms of species comparison, Rowenczyk et al. showed a general stimulating effect on the growth of diatoms, cyanobacteria, and green microalgae after 96 h of combined exposure to model PSNPs and humic acids, with only a marginal effect on photosynthesis [109].

Baudrimont et al. were the first to compare the ecotoxicological effects of nanoplastics obtained from virgin PE as a reference material and from weathered PE from the North Atlantic gyre as environmentally relevant one. Growth inhibition tests (48 h) were conducted both with the freshwater green alga Scenedemus subspicatus and the marine diatom Thalassiosira weissiflogii at higher concentrations of NPs (1, 10, 100, 1000, and 10 000 µg/L). The results showed that environmentally relevant PENPs were more harmful for algal growth (S. subspicatus > T. weissiflogii) than reference ones, suggesting that contaminants such as metals adsorbed onto nanoplastics from environmental samples contributed to their toxicity [110]. Those outcomes were associated with the different aggregation of NPs observed in freshwater (less aggregates and higher availability) vs. sea water (stronger aggregation and lower toxicity) media. A prolonged incubation (38 d) of the microalga Chlorella vulgaris to PSNPs (at 250 mg/L) led to a significant size-dependent inhibition of algal growth (20 nm vs 50 and 500 nm), with bare NPs leading to the strongest effect compared to PS-COOH [111]. In addition, bare PSNPs significantly decreased Chlorophyll-a after 14 d and induced lactate dehydrogenase and production of ROS after 28 d.

A recent study on one of the main bloom-forming diatom Skeletonema marinoi revealed that after 15 d PS-COOH (at 1, 10, 50 mg/L) strongly attached onto cell surface did not affect cell viability but caused a significant reduction in capacity of marine diatoms to form long chains [112]. Although the latter one is not a standard parameter evaluated in algal toxicity tests, it uncovers potential ecological implications, since the shortening in diatoms chains and consequent change in their buoyancy has repercussions on whole marine ecosystems.

More accurate representations of environmental scenarios consider the impact of nanoplastics on microalgae in combination with other stressors, such as ocean acidification or other relevant contaminants. Yang et al. investigated the combined effects of PSNPs on S. obliquus through a multiplexed experiment with more than 2000 combinations of environmental variables (i.e., CO_2, light, temperature) to represent future climate change scenarios [113]. Their findings showed several interactive effects among the disturbances that modulated the toxic potential of PSNPs, underlining the relevance of multi-stressors experiments for environmental risk assessment purposes.

The above-mentioned studies underline that in the natural environment, nanoplastics are likely to interact with microalgae, resulting in both positive and negative effects at the organism and population level, but more importantly altering interactions between species and energy fluxes in aquatic and marine environments, with consequent ecological harm.

The following studies also examined the toxicity of nanoplastics.

- The effects of nanoplastics on primary consumers have widely been investigated through standard toxicity tests with model organisms, mainly including rotifers (Brachionus spp.) and microcrustaceans belonging to Brachiopoda (e.g., Daphnia spp. and Artemia spp.) and Copepoda (e.g., Tigriopus spp.), as reviewed in [20, 93].
- The first studies on aquatic and marine zooplankton mostly focused on short-term experiments (6–48 h) to address the immobilization/lethality and biodistribution of nanoplastics through passive ingestion/egestion, and retention time in the digestive tract. Long-term tests ($\geq$ 14 d) had realistic exposure times to explore the toxic potential of nanoplastics. Several sub-lethal

end-points were studied, and macroscopic observations were combined with molecular and biochemical analyses [114, 115].
- In fresh water microcrustaceans, such as Daphnia spp., decrease in fecundity, reproduction efficiency and offspring number has been reported [20] and further confirmed upon transcriptomic analysis [116] and significant lifetime exposure (103 d) to PSNPs [117].
- In general, the toxicity of nanoplastics on zooplankton has been found to be functionalization- and concentration-dependent and is inversely correlated with the size of the particles, with the most striking effects observed for NPs with nominal size < 100 nm.
- Model neutral and negatively charged NPs (such as PS-COOH) usually did not exhibit acute toxicity, although they accumulated in the organisms, which could suggest trophic transfer upon predation on zooplankton if continuously exposed in the aquatic environment. Conversely, positively charged nanoplastics (i.e., PS-NH_2) were found to alter physiological (e.g., molting) and behavioral (e.g., swimming, escape response) traits in zooplankton and to cause oxidative stress and neurotoxicity [118], reviewed in [93] (Figure 13.1).

Concerning marine meroplankton, studies have focused on early larval stages of species having both key ecological role and high commercial value, such as the sea urchin Paracentrotus lividus [119, 120], the mussel Mytilus spp [121, 122]. and the Pacific oyster Crassostrea gigas [123, 124]. Overall, these studies point out that the toxicity of nanoplastics depends on their functionalization and surface charge, which also determines NP behavior and fate in sea water. In particular, PS-NH_2 led to severe malformations in embryos of marine invertebrates, even causing developmental delay and arrest. In contrast, PS-COOH were mostly associated with limited or no effect apart from ingestion/excretion, as shown for holoplankton. An interesting outcome in developing larvae was the altered shell mineralization and morphology observed in D-veliger larvae of bivalve species [121, 124], which could hamper larval survival and settlement. Recently, an embryotoxicity study was conducted on the ascidian Ciona robusta, a common inhabitant of coastal benthic communities in the Mediterranean Sea. This organism is also appointed as an early chordate model [125]. Results indicate no alteration in the phenotype of larvae after 22 h of exposure to PS-COOH up to 100 mg/L, while PS-NH_2 (2 – 15 mg/L) caused dose-dependent embryotoxicity including inhibition of hatching, with potential consequences for the recruitment of this species in contaminated marine environments.

Promising insights into the mode of action of nanoplastics are now achieved using -omic approaches. For example, by transcriptomic analysis, Jeong et al. unveiled the toxicity pathways of plain PSNPs (50 nm) and the water-accommodated fraction of crude oil on the rotifer B. koreanus after 7 d exposure. Data from differentially expressed genes supported an adverse outcome pathway and indicated synergistic effects, with alteration in rotifer reproduction and change in energy allocation as a potential defense mechanism [126].

Using model organisms in ecotoxicity testing offers some obvious advantages for environmental risk assessment, such as standardization and reproducibility of the experiments and well-known characterization of the biological responses of a species with respect to reference contaminants. However, this approach has limitations in terms of ecological relevance, especially for vulnerable and less studied ecosystems, such as remote polar regions.

In this regard, Bergami et al. were the first to report the effects of PSNPs on a keystone species of the Southern Ocean, the Antarctic krill Euphausia superba. Short-term (48 h) incubations of krill juveniles showed impairments in krill motility and physiology, suggesting spin-off effects for the Antarctic food web, which mostly relies on krill abundance. Moreover, agglomerates of PSNPs were found incorporated in krill fecal pellets, altering microbial community, and decreasing their sinking rate in the water column. These observations underpin repercussions at the ecosystem level, since lower sinking rates of krill fecal pellets might lead to a higher remineralization of this biogenic material in the ocean and consequently loss of organic C buried in the deep sea [127]. Predicted increasing levels of nanoplastics might act in concert with other environmental disturbances and disrupt ocean ecosystem functions and services [128].

Compared to the literature available on plankton, current knowledge on the impacts of nanoplastics on benthic organisms is still limited, reviewed in [93] (Figure 13.2). Most of the research focuses on filter-feeders and especially marine mussels (Mytilus spp.) [129] because they are known to accumulate pollutants in the marine environment and have recently been proposed as sentinel species for the biomonitoring of microplastic worldwide [130, 131]. Marine mussels have also been used in laboratory experiments to demonstrate the trophic transfer of model PSNPs (< 500 nm) upon predation by crabs [132].

First in vivo studies on marine mussels showed uptake/release of agglomerates of PSNPs and release of pseudofeces upon short-term exposures, reviewed in [93]. Likewise, by exposing the freshwater clam Corbicula fluminea to a mixture of microalgae (S. subspicatus) and environmentally relevant PE NPs (at 1000 µg/L) for 48 h, Baudrimont et al. reported no effect on bivalves' filtration activity, but an altered production of (pseudo)feces [110].

In the Mediterranean mussel Mytilus galloprovincialis, Auguste et al. reported that a discontinuous exposure to PS-NH$_2$ (10 µg/L) induced immune stimulation to counteract the external challenge and maintain homeostasis [133]. Capolupo et al. further exposed *M. galloprovincialis* to low concentrations (1.5, 15, and 150 ng/L) of PSNPs for 21 d, showing in a suite of biomarkers related to immunological functions, oxidative stress, and neurotoxicity [134]. The authors

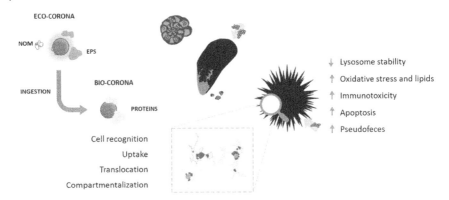

Figure 13.2 Interactions and effects of model nanoplastics to benthic organisms: eco- and bio-corona formation (left) and impact (right), based on the cited literature. Negative effects are listed together with red plain arrows. Not to scale.

thus demonstrated that long-term exposure to environmentally relevant concentrations of PSNPs could have a significant impact on the health status of marine mussels.

The bridge between in vitro and in vivo studies with benthic organisms is represented by the foraminifer Ammonia parkinsoniana, a single-cell marine eukaryote living in marine sediments. Ciacci et al. have shown a fast internalization and compartmentalization of PSNPs (1 mg/L) within 24 h, with consequent increase in ROS and neutral lipid levels [135]. However, further studies are necessary to elucidate the fate of nanoplastics in bottom-dwelling species.

Investigation of the immune cells of marine benthic organisms provides accurate information on the uptake and sub-cellular localization of nanoplastics and the generated immune response towards such external challenges. Haemocytes and coelomocytes obtained from the mussel M. galloprovincialis and the sea urchin P. lividus, respectively, are the most common organisms used in in vitro studies [100, 101, 136–138]. Sea urchins are excellent model organisms, as they have both key ecological functions as grazers in benthic communities, and a complex innate immunity to counteract external challenges.

Short-term cultures (1 – 24 h) of P. lividus coelomocytes to model PSNPs revealed that 10 and 25 mg/L doses of PS-NH$_2$ elicited dose-dependent immunotoxicity and affected the activity of ATP-binding cassette transporters [101]. A dose of PS-COOH did not cause severe effects and was quickly eliminated by phagocytes, although a 25 mg/L dose of PS-COOH did alter P. lividus phagocytic capacity and destabilized lysosomal membranes [138]. Similarly, a clear immune response against PSNPs was observed in coelomocytes of the Antarctic sea urchin Sterechinus neumayeri and supported by the modulation of genes involved in antioxidant activity and apoptosis [137].

13.3 Interactions between Nanoplastics and Aquatic Organisms | 425

Most of the studies of nanoplastics in fish are concentrated on fresh water and brackish species (i.e., zebrafish, Japanese medaka, common goby, Crucian carp, red tilapia) rather than marine species (e.g., sea bass) and more on the embryo/larval stages and less on juveniles or adults. D. rerio is the model species most investigated; either upon exposure to nanoplastics during embryo development, or on adults for nanoplastic uptake and distribution in organs and tissues (Figure 13.3).

Accumulation of nanoplastics in target tissues has been demonstrated in zebrafish (D. rerio) embryos/larvae associated with behavioral alterations, oxidative stress, and a reduction in acetylcholinesterase activity (AChE) [139, 140]. Similarly, a decrease in AChE activity is reported by Ding et al. upon exposure to PSNPs (100 nm) of the red tilapia Oreochromis niloticus (10–610 µg/L) [141]. Although the nanometer size of PS (50 nm) has been shown to readily penetrate the chorion of developing zebrafish embryos and mostly accumulated in lipid-rich regions such as the yolk lipids, only marginal effects on the survival, hatching rate, developmental abnormalities until cell death were observed. However, an exacerbation of Au ion-induced toxicity by PSNPs was observed in terms of the inflammatory response, oxidative stress up to mortality, and embryo deformations in zebrafish upon PSNPs bioaccumulation [142]. Therefore, the combined exposure to PSNPs and metal ions seems to induce a toxicity-dependent effect as a result of their nanoscale dimension and associated properties.

Sökmen et al. reported that upon microinjection into yolk sac of zebrafish embryos (at 120 h), PSNPs (20 nm) accumulate in the brain and cause oxidative DNA damage and body malformations in zebrafish embryos [143]. Brain accumulation has been

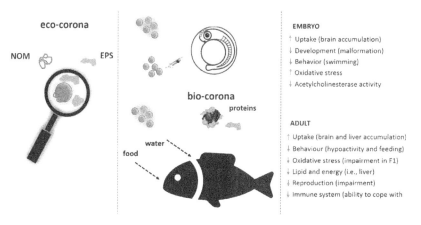

Figure 13.3 Interactions and effects of model nanoplastics to fish (middle), including fish embryos (right), based on the cited literature. Positive and negative effects are reported as green dotted and red plain arrows, respectively. Not to scale.

shown in adults of Crucian carp (Carassius carassius) fed with PSNPs (~130 mg of particles per feeding) with associated changes in brain morphology and organism behavior (lower activity and longer feeding time) [144]. Changes in locomotive activity have also been reported upon exposure to PSNPs (51 nm at 5×10^{-3} mg/L) upon trophic transfer in rice-fish Oryzias sinensis and dark chub Zacco temminckii; the latter also show abnormal liver histological patterns [145].

Nanoplastics can be ingested by fish through gills and transferred via the blood to several organs, including the brain [146]. The blood-brain barrier is established around 72 hours post fertilization (hpf) in the zebrafish [147]. Therefore, the brain could be a significant target for nanoplastics delivered in larger amounts to zebrafish larvae in the early stages. The ability of PSNPs to pass into the yolk sac and migrate through the gastrointestinal tract to the gallbladder, liver, pancreas, heart, and brain has been demonstrated [148]. It has also been shown that the PSNPs cause lower locomotor activity (hypoactivity) at 144 hpf.

Maternal transfer of PSNPs to offspring has been shown in association with antioxidant system impairment in tissues of adults and F1 larvae and their development [149]. Lu et al. reported that labelled-PS particles of various sizes, including NPs (20 μm, 5 μm, and 70 nm) distribute and accumulate differently in the organs of adult specimens in 7 d and cause histological changes and oxidative stress, disruption of lipid and energy in the liver [150]. Disruption of lipid metabolism has also been reported upon trophic transfer of PSNPs (24–48 nm) from algae to fish (Scenedesmus sp., D. magna, C. carassius) [151, 152]. More recently, Monikh et al. described a parental, trophic transfer and reproductive toxicity of iron oxide-doped PSNPD (Fe-PS-NPD, 270 nm) and Europium (Eu)-doped PSNPD (Eu-PS-NPD, 640 nm) as a function of NPD size being the smallest ones more hazardous [153].

Induction of oxidative stress has been largely investigated, but a multi-markers approach has been suggested to assess potential consequences as tissue damage and disruption of cellular functions. Parenti et al. revealed a protective role of superoxide dismutase (SOD) towards excessive ROS production in zebrafish upon exposure to nanoplastics as a mechanism against oxidative damage [154]. Similar findings were reported by Ding et al. on red tilapia O. niloticus where neither ROS production or malondialdehyde (MDA) were affected upon exposure to PSNPs (500 nm at 1×10^{-3} mg/mL and 100 nm $1 \times 10^{-6}/10^{-4}$ mg/mL respectively) [141]. Interactions of nanoplastics with natural colloids such as organic matter might also affect their impact on fish. Co-exposure to PSNPs (50–100 nm, $1 \times 10^{-6}/10^{-3}$ mg/mL) and fulvic acid caused a synergistic impact on ROS levels in zebrafish [155].

The innate immune system in fish is another target for PSNPs. Fathead minnow (Pimephales promelas) neutrophils in vitro exposed to PSNPs (41 nm, 100 mg/L, 2 h) show a significant increase in degranulation of primary granules and extracellular trap release thus affecting their ability to cope with disease [156]. An activation of the complement system in zebrafish after injection of 700 nm PSNPs at 5000 mg/L [157] and an upregulation of proinflammatory cytokines after both injection (1 mg/L) and waterborne exposure (10 mg/L) to 20 nm PSNPs documented the interaction of nanoplastics on teleost immune systems [142, 158]. Similarly, a synergistic effect on the expression of immune genes was also observed in zebrafish after 12 h exposure to 5 mg/L of 50 nm PSNPs, followed by a challenge with a viral stimulus [159].

Few studies addressed the impact of nanoplastics in marine fish species, although similar biological effects of freshwater species are documented as alteration in feeding behavior, decrease in protein and lipid content, and effects on immune system (Sebastes schlegelii 0.1 mg/L of PSNPs 500 nm; Dicentrachus labrax and Sparus aurata 0.001–10 mg/L of PMMA-NPs 45 nm 96 h waterborne) [160–162]. As far as in vitro studies, combined exposure with PSNP (100 nm) and drugs change their lethality in fish cell lines from gilthead sea bream (S. aurata) and European sea bass (D. labrax) [163].

13.4 Ingestion of Nanoplastics in Aquatic Organisms

Since the first hypothesis by Cózar et al. that plastic litter in the oceans could be prone to continuous weathering and degradation even below 1 μm, researchers have looked for nanoplastics in environmental samples [15]. However, the detection of the smallest fraction of plastic debris has proven to be challenging because of the organic nature of plastic polymers and their low concentrations in complex matrices. Therefore, the detection of nanoplastics in aquatic organisms has limped along with little success, comparable to looking for a needle in a haystack.

Transmission and Scanning Electron Microscopy (TEM and SEM) techniques have been employed to detect the interplay between model nanoplastics and microalgae [164–167] and to show morphological alterations in cells and/or tissues after exposure to NPs [100, 111]. For example, Murano et al. observed changes in the ultrastructure of sea urchin phagocytes following short-term exposure (4 h) to PS-COOH NPs (50 nm). In SEM and TEM images, treated cells displayed a narrowed structure, damaged membranes and filopodia, and abnormal nuclei compared to the control group [138].

The latest advancements in electron microscopy techniques are useful for examining biological samples without any preparation. Thus, we can observe the natural interactions with high-quality images, clearly valuable for impact assessment purposes. Studies that have used these techniques include the following:

- Variable Pressure SEM was used by Bergami et al. to capture images of fecal pellets of the Antarctic krill E. superba collected from the vials in which krill juveniles were exposed to PSNPs. SEM imaging revealed alteration in the peritrophic membrane of FPs from krill exposed to PSNPs, indirectly showing ingestion and egestion processes [127].
- Similarly, in vitro studies on trafficking of nanoparticles in human cells [168–170], research on nanoplastics and aquatic organisms largely adopted synthesized polymeric NPs coupled with fluorescent dyes as a proxy for nano-sized plastics interacting with aquatic organisms.
- Nanoparticle uptake can easily cross plasma membranes and their surface charges might also have a role. Amino-modified PSNPs, with a positive surface charge, are more prone to cross cell membranes and to damage them as a function of protein-corona formation [169, 171].
- Using fluorescently labelled nanoplastics coupled with microscopy techniques (optical fluorescent/laser scanning confocal microscopy, flow cytometry) was an effective strategy to follow the uptake and distribution of NPs in a range of species, from
    - adsorption onto microalgae [112, 165, 172],
    - ingestion, retention and excretion in zooplankton [119, 127, 165],
    - uptake, localization, and clearance (or removal) at the cellular level [135, 138],
- Sendra et al. also reported the presence of fluorescently labelled PSNPs in different tissues (gills, muscle, digestive gland) of the mussel M. galloprovincialis after short-term exposure (3 and 24 h), with size-dependent translocation into the circulatory system and uptake by hemocytes [173].

The above-mentioned pioneer studies have contributed to our understanding of the bioavailability, uptake, and fate of nanoplastics in biological systems, raising the concerns over potential repercussions on aquatic ecosystems. Innovative uses of dyes, as in the following studies, have helped researchers to trace nanoplastics through ingestion by organisms:

- Alternative testing materials to PS nanospheres may better represent environmentally relevant nanoplastics with various shapes, such as nano-sized PET and PE synthesized through dissolution and reprecipitation of polymeric flakes and labelled with Nile red for biodistribution assays [35, 174].
- However, in some recent studies, using nanoplastics with fluorescent dyes resulted in artifacts of the biodistribution of nanoplastics from dye leaching. These studies thus failed to target the actual presence and accumulation sites within the organisms [175, 176]. This effect also occurred in freshwater species, such as D. magna, and seems to be mostly related to the mechanism that

fluorescent dyes adsorbed onto the nanoplastic surface rather than inserted in their polymeric core.
- The stability of PS-COOH fluorophores in sea urchin coelomic fluid has recently been demonstrated [138]. Overall, dye-labeled nanoplastics have some major limitations related to possible dye leaching but also high detection limits of the fluorescent signal associated with the nanoplastics, which are thus tested at high concentrations, in the range of µg/mL. To overcome these issues, promising alternatives for proxies of nanoplastics have been developed, including metal-doped [177] and radiolabeled NPs [178]. For example, Al-Sid-Cheikh et al. have studied the tissue distribution and depuration kinetics of ^{14}C-radiolabeled PSNPs in the scallop Pecten maximus at realistic concentrations (15 µg/L), successfully detecting PSNPs in the different tissues of the scallop at various time-points by liquid scintillation counting and quantitative autoradiography. The authors demonstrated that PSNPs were still found in the hepatopancreas after 8 d depuration, suggesting bioaccumulation in the species upon chronic exposure [179].

Future research on a wide range of aquatic organisms from urban waters, rivers up to coastal environments, and the open ocean will allow us to better understand biointeractions with nanoplastics, their fate, and their impacts on biological systems.

13.5 Concluding Remarks and Future Recommendation

Scientists have defined several nanoplastic behaviors and identified major impacts and toxicity mechanisms for some key aquatic species (Figure 13.4). However, some knowledge gaps remain, and these are shaping the future studies in this

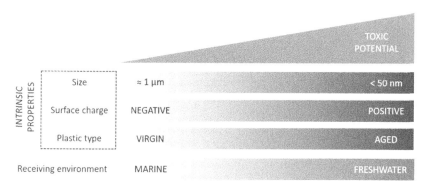

Figure 13.4 Schematic of the toxic potential of nanoplastics in natural waters based on their intrinsic and extrinsic properties, with reference to the literature cited in the main text.

growing research field. Future laboratory experiments should be carried out at environmentally realistic concentrations (< 1 µg/L), for example using metal-doped nanoplastics [177, 180] or radiolabeled ones [179, 181] that can be traced both in complex matrices and in the biota, allowing for a more reliable environmental risk assessment.

The current mismatch between microplastics field data and laboratory results is because most experiments so far having involved microfibers and PENPs that may not have been realistic models of the most common morphologies of environmental microplastics and nanoplastics needed to study the real toxic effects on aquatic species. Nanoplastic risk assessment based on laboratory studies are still not fully relevant for describing real exposure scenarios, because of a lack of testing environmentally realistic exposure conditions and media (i.e., predicted environmental concentrations, natural waters) and factors that affect nanoplastic behavior, fate, and toxicity (i.e., NOM, salinity etc.).

From an ecological perspective, long-term and chronic even unconventional endpoints should be preferred as study measures, over acute standard ecotoxicity tests. Multi-species studies should be encouraged to understand the impacts of nanoplastics over key ecological processes, such as algal bloom formation and particulate organic carbon flux.

"Omics" approaches are promising tools to determine signaling cascade and toxicity mechanisms triggered by anthropogenic nanoplastics in biological systems including the promising predictive tool of the adverse outcome pathway (AOP). Finally, gaining knowledge on bio- and eco-corona formation will be crucial to unveil the complex nano-bio interactions that occur in urban and natural systems. Their study and consideration in ecotoxicity studies are of paramount importance to obtain meaningful outcomes and extend them to real environmental scenarios.

Funding

The following funded projects partly supported this work: *Plastics in Antarctic Environment* (PLANET) and *Polymeric nanoparticles in the marine environment and Antarctic organisms* (NANOPANTA), both funded by the Italian National Antarctic Program (PNRA: 14_00090; 16_00075).

Acknowledgements

The authors acknowledge Christa Bedwin (University of British Columbia, Canada) for the English revision of the manuscript.

Competing Interests

The authors declare that there is no competing interest.

References

1 Lambert, S., Sinclair, C.J., Bradley, E.L. et al. (2013). Effects of environmental conditions on latex degradation in aquatic systems. *Science of the Total Environment* 447: 225–234.
2 Andrady, A.L. (2017). The plastic in microplastics: A review. *Marine Pollution Bulletin* 119 (1): 12–22.
3 Velev, O.D. and Kaler, E.W. (1999). In situ assembly of colloidal particles into miniaturized biosensors. *Langmuir* 15 (11): 3693–3698.
4 Guterres, S.S., Alves, M.P., and Pohlmann, A.R. (2007). Polymeric nanoparticles, nanospheres and nanocapsules, for cutaneous applications. *Drug Target Insights* 2: 147–157.
5 Leslie, H.A. (2014). Review of microplastics in cosmetics. *IVM Institute for Environmental Studies* 476: 33.
6 Hernandez, L.M., Xu, E.G., Larsson, H.C.E. et al. (2019). Plastic teabags release billions of microparticles and nanoparticles into tea. *Environmental Science and Technology* 53 (21): 12300–12310.
7 Jiménez-Fernández, E., Ruyra, A., Roher, N. et al. (2014). Nanoparticles as a novel delivery system for vitamin C administration in aquaculture. *Aquaculture* 432: 426–433.
8 Birch, Q.T., Potter, P.M., Pinto, P.X. et al. (2020). Sources, transport, measurement and impact of nano and microplastics in urban watersheds. *Reviews in Environmental Science and Bio/Technology* 19 (2): 275–336.
9 Hartmann, N.B., Hüffer, T., Thompson, R.C. et al. (2019). Are we speaking the same language? Recommendations for a definition and categorization framework for plastic debris. *Environmental Science and Technology* 53 (3): 1039–1047.
10 Zhang, H., Kuo, Y.Y., Gerecke, A.C. et al. (2012). Co-release of hexabromocyclododecane (HBCD) and nano- and microparticles from thermal cutting of polystyrene foams. *Environmental Science and Technology* 46 (20): 10990–10996.
11 Kuo, Y.Y., Zhang, H., Gerecke, A.C. et al. (2014). Chemical composition of nanoparticles released from thermal cutting of polystyrene foams and the associated isomerization of hexabromocyclododecane (HBCD) diastereomers. *Aerosol and Air Quality Research* 14 (4): 1114–1120.
12 Lambert, S. and Wagner, M. (2016). Formation of microscopic particles during the degradation of different polymers. *Chemosphere* 161: 510–517.

13 Davidson, T.M. (2012). Boring crustaceans damage polystyrene floats under docks polluting marine waters with microplastic. *Marine Pollution Bulletin* 64 (9): 1821–1828.

14 Lebreton, L. and Andrady, A. (2019). Future scenarios of global plastic waste generation and disposal. *Palgrave Communications* 5 (1): 1–11.

15 Cózar, A., Echevarría, F., González-Gordillo, J.I. et al. (2014). Plastic debris in the open ocean. *Proceedings of the National Academy of Sciences of the United States of America* 111 (28): 10239–10244.

16 Gigault, J., Pedrono, B., Maxit, B. et al. (2016). Marine plastic litter: The unanalyzed nano-fraction. *Environmental Science: Nano* 3 (2): 346–350.

17 Ter Halle, A., Jeanneau, L., Martignac, M. et al. (2017). Nanoplastic in the North Atlantic subtropical gyre. *Environmental Science and Technology* 51 (23): 13689–13697.

18 Schirinzi, G.F., Llorca, M., Seró, R. et al. (2019). Trace analysis of polystyrene microplastics in natural waters. *Chemosphere* 236: 124321.

19 Zhu, L., Zhao, S., Bittar, T.B. et al. (2020). Photochemical dissolution of buoyant microplastics to dissolved organic carbon: Rates and microbial impacts. *Journal of Hazardous Materials* 383: 121065.

20 Chae, Y. and An, Y.J. (2017). Effects of micro- and nanoplastics on aquatic ecosystems: Current research trends and perspectives. *Marine Pollution Bulletin* 124 (2): 624–632.

21 Lenz, R., Enders, K., and Nielsen, T.G. (2016). Microplastic exposure studies should be environmentally realistic. *Proceedings of the National Academy of Sciences* 113 (29): E4121–E4122.

22 Jakubowicz, I., Enebro, J., and Yarahmadi, N. (2021). Challenges in the search for nanoplastics in the environment—A critical review from the polymer science perspective. *Polymer Testing* 93: 106953.

23 Gigault, J., Ter Halle, A., Baudrimont, M. et al. (2018). Current opinion: What is a nanoplastic? *Environmental Pollution* 235: 1030–1034.

24 Koelmans, A.A., Besseling, E., and Shim, W.J. (2015). Nanoplastics in the aquatic environment. Critical review. In: *Marine Anthropogenic Litter* (ed. M. Bergmann, L. Gutow, and M. Klages). Springer International Publishing.

25 Gigault, J., El Hadri, H., Nguyen, B. et al. (2021). Nanoplastics are neither microplastics nor engineered nanoparticles. *Nature Nanotechnology* 16 (5): 501–507.

26 Anagnosti, L., Varvaresou, A., Pavlou, P. et al. (2021). Worldwide actions against plastic pollution from microbeads and microplastics in cosmetics focusing on European policies. Has the issue been handled effectively? *Marine Pollution Bulletin* 162: 111883.

27 Caputo, F., Vogel, R., Savage, J. et al. (2021). Measuring particle size distribution and mass concentration of nanoplastics and microplastics: Addressing some analytical challenges in the sub-micron size range. *Journal of Colloid and Interface Science* 588: 401–417.

28 Nguyen, B., Claveau-Mallet, D., Hernandez, L.M. et al. (2019). Separation and analysis of microplastics and nanoplastics in complex environmental samples. *Accounts of Chemical Research* 52 (4): 858–866.

29 Billuart, G., Bourgeat-Lami, E., Lansalot, M. et al. (2014). Free radical emulsion polymerization of ethylene. *Macromolecules* 47 (19): 6591–6600.

30 Ishii, H., Kuwasaki, N., Nagao, D. et al. (2015). Environmentally adaptable pathway to emulsion polymerization for monodisperse polymer nanoparticle synthesis. *Polymer* 77: 64–69.

31 Pessoni, L., Veclin, C., El Hadri, H. et al. (2019). Soap- and metal-free polystyrene latex particles as a nanoplastic model. *Environmental Science: Nano* 6 (7): 2253–2258.

32 Telford, A.M., Hawkett, B.S., Such, C. et al. (2013). Mimicking the wettability of the rose petal using self-assembly of waterborne polymer particles. *Chemistry of Materials* 25 (17): 3472–3479.

33 Schmitt, V., Destribats, M., and Backov, R. (2014). Colloidal particles as liquid dispersion stabilizer: Pickering emulsions and materials thereof. *Comptes Rendus Physique* 15 (8): 761–774.

34 Pikuda, O., Xu, E.G., Berk, D. et al. (2018). Toxicity assessments of micro-and nanoplastics can be confounded by preservatives in commercial formulations. *Environmental Science and Technology Letters* 6 (1): 21–25.

35 Balakrishnan, G., Déniel, M., Nicolai, T. et al. (2019). Towards more realistic reference microplastics and nanoplastics: Preparation of polyethylene micro/nanoparticles with a biosurfactant. *Environmental Science: Nano* 6 (1): 315–324.

36 Lambert, S. and Wagner, M. (2016). Characterisation of nanoplastics during the degradation of polystyrene. *Chemosphere* 145: 265–268.

37 Magrì, D., Sánchez-Moreno, P., Caputo, G. et al. (2018). Laser ablation as a versatile tool to mimic polyethylene terephthalate nanoplastic pollutants: Characterization and toxicology assessment. *ACS Nano* 12 (8): 7690–7700.

38 Stephens, B., Azimi, P., El Orch, Z. et al. (2013). Ultrafine particle emissions from desktop 3D printers. *Atmospheric Environment* 79: 334–339.

39 Rodríguez-Hernández, A.G., Chiodoni, A., Bocchini, S. et al. (2020). 3D printer waste, a new source of nanoplastic pollutants. *Environmental Pollution* 267: 115609.

40 Alimi, O.S., Farner Budarz, J., Hernandez, L.M. et al. (2018). Microplastics and nanoplastics in aquatic environments: Aggregation, deposition, and enhanced contaminant transport. *Environmental Science and Technology* 52 (4): 1704–1724.

41 Wang, M.-L., Chang, R.-Y., and Hsu, C.-H. (David) (ed.) (2018). *Molding Simulation: Theory and Practice*. Munich: Hanser Publishers.

42 Wang, Z., Taylor, S.E., Sharma, P. et al. (2018). Poor extraction efficiencies of polystyrene nano- and microplastics from biosolids and soil. *PLOS ONE* 13 (11): e0208009.

43 Bianco, A., Sordello, F., Ehn, M. et al. (2020). Degradation of nanoplastics in the environment: Reactivity and impact on atmospheric and surface waters. *Science of the Total Environment* 742: 140413.

44 Zhang, Y., Kang, S., Allen, S. et al. (2020). Atmospheric microplastics: A review on current status and perspectives. *Earth-Science Reviews* 203: 103118.

45 Mintenig, S.M., Bäuerlein, P.S., Koelmans, A.A. et al. (2018). Closing the gap between small and smaller: Towards a framework to analyse nano- and microplastics in aqueous environmental samples. *Environmental Science: Nano* 5 (7): 1640–1649.

46 Kay, P., Hiscoe, R., Moberley, I. et al. (2018). Wastewater treatment plants as a source of microplastics in river catchments. *Environmental Science and Pollution Research* 25 (20): 20264–20267.

47 Sun, J., Dai, X., Wang, Q. et al. (2019). Microplastics in wastewater treatment plants: Detection, occurrence and removal. *Water Research* 152: 21–37.

48 Meijer, L.J.J., van Emmerik, T., van der Ent, R. et al. (2021). More than 1000 rivers account for 80% of global riverine plastic emissions into the ocean. *Science Advances* 7 (18): eaaz5803.

49 Wahl, A., Le Juge, C., Davranche, M. et al. (2021). Nanoplastic occurrence in a soil amended with plastic debris. *Chemosphere* 262: 127784.

50 Davranche, M., Lory, C., Juge, C.L. et al. (2020). Nanoplastics on the coast exposed to the North Atlantic gyre: Evidence and traceability. *NanoImpact* 20: 100262.

51 Rist, S., Carney Almroth, B., Hartmann, N.B. et al. (2018). A critical perspective on early communications concerning human health aspects of microplastics. *Science of the Total Environment* 626: 720–726.

52 Hernandez, L.M., Yousefi, N., and Tufenkji, N. (2017). Are there nanoplastics in your personal care products? *Environmental Science and Technology Letters* 4 (7): 280–285.

53 Bundschuh, M., Filser, J., Lüderwald, S. et al. (2018). Nanoparticles in the environment: Where do we come from, where do we go to? *Environmental Sciences Europe* 30 (1): 6.

54 Hanna, S.K., Miller, R.J., Zhou, D. et al. (2013). Accumulation and toxicity of metal oxide nanoparticles in a soft-sediment estuarine amphipod. *Aquatic Toxicology* 142–143: 441–446.

55 Sun, T.Y., Bornhöft, N.A., Hungerbühler, K. et al. (2016). Dynamic probabilistic modeling of environmental emissions of engineered nanomaterials. *Environmental Science and Technology* 50 (9): 4701–4711.

56 Windsor, F.M., Durance, I., Horton, A.A. et al. (2019). A catchment-scale perspective of plastic pollution. *Global Change Biology* 25 (4): 1207–1221.
57 Pradel, A., El Hadri, H., Desmet, C. et al. (2020). Deposition of environmentally relevant nanoplastic models in sand during transport experiments. *Chemosphere* 255: 126912.
58 Hotze, E.M., Phenrat, T., and Lowry, G.V. (2010). Nanoparticle aggregation: Challenges to understanding transport and reactivity in the environment. *Journal of Environmental Quality* 39 (6): 1909–1924.
59 Singh, N., Tiwari, E., Khandelwal, N. et al. (2019). Understanding the stability of nanoplastics in aqueous environments: Effect of ionic strength, temperature, dissolved organic matter, clay, and heavy metals. *Environmental Science: Nano* 6 (10): 2968–2976.
60 Pradel, A., Ferreres, S., Veclin, C. et al. (2021). Stabilization of fragmental polystyrene nanoplastic by natural organic matter: Insight into mechanisms. *Environmental Science and Technology Water* 1 (5): 1198–1208.
61 Wu, X., Lyu, X., Li, Z. et al. (2020). Transport of polystyrene nanoplastics in natural soils: Effect of soil properties, ionic strength and cation type. *Science of the Total Environment* 707: 136065.
62 Cai, L., Hu, L., Shi, H. et al. (2018). Effects of inorganic ions and natural organic matter on the aggregation of nanoplastics. *Chemosphere* 197: 142–151.
63 Chen, C.S., Le, C., Chiu, M.H. et al. (2018). The impact of nanoplastics on marine dissolved organic matter assembly. *Science of the Total Environment* 634: 316–320.
64 Venel, Z., Tabuteau, H., Pradel, A. et al. (2021). Using a lab-on-a-chip: Where does the nanoscale fraction of plastic debris accumulate??? *Environmental Science and Technology* 55: 3001–3008.
65 Bhagat, J., Nishimura, N., and Shimada, Y. (2021). Toxicological interactions of microplastics/nanoplastics and environmental contaminants: Current knowledge and future perspectives. *Journal of Hazardous Materials* 405: 123913.
66 Huang, W., Song, B., Liang, J. et al. (2021). Microplastics and associated contaminants in the aquatic environment: A review on their ecotoxicological effects, trophic transfer, and potential impacts to human health. *Journal of Hazardous Materials* 405: 124187.
67 Reichel, J., Graßmann, J., Knoop, O. et al. (2021). Organic contaminants and interactions with micro-and nano-plastics in the aqueous environment: Review of analytical methods. *Molecules* 26 (4): 1164.
68 Wang, W. and Wang, J. (2018). Different partition of polycyclic aromatic hydrocarbon on environmental particulates in freshwater: Microplastics in comparison to natural sediment. *Ecotoxicology and Environmental Safety* 147: 648–655.
69 Teuten, E.L., Rowland, S.J., Galloway, T.S. et al. (2007). Potential for plastics to transport hydrophobic contaminants. *Environmental Science and Technology* 41 (22): 7759–7764.

70 Rochman, C.M., Hoh, E., Hentschel, B.T. et al. (2013). Long-term field measurement of sorption of organic contaminants to five types of plastic pellets: Implications for plastic marine debris. *Environmental Science and Technology* 47 (3): 1646–1654.

71 Hahladakis, J.N., Velis, C.A., Weber, R. et al. (2018). An overview of chemical additives present in plastics: Migration, release, fate and environmental impact during their use, disposal and recycling. *Journal of Hazardous Materials* 344: 179–199.

72 Bridson, J.H., Gaugler, E.C., Smith, D.A. et al. (2021). Leaching and extraction of additives from plastic pollution to inform environmental risk: A multidisciplinary review of analytical approaches. *Journal of Hazardous Materials* 414: 125571.

73 Town, R.M. and Van Leeuwen, H.P. (2020). Uptake and release kinetics of organic contaminants associated with micro-and nanoplastic particles. *Environmental Science and Technology* 54 (16): 10057–10067.

74 Abdolahpur Monikh, F., Vijver, M.G., Guo, Z. et al. (2020). Metal sorption onto nanoscale plastic debris and trojan horse effects in *Daphnia magna*: Role of dissolved organic matter. *Water Research* 186: 116410.

75 Davranche, M., Veclin, C., Pierson-Wickmann, A.-C. et al. (2019). Are nanoplastics able to bind significant amount of metals? The lead example. *Environmental Pollution* 249: 940–948.

76 Mao, R., Lang, M., Yu, X. et al. (2020). Aging mechanism of microplastics with UV irradiation and its effects on the adsorption of heavy metals. *Journal of Hazardous Materials* 393: 122515.

77 Barreto, A., Santos, J., Amorim, M.J.B. et al. (2021). Polystyrene nanoplastics can alter the toxicological effects of Simvastatin on *Danio rerio*. *Toxics* 9 (3): 44.

78 Chen, Q., Yin, D., Jia, Y. et al. (2017). Enhanced uptake of BPA in the presence of nanoplastics can lead to neurotoxic effects in adult zebrafish. *Science of the Total Environment* 609: 1312–1321.

79 Parenti, C.C., Magni, S., Ghilardi, A. et al. (2021). Does triclosan adsorption on polystyrene nanoplastics modify the toxicity of single contaminants? *Environmental Science: Nano* 8 (1): 282–296.

80 Yilimulati, M., Wang, L., Ma, X. et al. (2021). Adsorption of ciprofloxacin to functionalized nano-sized polystyrene plastic: Kinetics, thermochemistry and toxicity. *Science of the Total Environment* 750: 142370.

81 Xiong, Y., Zhao, J., Li, L. et al. (2020). Interfacial interaction between micro/nanoplastics and typical PPCPs and nanoplastics removal via electrosorption from an aqueous solution. *Water Research* 184: 116100.

82 Zhang, H., Liu, F., Wang, S. et al. (2020). Sorption of fluoroquinolones to nanoplastics as affected by surface functionalization and solution chemistry. *Environmental Pollution* 262: 114347.

83 Cortés-Arriagada, D. (2021). Elucidating the co-transport of bisphenol A with polyethylene terephthalate (PET) nanoplastics: A theoretical study of the adsorption mechanism. *Environmental Pollution* 270: 116192.

84 Liu, J., Ma, Y., Zhu, D. et al. (2018). Polystyrene nanoplastics-enhanced contaminant transport: Role of irreversible adsorption in glassy polymeric domain. *Environmental Science and Technology* 52 (5): 2677–2685.

85 Liu, J., Zhang, T., Tian, L. et al. (2019). Aging significantly affects mobility and contaminant-mobilizing ability of nanoplastics in saturated loamy sand. *Environmental Science and Technology* 53 (10): 5805–5815.

86 Liu, L., Fokkink, R., and Koelmans, A.A. (2016). Sorption of polycyclic aromatic hydrocarbons to polystyrene nanoplastic. *Environmental Toxicology and Chemistry* 35 (7): 1650–1655.

87 Velzeboer, I., Kwadijk, C., and Koelmans, A.A. (2014). Strong sorption of PCBs to nanoplastics, microplastics, carbon nanotubes, and fullerenes. *Environmental Science and Technology* 48 (9): 4869–4876.

88 Jiang, R., Lin, W., Wu, J. et al. (2018). Quantifying nanoplastic-bound chemicals accumulated in Daphnia magna with a passive dosing method. *Environ Science: Nano* 5 (3): 776–781.

89 Ma, Y., Huang, A., Cao, S. et al. (2016). Effects of nanoplastics and microplastics on toxicity, bioaccumulation, and environmental fate of phenanthrene in fresh water. *Environmental Pollution* 219: 166–173.

90 Lin, W., Jiang, R., Xiao, X. et al. (2020). Joint effect of nanoplastics and humic acid on the uptake of PAHs for *Daphnia magna*: A model study. *Journal of Hazardous Materials* 391: 122195.

91 Trevisan, R., Uzochukwu, D., and Di Giulio, R.T. (2020). PAH sorption to nanoplastics and the trojan horse effect as drivers of mitochondrial toxicity and PAH localization in zebrafish. *Frontiers in Environmental Science* 8. doi: 10.3389/fenvs.2020.00078.

92 Trevisan, R., Voy, C., Chen, S. et al. (2019). Nanoplastics decrease the toxicity of a complex PAH mixture but impair mitochondrial energy production in developing zebrafish. *Environmental Science and Technology* 53 (14): 8405–8415.

93 Corsi, I., Bergami, E., and Grassi, G. (2020). Behaviour and bio-interactions of anthropogenic particles in marine environment for a more realistic ecological risk assessment. *Frontiers in Environmental Science* 8 (60): 1–21.

94 Cedervall, T., Lynch, I., Foy, M. et al. (2007). Detailed identification of plasma proteins adsorbed on copolymer nanoparticles. *Angewandte Chemie - International Edition* 46 (30): 5754–5756.

95 Wan, S., Kelly, P.M., Mahon, E. et al. (2015). The "sweet" Side of the protein corona: Effects of glycosylation on nanoparticle-cell interactions. *ACS Nano* 9 (2): 2157–2166.

96 Lara, S., Alnasser, F., Polo, E. et al. (2017). Identification of receptor binding to the biomolecular corona of nanoparticles. *ACS Nano* 11 (2): 1884–1893.
97 Winzen, S., Schoettler, S., Baier, G. et al. (2015). Complementary analysis of the hard and soft protein corona: Sample preparation critically effects corona composition. *Nanoscale* 7 (7): 2992–3001.
98 Canesi, L. and Corsi, I. (2016). Effects of nanomaterials on marine invertebrates. *Science of the Total Environment* 565: 933–940.
99 Canesi, L., Balbi, T., Fabbri, R. et al. (2017). Biomolecular coronas in invertebrate species: Implications in the environmental impact of nanoparticles. *NanoImpact* 8: 89–98.
100 Canesi, L., Ciacci, C., Fabbri, R. et al. (2016). Interactions of cationic polystyrene nanoparticles with marine bivalve hemocytes in a physiological environment: Role of soluble hemolymph proteins. *Environmental Research* 150: 73–81.
101 Marques-Santos, L.F., Grassi, G., Bergami, E. et al. (2018). Cationic polystyrene nanoparticle and the sea urchin immune system: Biocorona formation, cell toxicity, and multixenobiotic resistance phenotype. *Nanotoxicology* 12 (8): 847–867.
102 Grassi, G., Landi, C., Della Torre, C. et al. (2019). Proteomic profile of the hard corona of charged polystyrene nanoparticles exposed to sea urchin: *Paracentrotus lividus* coelomic fluid highlights potential drivers of toxicity. *Environmental Science: Nano* 6 (10): 2937–2947.
103 Kihara, S., Köper, I., Mata, J.P. et al. (2021). Reviewing nanoplastic toxicology: It's an interface problem. *Advances in Colloid and Interface Science* 288: 102337.
104 Troost, T.A., Desclaux, T., Leslie, H.A. et al. (2018). Do microplastics affect marine ecosystem productivity? *Marine Pollution Bulletin* 135: 17–29.
105 Reichelt, S. and Gorokhova, E. (2020). Micro- and nanoplastic exposure effects in microalgae: A meta-analysis of standard growth inhibition tests. *Frontiers in Environmental Sciences* 8. doi: 10.3389/fenvs.2020.00131.
106 Gao, G., Zhao, X., Jin, P. et al. (2021). Current understanding and challenges for aquatic primary producers in a world with rising micro- and nano-plastic levels. *Journal of Hazardous Materials* 406: 124685.
107 Schampera, C., Wolinska, J., Bachelier, J.B. et al. (2021). Exposure to nanoplastics affects the outcome of infectious disease in phytoplankton. *Environmental Pollution* 277: 116781.
108 Gomes, T., Almeida, A.C., and Georgantzopoulou, A. (2020). Characterization of cell responses in *Rhodomonas baltica* exposed to PMMA nanoplastics. *Science of the Total Environment* 726: 138547.
109 Rowenczyk, L., Leflaive, J., Clergeaud, F. et al. (2021). Heteroaggregates of polystyrene nanospheres and organic matter: Preparation, characterization and evaluation of their toxicity to algae in environmentally relevant conditions. *Nanomaterials* 11 (2): 1–15.

110 Baudrimont, M., Arini, A., Guégan, C. et al. (2020). Ecotoxicity of polyethylene nanoplastics from the North Atlantic oceanic gyre on freshwater and marine organisms (microalgae and filter-feeding bivalves). *Environmental Science and Pollution Research* 27 (4): 3746–3755.

111 Hazeem, L.J., Yesilay, G., Bououdina, M. et al. (2020). Investigation of the toxic effects of different polystyrene micro-and nanoplastics on microalgae *Chlorella vulgaris* by analysis of cell viability, pigment content, oxidative stress and ultrastructural changes. *Marine Pollution Bulletin* 156: 111278.

112 Bellingeri, A., Bergami, E., Grassi, G. et al. (2019). Combined effects of nanoplastics and copper on the freshwater alga *Raphidocelis subcapitata*. *Aquatic Toxicology* 210: 179–187.

113 Yang, Y., Guo, Y., Brien, A.O. et al. (2020). Biological responses to climate change and nanoplastics are altered in concert: Full-factorial screening reveals effects of multiple stressors on primary producers. *Environmental Science and Technology* 54 (4): 2401–2410.

114 Bergami, E., Pugnalini, S., Vannuccini, M.L. et al. (2017). Long-term toxicity of surface-charged polystyrene nanoplastics to marine planktonic species *Dunaliella tertiolecta* and *Artemia franciscana*. *Aquatic Toxicology* 189: 159–169.

115 Varó, I., Perini, A., Torreblanca, A. et al. (2019). Time-dependent effects of polystyrene nanoparticles in brine shrimp *Artemia franciscana* at physiological, biochemical and molecular levels. *Science of the Total Environment* 675: 570–580.

116 Zhang, W., Liu, Z., Tang, S. et al. (2020). Transcriptional response provides insights into the effect of chronic polystyrene nanoplastic exposure on *Daphnia pulex*. *Chemosphere* 238: 124563.

117 Kelpsiene, E., Torstensson, O., Ekvall, M.T. et al. (2020). Long-term exposure to nanoplastics reduces life-time in *Daphnia magna*. *Scientific Reports* 10 (1): 1–7.

118 Saavedra, J., Stoll, S., and Slaveykova, V.I. (2019). Influence of nanoplastic surface charge on eco-corona formation, aggregation and toxicity to freshwater zooplankton. *Environmental Pollution* 252: 715–722.

119 Della Torre, C., Bergami, E., Salvati, A. et al. (2014). Accumulation and embryotoxicity of polystyrene nanoparticles at early stage of development of sea urchin embryos *Paracentrotus lividus*. *Environmental Science and Technology* 48 (20): 12302–12311.

120 Pinsino, A., Bergami, E., Della Torre, C. et al. (2017). Amino-modified polystyrene nanoparticles affect signalling pathways of the sea urchin (*Paracentrotus lividus*) embryos. *Nanotoxicology* 11 (2): 201–209.

121 Balbi, T., Camisassi, G., Montagna, M. et al. (2017). Impact of cationic polystyrene nanoparticles (PS-NH$_2$) on early embryo development of *Mytilus galloprovincialis*: Effects on shell formation. *Chemosphere* 186: 1–9.

122 Rist, S., Baun, A., Almeda, R. et al. (2019). Ingestion and effects of micro- and nanoplastics in blue mussel (*Mytilus edulis*) larvae. *Marine Pollution Bulletin* 140: 423–430.

123 Cole, M. and Galloway, T.S. (2015). Ingestion of nanoplastics and microplastics by Pacific oyster larvae. *Environmental Science and Technology* 49 (24): 14625–14632.

124 Tallec, K., Huvet, A., Di Poi, C. et al. (2018). Nanoplastics impaired oyster free living stages, gametes and embryos. *Environmental Pollution* 242: 1226–1235.

125 Eliso, M.C., Bergami, E., Manfra, L. et al. (2020). Toxicity of nanoplastics during the embryogenesis of the ascidian *Ciona robusta* (Phylum Chordata). *Nanotoxicology* 14 (10): 1415–1431.

126 Jeong, C.-B., Kang, H.-M., Byeon, E. et al. (2021). Phenotypic and transcriptomic responses of the rotifer *Brachionus koreanus* by single and combined exposures to nano-sized microplastics and water-accommodated fractions of crude oil. *Journal of Hazardous Materials* 416: 125703.

127 Bergami, E., Manno, C., Cappello, S. et al. (2020). Nanoplastics affect moulting and faecal pellet sinking in Antarctic krill (*Euphausia superba*) juveniles. *Environment International* 143: 105999.

128 Galgani, L. and Loiselle, S.A. (2021). Plastic pollution impacts on marine carbon biogeochemistry. *Environmental Pollution* 268: 115598.

129 Sendra, M., Sparaventi, E., Novoa, B. et al. (2021). An overview of the internalization and effects of microplastics and nanoplastics as pollutants of emerging concern in bivalves. *Science of the Total Environment* 753: 142024.

130 Chen, C.-Y., Lu, T.-H., Yang, Y.-F. et al. (2021). Marine mussel-based biomarkers as risk indicators to assess oceanic region-specific microplastics impact potential. *Ecological Indicators* 120: 106915.

131 Li, J., Lusher, A.L., Rotchell, J.M. et al. (2019). Using mussel as a global bioindicator of coastal microplastic pollution. *Environmental Pollution* 244: 522–533.

132 Farrell, P. and Nelson, K. (2013). Trophic level transfer of microplastic: *Mytilus edulis* (L.) to *Carcinus maenas* (L.). *Environmental Pollution* 177: 1–3.

133 Auguste, M., Balbi, T., Ciacci, C. et al. (2020). Shift in immune parameters after repeated exposure to nanoplastics in the marine bivalve *Mytilus*. *Frontiers in Immunology* 11: 1–11.

134 Capolupo, M., Valbonesi, P., and Fabbri, E. (2021). A comparative assessment of the chronic effects of micro- and nano-plastics on the physiology of the mediterranean mussel *Mytilus galloprovincialis*. *Nanomaterials* 11 (3): 649.

135 Ciacci, C., Grimmelpont, M.V., Corsi, I. et al. (2019). Nanoparticle-biological interactions in a marine benthic foraminifer. *Scientific Reports* 9 (1): 19441.

136 Canesi, L., Ciacci, C., Bergami, E. et al. (2015). Evidence for immunomodulation and apoptotic processes induced by cationic polystyrene nanoparticles in the

hemocytes of the marine bivalve *Mytilus*. *Marine Environmental Research* 111: 34–40.

137 Bergami, E., Krupinski Emerenciano, A., González-Aravena, M. et al. (2019). Polystyrene nanoparticles affect the innate immune system of the Antarctic sea urchin *Sterechinus neumayeri*. *Polar Biology* 42: 743–757.

138 Murano, C., Bergami, E., Liberatori, G. et al. (2021). Interplay between nanoplastics and the immune system of the mediterranean sea urchin *Paracentrotus lividus*. *Frontiers in Marine Science* 8: 1–16.

139 Chen, Q., Gundlach, M., Yang, S. et al. (2017). Quantitative investigation of the mechanisms of microplastics and nanoplastics towards zebrafish larvae locomotor activity. *Science of the Total Environment* 584–585: 1022–1031.

140 Van Pomeren, M., Brun, N.R., Peijnenburg, W.J.G.M. et al. (2017). Exploring uptake and biodistribution of polystyrene (nano)particles in zebrafish embryos at different developmental stages. *Aquatic Toxicology* 190: 40–45.

141 Ding, J., Zhang, S., Razanajatovo, R.M. et al. (2018). Accumulation, tissue distribution, and biochemical effects of polystyrene microplastics in the freshwater fish red tilapia (*Oreochromis niloticus*). *Environmental Pollution* 238: 1–9.

142 Lee, W.S., Cho, H.-J., Kim, E. et al. (2019). Bioaccumulation of polystyrene nanoplastics and their effect on the toxicity of Au ions in zebrafish embryos. *Nanoscale* 11 (7): 3173–3185.

143 Sökmen, T.Ö., Sulukan, E., Türkoğlu, M. et al. (2020). Polystyrene nanoplastics (20 nm) are able to bioaccumulate and cause oxidative DNA damages in the brain tissue of zebrafish embryo (*Danio rerio*). *NeuroToxicology* 77: 51–59.

144 Mattsson, K., Johnson, E.V., Malmendal, A. et al. (2017). Brain damage and behavioural disorders in fish induced by plastic nanoparticles delivered through the food chain. *Scientific Reports* 7 (1): 1–7.

145 Chae, Y., Kim, D., Kim, S.W. et al. (2018). Trophic transfer and individual impact of nano-sized polystyrene in a four-species freshwater food chain. *Scientific Reports* 8 (1): 1–11.

146 Raftis, J.B. and Miller, M.R. (2019). Nanoparticle translocation and multi-organ toxicity: A particularly small problem. *Nano Today* 26: 8–12.

147 Xie, J., Farage, E., Sugimoto, M. et al. (2010). A novel transgenic zebrafish model for blood-brain and blood-retinal barrier development. *BMC Developmental Biology* 10 (1): 76.

148 Pitt, J.A., Kozal, J.S., Jayasundara, N. et al. (2018). Uptake, tissue distribution, and toxicity of polystyrene nanoparticles in developing zebrafish (*Danio rerio*). *Aquatic Toxicology* 194: 185–194.

149 Pitt, J.A., Trevisan, R., Massarsky, A. et al. (2018). Maternal transfer of nanoplastics to offspring in zebrafish (*Danio rerio*): A case study with nanopolystyrene. *Science of the Total Environment* 643: 324–334.

150 Lu, Y., Zhang, Y., Deng, Y. et al. (2016). Uptake and accumulation of polystyrene microplastics in zebrafish (*Danio rerio*) and toxic effects in liver. *Environmental Science and Technology* 50 (7): 4054–4060.

151 Cedervall, T., Hansson, L.A., Lard, M. et al. (2012). Food chain transport of nanoparticles affects behaviour and fat metabolism in fish. *PLoS ONE* 7 (2): e32254.

152 Mattsson, K., Ekvall, M.T., Hansson, L.A. et al. (2015). Altered behavior, physiology, and metabolism in fish exposed to polystyrene nanoparticles. *Environmental Science and Technology* 49 (1): 553–561.

153 Abdolahpur Monikh, F., Chupani, L., Vijver, M.G. et al. (2021). Parental and trophic transfer of nanoscale plastic debris in an assembled aquatic food chain as a function of particle size. *Environmental Pollution* 269: 116066.

154 Parenti, C.C., Ghilardi, A., Della Torre, C. et al. (2019). Evaluation of the infiltration of polystyrene nanobeads in zebrafish embryo tissues after short-term exposure and the related biochemical and behavioural effects. *Environmental Pollution* 254: 112947.

155 Liu, Y., Wang, Z., Wang, S. et al. (2019). Ecotoxicological effects on *Scenedesmus obliquus* and *Danio rerio* Co-exposed to polystyrene nano-plastic particles and natural acidic organic polymer. *Environmental Toxicology and Pharmacology* 67: 21–28.

156 Greven, A.-C., Merk, T., Karagöz, F. et al. (2016). Polycarbonate and polystyrene nanoplastic particles act as stressors to the innate immune system of fathead minnow (*Pimephales promelas*). *Environmental Toxicology and Chemistry* 35 (12): 3093–3100.

157 Veneman, W.J., Spaink, H.P., Brun, N.R. et al. (2017). Pathway analysis of systemic transcriptome responses to injected polystyrene particles in zebrafish larvae. *Aquatic Toxicology* 190: 112–120.

158 Brun, N.R., Koch, B.E.V., Varela, M. et al. (2018). Nanoparticles induce dermal and intestinal innate immune system responses in zebrafish embryos. *Environmental Science: Nano* 5 (4): 904–916.

159 Brandts, I., Garcia-Ordoñez, M., Tort, L. et al. (2020). Polystyrene nanoplastics accumulate in ZFL cell lysosomes and in zebrafish larvae after acute exposure, inducing a synergistic immune response: *In vitro* without affecting larval survival in vivo. *Environmental Science: Nano* 7 (8): 2410–2422.

160 Yin, L., Liu, H., Cui, H. et al. (2019). Impacts of polystyrene microplastics on the behavior and metabolism in a marine demersal teleost, black rockfish (*Sebastes schlegelii*). *Journal of Hazardous Materials* 380: 120861.

161 Brandts, I., Teles, M., Tvarijonaviciute, A. et al. (2018). Effects of polymethylmethacrylate nanoplastics on *Dicentrarchus labrax*. *Genomics* 110 (6): 435–441.

162 Brandts, I., Barría, C., Martins, M.A. et al. (2021). Waterborne exposure of gilthead seabream (*Sparus aurata*) to polymethylmethacrylate nanoplastics causes effects at cellular and molecular levels. *Journal of Hazardous Materials* 403: 123590.

163 Almeida, M., Martins, M.A., Soares, A.M.V. et al. (2019). Polystyrene nanoplastics alter the cytotoxicity of human pharmaceuticals on marine fish cell lines. *Environmental Toxicology and Pharmacology* 69: 57–65.

164 Bellingeri, A., Casabianca, S., Capellacci, S. et al. (2020). Impact of polystyrene nanoparticles on marine diatom Skeletonema marinoi chain assemblages and consequences on their ecological role in marine ecosystems. *Environmental Pollution* 262: 114268.

165 Bergami, E., Bocci, E., Vannuccini, M.L. et al. (2016). Nano-sized polystyrene affects feeding, behavior and physiology of brine shrimp *Artemia franciscana* larvae. *Ecotoxicology and Environmental Safety* 123: 18–25.

166 Bhattacharya, P., Lin, S., Turner, J.P. et al. (2010). Physical adsorption of charged plastic nanoparticles affects algal photosynthesis. *Journal of Physical Chemistry C* 114 (39): 16556–16561.

167 González-Fernández, C., Toullec, J., Lambert, C. et al. (2019). Do transparent exopolymeric particles (TEP) affect the toxicity of nanoplastics on Chaetoceros neogracile? *Environmental Pollution* 250: 873–882.

168 Bertoli, F., Garry, D., Monopoli, M.P. et al. (2016). The intracellular destiny of the protein corona: A study on its cellular internalization and evolution. *ACS Nano* 10 (11): 10471–10479.

169 Salvati, A., Åberg, C., dos Santos, T. et al. (2011). Experimental and theoretical comparison of intracellular import of polymeric nanoparticles and small molecules: Towards models of uptake kinetics. *Nanomedicine: Nanotechnology, Biology, and Medicine* 7 (6): 818–826.

170 Liu, Y., Li, W., Lao, F. et al. (2011). Intracellular dynamics of cationic and anionic polystyrene nanoparticles without direct interaction with mitotic spindle and chromosomes. *Biomaterials* 32 (32): 8291–8303.

171 Wang, F., Bexiga, M.G., Anguissola, S. et al. (2013). Time resolved study of cell death mechanisms induced by amine-modified polystyrene nanoparticles. *Nanoscale* 5 (22): 10868–10876.

172 Sendra, M., Staffieri, E., Yeste, M.P. et al. (2019). Are the primary characteristics of polystyrene nanoplastics responsible for toxicity and ad/absorption in the marine diatom *Phaeodactylum tricornutum*? *Environmental Pollution* 249: 610–619.

173 Sendra, M., Saco, A., Yeste, M.P. et al. (2019). Nanoplastics: From tissue accumulation to cell translocation into *Mytilus galloprovincialis* hemocytes. resilience of immune cells exposed to nanoplastics and nanoplastics plus *Vibrio splendidus* combination. *Journal of Hazardous Materials* 388: 121788.

174 Rodríguez-Hernández, A.G., Muñoz-Tabares, J.A., Aguilar-Guzmán, J.C. et al. (2019). A novel and simple method for polyethylene terephthalate (PET) nanoparticle production. *Environmental Science: Nano* 6 (7): 2031–2036.

175 Catarino, A.I., Frutos, A., and Henry, T.B. (2019). Use of fluorescent-labelled nanoplastics (NPs) to demonstrate NP absorption is inconclusive without adequate controls. *Science of the Total Environment* 670: 915–920.

176 Schür, C., Rist, S., Baun, A. et al. (2019). When fluorescence is not a particle: The tissue translocation of microplastics in *Daphnia magna* seems an artifact. *Environmental Toxicology and Chemistry* 38 (7): 1495–1503.

177 Mitrano, D.M., Beltzung, A., Frehland, S. et al. (2019). Synthesis of metal-doped nanoplastics and their utility to investigate fate and behaviour in complex environmental systems. *Nature Nanotechnology* 14 (4): 362–368.

178 Al-Sid-Cheikh, M., Rowland, S.J., Kaegi, R. et al. (2020). Synthesis of ^{14}C-labelled polystyrene nanoplastics for environmental studies. *Communications Materials* 1 (1): 97.

179 Al-Sid-Cheikh, M., Rowland, S.J., Stevenson, K. et al. (2018). Uptake, whole-body distribution, and depuration of nanoplastics by the scallop pecten maximus at environmentally realistic concentrations. *Environmental Science and Technology* 52 (24): 14480–14486.

180 Koelmans, A.A. (2019). Proxies for nanoplastic. *Nature Nanotechnology* 14 (4): 307–308.

181 Bourgeault, A., Cousin, C., Geertsen, V. et al. (2015). The challenge of studying TiO2 nanoparticle bioaccumulation at environmental concentrations: Crucial use of a stable isotope tracer. *Environmental Science and Technology* 49 (4): 2451–2459.

Index

Note: Page numbers for figures are in *italics*, tables are in **bold**.

a

abrasive damage   330
ABS *see* acrylonitrile butadiene styrene
acetate kinase (AK)   158–159, 165
acetylcholinesterase (AChE)   393, 425
acid treatments   18–25
acrylics   67, 151–154, 182, **198**, 245
acrylonitrile butadiene styrene
  (ABS)   19, **99**, 105
activated sludge (AS)   119–126,
  129–134, 137, 148–155, 158, 165
additives   326–330, 346–351, 356–363
adipates   356, 359
adsorption
  aging of microplastics   251
  algae   291, 297
  aquatic organisms   291, 297, 328–329
  chemicals associated with
    microplastics   346–363
  contaminant-microplastic
    interactions   375–391, 395–396
  degradation aspects   213–219,
    223–225
  drinking water systems   76

nanoplastics   414–418, 428
removal of microplastics   213–219,
  223–225
sample purification   22
sewage sludge   152
advanced drinking water treatment
  plants (ADWTP)   68–69, 77, 80
advanced oxidation processes   109
advanced wastewater
  treatments   122–123
aerobic digestion   124, 134–136
aerobic granular sludge   **125**, 126–127,
  129–132
AFM-IR *see* Atomic Force
  Microscopy-Infrared
Africa   181–196, 248, 254, 266
aging of microplastics
  algae   296
  chemicals associated with   350–357
  contaminant-microplastic
    interactions   388–390
  receiving waters   246, 251, 262, 266
agriculture practices
  chemicals   345–346, 361–362

Microplastics in Urban Water Management, First Edition. Edited by Bing-Jie Ni,
Qiuxiang Xu, and Wei Wei.
© 2023 John Wiley & Sons, Inc. Published 2023 by John Wiley & Sons, Inc.

drinking water systems 58–59
nanoplastics 411
receiving waters 245
sludge 160–162
wastewater treatment
plants 197–198
air flotation **95**, 107
AK *see* acetate kinase
algae 287–314, 321–324
antibiotics 296–300
contamination 288, 292, 296–303
ecological toxicity 288, 291–303
effects of microplastics 287–314,
321–324
growth 289–303
heavy metals 296, 300–301
metal contamination 296, 300–301
nanoplastics 419–427
population effects 289–293, **298**
research gap 302–303
toxicity 288, 291–303
transfer aspects 321–324
wastewater treatment
plants 287–288
alkaline treatments 18–25, 158
alkyd resins **151**, 154
aluminium-based salts 217–219
ammonia-oxidizing bacteria 129–131
amoxicillin 348–349, 376–377
anaerobic-anoxic-aerobic process **95–97**,
102, 119, 121
anaerobic digestion 124, **135**, 148,
155–165
anaerobic granular sludge **125**, 127–132
antagonistic toxicity 393
Antarctic 185–187, 191, 197–198
anthracene 348–351
antibiotic-resistant bacteria (ARB) 375,
383–385, 391–392, 395
antibiotic resistant genes (ARG) 259,
375, 383–385, 391–392, 395

antibiotics
algae 296–300
chemicals associated with
microplastics 349–352
contaminant-microplastic
interactions 375–378, 381–388,
391–392, 395
receiving waters 262, 296–300
antioxidants 356–359
aquatic environments
algae 287–303, 321–324
contaminant-microplastic
interactions 390–393
nanoplastics 407–430
receiving waters 243–267
wastewater treatment plants 91–93,
102–109
aquatic organisms
adsorbed pollutants 328–330
algae 287–314, 321–324
bacteria/microbials 316–317,
322–325, 329–330
chemicals 326–330
consumers 324–326
contaminant-microplastic
interactions 390–393
decomposers 322–323
discharge of microplastics 175,
193–199
ecological impacts/risks 251–256,
291–303, 318–330
food chains 251–256, 316, 319–330
growth 253, 288–303, 323–326,
390–391, 419–421
immune responses 326, 393, 423–427
ingestion of microplastics 320–329,
421–422, 426–430
nanoplastics 418–430
physical damage effects 329–330
plants 287–303, 316, 319–324,
327–330

producers 323–324
receiving waters 246, 251–256, 287–303, 315–343
sediments 316–320
toxicity 253, 288, 291–303, 322–330, 393, 418–426, 429–430
uptake of microplastics 213, 251–256, 421–422, 426–430
wastewater treatment plants 91–93, 102–109, 175, 193–199
ARB *see* antibiotic-resistant bacteria
Arctic 185–187, 190–191, 197–198
ARG *see* antibiotic resistant genes
Argentina **183**, **194**
AS *see* activated sludge
Atlantic ocean 185–186, 191, 195–198, 420–421
atmosphere 64–65, 263–264, 266, 411, 419
Atomic Force Microscopy-Infrared (AFM-IR) 28
attached-growth processes 122, 138
Attenuated total reflection (ATR) 28
Australia 97, 100–102, 161–162, **178**, **194**, **196**
Austria 247
automatic mapping **26**, 28–29

b

bacteria *see also* microbials
aquatic organisms 316–317, 322–325, 329–330
contaminant-microplastic interactions 375, 383–385, 391–395
receiving waters 256–260
wastewater treatment processes 108, 119–139
BaP *see* benzo(a)pyrenes
basket samplers 3
beaches
chemicals in microplastics 347–348, 353, 363

contaminant-microplastic interactions **379**
discharge of microplastics 186, 188–189
sample detection 25–26
benthic organisms 197–198, 423–424
benzo(a)pyrenes (BaP) 392–393
benzo(e)pyrenes (BeP) 387–388
biochemical oxygen demand (BOD) 121–122
biodegradable organic matters (BOD) 108
biofilms
aquatic organisms 316–317, 323–324, 328–329
characteristics 256–257
formation 256–259
genetic material transfer 258–259
receiving waters 245–246, 250–251, 256–262, 316–317, 323–324, 328–329
sewage sludge 148, 152–153, 164
wastewater treatment plants 108, 148, 152–153, 164
wastewater treatment processes 119–123, 128–129, 137–138
biological wastewater treatments 94–103, 119–139
biota
algae 292
chemicals associated with microplastics 360
contaminant-microplastic interactions 378, 383–385
emergence of microplastics 59–60
human health 265–267
nanoplastics 407–408, 430
sampling 1, 17–19, 27
sludge to soil transport 161
toxicity to aquatic organisms 325, 329
uptake in aquatic organisms 193
birds 253–254, 323
bismuth-based photocatalysis 229
bisphenol A (BPA)

algae   288, 302
chemicals associated with microplastics   346–347, 351, 362–363
contaminant-microplastic interactions   **382**
nanoplastics   415–416
removal and degradation aspects   213
wastewater treatment processes   120, 126, 128, 134
bivalves   252–253, 261–262, 321, 392, 422–423
BK *see* butyrate kinase
blood-brain barrier   423
BOD *see* biochemical oxygen demand; biodegradable organic matters
bonding
chemicals associated with microplastics   350–353, 356–360
contaminant-microplastic interactions   376, 380, 389
nanoplastics   414–416
purification   20
bongo nets   4
bottle caps   29, 57–58
bottled water   14–16, 29, 54, 57–58, 261–262
BPA *see* bisphenol A
Brazil   63, **183**, **194**, 260
Brownian motion   34–35
bulk collection   1, 4–7
butyrate kinase (BK)   158–159, 165

C

Canada
aquatic organisms   **317**
discharge of microplastics   **177**, **183**, 188, **190**
receiving waters   247, 354
removal of microplastics   **94**, **97**, 101
sewage sludge   152–153

Canary Islands   195
caprolactam (CPL)   156, 159
carbonyl index (CI)   20, 22, **22**, 226, 416
cellulose acetates   220
CFS *see* coagulation, flocculation, and sedimentation
charge neutralization   215–216–20, 218, 220
chemical exchange   415
chemical imaging   28–29
chemical oxidation   214, 229–232
chemical purification   20–25
chemicals associated with microplastics   345–372
additives   326–330, 346–351, 356–363
aquatic organisms   326–330
bisphenol A   346–347, 351, 362–363
consumer products   356–358
environmental impacts   358–363
phthalic acid esters   346, 348, 356, 359–362
receiving waters   326–330
release aspects   358–363
toxicity   326–330, 345–348, 358–360
chemisorption   351–352, 354, 356
China
algae   288, 297
aquatic organisms   317
chemicals associated with microplastics   346, 348, 354, 359–363
discharge of microplastics   176–189, 194–198
drinking water systems   58, 61–63, 66–67, 70–71, 74
receiving waters   247–250, 253, 258
sewage sludge treatments   160–162
wastewater treatment plants   93–107, 160–162, 176–189, 194–198
Chlorella pyrenoidosa   **299**
Chlorella vulgaris   **298**

chlorination   109
Chrysene   387
CI *see* carbonyl index
ciprofloxacin   297, 376
coagulation   68–69, **71–72**, 75–82, 109
coagulation, flocculation, and sedimentation (CFS)   75–76, 79, 214–223
coastal areas
   aquatic organisms   253–255
   chemicals associated with microplastics   354, 362
   discharge of microplastics   182–188, 195–197
   sludge   164
collection techniques
   drinking water samples   14–15
   freshwater samples   1–9, 12
   sludge samples   12
   wastewater samples   9–11
colloidal nanoplastics   408–410, 413–414
Colombia   181, 190
color of microplastics
   aging   251
   algae   296
   chemicals associated with microplastics   349–350, 357–358
   freshwater   249–250
   sample identification   25–27, 37
   sewage sludge   151–154
   wastewater treatment plants   92–100, 105
composition of microplastics
   contaminant interactions   383–384, 390–391
   drinking water systems   67–68, 74–75, 81–82
   nanoplastics   409, 412
   receiving waters   246–250, 255, 291–292, 295–296
   sampling   15, 20, 25–29
   wastewater treatments   102–106, 122–123, 136, 139, 151–152, 187
composting   163–164
constructed wetlands   **95**, 102
consumer products, chemicals in   356–358
consumers, aquatic organisms   324–326
container collection   9–10
contaminant-microplastic interactions   373–406
   adsorption   375–391, 395–396
   aging of microplastics   388–390
   antibiotics   375–378, 381–386, 388, 391–392, 395
   aquatic environments   390–393
   aquatic organisms   390–393
   bacteria   375, 383–385, 391–395
   environmental impacts   375–396
   heavy metals   375–381, 384–395
   human health   375–378, 381–386, 388, 391–396
   ionic strength   387–388, 413–416
   joint potential risks   390–396
   nanoplastics   414–418
   organic pollutants   375–376, 380–389, 395–396
   salinity   388
   selection techniques   383–385
   temperature   386–388
   weathering   388–390
contamination
   algae   288, 292, 296–303
   antibiotics   296–300
   freshwater resources   247–250
   quality control   36–37
   receiving waters   243–286
conventional unit operations/processes   119–123
cormorants   253–254
corona formation   419–420, 424, 428, 430

CPL *see* caprolactam
crabs   252, 322, 326, 423
crops *see* agricultural practices
cross-contamination   36–37
crustaceans   252–253, 321–322, 325–326, 417, 421–423
curing agents   356, 415
Cyanobacteria   257
Czech Republic   62–63, 69–74

d

Daphnia magna   321–322, 324, 327, 417
DBP *see* dibutyl phthalates
decomposers   322–323
deep-bed filtration   216, 221
degradation   211–242
   adsorption   213–219, 223–225
   CFS   214–223
   chemical oxidation   214, 229–231
   drinking water treatment plants   212, 215–221, 231–232
   electrocoagulation   214, 219–221
   filtration   214–216, 221–223
   flocculation   214–223
   future directions   231–232
   membrane separation   214, 222–223, 231
   nanoplastics   410–412, 427
   photocatalysis   214–215, 225–229, 232
   separation techniques   214–232
   wastewater treatment plants   212–216, 221–227, 232
dehydration   **151**, 153–155, 162–163
denitrification   120–133, 136
Denmark   62, 161
density separation   7–8, 14
DEP *see* diethyl phthalates
dermal contact   263–264, 320, 393–394
detection techniques   1–52
   identification   25–32
   purification   16–25
   quality control   36–37
   quantitative analysis   32–36
   sampling   1–32
dewatered/dewatering sludge   134–136, 148–149, 153–155, 160, 163
di-2-ethylhexyl phthalate   148, 158, 360–362
diatoms   **290**, 292, 420–421
dibutyl phthalates (DBP)   148, 348, 352, 360–362
dietary exposure, human health   246, 260–263
diethyl phthalates (DEP)   352, 360
differential scanning calorimetry (DSC)   31, **32**
digestion
   aerobic   124, 134–136
   contaminant-microplastic interactions   391–394
   detection techniques   6
   organic matter   17–23
   purification   17–23
   sediments   6
   wastewater treatments   124, 134–136
dimethyl phthalates   361–362
disc filters   **95**
discharge of microplastics   175, 180, 184–188, 191, 195–197
   aquatic organisms   175, 193–199
   chemicals   345–346, 360, 363
   effluent water   175–199
   freshwater environments   175, 180, 184, 188, 191, 193–195
   future directions   198–199
   lakes   175, 179–184, 188–195, 199
   marine resources   175, 180, 184–188, 191, 195–197
   oceans   175, 179–180, 184–192, 195–199
   receiving waters   179–192

rivers   175, 179–191, 194–195, 199
sediments   175, 180–182, 188–193, 197–199
wastewater treatment plants   175–209
diseases, human health   264–266
disinfection   **72**, 80, **94**, 109, 301
dispersed particles tracking   34–35
dissolved air flotation   **95**, 107
dissolved organic matter (DOM)   417
DLS *see* dynamic light scattering
dolphins   255–256
drinking water
  atmosphere   64–65
  bottled water   14–16, 29, 54, 57–58, 261–262
  CFS   75–76, 79
  coagulation   68–69, **71–72**, 75–82
  collection techniques   14–15
  composition dependency   74–75
  distribution of microplastics   64–67
  emergence of microplastics   57–60
  fibrous microplastics   66–67, 73
  filtration   68–70, **71–72**, 76–82
  flocculation   68–69, **71–72**, 75–82
  fragmented microplastics   66–67, 73–74
  freshwater environments   63–67, 81–82
  human health   246–247, 261–263
  labs-scale studies   78–80, 82
  microplastic sources   59
  morphological distribution/transformation   59–60, 66–67
  occurrence of microplastics   53–68, 80–82
  ozone treatments   69, 77–78, 82
  receiving waters   246–247, 261–263
  removal of microplastics   58–59, 68–82
  safe supply   68–82
  sampling techniques   14–16
  sedimentation   68–69, **71–72**, 75–78
  size of microplastics   65–66, 70–73
  tap water sampling   15–16
  toxicity aspects   54, 60, 80
  treatment processes   16, 59, 68–82
  weathered materials   60, 79–80
drinking water treatment plants (DWTP)
  degradation   212, 215–221, 231–232
  occurrence of microplastics   62–64
  removal of microplastics   58–59, 68–82, 212, 215–221, 231–232
DSC *see* differential scanning calorimetry
dust   318
dynamic light scattering (DLS)   33–35

e

eco-corona formation   419
ecological impacts/risks
  algae   288, 291–303
  aquatic organisms   251–256, 288, 291–303, 318–330
Ecuador   **181**
EDC *see* endocrine disruptors
EDS *see* energy-dispersive X-ray spectroscopy
effluent water
  aquatic organisms   318
  discharge of microplastics   175–199
  drinking water   15, 69–70, 73–78
  nanoplastics   411, 416
  wastewater treatment plants   92–110, 175–199
Egypt   **185**, **189**, 195–196
electrocoagulation   214, 219–221
electron microscopy techniques   427–428
electrostatic interactions   376, 386, 388, 416
elutriation   8

embryonic fish   422, 425–426
embryotoxicity   422
endocrine disruptors (EDC)   301, 380
energy-dispersive X-ray spectroscopy (EDS)   27
engineered nanomaterials (ENM)   412
environmental impacts
   algae   287–303, 321–324
   chemicals associated with microplastics   358–363
   contaminant-microplastic interactions   375–396
   discharge of microplastics   175, 180, 184, 188, 191, 193–195
   drinking water systems   63–67, 81–82
   nanoplastics   407–430
   receiving waters   243–267, 287
   wastewater treatment plants   91–93, 102–109, 175, 180, 184, 188, 191, 193–195
enzymes
   effects of microplastics   **121**, 128–134, 138–139
   enzymatic degradation   17–18
   human health   264
   sewage sludge   158–159, 165
   wastewater treatments   **121**, 128–134, 138–139, 158–159, 165
epoxy resins   62, 67, 362
Ethiopia   **189**, **194**
excretion, aquatic organisms   321–322, 325, 329–330
exposure duration, biofilms   258
extracellular polymeric substances (EPS)
   algae   291–295
   chemicals   351, 358
   drinking water   60
   effects of microplastics   120–139
   nanoparticles   426
   receiving waters   256, 291–295
   sampling   12–13
   sewage sludge   152, 157–160
   wastewater treatments   120–139, 152, 157–160
extraction
   sediments   6, 7–9
   sludge   14

f

Fathead minnow   426–427
Fenton's reagent   13–14, 17
fermentation   148, 155–158, 165
fibrous microplastics
   algae   287–288
   aquatic organisms   315–318, 321, 325
   drinking water systems   58–59, 66–67, 73
   nanoplastics   407, 430
   receiving waters   243–247, 250–251, 254, 263, 266–267
   sewage sludge   147–155, *161*, 165
   wastewater treatment plants   91–92, **94–100**, 101, 104–109, 147–155, *161*, 165
   wastewater treatment processes   120–123, 139
fill-and-empty cycles   15
films
   aquatic organisms   316–317, 323–324, 328–329
   receiving waters   245–246, 250–251, 256–262, 316–317, 323–324, 328–329
   sewage sludge   148, 152–153, 164
   wastewater treatment plants   **94–99**, 104–105, 108, 148, 152–153, 164
filtration
   aging of microplastics   296
   degradation aspects   214–216, 221–223
   drinking water systems   68–72, 76–82

removal of microplastics 68–72, 76–82, 214–216, 221–223
wastewater sampling 9–12
wastewater treatment plants **95**, 109
Finland **95**, 101–103, **177**, 185–190, **196**, 249
fish
  contaminant-microplastic interactions 392–393
  drinking water systems 66
  nanoplastics 416–427
  receiving waters 246, 260–262, 316, 319–330
  uptake of microplastics 3, 193–197
flame retardants 356–357, 414–415
flocculation 68–72, 75–82, 107–108, 124, 214–223
flocs, sludge samples 12–13
fluoranthene **382**, 387
fluoroquinolones 373, 376, 416
foams
  aquatic organisms 316
  chemicals 348–350, 353
  drinking water 63, 66–67
  receiving waters 246, 250
  wastewater treatment plants **94**, **97**, 104, 107
focal plane array (FPA) reflectance FTIR **26**, 28–29
food chains
  aquatic organisms 251–256, 316, 319–330
  di-2-ethylhexyl phthalate 360
  human health 246, 260–263
Fourier transform infrared (FTIR) **15**, 18–20, 26–30, 36–37
FPA *see* focal plane array reflectance FTIR
FPA micro-FTIR 28–29
fragmented microplastics
  aquatic organisms 315–325, 329–330

chemicals associated with microplastics 347–356
degradation 211, 229
drinking water systems 66–67, 73–74
nanoplastics 407–408, 411
receiving waters 243–247, 250, 266–267
removal of 73–74, 92–100, 104–106, 211, 229
sewage sludge 149, 152, 164
wastewater treatment plants 92–100, 104–106, 149, 152, 164, 182, 186
wastewater treatment processes 120–123
France **99**, 101, 177–178, 247, 261, 317
freshwater environments
  discharge of microplastics 175, 180, 184, 188, 191, 193–195
  drinking water systems 63–67, 81–82
  nanoplastics 416–423, 427–428
  receiving waters 247–250
  wastewater treatment plants 175, 180, 184, 188, 191, 193–195
freshwater organisms 193–195, **298**, 416–423, 427–428
freshwater sampling 1–9, 12
  collection 1–9, 12
  digestion 6
  extraction 6, 7–9
  grab sampling 3–5
  inspection 6
  nets 2–5, *4*, 12
  pump sampling 3–5
  sediments 5–9, 12
  separation 6, 7–9
  trawls 2–5, 12
Freundlich model 348–349, 352, 354, 415–416
FTIR *see* Fourier transform infrared
fungi 126, 246, 322–323

g

GAC *see* granular activated carbon
Gammarus species   253, 322, 324–325
gas chromatography   **26**, 30–32
genetic vectors   258–259, 375, 383–385, 391–392, 395
Germany   15, **177**, 181–184, **194**, 247–248, 258
Ghana   195–196
grab sampling   3–5
granular activated carbon (GAC) filtration   70, **71–72**, 76–77, 216, 221
granular microplastics   91, **95**, **99**, 104
gravity filters   **96**
growth
    algae   288–303
    aquatic organisms   253, 288–303, 323–326, 390–391, 419–421
    degradation/removal of microplastics   213
    wastewater treatment processes   119–123, 131–134, 137–139

h

HBCD *see* hexabromocyclododecane
HDPE *see* high-density polyethylene
heavy metals
    algae   296, 300–301
    aquatic organisms   251–252, 296, 300–302, 328
    chemicals associated with microplastics   354–355
    contaminant-microplastic interactions   375–381, 384–395
heterotrophic bacteria   128–129
hexabromocyclododecane (HBCD)   358
high concentration microplastics
    algae   287–288, 293
    drinking water systems   61
    wastewater treatments   93, 120, 124–127, 136, 181, 188
high-density polyethylene (HDPE)
    algae   289–290, 294–299
    aquatic organisms   289–290, 294–299, 328–329
    contaminant interactions   **379**
    degradation/removal   228–229
    drinking water   59
    sampling   15, 19
    wastewater treatment plants   **98**, 105, **190**
high temperature composting   163–164
Hong Kong   **94**, 101, 103, 300, 363, 379
horizontal gene transfer   124–126, 259, 384
hormones   301, 325
human health
    contaminant-microplastic interactions   375–378, 381–386, 388, 391–396
    dermal contact   263–264
    dietary exposure   246, 260–263
    diseases   264–266
    drinking water   246–247, 261–263
    food chains   246, 260–263
    immune responses   263–264
    ingestion of microplastics   211–216, 226–227, 231–232
    inhalation exposure   263–264
    nanoplastics   428
    receiving waters   246, 260–267
    toxicity aspects   263–266
Hungary   **181**, 182–183, **190**
hydrochloric acid   18–24
hydrogen, sewage sludge   158–159, 165
hydrogen peroxide   13–14, 17
hydrophobicity
    algae impacts   296–302
    aquatic organism impacts   251, 328–329
    biofilms   257–258
    chemicals associated with microplastics   347–352, 362–363

contaminant-microplastic
    interactions   374–376, 381, 387–389
  degradation/removal aspects   213, 224
  in drinking water   60
  nanoplastics   411, 416–417
  in receiving waters   264, 296–302,
    328–329
  wastewater treatments   123, 128,
    138, 155

i

identification techniques   25–32
  detection   25–32
  differential scanning calorimetry
    31, **32**
  FPA-reflectance   **26**, 28–29
  FTIR techniques   **26**, 28–30
  gas chromatography   **26**, 30–32
  mass spectrometry   **26**, 29–32
  microscopy   26–27
  pyro-GC/MS   **26**, 30–32, **32**
  Raman spectroscopy   **26**, 29–30
  size distribution   25–32
  spectroscopy   **26**, 27–32
  TED-GC-MS   30–31, **32**
  thermal analysis   **26**, 29–32
  thermogravimetric analysis   30–32
  vibration spectroscopy   **26**, 27–32
immune responses   263–264, 326, 393,
    423–427
incineration   164–165
India   62, **71**, 76, **183**, 188–189, 346
Indonesia   180–181, **189**
indoor air   263–264
influent water
  degradation/removal aspects   216
  drinking water treatment   59, 62–63,
    69–70, 76–77
  sampling   12–15, 29
  wastewater treatment   93–105, 109,
    149–152

ingestion of microplastics
  aquatic organisms   320–329, 421–422,
    426–430
  contaminant-microplastic
    interactions   393–395
  human health   211–216, 226–227, 231–232
  nanoplastics   421–422, 426–430
inhalation exposure   263–264, 393–394
inspection techniques   6, 35–37
internal deviation   36–37
invertebrates   198, 251–253, 316, 419, 422
ionic strength   387–388, 413–416
Iran   **97**, 101, 162–163
Israel   **94**, 101
Italy   **94**, 99–101, 177–181, 247–249, 261

j

Japan   317, **379**
jar tests   78, 215
joint potential contaminant
    risks   390–396
judgement errors   37

k

Kenya   183–184
Korea   **98**, **100**, 101–102, 346, 362
krill   423, 428

l

labs-scale studies
  additive leaching   327
  algae   302–303
  aquatic organisms   302–303, 327–328
  contaminant-microplastic
    interactions   380, 393, 395–396
  degradation aspects   221
  detection techniques   1–5
  drinking water systems   78–80, 82
  nanoplastics   423, 430
  removal of microplastics   78–80, 82,
    216, 220–222

wastewater treatment
    processes   137–138
lakes
    aquatic organisms   316–317
    chemicals associated with
        microplastics   348, 361–362
    discharge of microplastics   175,
        179–184, 188–195, 199
    drinking water systems   63
    receiving waters   248–254, 257, 261
    wastewater treatment plants   175,
        179–184, 188–195, 199
landfills   53, 57–59, 160–162, 165–166,
    411
Langmuir model   348–349, 354, 381
laser diffraction (LD)   33–35
laundries
    receiving waters   179, 243, 250, 315
    wastewater treatment plants   92, 104,
        154, 179
LD *see* laser diffraction
LDPE *see* low-density polyethylene
line microplastics   **99**
low-density plastics *see also individual
    plastics*
    detection   8, 11, 14, 17–19
    discharge of   154, 188, 197
    removal   221, 226–227
    sediment   317
    sewage sludge   154
    toxicity   327
    wastewater treatment plants   154,
        188, 197
low-density polyethylene (LDPE)
    aquatic organisms   328–329
    chemicals associated with   345
    degradation/removal   221, 226–227
    drinking water systems   59
    purification   19
    receiving waters   **189**, 328–329

separation technology   217–218
wastewater treatment plants   **99**

m

macrolides   376
magnesium hydroxide   219
magnetic extraction   214, 217–218
Malaysia   185–186, 354
mammals   254–256
manta nets   3–5, 12
marine life/resources
    discharge of microplastics   175, 180,
        184–188, 191, 195–197
    microalgae   **298–299**
    uptake of microplastics   195–197
    wastewater treatment plants   175,
        180, 184–188, 191, 195–197
mass, sample purification   20, **21**
mass spectrometry (MS)   **26**, 29–32
MBBR *see* moving bed biofilm reactors
MBR *see* membrane bioreactors
MCT *see* mercury telluride single-mode
    FTIR
Mediterranean Sea
    aquatic organisms   195–196, 255,
        260, **317**
    chemicals associated with
        microplastics   358
    nanoplastics   422–424
    wastewater treatments   185–190
megafauna   254–256
membrane bioreactors (MBR)   93–95,
    99, 108–109, 119–123, 137
membrane filtration   80, 109
membrane separation   214, 222–223, 231
mercury telluride (MCT) single-mode
    FTIR   28
meroplankton   422
mesh sieve devices   9–11
mesh sizes   12

Index | 457

metal ions   353–355, 389, 425
metals
  algae   296, 300–301
  aquatic organisms   296, 300–301, 328–329
  chemicals associated with microplastics   354–355, 357, 359
  contaminant-microplastic interactions   375–381, 384–395
methane   124–130, **135**, 155–156, 159
Mexico   **317**
microalgae   419–427
microbeads
  algae   287, 289, 293
  aquatic organisms   315–318, 321, 324–325, 329–330
  chemicals associated with microplastics   348, 350
  drinking water   66
  nanoplastics   407, 410, 416, 428
  receiving waters   246, 250, 315–318, 321, 324–325, 329–330
  sampling   25
  wastewater treatment plants   **94, 97–99**, 104, 109, 147–148, *161*
microbials
  anaerobic digestion   *157*, 159–160
  aquatic organisms   316–317, 322–325, 329–330
  contaminant-microplastic interactions   375, 383–385, 391–395
  effects of microplastics   119–139
  receiving waters   256–260, 316–317, 322–325, 329–330
  sewage sludge   *157*, 159–160
  wastewater treatments   108, 119–139, *157*, 159–160
microfiltration   222–223
micro-FTIR   28

microorganisms *see* microbials
microplastics, definitions   55–57
microscopy   26–27, *150*
minnows   426–427
mollusk species   252–253, 392
Mongolia   248, 257
morphological distribution/transformation, in drinking water   59–60, 66–67
moving bed biofilm reactors (MBBR)   119, 122–123
MS *see* mass spectrometry
mussels
  contaminant-microplastic interactions   392
  degradation/removal aspects   228
  nanoplastics   422–424
  receiving waters   252–253, 261–262, 321–322, 325–326
Mytilus galloprovincialis   422–424

n

nanofiltration   68–69, 80
nanoparticles (NP), definitions   408–409
nanoparticle tracking analysis (NTA)   33–36
nanoplastics   407–444
  algae   419–427
  aquatic environments   407–430
  aquatic organisms   418–430
  behaviour and fate   412–414
  contaminant interactions   414–418
  definitions   408–410
  degradation   410–412, 427
  environmental impacts   407–430
  fish   419–427
  formation pathways   410–411
  ingestion   421 422, 426–430
  microalgae   419–427

omics 422, 429–430
sources 411–412
temperature 412, 421
toxicity 418–426, 429–430
UV radiation 407, 410–412, 415–418
weathering 407, 411–414, 420, 427
nanoprecipitation 410
natural organic matter (NOM) 409–410, 413–415, 419, 430
Netherlands 61, **96**, 176–177
nets 2–5, 12
neuston nets 3
Nigeria 183–184, **189**
nitrates 127, 131–132
nitric acid 18–24
nitrification 120–133, 136
nitrite-oxidizing bacteria 132
nitrogen 108, 120–127, 136, 160
NOM *see* natural organic matter
nonylphenol **299**, 302, 327, 346, 357
norfloxacin 386
NP *see* nanoparticles
NTA *see* nanoparticle tracking analysis
nylon
  aging 354
  contaminant interactions 379
  degradation/removal 221
  drinking water systems 67–68
  nanoplastics 412
  purification 19
  receiving waters 248
  separation technology 221

o

oceans
  chemicals associated with microplastics 345–346
  degradation/removal of microplastics 211
  detection techniques 7, 30
  discharge of microplastics 175, 179–180, 184–192, 195–199
  nanoplastics 410–411, 421–424, 427–429
  receiving water contaminants 254
oil extraction 7–8
omics, nanoplastics 422, 429–430
organic matter purification 16–25
organic pollutants
  aquatic organisms 328–329
  chemicals associated with microplastics 347–353, 360–362
  contaminant-microplastic interactions 375–376, 380–389, 395–396
  detection 30
  drinking water systems 54, 57–59
  extraction 8–9
  nanoplastics 410, 413–417, 423, 426–427
  particle size analysis 35
  pressurized fluid extraction 8–9
  wastewater treatment processes 119–120, 134
oxidation ditch 119, 121
oysters 253, 326, 422
ozonation/ozone treatments 69, 77–78, 82, 109, 416–417

p

PA *see* polyamides
Pacific ocean 186, 191
packaging
  aquatic organisms 326–327
  chemicals associated with 359–362
  drinking water 14–15, 57–58, 67–68
  receiving waters 262, 326–327
  wastewater treatment plants 104–105, 179, 182
PAE *see* phthalates; phthalic acid esters
PAH *see* polycyclic aromatic hydrocarbons
Pakistan 180–181, 188–189
PAM *see* polyacrylamides

particle size distribution 33–36
particulate matter 55, 151–153
pathogen carriers 260
PBD *see* polybrominated diethers
PBDE *see* polybrominated diphenyl ethers
PC *see* polycarbonates
PCB *see* polychlorinated biphenyls
PE *see* polyethylene
PEC *see* predicted environmental concentrations
pellet microplastics
 aquatic organisms 315–318, 321, 324–325, 329–330
 chemicals associated with 348, 350
 drinking water 66
 receiving waters 246, 250, 315–318, 321, 324–325, 329–330
 sampling 25
 wastewater treatment plants **94, 97–99**, 104, 109
PENP *see* polymeric engineered nanoparticles
PE-PP *see* polypropylene copolymers
perchloric acid 19
perfluorinated alkyl substances (PFAS) 328–329
perfluorooctanesulfonamide (FOSA) 349
perfluorooctanesulfonate (PFOS) 349, 386, 390, 393
peroxidation 16–17
persistent organic pollutants (POP) 154, 349, 386
PES *see* polyesters
pesticides 301, 328–329, 380
PET *see* polyethylene terephthalates
PFAS *see* perfluorinated alkyl substances
PFOS *see* perfluorooctanesulfonate
pH 350–351, 354–355, 375, 386–389
pharmaceuticals 301, 328–329, 376, 380, 407, 414–416

phenanthrene (Phe) 346, 349, 353, 387, 389
phenolic resins **99**
phenols 348–349, 351–352, 357–359 *see also* bisphenol A
phosphorus 108, 120, 124, 127, 133, 160
photocatalysis 214–215, 225–229, 232
photooxidation degradation 296
phthalates (PAE) 120, 213, 346–348, 356, 359–362 *see also* polyethylene terephthalates
phthalic acid esters (PAE) 346, 348, 356, 359–362
phylum Proteobacteria 256–257
physical damage effects 329–330
physical transformation 59–60
physicochemical properties 23–24
physisorption 351–352, 354, 356
plankton nets 2–5, 12
plants
 aquatic organisms 287–303, 316, 319–324, 327–330
 contaminant-microplastic interactions 390–391
 receiving waters 287–303, 316, 319–324, 327–330
plasma membranes 428
plasticizers 148–149, 326–329, 356–360, 415
plasticulture 58–59
PMMA *see* polymethyl methacrylate
PO *see* polyolefins
Poland **181**
polar regions 185–187, 190–191, 197–198
polyacrylamides (PAM) 59, 68, 75–79, 82, 217–219
polyamides (PA)
 contaminant interactions 376–379, 381, 388
 degradation/removal 221–222
 drinking water 67–68

identification techniques 29
purification 19–25
sewage sludge 148, **151**, 152–154, 156
wastewater treatment plants **95–100**, 104–105, 177–182, 185–191, 194–198
polyamides (PA) receiving waters 245, 248–251
polybrominated diethers (PBD) 328–329, 345–346, 414
polybrominated diphenyl ethers (PBDE) 345, 414
polycarbonates (PC) 67–68, 362–363
polychlorinated biphenyls (PCB) 328–329, 347–349, 387–388, 392, 414, 417
polycyclic aromatic hydrocarbons (PAH) 328–329, 347–349, 414–415, 417
polyesters (PES)
  algae 295
  contaminant interactions **379**
  drinking water 62, 67–68
  receiving waters 245
  removal/degradation 216, 218–219
  sewage sludge 148, **151**, 153, 155–156
  wastewater treatment plants **94–100**, 104–105, 176–185, 191, 194–196
  wastewater treatment processes 120–121, 125–132, **135**, 148, 151–156
polyethylene (PE)
  algae 289–290, 293–299
  aquatic organisms 289–290, 293–299, 316, 319, 327–330
  chemicals associated with 347–361
  contaminant interactions 376–382, 386–388
  drinking water 62, 67–68, 74–80
  identification techniques 27, 29–31
  nanoplastics 410–412, 417–423, 428
  particle size distribution 34–35
  receiving waters 245–249, 257, 262, 289–290, 293–299, 316, 319, 327–330
  removal/degradation 213, 216–222, 227–230
  sample purification treatments 17, 20–25
  sewage sludge 148, 152, 155–158, 163–164
  wastewater treatment plants **94–100**, 104–105, 148, 155–158, 176–198
  wastewater treatment processes 120–136
polyethylene terephthalates (PET)
  contaminant interactions **378–380, 382**, 388
  drinking water 62, 67–68, 74–75
  nanoplastics 411–412, 416, 428
  receiving waters 245–253, 257–258, 261
  removal/degradation 216, 218
  sample purification treatments 17, 19–25
  sewage sludge 151–154, 157–158
  wastewater treatment plants **95–100**, 104–105, 151–154, 157–158, 177–198
polyhydroxyalkanoates 346, 350
polylactic acids **382**, 411
polymeric engineered nanoparticles (PENP) 409–412, 415, 418–421, 430
polymeric nanoparticles 408–412, 415, 418–421, 428, 430
polymerization 67, 326, 356, 410
polymethyl methacrylate (PMMA)
  chemicals associated with 354–356
  degradation/removal 220
  nanoplastics 420, 427

receiving waters   248–249
  sample purification effects   20–25
  wastewater treatment plants   182, **190**
polyolefins (PO)   **151**, 154
polypropylene copolymers
    (PE-PP)   178–179
polypropylene (PP)
  algae   289–290, 293–297
  aquatic organisms   316, 327–330
  chemicals associated with   347–348,
    354, 359
  contaminant interactions   **377–380**,
    382, 388
  degradation/removal   220, 227
  discharge of   176–198
  drinking water   62, 67–68, 74–76
  particle size distribution   34–35
  purification   17, 19–25
  receiving waters   245–251, 257–258,
    261, 289–290, 293–297, 316,
    327–330
  sewage sludge   151–153, 157
  wastewater treatment plants
    93–100, **94–100**, 151–153, 157,
    176–198
  wastewater treatment processes
    120–125, 129–132, 135–136
polystyrene (PS)
  algae   290–294, 297–301
  aquatic organisms   290–294, 297–301,
    328–330
  chemicals associated with   347–358
  contaminant interactions   376–382,
    388
  drinking water   62, 67–68, 74–75, 80
  identification techniques   29–31
  nanoplastics   410–411, 415–419,
    421–429
  particle size distribution   34–35
  purification   17, 19–25

  receiving waters   245–253, 257, 265,
    290–294, 297–301, 328–330
  removal/degradation   213, 216–219,
    227–228
  wastewater treatment plants   **95–100**,
    105, **151**, 154–159, 164, 177–191,
    194–195, 198
  wastewater treatment processes
    120–132, 135–136
polytetrafluorethylene (PTFE)   19, 68, 115
polyurethanes (PUR)   67–68, **94–95**,
    100, 105, 247, 327
polyvinyl chlorides (PVC)
  algae   289–290, 293–295, 299
  aquatic organisms   327–330
  chemical oxidation   230–231
  chemicals associated with   347–354,
    358–360
  contaminant interactions   **377–379**,
    381–382, 388
  drinking water   67–68, 74
  identification techniques   29
  purification   17, 19
  receiving waters   245–253, 257,
    262–264, 289–290, 293–295, 299,
    327–330
  removal/degradation   230–231
  sewage sludge   148, 155–158
  wastewater treatment plants   **95**,
    98–100, 105, 176, 183–192,
    197–198
  wastewater treatment
    processes   120–138
ponds   316–317
POP *see* persistent organic pollutants
population effects, algae   289–293, **298**
pore sizes   12
Portugal   194–196, 247
potassium   160
potassium hydroxide   13–14, 19

potentiating effects, aquatic organisms   393
PP *see* polypropylene
precipitation, drinking water systems   76
predicted environmental concentrations (PEC)   99, 348, 408
pressurized fluid extraction   7–9
pretreatments, sludge samples   12–14
primary microplastics
   chemicals associated with   346–347, 360
   definitions   56
   receiving waters   243–245, 248–250, 266
   wastewater treatment plants   91
primary sludge   150–154, 165
primary treatments   **96**, 106–110, 150–153
principle components analysis   22–25
producers, aquatic organisms   323–324
protein-corona formation   419
Proteobacteria   256–257
PS *see* polystyrene
pseudo-second-order models   349–350
PTFE *see* polytetrafluoroethylene
pump sampling   3–5, 9–11, 36
PUR *see* polyurethanes
purification   16–25
   acid treatments   18–25
   alkaline treatments   18–25
   detection techniques   16–25
   digestion   17–23
   enzymatic degradation   17–18
   hydrochloric acid   18–24
   microplastic property effects   20–25
   nitric acid   18–24
   physicochemical properties   23–24
   property effects   20–25
   quantitative indicators   22–25
   sodium hydroxide   19, 21–25
   wet peroxidation   16–17

PVC *see* polyvinyl chlorides
pyrenes   **382**, 387
pyrolysis-gas chromatography-mass spectrometry (pyro-GC/MS)   **26**, 30–32

q

quality control   36–37
quantitative analysis/indicators   22–25, 32–36

r

Raman spectroscopy   **26**, 29–30
rapid sand filters   **95**, 216, 221–222
raw water   59, 63, 68–77, 81, 216
rayon   **97–98**, 176–181, 194–196
reactive oxygen species (ROS)
   contaminant-microplastic interactions   393
   degradation of microplastics   225–226
   nanoplastics   420–421, 424, 426
   sludge-microplastic interactions   127, 156–159, 165
   toxicity of microplastics   264–265
   wastewater treatment processes   127, 133, 156–159, 165
receiving waters
   adsorbed pollutants   328–330
   algae   287–314, 321–324
   aquatic organisms   246, 251–256, 287–303, 315–343
   bacteria   316–317, 322–325, 329–330
   biofilms   245–246, 250–251, 256–262, 316–317, 323–324, 328–329
   chemicals   326–330
   contamination   243–286
   discharge of microplastics   179–192
   ecological impacts/risks   251–256, 291–303, 318–330
   effects of microplastics   243–343
   food chains   251–256, 316, 319–330

Index | 463

important sources 179–192
microbials 316–317, 322–325, 329–330
physical damage effects 329–330
plants 316, 319–324, 327–330
sediments 316–320
toxicity 253, 288, 291–303, 322–330
reclamation plants **95**, 105–106
recovery experiment design 36–37
release aspects, chemicals 358–363
removal of microplastics
  advanced drinking water treatment plants 68–69, 77, 80
  biological processes **94, 96–100**, 101–103, 106–109
  CFS 75–76, 79, 214–223
  chemical oxidation 214, 229–231
  coagulation 68–69, **71–72**, 75–82
  composition dependency 74–75
  degradation 211–242
  drinking water 58–59, 68–82, 212, 215–221, 231–232
  electrocoagulation 214, 219–221
  filtration 68–70, **71–72**, 76–82, 214–216, 221–223
  flocculation 68–72, 75–82, 107–108, 214–223
  future directions 231–232
  labs-scale studies 78–80, 82
  membrane separation 214, 222–223, 231
  photocatalysis 214–215, 225–229, 232
  primary treatments **96**, 106–110, 150–153
  secondary treatments 94–103, 106–109, 151–153
  sedimentation 68–69, **71–72**, 75–78, 214–223
  separation techniques 214–232

sewage sludge 148–155, 162–166
size dependency 70–73
sludge 148–155, 162–166
tertiary treatments 93–101, 106, 108–109
type dependency 73–74
wastewater treatment plants 92–110, 148–155, 162–166, 212–216, 221–227, 232
wastewater treatment processes 119–120
resource recovery, sewage sludge 148, 155
reverse osmosis 222–223
rivers
  aquatic organisms 316–318
  contamination 247–250
  discharge of microplastics 175, 179–191, 194–195, 199
  drinking water systems 62–63, **71**
  receiving waters 247–250, 316–318
  wastewater treatment plants 175, 179–191, 194–195, 199
ROS *see* reactive oxygen species
Rose Bengal 13–14
rotating biological contactors 122
Ruttner samplers 9

s

SAF *see* submerged aerated filters
safe supply, drinking water 68–82
salinity 347, 350–351, 355, 388
salts, contamination 261–262
sampling 1–16
  collection 1–16
  detection techniques 1–32
  drinking water samples 14–16
  freshwater 1–9, 12
  identification 25–32
  purification 16–25
  separation 1–32
  size distribution 2–5, 9–12

sludge samples   12–14
visual techniques   25–26
wastewater samples   9–12
sand filtration   76–77, **95**, 109, 216, 221–222
Saudi Arabia   182–184
SBR *see* sequencing batch reactors
scanning electron microscopy (SEM)   27
SCFA *see* short-chain fatty acids
Scotland   **96**, 177, **185**, 187
seabirds   253–254
seals   255–256
sea salt   261–262
sea water
  algae   292
  aquatic organisms   251, 259, 292, 321, 329
  chemicals associated with microplastics   354–355, 362
  contaminant-microplastic interactions   376–377, 385–386, 390
  degradation/removal of microplastics   211
  detection techniques   7, 30
  discharge of microplastics   175, 179–180, 184–192, 195–199
  drinking water   54, 59
  nanoplastics   410–411, 421–424, 427–429
  receiving waters   179, 187, 254, 257, 260–262, 315–317
  sampling   18, 34
  wastewater treatment plants   175, 179–180, 184–192, 195–199
secondary microplastics   56, 91–92, 243–245, 266, 346–347
secondary treatments   94–103, 106–109, **151**, 152–153
sedimentation
  degradation   214–223

drinking water systems   68–69, **71–72**, 75–78
removal of microplastics   68–69, **71–72**, 75–78, 214–223
sewage sludge   151–152
wastewater treatment plants   151–152
wastewater treatment processes   120, 137
sediments
  aquatic organisms   316–320
  density separation   7–8
  digestion   6
  discharge of microplastics   175, 180–182, 188–193, 197–199
  elutriation
  extraction   6, 7–9
  freshwater samples   5–9, 12
  inspection   6
  oil extraction   7–8
  pressurized fluid extraction   7–9
  receiving waters   316–320
  separation   6, 7–9
  visual separation   7–8
  wastewater treatment plants   175, 180–182, 188–193, 197–199
selective collection, freshwater samples   1, 5
self-regulation, algae   291–292
SEM *see* scanning electron microscopy
separation techniques
  degradation   214–232
  freshwater samples   6, 7–9
  removal of microplastics   214–232
  sampling   6–32
  sediments   6, 7–9
sequencing batch reactors (SBR)   119, 121–122
sewage sludge
  anaerobic digestion   148, 155–165
  biofilms   148, 152–153, 164
  composting   163–164

dehydration **151**, 153–155, 162–163
effects of microplastics 155–160
enzymes 158–159, 165
fermentation 148, 155–158, 165
fibrous microplastics 147–155, *161*, 165
films 148, 152–153, 164
high temperature composting 163–164
hydrogen 158–159, 165
incineration 164–165
microbials *157*, 159–160
occurrence of microplastics 147–173
primary treatments 150–153
receiving waters 245
removal of microplastics 148–155, 162–166
secondary treatments **151**, 152–153
short-chain fatty acids 156–159
toxicity aspects 148–149, 156–158, 165
transportation of microplastics 160–162, 165–166
wastewater treatment plants 147–173
shape of microplastics
 aquatic organisms 320–323
 chemicals associated with 350
 discharge of microplastics 192
 sewage sludge **151**
 wastewater treatment plants 92, **94–100**, 103–104
 wastewater treatment processes 120–123, 126, 137–139
sharks 255–256
sheet microplastics **94**, **98**, **100**
short-chain fatty acids (SCFA) 156–159
size of microplastics
 aquatic organisms 320–323
 biofilms 257
 discharge of microplastics 192
 drinking water systems 65–66, 70–73

identification techniques 25–32
plastic-particulate matter 55
purification effects 20, **21**
quantitative analysis 32–36
sampling techniques 2–5, 9–12
toxic effects on algae **294**, 295
wastewater treatment 92–104, 107–109, 121–129, 136–139
Skeletonema costatum **298–299**
Skeletonema marinoi 421
skin contact, human health 263–264
sludge
 collection techniques 12
 dehydration **151**, 153–155, 162–163
 dewatered/dewatering 134–136, 148–149, 153–155, 160, 163
 effects of microplastics 119–139
 extraction 14
 Fenton's reagent 13–14
 hydrogen peroxide 13–14
 landfills 160–162, 165–166
 microplastic interactions 123–128
 occurrence of microplastics 147–173
 organic matter 12–14
 potassium hydroxide 13–14
 pretreatments 12–14
 receiving waters 245
 removal of microplastics 148–155, 162–166
 sampling techniques 12–14
 sodium hydroxide 13–14
 soils 160–162
 stabilization/stabilizers 134–136, 153–155, 163
 thickening 162–163
 transportation 160–162, 165–166
 wastewater treatment 93, 103–110, 119–139, 147–173
sodium hydroxide 13–14, 19, 21–25
soil

chemicals associated with
microplastics   361–363
nanoplastics   412
sludge   160–162, 165
uptake of microplastics   197–198
solubility   134, 148, 156–157, 352–353, 358
sorption *see* adsorption
South Korea   **98**, **100**, 101–102
Spain   96–102, 160, 181–182, 188–190, 197–198, 261
spatial distribution of microplastics   64, 154
specific surface area
algae   296, 300
contaminant-microplastic interactions   374, 383, 389
degradation/removal   213, 217, 224, 229
sludge   135
spectroscopic identification techniques   **26**, 27–32
spherical microplastics
algae   287, 289, 293
aquatic organisms   315–318, 321, 324–325, 329–330
chemicals associated with   348, 350
drinking water   66, 74
nanoplastics   407, 410, 416, 428
receiving waters   246, 250, 315–318, 321, 324–325, 329–330
sampling   25
sludge   147–148, *161*
wastewater treatment   **94–95**, **97–100**, 104, 109, 147–148, *161*
sponge materials   223–225
stabilization/stabilizers
aquatic organisms   327
chemical additives   356–359
nanoplastics   413–415, 419
sludge   134–136, 153–155, 163

stacked mesh sieve devices   9–11
stereomicrographs   *150*
straining, drinking water systems   76
submerged aerated filters (SAF)   122
sulfadiazine   376
sulfamethazine   297
sulfamethoxazole   297, 376, **378**, 388, 391
sulfuric acid   19
surface filtration   11–12
surface waters   247–250
surfactants   410
suspended-growth processes   120–122
Sweden   **94–95**, 101, 245, 361
sweep flocculation   215, 218
synergistic effects   393

t

tandem nets   3
tap water   15–16, 62, 67, 78
teabags   412
TED-GC-MS *see* thermal extraction-desorption gas chromatography-mass spectroscopy
temperature
chemicals associated with microplastics   348–351, 356–359, 362
composting   163–164
contaminant-microplastic interactions   386–388
degradation/removal aspects   224, 228–232
nanoplastics   412, 421
sampling   13, 17–19, 24, 30–32
terrestrial environments
algae   287
receiving waters   243–245, 287, 315
sludge   149, 165
tertiary treatments   93–101, 106, 108–109

tetracycline   297, 376–378, 388
Tetraselmis chuii   **298**
textiles *see* fibrous microplastics
TGA *see* thermogravimetric analysis
Thailand   66, **99**, 101, 194
thermal analysis   **26**, 29–32
thermal extraction-desorption gas chromatography-mass spectroscopy (TED-GC-MS)   30–31, **32**
thermogravimetric analysis (TGA)   30–32
thermoplastics   67–68, 356
thermoset plastics   67–68
thickening, sludge   162–163
titanium dioxide-based photocatalysis   227–229
toxicity
 algae   288, 291–303
 aquatic organisms
  effects of microplastics   393
  nanoplastics   418–426, 429–430
  receiving waters   253, 288, 291–303, 322–330
 chemicals associated with microplastics   326–330, 345–348, 358–360
 drinking water systems   54, 60, 80
 human health   263–266
 nanoplastics   418–426, 429–430
 receiving waters   253, 263–266, 288, 291–303, 322–330
 sewage sludge   148–149, 156–158, 165
 wastewater treatment plants   148–149, 156–158, 165
transportation of microplastics
 aquatic organisms   253, 259–260, 293–294, 315–324
 chemicals associated with   345, 358
 contaminant interactions   378 380, 390–396
 drinking water   64–65, 81

landfills/sludge/soil   160–162, 165–166
nanoplastics   412–418, 424
wastewater treatments   180, 184–185, 245–247
traps   2–3, *4*
trawls   2–5, 12
trickling filters   122
triclosan   **299**, 301, 393, 416
trimethoprim   376–377
trophic transfer of microplastics   322
Tunisia   188, **189–190**
Turkey   **96–97**, 101–102, 152–153, 177–178

u

UK *see* United Kingdom
ultrafiltration   68–69, 80, 109, 122, 222–223
ultraviolet UV radiation
 aging   20, 251
 contaminant-microplastic interactions   388–389
 drinking water systems   56, 60, 72, 79
 nanoplastics   407, 410–412, 415–418
 photocatalysis degradation   227–228
 purification   20
 receiving waters   258
 toxicity sources   327
 wastewater treatment plants   **94**, 105, 109
United Kingdom (UK)
 receiving waters   247, 253, 261
 wastewater treatment plants   **96–97**, 101, 177–179, **181**, 188, **190**, 194
United States (USA)
 aquatic organisms   **317**
 chemicals associated with microplastics   346, 359–360
 discharge of microplastics   177–178, 181–188, 190–191, 194–195
 drinking water   61

receiving waters   247
sewage sludge   148, 160, 162
wastewater treatment plants   **96**, 99–101, 148, 160–162, 177–178, 181–191, 194–195

V

Van der Waals forces   351–352, 360, 414
veterinary antibiotics   262
vibration spectroscopy   **26**, 27–32
Vietnam   247
visual techniques
  sample identification   25–26
  sediment separation   7–8
Vis–NIR (visible–near infrared) spectroscopy   29
volume-reduced collection   1, 3–5

W

waste-activated sludge (WAS)   148–149, 152–155, 395
wastewater samples   9–12
  collection   9–11
  container collection   9–10
  filtration   9–12
  mesh sieve devices   9–11
  pumping sampling   9–11
  surface filtration   11–12
wastewater treatment plants (WWTP)
  see also wastewater treatments
  algae   287–288
  anaerobic-anoxic-aerobic processes   **95–97**, 102
  anaerobic digestion   124, **135**, 148, 155–165
  aquatic organisms   91–93, 102–109, 175, 193–199
  aquatic systems   91–93, 102–109
  biofilms   148, 152–153, 164
  biological processes   **94**, **96–100**, 101–103, 106–109
  color   92, **94–100**, 105
  composting   163–164
  degradation   212–216, 221–227, 232
  dehydration   **151**, 153–155, 162–163
  dewatered sludge   148–149, 153–155, 160, 163
  discharge of microplastics   175–209
  disinfection   **94**, 109
  effluent water   92–110, 175–199
  enzymes   158–159, 165
  fermentation   148, 155–158, 165
  films   **94–99**, 104–105, 108, 148, 152–153, 164
  filtration   **95**, 109
  flocculation   107–108
  FPA-micro-FTIR   29
  freshwater environments   175, 180, 184, 188, 191, 193–195
  future directions   198–199
  granular microplastics   91, **95**, **99**, 104
  high temperature composting   163–164
  hydrogen   158–159, 165
  identification techniques   29
  incineration   164–165
  influent water   93–95, **94–100**, 109
  lakes   175, 179–184, 188–195, 199
  landfills   162
  marine resources   175, 180, 184–188, 191, 195–197
  membrane bioreactors   93–95, **99**, 108–109
  microbials   *157*, 159–160
  nanoplastics   407, 411, 414, 418
  occurrence of microplastics   91–117, 147–173
  oceans   175, 179–180, 184–192, 195–199

primary sludge   150–154, 165
primary treatments   **96**, 106–107, 109–110, 150–153
receiving waters   243–250, 260
reclamation plants   **95**, 105–106
removal of microplastics   92–110, 148–155, 162–166, 212–216, 221–227, 232
rivers   175, 179–191, 194–195, 199
sand filtration   **95**, 109
secondary treatments   **94**, **96–100**, 101–103, 106–109, **151**, 152–153
sediments   175, 180–182, 188–193, 197–199
sewage sludge   147–173, *157*, 162–166
short-chain fatty acids   156–159
sludge   93, 103–110, 147–173
tertiary treatments   93–101, 106, 108–109
toxicity   148–149, 156–158, 165
transportation of microplastics   160–162, 165–166
UV treatments   **94**, 105, 109
water reclamation plants   **95**, 105–106
wastewater treatments
  activated sludge   119–126, 129–134, 137, 148–155, 158, 165
  aerobic digestion   124, 134–136
  aerobic granular sludge   **125**, 126–127, 129–132
  ammonia-oxidizing bacteria   129–131
  anaerobic-anoxic-aerobic process   **95–97**, 102, 119, 121
  anaerobic granular sludge   **125**, 127–132, *130*
  attached-growth processes   122, 138
  bacteria   108, 119–139
  biofilms   119–123, 128–129, 137–138
  biological processes   94–103, 119–139
  bisphenol A   120, 126, 128, 134
  composition of microplastics   102–106, 122–123, 136, 139, 151–152, 187
  conventional units   119–123
  digestion   124, 134–136, **135**
  effects of microplastics   119–145
  enzymes   **121**, 128–134, 138–139
  extracellular polymeric substances   120–139
  heterotrophic bacteria   128–129
  labs-scale studies   137–138
  membrane bioreactors   119–123, 137
  methane   124–130, **135**
  microbials   119–139
  moving bed biofilm reactors   119, 122–123
  nitrates   127, 131–132
  nitrite-oxidizing bacteria   132
  nitrogen   120–127, 136
  oxidation ditch   119, 121
  performance   120–124, 127–129, 132–139
  phosphorus   120, 124, 127, 133
  removal rates   119–120
  sedimentation   120, 137
  sequencing batch reactors   119, 121–122
  sludge   119–139
  suspended-growth processes   120–122
  ultrafiltration   122
waterbirds   253–254
waterfalls   11–12
water reclamation plants   **95**, 105–106
weathering
  contaminant-microplastic interactions   388–390
  drinking water systems   79–80
  microplastic sources   60

nanoplastics 407, 411–414, 420, 427
receiving waters 243, 250–251
wet peroxidation (WPO) 16–17
whales 255–256
world plastics production 243, *244*
WWTP *see* wastewater treatment plants

z

zebrafish 229, 232, 321, 325, 416, 425–426
zeta potentials 22, *24*
zinc oxide-based photocatalysis 226–227
zooplankton **196**, 319, 392–393, 420–422, 428

Printed and bound by CPI Group (UK) Ltd, Croydon, CR0 4YY
29/12/2022